Lecture Notes in Computer Science 16036

The series Lecture Notes in Computer Science (LNCS), including its subseries Lecture Notes in Artificial Intelligence (LNAI) and Lecture Notes in Bioinformatics (LNBI), has established itself as a medium for the publication of new developments in computer science and information technology research, teaching, and education.

LNCS enjoys close cooperation with the computer science R & D community, the series counts many renowned academics among its volume editors and paper authors, and collaborates with prestigious societies. Its mission is to serve this international community by providing an invaluable service, mainly focused on the publication of conference and workshop proceedings and postproceedings. LNCS commenced publication in 1973.

Sang-Ki Ko · Florin Manea
Editors

Developments in Language Theory

29th International Conference, DLT 2025
Seoul, South Korea, August 19–22, 2025
Proceedings

Springer

Editors
Sang-Ki Ko
University of Seoul
Seoul, Korea (Republic of)

Florin Manea
University of Göttingen
Göttingen, Germany

ISSN 0302-9743 ISSN 1611-3349 (electronic)
Lecture Notes in Computer Science
ISBN 978-3-032-01474-0 ISBN 978-3-032-01475-7 (eBook)
https://doi.org/10.1007/978-3-032-01475-7

This Springer imprint is published by the registered company Springer Nature Switzerland AG
The registered company address is: Gewerbestrasse 11, 6330 Cham, Switzerland

Preface

The 29th International Conference on Developments in Language Theory (DLT 2025) was held from August 19 to August 22, 2025 in Seoul, South Korea, at the Centennial Memorial Hall of the University of Seoul. It was organised by the Computational Intelligence and Data Analytics Laboratory of the University of Seoul.

The DLT conference series provides a forum for presenting current developments in formal languages and automata. Its scope is very general and includes, among others, the following topics and areas: grammars, acceptors and transducers for words, trees and graphs; algebraic theories of automata; algorithmic, combinatorial and algebraic properties of words and languages; relations between formal languages and artificial neural networks; variable-length codes; symbolic dynamics; cellular automata; groups and semigroups generated by automata; polyominoes and multi-dimensional patterns; decidability questions; image manipulation and compression; efficient text algorithms; relationships to cryptography, concurrency, complexity theory and logic; bio-inspired computing; and quantum computing.

Following its establishment by Grzegorz Rozenberg and Arto Salomaa in Turku (1993), a DLT conference was held every other year in Magdeburg (1995), Thessaloniki (1997), Aachen (1999) and Vienna (2001). From 2001 onwards, DLT conferences have been organised yearly, typically alternating between locations within and outside of Europe. The locations of DLT conferences since 2002 were: Kyoto (2002), Szeged (2003), Auckland (2004), Palermo (2005), Santa Barbara (2006), Turku (2007), Kyoto (2008), Stuttgart (2009), London (2010), Milan (2011), Taipei (2012), Marne-la-Vallée (2013), Ekaterinburg (2014), Liverpool (2015), Montréal (2016), Liège (2017), Tokyo (2018), Warsaw (2019), Porto (2021), Tampa (2022), Umeå (2023) and Göttingen (2024); since DLT 2001, the proceedings have been published in the Lecture Notes in Computer Science series of Springer. DLT 2020 was initially planned to be held in Tampa, Florida, but was unfortunately cancelled due to the COVID-19 pandemic (instead Tampa hosted DLT 2022). The accepted papers of DLT 2020 were nevertheless published as a volume of Lecture Notes in Computer Science, and the authors of these papers could present their work at DLT 2021 in Porto. After DLT 2024 broke from the rule of alternating European conferences with ones organised outside Europe, DLT 2025 is once again organised outside Europe, being held in South Korea for the first time.

At the end of January, we received the bad news that Arto Salomaa passed away. As we mentioned, Prof. Salomaa was one of the founders of the conference. His scientific contributions are fundamental in automata and formal language theory and in numerous other areas of theoretical computer science. He will be greatly missed by the DLT community, but he will continue to be present in our research and studies, with the influence and inspiration that comes from his seminal findings. A commemoration of Prof. Salomaa, with a talk given by Jarkko Kari (University of Turku), was held during the conference.

In 2018, the DLT conference series instituted the Salomaa Prize to honour the work of the late Arto Salomaa, and to increase the visibility of research on automata and formal language theory. The prize is funded by the University of Turku. The ceremony for the Salomaa Prize 2025 took place during the conference. We congratulate the winner Georg Zetzsche, from the Max Planck Institute for Software Systems in Kaiserslautern.

This volume contains invited contributions as well as the accepted papers of DLT 2025. There were 38 submissions. Each submission was single-blindly reviewed by at least three experts in the field and thoroughly discussed by the Program Committee. In the end, the Program Committee decided to accept 18 of these submissions for presentation at the conference. We would like to thank the members of the Program Committee, and all external reviewers, for their hard work in evaluating the papers and for valuable comments that led to the selection of the contributed papers.

There were four invited talks, which were presented by:

- Kyungmin Bae (POSTECH, South Korea): Robust Model Checking for Signal Temporal Logic
- Christoph Haase (University of Oxford, UK): Formal Languages and Arithmetic Theories: Recent Results and Open Problems
- Shunsuke Inenaga (Kyushu University, Japan): Recent Developments on the CDAWGs: Algorithms and Combinatorics
- Dora Giammarresi (University of Rome Tor Vergata, Italy): Partial Cubes and Fibonacci Dimension: Insights and Perspectives.

We would like to express our sincere thanks to the four invited speakers, to the presenters of contributed talks and to all authors of submitted papers for their contributions to the success of DLT 2025.

The EasyChair conference system provided excellent support in the selection of the papers and the preparation of these proceedings. Special thanks are due to the Lecture Notes in Computer Science team at Springer for having granted us the opportunity to publish these proceedings in the series, and also for their help during the process. The registration of the conference was implemented using the conference management systems provided by the Korean Institute of Information Scientists and Engineers.

Most importantly, the DLT 2025 conference would not have been possible without the financial support of the University of Seoul and the National Research Foundation of Korea (NRF) through the grant funded by MSIT (project number RS-2024-00456065). We express our deepest thanks to the NRF for this support.

We are grateful, as well, to the members of the Organizing Committee: Carl-Fredrik Nyberg-Brodda (Korea Institute for Advanced Study), Seongmin Kim (University of Seoul), Suyoung Park (University of Seoul), Aditi (University of Seoul), Gunhee Joe (University of Seoul), Miru Hong (University of Seoul) and Geoungjoo Lee (University of Seoul) for their help in the organization of the conference.

June 2025

Sang-Ki Ko
Florin Manea

Organization

Program Committee Co-Chairs

Sang-Ki Ko	University of Seoul, South Korea
Florin Manea	University of Göttingen, Germany

Program Committee

Valérie Berthé	CNRS IRIF, France
Marie-Pierre Béal	Université Gustave Eiffel, France
Emilie Charlier	University of Liège, Belgium
Erzsébet Csuhaj-Varjú	Eötvös Loránd University, Hungary
Joel Day	Loughborough University, UK
Szilárd Zsolt Fazekas	Akita University, Japan
Pamela Fleischmann	Christian Albrechts University of Kiel, Germany
Yo-Sub Han	Yonsei University, South Korea
Sang-Ki Ko (Chair)	University of Seoul, South Korea
Florin Manea (Chair)	University of Göttingen, Germany
Ian McQuillan	University of Saskatchewan, Canada
Nelma Moreira	University of Porto, Portugal
Carl-Fredrik Nyberg-Brodda	Korean Institute for Advanced Study, South Korea
Giovanni Pighizzini	University of Milan, Italy
Igor Potapov	University of Liverpool, UK
Markus Schmid	Humboldt-Universität zu Berlin, Germany
Marinella Sciortino	University of Palermo, Italy
Ryoma Sin'ya	Akita University, Japan
Taylor Smith	St. Francis Xavier University, Canada
Bianca Truthe	Justus-Liebig-Universität Gieß en, Germany
Hsu-Chun Yen	National Taiwan University, Taiwan

Additional Reviewers

Abuzer Yakaryilmaz
Andrew Ryzhikov
Andreas Malcher
Annika Huch
Attila Kiss
Beatrice Palano
Bow-Yaw Wang
Brennan Lockinger

Bruno Guillon
Carlo Mereghetti
Christian Rauch
Djamal Belazzougui
Dominik D. Freydenberger
Dominik Köppl
Dominique Perrin
Duncan Adamson
Elena Di Lavore
Emmanuel Filiot
Flavio D'Alessandro
Francesco Dolce
Gabriele Fici
Galina Jiraskova
Ghajendran Poovanandran
Giuseppe Romana
Guilherme Duarte
Henning Bordihn
Henning Fernau
Henning Urbat
Hideo Bannai
Indhumathi Raman
Ismaël Jecker
Jaehwan Jeong
Jean-Éric Pin
Julien Leroy
Jozef Jirasek
Kai Salomaa
Liat Peterfreund
Luís Cunha
Markus Holzer
Markus Whiteland
Martin Kutrib
Massimiliano Goldwurm
Matthias Wendlandt
Max Wiedenhöft
Michal Hospodár
Michel Rigo
Narad Rampersad
Noam Zeilberger
Olga Martynova
Özlem Salehi
Pascal Caron
Paweł Gawrychowski
Rogério Reis
Robert Mercaş
Sam Thompson
Sabine Broda
Sabrina Mantaci
Sergiu Ivanov
Stefan Hoffmann
Stavros Konstantinidis
Takuya Mieno
Tomasz Kociumaka
Tomoyuki Yamakami
Tore Koß
Victor Marsault
Wen Chean Teh
Yuto Nakashima
Zbyněk Křivka
Zsuzsanna Lipták

Robust Model Checking for Signal Temporal Logic

Kyungmin Bae

Pohang University of Science and Technology, Pohang, South Korea
kmbae@postech.ac.kr

Abstract. Signal Temporal Logic (STL) is a temporal logic for specifying linear-time properties of continuous signals. It is widely used in monitoring and testing hybrid systems that exhibit both discrete and continuous behavior. However, model checking for such systems has mostly focused on invariants and reachability, due to the difficulty of handling uncountably many continuously evolving signals. In this talk, I introduce a bounded model checking approach for general STL properties of hybrid systems. Our method is built on a novel logical foundation for STL, including: (i) *syntactic separation* [1], which decomposes STL formulas into parts that depend exclusively on separate signal segments; (ii) *signal discretization* [2], which abstracts continuous signals into discrete elements; and (iii) *epsilon-strengthening* [3, 4], which reduces robust STL checking to Boolean STL checking. This foundation enables STL model checking—up to given bounds and robustness thresholds—to be reduced to the satisfiability of a first-order formula, which is decidable whenever the reachability problem for the underlying hybrid system is decidable. Based on this reduction, we develop a pioneering bounded model checker for hybrid systems, called STLmc [4].

Keywords. Model checking · Signal temporal logic · Hybrid systems · Satisfiability modulo theories

References

1. Bae, K., Lee, J.: Bounded model checking of signal temporal logic properties using syntactic separation. Proc. ACM Program. Lang. **3**(POPL), 51:1–51:30 (2019)
2. Lee, J., Yu, G., Bae, K.: Efficient SMT-based model checking for signal temporal logic. In: ASE, pp. 343–354. IEEE (2021)
3. Lee, J., Yu, G., Bae, K.: SMT-based robust model checking for signal temporal logic. Sci. Comput. Program. **246**, 103332 (2025)
4. Yu, G., Lee, J., Bae, K.: STLmc: robust STL model checking of hybrid systems using SMT. In: CAV. LNCS, vol. 13371, pp. 524–537. Springer, Cham (2022)

This work was partially supported by the NRF grant (RS-2021-NR060080) and the IITP grant (RS-2024-00439856), both funded by the South Korean government (MSIT).

Recent Developments on the CDAWGs: Algorithms and Combinatorics

Shunsuke Inenaga

Department of Informatics, Kyushu University, Fukuoka, Japan
inenaga.shunsuke.380@m.kyushu-u.ac.jp

Abstract. The *Compact directed acyclic word graph* (*CDAWG*) [Blumer et al. 1987] of a string T, denoted $\mathsf{CDAWG}(T)$, is an edge-labeled DAG that is obtained by merging isomorphic subtrees of the suffix tree [Weiner 1973] of the same string T. In other words, $\mathsf{CDAWG}(T)$ is the minimal partial DFA that recognizes all the substrings of T. CDAWGs are a fundamental string data structure with applications in text pattern searching, data compression, and pattern discovery. For instance, $\mathsf{CDAWG}(T)$ allows for matching patterns and computing *maximal exact matches* (*MEMs*) in optimal $O(m)$ time with $O(\mathsf{e}(T))$ space, where m is the pattern length and $\mathsf{e}(T)$ is the number of edges in $\mathsf{CDAWG}(T)$.

It is known that there is a one-to-one correspondence between the internal nodes of $\mathsf{CDAWG}(T)$ and the *maximal repeats* in T, and that there is a one-to-one correspondence between the edges of $\mathsf{CDAWG}(T)$ and the right-extensions of the maximal repeats in T. Hence, the number e of the edges of the CDAWG tends to be small when the string contains a small number of distinct maximal repeats (and their right extensions). In the best case, e can be $O(\log n)$ for some highly repetitive strings of length n, including for the Fibonacci words and Thue-Morse words. In this talk, after revisiting the basic notions and properties, we review some of the recent developments on CDAWGs in both algorithms and combinatorics perspectives, including the following:

- $\mathsf{CDAWG}(T)$ can be stored in $O(\mathsf{e}(T))$ space, without storing the input string T [Belazzougui & Cunial 2017, Inenaga 2024].
- $\mathsf{CDAWG}(T)$ can compute *unusual words* for the input string T, including the *minimal unique substrings* (*MUSs*), *minimal rare words* (*MRWs*) and *minimal absent words* (*MAWs*) in optimal time with $O(\mathsf{e}(T))$ working space [Inenaga et al. 2024, Mieno & Inenaga 2025].
- *Sensitivity*, the maximum increase in the size after an edit operation, is only a constant for CDAWGs. Namely, $\mathsf{e}(T')/\mathsf{e}(T) = O(1)$ holds for any string T and any string T' whose edit distance from T is $O(1)$ [Hamai et al. 2025].
- Matching upper and lower bounds $\mathsf{e}(T^R)/\mathsf{e}(T) = \Theta(\sqrt{n})$, where T is any string of length n and T^R is the reversal of T [Inenaga & Kosolobov, 2024].

Keywords. CDAWGs · compressed suffix automata · string data structures · maximal repeats · compression sensitivity

Contents

Formal Languages and Arithmetic Theories: Recent Results and Open Problems

Christoph Haase[1(✉)] and Mikhail Starchak[2]

[1] Department of Computer Science, University of Oxford, Oxford, UK
christoph.haase@cs.ox.ac.ukl

[2] Saint Petersburg Petersburg State University, Saint Petersburg, Russia

Abstract. This paper surveys recent developments at the intersection of formal language theory and logical theories of arithmetic, focusing particularly on Büchi and Semënov arithmetic. We explore the connections between sets of integers defined by logical formulas and their representation as formal languages, and present some key results concerning the expressive power and algorithmic properties of these arithmetic theories. We also outline several open problems that arise from the current body of literature.

Keywords: Presburger arithmetic · Büchi arithmetic · exponentiation · Semënov arithmetic · string constraints · automatic structures

1 Introduction

This paper accompanies an invited talk of the first author at the 29th International Conference on Developments in Language Theory in Seoul, South Korea. One goal of this paper is to non-exhaustively survey some results at the intersection of formal language theory and logical theories of arithmetic that have been obtained by the authors and others. At this intersection, we look at how to relate sets of sets of integers to classes of formal languages studied in formal language theory. By viewing a word over the alphabet $\Sigma = \{0, \ldots, p-1\}$ as a base-p encoding of a natural number, we can, for instance, ask what logical theory of arithmetic characterizes a certain class of formal languages. Conversely, we can study what classes of formal languages can capture sets of numbers definable in (fragments of) some logical theory of arithmetic. One of the most prominent of such correspondences and starting point of this paper is the Büchi-Bruyère theorem, showing that Büchi arithmetic fully characterizes the class of regular languages. In the second part of this article, we focus on Semënov arithmetic and its connection to deciding string constraints.

As another goal of this paper, we also distill a list of open problems arising from those existing results. Most of them lie at the intersection of formal language theory and logical theories of arithmetic, and we believe that approaching

S.-K. Ko and F. Manea (Eds.): DLT 2025, LNCS 16036, pp. 1–14, 2026.
https://doi.org/10.1007/978-3-032-01475-7_1

them will lead to new developments in language theory. We also discuss some open problems related to the computational complexity of logical theories of arithmetic.

2 Preliminaries

A *non-deterministic finite automaton (NFA)* is a tuple $A = (Q, \Sigma, \delta, q_0, F)$, where Q is a finite set of states, Σ is a finite alphabet, $\delta \subseteq Q \times \Sigma \times Q$ is the transition relation, $q_0 \in Q$ is the initial state and $F \subseteq Q$ is the set of final states. If for every $q \in Q$ and $a \in \Sigma$ there is at most one $r \in Q$ such that $(q, a, r) \in \delta$ then we call A a deterministic finite automaton (DFA). Throughout this paper, we assume that $\Sigma = \{0, \ldots, p-1\}$ for some $p \geq 2$. Given $a \in \Sigma$, we write $q \xrightarrow{a} r$ whenever $(q, a, r) \in \delta$. A word $w = a_1 a_2 \cdots a_n \in \Sigma^*$ has a run from q to r in A if $q_1 \xrightarrow{a_1} q_2 \xrightarrow{a_2} \cdots \xrightarrow{a_n} q_{n+1}$ such that $q = q_1$ and $r = q_{n+1}$. We say that w is accepted by A if w has a run from q_0 to q_f in A for some $q_f \in F$. The *language* $L(A)$ *of* A is the set of all words accepted by A. We can also view an NFA as a *Büchi automaton* accepting infinite words $w \in \Sigma^\omega$ by changing the acceptance conditions. Formally, $w = a_1 a_2 a_3 \cdots \in \Sigma^\omega$ is accepted by the Büchi automaton A if $q_0 \xrightarrow{a_1} q_1 \xrightarrow{a_2} q_2 \xrightarrow{a_3} \cdots$ and there is some $q \in F$ such that $q = q_i$ for infinitely many i.

For a language $L \subseteq \Sigma^*$, we define its density function $\#_L \colon \mathbb{N} \to \mathbb{N}$ as

$$\#_L(n) := \#(L \cap \Sigma^n),$$

where $\#$ is the cardinality function. We can then speak of some L having, e.g., polynomial density if $\#_L = O(p(n))$ for some polynomial $p \colon \mathbb{N} \to \mathbb{N}$.

Given $w = a_0 \cdots a_n \in \Sigma^*$, if read in base p, w encodes a number $[\![w]\!]$ in most significant digit first representation:

$$[\![w]\!] := \sum_{i=0}^{n} p^i a_i .$$

We naturally lift this definition to languages $L \subseteq \Sigma^*$ and define a set of non-negative integers $[\![L]\!] = \{[\![w]\!] \in \mathbb{N} : w \in L\}$. Note that $[\![\cdot]\!]$ is not injective due to the possibility of trailing zeros. For $b \in \mathbb{N} \setminus \{0\}$, we define $[\![b]\!]^{-1}$ to be the unique $w = a_n \cdots a_0 \in \Sigma^*$ such that $b = [\![w]\!]$ and $a_n \neq 0$, and $[\![0]\!]^{-1} := 0$.

We often consider NFA that read words over Σ synchronously. The alphabet of an NFA reading k words synchronously consequently is Σ^k. Given a word

$$w = \begin{bmatrix} a_{n,1} \\ \vdots \\ a_{n,k} \end{bmatrix} \cdots \begin{bmatrix} a_{1,1} \\ \vdots \\ a_{1,k} \end{bmatrix} \begin{bmatrix} a_{0,1} \\ \vdots \\ a_{0,k} \end{bmatrix} \in (\Sigma^k)^*,$$

we let $\pi_i(w) := a_{n,i} \cdots a_{1,i} a_{0,i}$ be the word encoded in the i-th component of w. Such a word w encodes a k-tuple of numbers: $[\![w]\!] := ([\![\pi_1(w)]\!], \ldots, [\![\pi_k(w)]\!]) \in \mathbb{N}^k$, and for $L \subseteq (\Sigma^k)^*$ we define $[\![L]\!] := \{[\![w]\!] \in \mathbb{N} : w \in L\}$. Finally, the size of an NFA A over Σ^k is defined as $|A| :- |Q| + |Q|^2 \cdot p^k$.

We now turn towards introducing logical theories of arithmetic as relational first-order structures. *Presburger arithmetic* commonly refers to the first-order theory of the structure $\langle \mathbb{N}, 0, 1, +, \leq \rangle$ in which all constants and relation symbols are interpreted in their natural semantics. *Büchi arithmetic of base p* is the first-order theory of the extended structure $\langle \mathbb{N}, 0, 1, +, V_p, \leq \rangle$. Here, V_p is a binary predicate such that $V_p(x, y)$ holds whenever x is the largest power of p dividing y without remainder. Observe that using the V_p predicate, we can express the unary predicate $p^{\mathbb{N}}$ asserting that a number is a power of p: $p^{\mathbb{N}}(x)$ holds if and only if $V_p(x, x)$ holds. *Semënov arithmetic* is the first-order theory of the structure $\langle \mathbb{N}, 0, 1, +, \exp_p, \leq \rangle$, where $\exp_p$ is a binary *exponentiation* predicate such that $\exp_p(x, y)$ holds whenever $x = p^y$. For any formula $\Phi(x_1, \ldots, x_n)$ in a logical theory of arithmetic open in the variables $x_1, \ldots, x_k$, we denote by $[\![\Phi]\!] \subseteq \mathbb{N}^k$ its set of satisfying assignments.

Given a class of languages $\mathcal{L}$, we say that $\mathcal{L}$ is characterized by a logical theory whenever for any $L \in \mathcal{L}$ there exists a formula $\Phi_L(x)$ such that $[\![L]\!] = [\![\Phi_L]\!]$.

3 Büchi Arithmetic and Regular Languages

Büchi's seminal work [9] introduced two central ideas. First, Büchi showed how to decide Büchi arithmetic by a reduction to the language non-emptiness problem for NFA. To this end, he showed that every k-ary predicate of Büchi arithmetic can be implemented by a DFA over the alphabet Σ^k. Hence, implementing, e.g., the ternary addition predicate " + " requires defining an NFA A over Σ^3 such that $w \in L(A)$ if and only if $\pi_1(w) + \pi_2(w) = \pi_3(w)$. An example of such a DFA with $\Sigma = \{0, 1\}$ is given in Fig. 1

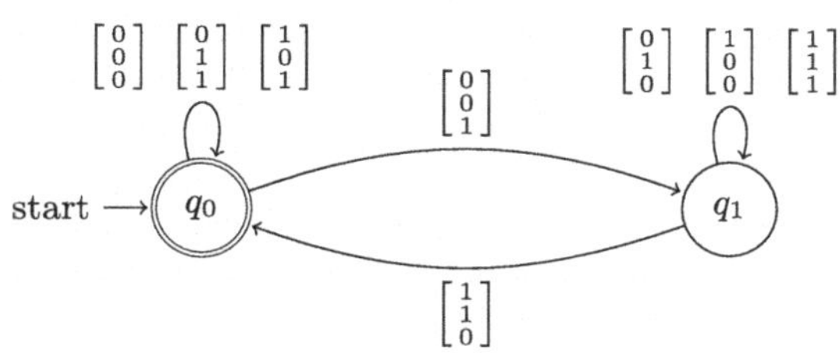

Fig. 1. A DFA for the ternary addition predicate.

With a DFA for addition at hand and observing that constructing a DFA for the V_p predicate is fairly straight forward, it is now not difficult to see how Büchi arithmetic can be decided: For a given formula $\Phi(x_1, \ldots, x_k)$, construct by structural induction on Φ an NFA for all atomic formulas, and then use the closure of regular languages under inverse homomorphisms together with closure under intersection for conjunctions; union for disjunctions; complement for negations; and homomorphisms for existential quantification (universal quantifiers can be dealt with using the standard identity $\forall x\, \Phi \equiv \neg\exists x\, \neg\Phi$). Ultimately, we arrive at an NFA A_Φ over the alphabet Σ^k such that $[\![L(A_\Phi)]\!] = [\![\Phi]\!]$. Note that since regular languages are closed under reversal, the choice between least and most significant digit first encodings does not matter, and different authors prefer different encodings.

The algorithmics of Büchi's approach to deciding Büchi arithmetic have been investigated from the 1990s onwards. One challenge is that deciding a sentence with existential and universal quantifier alternations leads to an exponential blow-up in the size of an NFA obtained in the construction outlined above due to repeated complementation steps. For Büchi arithmetic, this is optimal since a non-elementary TOWER lower bound can easily be shown e.g. by a reduction from WS1S [27,36]. However, it was an open problem for a long time whether Büchi's approach is elementary when given formulas of Presburger arithmetic as input. This was resolved in the affirmative by Durand-Gasselin and Habermehl in 2012 [16]. The question of whether the approach of eliminating a universal quantifier via the identity $\forall x\, \Phi \equiv \neg\exists x\, \neg\Phi$ is optimal, as it involves two complementation steps, has recently been answered [20]. There, the authors show that the smallest NFA may inherently grow doubly exponential when projecting away a universal quantifier.

There are at least three open questions on the algorithmics of Büchi's approach and Büchi arithmetic that the authors of this article can identify as open and worthwhile to pursue. Given a system of equations over the natural numbers $A \cdot x = b$, it is well known that if one exists, there is a solution x^* in which the bit length of all components of x^* is polynomial in A and b [6,18]. By the construction above, we can alternatively represent all solutions to $A \cdot x = b$ as the language of the intersection of m DFA $A_1, \ldots, A_m$ similar to those displayed in Fig. 1 for some m which is polynomial in the size of the description of A and b. In general, the shortest word in the intersection of m DFA, all of size at least s, may be of length $\Omega(s^m)$. However, the existing results on small solutions established for integer linear programming suggest that the DFA displayed in Fig. 1 possess some properties that guarantee the existence of a shortest word in the intersection of $L(A_1), \ldots, L(A_m)$ of length polynomial in the description $A_1, \ldots, A_m$. This observation gives rise to the following open problem.

Open Problem 1. Is there a syntactic subclass $\mathcal{A}$ of DFA that includes the DFA for the addition predicate such that for any $A_1, \ldots, A_m \in \mathcal{A}$, if their intersection $L = \bigcap_{i=1}^m L(A_i)$ is non-empty then L contains a word of length polynomial in $|A_1| + \cdots + |A_m|$?

The second open problem pertains the aforementioned elimination of a universal quantifier by application of the identity $\forall x\, \Phi \equiv \neg\exists x\, \neg\Phi$. As mentioned above, this approach incurs a doubly exponential blow-up but is, in general, optimal [20]. However, recent work has shown that for certain ω-regular structures, double complementation can in fact be avoided [5], and that the Büchi automaton resulting from eliminating a universal quantifier is only of exponential size. Similar observations have been made for the elimination of universal quantifiers in quantified integer programming [12]. Given a DFA A over Σ^{k+1}, encoding assignments of $k+1$ variables, eliminating a universally quantified variable computes the universal projection of $L(A)$

$$\{(w_1,\dots,w_k)\in(\Sigma^k)^* : (w_1,\dots,w_k)\otimes w_{k+1}\in L(A) \text{ for all } w_{k+1}\in\Sigma^*\},$$

where $\otimes$ is the convolution operator that ensures that w_{k+1} and all other w_i have the same length by padding with zeros appropriately.

Open Problem 2. Is there a syntactic subclass $\mathcal{A}$ of DFA such that for any $A \in \mathcal{A}$ over Σ^{k+1}, the universal projection of $L(A)$ to Σ^k can be represented by an NFA of size exponential in $|A|$?

The final open problem on the algorithmics of Büchi arithmetic pertains the complexity of Büchi arithmetic itself. It is known that its existential fragment is NP-complete [13,19], and it easily seen that the logic is TOWER-complete by a reduction from WS1S. However, despite some initial efforts [21], the precise complexity of Büchi arithmetic with a fixed quantifier alternation prefix is still an open problem.

Open Problem 3. Determine the complexity of Büchi arithmetic with a fixed number of quantifier alternations.

Returning now to Büchi's seminal article [9], he also claimed that regular langugages are definable in arithmetic, even in the restricted structure of Büchi arithmetic $\langle \mathbb{N}, 0, 1, +, p^{\mathbb{N}}, \leq\rangle$. That is, for any regular language $L \subseteq \Sigma^*$, Büchi claimed that there is a $\Phi_L(x)$ in that structure such that $[\![L]\!] = [\![\Phi_L]\!]$. A flaw in Büchi's argument was pointed out by McNaughton [26], and Semënov later observed [30] that $p^{\mathbb{N}}$ cannot be used to obtain the result Büchi was hoping for. To this end, Semënov constructed a complete description of the expressive power of that structure. An important role here play the languages $\Sigma_{\ell,m,c}$, which are defined for every fixed triple $\ell, m, c \in \mathbb{N}$ as follows. If $\ell \neq 0$ and $m \neq 0$, then $\Sigma_{\ell,m,c}$ is the set of all words $w = [\![b]\!]^{-1}$ such that $b \in \mathbb{N}$ is congruent to c modulo m and the length of w is divisible by ℓ. Otherwise, we define $\Sigma_{0,m,c} = \Sigma_{\ell,0,c} := \{\epsilon\}$ for every $\ell, m, c \in \mathbb{N}$. For example, in the case of decimal encodings of non-negative integers, the language $\Sigma_{2,25,3}$ is composed of the words $28, 53, 78, 1003, 1028$, *etc.*. The desired characterization is given by the next theorem.

Theorem 1 (*[30]*). *For every language $L \subseteq \Sigma^*$ there is a formula Φ_L in the structure $\langle \mathbb{N}, 0, 1, +, p^{\mathbb{N}}, \leq\rangle$ with $[\![L]\!] = [\![\Phi_L]\!]$ if and only if L is a finite union of languages of the form $v_0 w_1^* v_1 w_2^* v_2 \dots w_n^* v_n \Sigma_{\ell,m,c}$ for some fixed words $v_i, w_i \in \Sigma^*$ and non-negative integers ℓ, m, c.*

The proof in one direction is based on a quantifier elimination approach. The variables in $\Phi_L(x)$ are split into two sorts: for non-negative integers and for non-negative integer powers of p. Semënov shows that $\Phi_L(x)$ is equivalent to an existential formula $\exists(p^{\alpha_1})\dots\exists(p^{\alpha_n}) : \Psi_L(p^{\alpha_1},\dots,p^{\alpha_n},x)$, where p^{α_i} is an alias for a variable y_i with an extra constraint $p^{\mathbb{N}}(y_i)$. To give some insight into the main idea of the proof, let us assume that $p = 10$ and that there are only two variables for powers of 10 with $\Delta \equiv (10^2 < 10^{\alpha_1}) \wedge (10^2 \cdot 10^{\alpha_1} < 10^{\alpha_2})$. Consider a pair of inequalities Γ defined as

$$3 \cdot 10^{\alpha_2} + 17 \cdot 10^{\alpha_1} - 1 \leq x \leq 4 \cdot 10^{\alpha_2} - 22 \cdot 10^{\alpha_1} + 13$$

in the formula $\Phi_{E1}(x) \equiv \exists(10^{\alpha_1})\exists(10^{\alpha_2}) : \Delta \wedge \Gamma$. This formula expresses the fact that x (when considered in its decimal representation) is such that

$$3\,0^{\beta_2}\,16\,9^{\beta_1}\,99 \leq x \leq 3\,9^{\beta_2}\,78\,0^{\beta_1}\,13,$$

where $\beta_1 = \alpha_1 - 2$ and $\beta_2 = \alpha_2 - \alpha_1 - 2$. The existence of the representation from Theorem 1 for the formula Φ_{E1} is not difficult to show, and we leave the proof as an exercise for the interested reader. For the converse direction, let us only notice that the class of all languages $\Sigma_{\ell,m,c}$ can easily be characterized by existential formulas. For example, $[\![\Sigma_{2,2,1}]\!] = [\![\Phi_{2,2,1}]\!]$ for the formula $\Phi_{2,2,1}(x)$ defined as

$$\exists u \exists v \exists w : 10^{\mathbb{N}}(u) \wedge (u \leq x) \wedge (x+1 \leq 10 \cdot u) \wedge (99 \cdot v = 10 \cdot u - 1) \wedge (2 \cdot w = x - 1).$$

Theorem 1 allows to conclude that there is no formula Φ in the structure $\langle \mathbb{N}, 0, 1, +, p^{\mathbb{N}}, \leq \rangle$ with $[\![\Phi]\!] = [\![(01 \cup 10)^*]\!]$. In fact, the predicate $p^{\mathbb{N}}$ is in a sense the weakest regular predicate which cannot be defined in Presburger arithmetic. In 1997, Bès [4] proved that for every language $L \subseteq (\Sigma^k)^*$ such that $[\![L]\!]$ is not definable in Presburger arithmetic, the predicate $p^{\mathbb{N}}$ is definable in the structure $\langle \mathbb{N}, 0, 1, +, [\![L]\!], \leq \rangle$, where by $[\![L]\!]$ stands for the unary predicate $x \in [\![L]\!]$.

An alternative proof of McNaughton's conjecture (which was obtained in 1984 [31,32]) uses a decidability argument. Semënov arithmetic is decidable and model-complete, i.e., any formula of Semënov arithmetic is logically equivalent to an existential formula. However, being endowed with additional regular predicates, this theory becomes undecidable. Since $p^{\mathbb{N}}$ is trivially definable in the existential fragment of Semënov arithmetic, it follows that $\langle \mathbb{N}, 0, 1, +, p^{\mathbb{N}}, \leq \rangle$ is insufficiently expressive to capture all regular languages. Let us conclude the discussion of that structure with an interesting open problem.

Open Problem 4. Does the first-order theory of the structure $\langle \mathbb{N}, 0, 1, +, p^{\mathbb{N}}, \leq \rangle$ characterize all regular predicates definable in Semënov arithmetic in base p?

Büchi arithmetic is already expressive enough to characterize the class of all regular languages. From the classical results of Bruyère [7,8] and Villemaire [37] we know that runs of NFA can be arithmetized by formulas of Büchi arithmetic with two quantifier alternations. It was later observed that Villemaire's construction can be modified to establish that one quantifier alternation is sufficient for expressive completeness. The existential fragment of Büchi arithmetic is, however, strictly less expressive: the main result of [22] is the existence of a regular language that cannot be characterized by any existential formula in that theory. To this end, the authors established a density argument and showed that for any formula of existential Büchi arithmetic $\Phi(x)$ and the language of its encodings L_Φ, either $d_{L_\Phi}(n) = O(n^c)$ for some constant $c > 0$, or $d_{L_\Phi}(n) \geq c \cdot p^n$ for infinitely many n. Observe that $d_{(01 \cup 10)^*}(n)$ is either equal to zero or $\Theta(2^{n/2})$, and hence $(01 \cup 10)^*$ is not definable in existential Büchi arithmetic. In summary, we have the following fundamental result.

Theorem 2 *The class of regular languages over $(\Sigma^k)^*$ is characterized by Büchi arithmetic with one quantifier alternation. However, the language $(01 \cup 10)^*$ is not definable in the existential fragment Büchi arithmetic.*

An immediate follow-up question is whether we can we completely characterize the expressive power of the existential fragment of Büchi arithmetic of base p. As a starting point in attempts to find an answer, we could appeal to Theorem 1 on the expressive power of Presburger arithmetic with $p^{\mathbb{N}}$. Since every formula in that theory is equivalent to an existential formula, the finite unions from Theorem 1 can also be characterized by existential Büchi arithmetic of base p. However, in contrast to $p^{\mathbb{N}}$, the predicate V_p allows specifying arbitrary positions for the languages $\Sigma_{\ell,m,c}$ in the p-ary expansions of non-negative integers. For example, the following formula describes the set $[\![\Sigma_{2,1,0}29\Sigma_{2,2,1}]\!]$ when $p = 10$:

$$\exists(10^{\alpha})\,\exists u\,\exists v\,\exists r : V_{10}(v,\, u) \wedge (100 \cdot 10^{\alpha} \leq v) \wedge (x = u + 29 \cdot 10^{\alpha} + r) \wedge \Phi_{2,2,1}(u+1) \wedge (r + 1 \leq 10^{\alpha}) \wedge \Phi_{2,2,1}(r).$$

This formula states that for an integer 10^{α}, the remainder r of the division of x by 10^{α} is such that $r \in [\![\Sigma_{2,2,1}]\!]$; the quotient of the division of x by $100 \cdot 10^{\alpha}$ can be any positive integer whose decimal expansion has even length. As it turns out, by allowing this freedom for the languages $\Sigma_{\ell,m,c}$, it becomes possible to completely characterize the expressive power of the existential Büchi arithmetic.

For a pair of positive integers ℓ and m, define a class $\mathcal{C}_{\ell,m}$ as the union of the following languages:

- $\{w\}$ and w^* for every word $w \in \Sigma$ of length at most ℓ;
- $\Sigma_{\ell',m',c'}$ for every $\ell' \leq \ell$, $m' \leq m$, $c' \in [0..m'-1]$.

Let us also say that a regular language $L \subseteq \Sigma$ has a $\{\cdot, \cup\}$-representation over some class of languages $\mathcal{C}$ if L can be obtained by a finite number of applications of concatenation and union to $\mathcal{C}$. The main result of [34] is the next theorem.

Theorem 3 (*[34]*). *For any language $L \subseteq \Sigma^*$ there is a formula Φ_L of existential Büchi arithmetic with $[\![L]\!] = [\![\Phi_L]\!]$ if and only if there exist positive integers ℓ and m such that L has $\{\cdot, \cup\}$-representation over $\mathcal{C}_{\ell,m}$.*

The proof is based on quantifier elimination for a slightly modified existential theory: the predicate V_p is replaced by the function "$\bmod_p$", which computes the remainder of the division by a power of p, instead of V_p. This replacement does not change the expressive power of the theory. Every formula $\Phi(x)$ is equivalent to an existential formula $\exists(p^{\alpha_1}) \ldots \exists(p^{\alpha_n}) : \Psi(p^{\alpha_1}, \ldots, p^{\alpha_n}, x)$, but now the function "$\bmod_p$" splits the base p representation of x into independent fragments (similar to the variables u and r in the example above), each of which can be described using Theorem 1. As a consequence of Theorem 3, the density argument again rules out the possibility of expressing the set $[\![(01 \cup 10)^*]\!]$ in existential Büchi arithmetic. This theorem also reveals a remarkable relationship between the decidability problems in arithmetic theories and in the theory of regular languages. The next open problem is formulated in [22].

Open Problem 5. Is it decidable whether for a given regular language $L \subseteq \Sigma^*$, $[\![L]\!]$ is definable in the existential fragment of Büchi arithmetic of base p?

In 1983, Hashiguchi proved the so-called representation theorem on regular languages [23]. Given a set of operations $\Omega \subseteq \{\cdot, \cup, *\}$ and a finite class of regular languages $\mathcal{C}$, it is a decidable problem to determine whether a regular language L can be obtained by a finite number of applications of operations from Ω to the languages in $\mathcal{C}$. In the light of [23, Theorem 6.1] for $\Omega = \{\cdot, \cup\}$, to prove that Open Problem 4 is decidable, one could attempt to construct an upper bound on ℓ and m depending on the number of states of the minimal DFA recognizing the input language L.

Another important question in connection with Theorem 2 is whether there is a "natural" regular language $L \subseteq (\Sigma^k)^*$ such that the class of regular languages is characterized by the existential theory of the structure $\langle \mathbb{N}, 0, 1, +, [\![L]\!], \leq \rangle$? The arithmetization of NFA by Villemaire [37] uses a unique bounded universal quantifier, and this fact suggests turning to the existential encoding of computations of Turing machines from the seminal proof of the undecidability of the Hilbert's 10th Problem. A successful candidate for the regular language L comes from the papers by Matiyasevich [25] and his joint paper with Jones [24]: the ternary bitwise minimum operation of base p, which will be denoted by $\&_p$. Its semantics is natural, for example, $(12345 \&_{10} 54321 = x)$ is true only when $x = 12321$; for simplicity, we sometimes treat $\&_{10}$ as a function. The DFA recognizing $\&_p$ is easy to construct, and from the WS1S viewpoint, this predicate in base 2 corresponds to the intersection of finite subsets of $\mathbb{N}$. Whereas all recursively enumerable languages can be characterized by the first-order theory of the structure $\langle \mathbb{N}, 0, 1, +, \cdot, \exp, \&_2, \leq \rangle$ with the binary exponentiation function exp, for the regular languages there is the following theorem.

Theorem 4 *([33]). The class of regular languages over $(\Sigma^k)^*$ is characterized by the existential first-order theory of the structure $\langle \mathbb{N}, 0, 1, +, \&_p, \leq \rangle$.*

Let us now consider an example illustrating the idea underlying the proof of Theorem 4 where, for simplicity, we take $p = 2$ and omit the subscript 2. An extra symbol $\subseteq$ is defined by the formula $x \subseteq y \iff x \,\&\, y = x$ specifying that every '1' in the binary expansion of x occurs at the same position in the binary expansion of y. The DFA for the language $(01 \cup 10)^*$ and the computation table on input word 10100110 are given below.

Each blank cell in Fig. 2(b) contains '0', and we treat each row as a binary representation of the variable written next to the left border. The positions of '1' in the variables q_0, q_1, q_2 describe a run in the DFA. Now consider the following existential definition.

$$x \in [\![(01 \cup 10)^*]\!] \iff \exists t\, \exists \widetilde{x}\, \exists q_0 \exists q_1 \exists q_2 \Big[t \,\&\, (t-1) = 0 \wedge (x + \widetilde{x} = t - 1) \wedge \quad (1)$$

$$(q_0 \,\&\, q_1 = 0) \wedge (q_0 \,\&\, q_2 = 0) \wedge (q_1 \,\&\, q_2 = 0) \wedge (q_0 + q_1 + q_2 = 2t - 1) \wedge \quad (2)$$

$$(q_0 \,\&\, t = t) \wedge (q_0 \,\&\, 1 = 1) \wedge \quad (3)$$

$$(q_0 \,\&\, 2 \cdot x) \subseteq 2 \cdot q_2 \wedge (q_1 \,\&\, 2 \cdot x) \subseteq 2 \cdot q_0 \wedge (q_2 \,\&\, 2 \cdot x) \subseteq 0 \wedge \quad (4)$$

$$(q_0 \,\&\, 2 \cdot \widetilde{x}) \subseteq 2 \cdot q_1 \wedge (q_1 \,\&\, 2 \cdot \widetilde{x}) \subseteq 0 \wedge (q_2 \,\&\, 2 \cdot \widetilde{x}) \subseteq 2 \cdot q_0 \Big]. \quad (5)$$

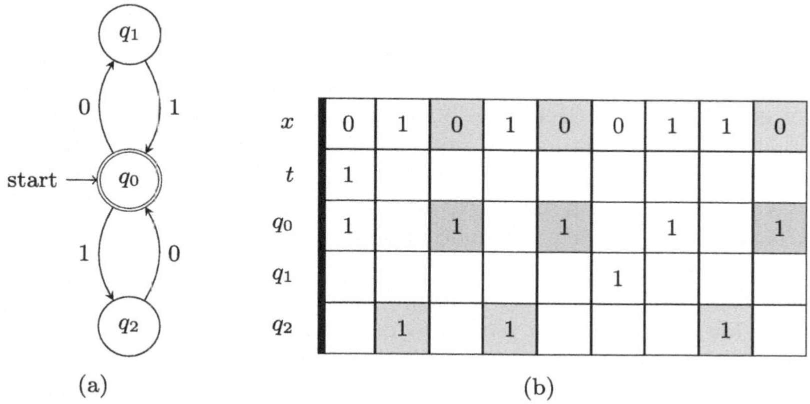

Fig. 2. DFA for the language $(01 \cup 10)^*$ and a run on input word 10100110.

Here, the conjunction in (1) says that the variable t is a power of 2 which is greater than $x + 1$ and the variable $\widetilde{x}$ corresponds to the two's-complement of x. The next line of the formula expresses the fact that the variables q_0, q_1, q_2 describe a sequence of states of length $\log_2(t) + 1$. Due to conjunction (3), this sequence should start and end in the same state q_0. Lines (4) and (5) describe the transition relation. For example, the last constraint specifies that every time the computation is in state q_2 and reads '0', the next state is q_0. The proof of Theorem 4 is a straightforward generalization of this example.

The language $(01 \cup 10)^*$ itself has the same property as the bitwise minimum: the existential theory of the structure $\langle \mathbb{N}, 0, 1, +, [\![(01 \cup 10)^*]\!], \leq \rangle$ characterizes all regular languages. We conclude this section with an important open problem.

Open Problem 6. Is it decidable whether for a given pair of regular languages L_1 and L_2 the set $[\![L_1]\!]$ is definable in the existential first-order theory of the structure $\langle \mathbb{N}, 0, 1, +, [\![L_2]\!], \leq \rangle$?

Note that a positive answer to this problem will give us an algorithm to determine whether a regular language L' preserves the property of expressive completeness of the existential theory of $\langle \mathbb{N}, 0, 1, +, [\![L']\!], \leq \rangle$: It would be sufficient to apply such an algorithm to the pair of languages $((01 \cup 10)^*, L')$.

4 Semënov Arithmetic and String Constraints

In this section we review some results on the decidability and complexity of Semënov arithmetic, and, as before, identify some open problems. We also put a particular emphasis on the connection between Semënov arithmetic and deciding string constraints.

The decidability of Semënov arithmetic was first shown by Semënov using quantifier elimination [32]. A major underlying conceptual contribution was

Semënov's use of almost periodic sequences [28], which enabled him to show an even more general result. He showed that the first-order theory of any extension of Presburger arithmetic with sparse functions such as exponentiation or the factorial function is decidable. Soon after, Cherlin and Point gave a more traditional quantifier elimination procedure for Semënov arithmetic [11]. Point also observed that Semënov arithmetic has non-elementary complexity as it can encode a pairing function [29], i.e., an injective map from $\mathbb{N} \times \mathbb{N}$ to $\mathbb{N}$, and any extension of Presburger arithmetic that can encode such functions has non-elementary complexity [14].

The existential fragment of Semënov arithmetic and its extensions have received a lot of attention in recent years, both from the point of view of its computational complexity and its applications in e.g. program verification. In 2024, Chistikov, Mansutti and Starchak developed a quantifier-elimination procedure yielding NP-completeness for existential Semënov arithmetic extended with Büchi's V_p predicate [13]. This low complexity is certainly surprising given that smallest solutions and solution sets of Semënov arithmetic can grow non-elementary, as, e.g., shown by the following family of formulas:

$$\Phi_n \equiv (x_1 = 1) \wedge \bigwedge_{1<i<n} \exp_2(x_{i+1}, x_i) \wedge (y < x_n) .$$

This family of formulas can also be used to show that, unsuprisingly, Semënov arithmetic is not an automatic structure and cannot be decided using finite automata as illustrated for Büchi arithmetic in the previous section. Just as for Büchi arithmetic, the complexity of Senënov arithmetic with a fixed number of quantifier alternations is an open problem [2].

Open Problem 7. Determine the complexity of deciding Semënov arithmetic with a fixed number of quantifier alternations.

Semënov arithmetic has an interesting connection to deciding string constraints. Let $\Sigma = \{0, 1\}$. Following [3], the *theory of enriched string constraints* T_{lnc} is the first-order theory of the two-sorted structure

$$\langle \Sigma^*, \mathbb{N}; \{w\}_{w \in \Sigma^*}, \cdot, \mathit{len}, \mathit{sn}, \{R_i\}_{i \in \mathbb{N}}, 0, 1, + \rangle,$$

where

- the binary function $\cdot$ over Σ^* is the string concatenation operator,
- the unary function $\mathit{len}\colon \Sigma^* \to \mathbb{N}$ returns on input w the length $|w|$ of w,
- the unary function $\mathit{sn}\colon \Sigma^* \to \mathbb{N}$ on input u returns $[\![u]\!]$, and
- $R_0, R_1, \ldots \subseteq \Sigma^*$ is an enumeration of all regular languages.

The remaining predicates, constant and function symbols are defined in their standard semantics. In all generality, the existential fragment of T_{lnc} is undecidable [3]. However, various fragments of T_{lnc} have been studied in the literature, in particular due to their applications in the verification of string-manipulating programs [3].

Of particular interest to us are the existential fragments T_{REln} and T_{ln}, which are obtained from T_{lnc} by omitting the concatenation operator (T_{REln}) and additionally omitting regular language membership (T_{ln}). As observed in [15], it is actually not difficult to see that existential T_{ln} reduces to existential Semënov arithmetic. The idea is to define a bijection $h\colon \{0,1\}^* \to \mathbb{N} \setminus \{0\}$ as $h(s) := [\![1s]\!]$. The function $sn(s,x)$ can then be "implemented" by the formula

$$\Phi_{sn}(s,x) \equiv \exists y \exists z_1 \exists z_2 : \exp_2(z_1,y) \wedge (z_1 \le s) \wedge \\ \exp_2(z_2, y+1) \wedge (s \le z_2) \wedge (x = s - z_1)\,.$$

Likewise, $len(s,x)$ computes the floored logarithm of s and can hence be "implemented" by

$$\Phi_{ln}(s,x) \equiv \exists y_1 \exists y_2 : \exp_2(y_1,x) \wedge (y_1 \le s) \wedge \exp_2(y_2, x+1) \wedge (s < y_2)\,.$$

Together with the results from [13], we get that existential T_{ln} is NP-complete.

Due to the presence of regular membership predicates, existential T_{REln} does not directly reduce to existential Semënov arithmetic. As a first step, a dedicated decision procedure for existential T_{REln} running in 3-EXP was given in [38]. This upper bound was independently improved to EXPSPACE in [15] by a reduction to the existential fragment of *generalized Semënov arithmetic*, which is obtained by adding regular membership predicates for every regular language $L \subseteq (\Sigma^k)^*$ for all $k > 0$ to the structure of Semënov arithmetic. As expected, those predicates allow for testing whether the p-ary expansion of a finite number of variables is contained in a regular language. While the discussion in the previous section of this paper shows that the full first-order theory of generalized Semënov arithmetic is undecidable, the existential fragment of generalized Semënov arithmetic was shown decidable in EXPSPACE in [15] by a reduction to the language emptiness problem in a restricted class of affine vector addition systems with states. More recently, an alternative proof of EXPSPACE membership of existential generalized Semënov arithmetic based on quantifier elimination interlinked with the finite automata approach for Büchi arithmetic has been established by the second author, together with a PSPACE lower bound for the fixed regular languages such as $(01 \cup 10)^*$ defined in the previous section [35]. At present, the precise complexities of both T_{REln} and existential generalized Semënov arithmetic are unknown; their complexities may not coincide since, informally speaking, T_{REln} cannot stack up towers of exponentials.

Open Problem 8. Determine the complexity of deciding the existential theories of T_{REln} and generalized Semënov arithmetic.

Another problem that appears worthwhile studying is the decidability of the existential fragment of T_{nc} in which only the concatenation operator and sn function are present. It may well be that this requires a more direct proof rather than a reduction to a theory of arithmetic since, at present, no good candidate logic for encoding the concatenation operator seems to exist.

Open Problem 9. Determine the decidability of the existential theory of T_{nc}.

We close this section by mentioning that tool support for Semënov arithmetic and its fragments has recently been developed. This includes support for integer exponentiation in CVC, the tool SwInE [17] that uses incremental linearization to decide integer arithmetic with exponentiation, the tool ExpSolver accompanying [38], and the tool ChrobELIAS [1] by the second author and collaborators that is based on the automata-based techniques described in [35] and can be used to decide existential formulas of generalized Semënov arithmetic.

Acknowledgments. The first author is supported by the European Research Council (ERC) under the European Union's Horizon 2020 research and innovation programme (Grant agreement No. 852769, ARiAT).

References

1. Belyanin, G., Frolov, T., Starchak, M.: ChrobELIAS: chrobak normal form-based exponential linear integer arithmetic solver (2025). https://github.com/Chrobelias/Chrobelias. Accessed 8 June 2025
2. Benedikt, M., Chistikov, D., Mansutti, A.: The complexity of Presburger arithmetic with power or powers. In: Proceedings of the ICALP. Leibniz International Proceedings in Informatics (LIPIcs), vol. 261, pp. 112:1–112:18. Schloss Dagstuhl – Leibniz-Zentrum für Informatik (2023). https://doi.org/10.4230/LIPIcs.ICALP.2023.112
3. Berzish, M., et al.: Towards more efficient methods for solving regular-expression heavy string constraints. Theor. Comput. Sci. **943**, 50–72 (2023). https://doi.org/10.1016/j.tcs.2022.12.009
4. Bès, A.: Undecidable extensions of Büchi arithmetic and Cobham-Semënov theorem. J. Symb. Log. **62**(4), 1280–1296 (1997). https://doi.org/10.2307/2275643
5. Boigelot, B., Fontaine, P., Vergain, B.: Universal first-order quantification over automata. In: Proceedings of the CIAA. LNCS, vol. 14151, pp. 91–102. Springer (2023). https://doi.org/10.1007/978-3-031-40247-0_6
6. Borosh, I., Treybig, L.B.: Bounds on positive integral solutions of linear Diophantine equations. Proc. Am. Math. Soc. **55**(2), 299–304 (1976). https://doi.org/10.2307/2041711
7. Bruyère, V.: Entiers et automates finis. Mémoire de fin d'études (1985)
8. Bruyère, V., Hansel, G., Michaux, C., Villemaire, R.: Logic and p-recognizable sets of integers. Bull. Belg. Math. Soc. Simon Stevin **1**(2), 191–238 (1994). https://doi.org/10.36045/bbms/1103408547
9. Büchi, R.J.: Weak second-order arithmetic and finite automata. Math. Logic Quart. **6**(1–6), 66–92 (1960). https://doi.org/10.1002/malq.19600060105
10. Büchi, R.J.: On a decision method in restricted second order arithmetic. In: Nagel, E., Suppes, P., Tarski, A. (eds.) Logic, methodology and philosophy of science, Proceedings of the 1960 International Congress, pp. 1–11. Stanford University Press, Stanford, Calif (1962)
11. Cherlin, G., Point, F.: On extensions of Presburger arithmetic. In: Proceedings of the fourth Easter conference on model theory in Groß Köris, pp. 17–34. Seminarberichte 86, Humboldt University, Berlin, Groß Köris, Germany (1986)
12. Chistikov, D., Haase, C.: On the complexity of quantified integer programming. In: Proceedings of the ICALP. LIPIcs, vol. 80, pp. 94:1–94:13. Schloss Dagstuhl - Leibniz-Zentrum für Informatik (2017). https://doi.org/10.4230/LIPICS.ICALP.2017.94
13. Chistikov, D., Mansutti, A., Starchak, M.R.: Integer linear-exponential programming in NP by quantifier elimination. In: Proceedings of the ICALP. LIPIcs, vol. 297, pp. 132:1–132:20 (2024). https://doi.org/10.4230/LIPIcs.ICALP.2024.132

14. Compton, K.J., Henson, C.W.: A uniform method for proving lower bounds on the computational complexity of logical theories. Ann. Pure Appl. Logic **48**(1), 1–79 (1990). https://doi.org/10.1016/0168-0072(90)90080-L
15. Draghici, A., Haase, C., Manea, F.: Semënov arithmetic, affine VASS, and string constraints. In: Proceedings of the STACS, pp. 29:1–29:19. Leibniz International Proceedings in Informatics (LIPIcs) (2024). https://doi.org/10.4230/LIPIcs.STACS.2024.29
16. Durand-Gasselin, A., Habermehl, P.: Ehrenfeucht-fraïssé goes elementarily automatic for structures of bounded degree. In: Proceedings of the STACS. LIPIcs, vol. 14, pp. 242–253. Schloss Dagstuhl - Leibniz-Zentrum für Informatik (2012). https://doi.org/10.4230/LIPICS.STACS.2012.242
17. Frohn, F., Giesl, J.: Satisfiability modulo exponential integer arithmetic. In: Proceedings of the IJCAR, pp. 344–365. Springer (2024). https://doi.org/10.1007/978-3-031-63498-7_21
18. von zur Gathen, J., Sieveking, M.: A bound on solutions of linear integer equalities and inequalities. Proc. Am. Math. Soc. (1978). https://doi.org/10.1090/S0002-9939-1978-0500555-0
19. Guépin, F., Haase, C., Worrell, J.: On the existential theories of Büchi arithmetic and linear p-Adic fields. In: Proceedings of the LICS, pp. 1–10 (2019). https://doi.org/10.1109/LICS.2019.8785681
20. Haase, C., Piórkowski, R.: Universal quantification makes automatic structures hard to decide. In: Proceedings of the CONCUR 2023. LIPIcs, vol. 279, pp. 13:1–13:17. Schloss Dagstuhl - Leibniz-Zentrum für Informatik (2023). https://doi.org/10.4230/LIPICS.CONCUR.2023.13
21. Haase, C., Piórkowski, R.: Universal quantification makes automatic structures hard to decide. CoRR (2024https://doi.org/10.48550/ARXIV.2306.10432
22. Haase, C., Różycki, J.: On the expressiveness of Büchi arithmetic. Presented at the (2021). https://doi.org/10.1007/978-3-030-71995-1_16
23. Hashiguchi, K.: Representation theorems on regular languages. J. Comput. Syst. Sci. **27**(1), 101–115 (1983). https://doi.org/10.1016/0022-0000(83)90031-4
24. Jones, J.P., Matijasevich, Y.V.: Register machine proof of the theorem on exponential diophantine representation of enumerable sets. J. Symb. Log. **49**(3), 818–829 (1984). https://doi.org/10.2307/2274135
25. Matiyasevich, Yu.V.: A new proof of the theorem on exponential diophantine representation of enumerable sets (in Russian). Zapiski Nauchnykh Seminarov LOMI **60**, 75–92 (1976). https://doi.org/10.1007/BF01693980, English translation: J. Soviet Math. **14**(5), 1475–1486 (1980)
26. McNaughton, R.: Review of [9] and [10]. J. Symb. Log. **28**(1), 100–102 (1963). https://doi.org/10.2307/2271342
27. Meyer, A.R.: Weak monadic second order theory of succesor is not elementary-recursive. In: Logic Colloquium. Lecture Notes in Mathematics, vol. 453, pp. 132–154. Springer Berlin Heidelberg (1975https://doi.org/10.1007/BFb0064872
28. Muchnik, A., Semenov, A., Ushakov, M.: Almost periodic sequences. Theor. Comput. Sci. **304**(1), 1–33 (2003). https://doi.org/10.1016/S0304-3975(02)00847-2
29. Point, F.: On the expansion of $(\mathbb{N}, +, 2^x)$ of Presburger arithmetic (2010). https://u.osu.edu/friedman.8/files/2014/01/0AppB072710-1mnlwyn.pdf, unpublished manuscript
30. Semënov, A.L.: On certain extensions of the arithmetic of addition of natural numbers. Math. USSR-Izvestiya **15**(2), 401–418 (1980). https://doi.org/10.1070/im1980v015n02abeh001252

31. Semënov, A.L.: Logical theories of one-place functions on integers. Habilitation Thesis (1984)
32. Semënov, A.L.: Logical theories of one-place functions on the set of natural numbers. Math. USSR-Izvestiya **22**(3), 587–618 (1984). https://doi.org/10.1070/im1984v022n03abeh001456
33. Starchak, M.R.: On the existential arithmetics with addition and bitwise minimum. In: Proc. FoSSaCS. LNCS, vol. 13992, pp. 176–195. Springer International Publishing (2023). https://doi.org/10.1007/978-3-031-30829-1_9
34. Starchak, M.R.: Existential definability of unary predicates in Büchi arithmetic. In: Proc. Computability in Europe (CiE). LNCS, vol. 14773, pp. 218–232. Springer Nature Switzerland (2024).
35. Starchak, M.R.: Quantifier elimination for regular integer linear-exponential programming. In: Proceedings of the LICS (2025). to appear
36. Villemaire, R.: Joining k- and l-recognizable sets of natural numbers. In: Proceedings of the STACS, pp. 83–94. Springer Berlin Heidelberg (1992). https://doi.org/10.1007/3-540-55210-3_175
37. Villemaire, R.: The theory of $\langle N, +, V_k, V_l \rangle$ is undecidable. Theor. Comput. Sci. **106**(2), 337–349 (1992). https://doi.org/10.1016/0304-3975(92)90256-F
38. Wu, H., Chen, Y.F., Wu, Z., Xia, B., Zhan, N.: A decision procedure for string constraints with string/integer conversion and flat regular constraints. Acta Inform. (2023). https://doi.org/10.1007/s00236-023-00446-4

Partial Cubes and Fibonacci Dimension: Insights and Perspectives

Marcella Anselmo[1], Dora Giammarresi[2](✉), Maria Madonia[3], and Sabrina Mantaci[4]

[1] Dipartimento di Informatica, Università di Salerno, Fisciano, SA, Italy
manselmo@unisa.it
[2] Dipartimento di Matematica, Università di Roma "Tor Vergata", Roma, Italy
giammarr@mat.uniroma2.it
[3] Dipartimento di Matematica e Informatica, Università di Catania, Catania, Italy
maria.madonia@unict.it
[4] Dipartimento di Matematica e Informatica, Università di Palermo, Palermo, Italy
sabrina.mantaci@unipa.it

Abstract. A *partial cube* is a graph G that can be isometrically embedded into a hypercube Q_k, with the minimum of such k called the *isometric dimension*, $\text{idim}(G)$, of G. A *Fibonacci cube* Γ_k excludes strings containing 11 from the vertices. Any partial cube G embeds into some Γ_d, defining *Fibonacci dimension*, $\text{fdim}(G)$, as the minimum of such d. It holds $\text{idim}(G) \leq \text{fdim}(G) \leq 2 \cdot \text{idim}(G) - 1$. While $\text{idim}(G)$ is computable in polynomial time, check whether $\text{idim}(G) = \text{fdim}(G)$ is NP-complete. We survey the properties of partial cubes and Generalized Fibonacci Cubes and present a new family of graphs G for which $\text{idim}(G) = \text{fdim}(G)$. We conclude with some open problems.

Keywords: Partial Cubes · Fibonacci Cubes · Isometric words

1 Introduction

The k-dimensional cube Q_k, or simply k-cube or hypercube, is defined as an undirected graph with 2^k vertices, each corresponding to (and labeled with) a binary string of length k, representing positions in the k dimensional space. The distance between two vertices in Q_k is given by the *Hamming distance*, that is, the number of differing bits between the corresponding binary labels. Then Q_k generalizes the notion of a cube to higher dimensions.

Partially supported by INdAM-GNCS Project 2025 - CUP E53C24001950001, FARB Project ORSA240597 of University of Salerno, TEAMS Project and PNRR MUR Project PE0000013-FAIR University of Catania, PNRR MUR Project ITSERR CUP B53C22001770006 and FFR fund University of Palermo, MUR Excellence Department Project MatMod@TOV, CUP E83C23000330006, awarded to the Dept of Mathematics, University of Rome Tor Vergata.

S.-K. Ko and F. Manea (Eds.): DLT 2025, LNCS 16036, pp. 15–29, 2026.
https://doi.org/10.1007/978-3-032-01475-7_2

A graph G is called a *partial cube* if it can be isometrically embedded into a k-cube. This implies that there exists a set of binary strings S to be used to label the vertices in such a way that the graph-theoretic distance between any two vertices equals the Hamming distance between their labels. The minimum k for which such embedding is possible is called the *isometric dimension* of the graph G and is denoted by $\text{idim}(G)$. If G is not a partial cube then $\text{idim}(G) = \infty$. Partial cubes naturally arise in several domains, including computational geometry, order theory, interconnection networks, and theoretical chemistry [25].

In 1993, Hsu [27] introduced the *Fibonacci cube* Γ_k, a subgraph of the k-cube Q_k induced by all binary strings of length k that avoid the factor 11. Notably, Γ_k is itself a partial cube, so the Hamming distance remains valid as a graph distance metric.

In 2011, Cabello, Eppstein, and Klavžar [15] introduced the *Fibonacci dimension* of a graph G, denoted $\text{fdim}(G)$, defined as the smallest integer d such that G can be isometrically embedded into the Fibonacci cube Γ_d. It is not difficult to show that every partial cube can also be isometrically embedded in a Fibonacci cube and that $\text{idim}(G) \leq \text{fdim}(G)$. The problem of characterizing partial cubes with minimal Fibonacci dimension is worth investigating [5,15].

A different approach to partial cubes was introduced in 2012 by Ilić, Klavžar, and Rho [28], who considered special subgraphs of Q_k obtained by selecting only the vertices whose labels avoid a given binary string f as a factor. These graphs, denoted $Q_k(f)$, are partial cubes when the string f satisfies certain structural properties and are referred to as *Generalized Fibonacci Cubes*.

In this paper, we introduce the main definitions of partial cubes and survey related results from both the algorithmic and the combinatorics on words sides, highlighting the rich interplay between these two viewpoints within this special class of graphs. We conclude with a discussion of related work and research directions.

2 Partial Cubes

Let $G = (V(G), E(G))$ be a graph. The distance of two nodes $u, v \in V(G)$, denoted by $\text{dist}_G(u, v)$, is the length of the shortest path that connects u and v in G. A subgraph G' of G is defined by a subset of selected vertices of G, $V(G') \subseteq V(G)$, together with all the edges of G that connect the selected vertices. A subgraph G' of a graph G is an *isometric subgraph* of G if for any $u, v \in V(G')$, $\text{dist}_{G'}(u, v) = \text{dist}_G(u, v)$.

Let G, G'' be two graphs. An *isometric embedding* β *of* G'' *in* G is a map $\beta : V(G'') \to V(G)$ which preserves distances, i.e., for any $u, v \in V(G'')$, $\text{dist}_{G''}(u, v) = \text{dist}_G(\beta(u), \beta(v))$. If an isometric embedding β of G'' in G does exist, then we say that G'' *can be isometrically embedded into* G. Note that this is equivalent to saying that G'' is isomorphic to an isometric subgraph G' of G.

The k-*dimensional cube*, or simply the *hypercube* Q_k, generalizes in k dimensions the common notion of cube. It is a graph with 2^k vertices, each associated with a binary word of length k, that describes its position in the k-dimensional grid in such a way that the two words associated with two adjacent vertices differ only in one position. Recall that, given two binary strings x and y of equal length, the *Hamming distance* between x and y, indicated by $\text{dist}_H(x, y)$, is the number of positions where the two strings differ. Then, more formally, the k-cube Q_k is a graph with 2^k vertices, equipped with a labeling function $\lambda_k : V(Q_k) \rightarrow \{0,1\}^k$ such that two vertices u and v are adjacent if and only if $\text{dist}_H(\lambda_k(u), \lambda_k(v)) = 1$. This implies that the distance between any two vertices u and v is simply given by the Hamming distance between the corresponding labels, i.e. $\text{dist}_{Q_k}(u, v) = \text{dist}_H(\lambda_k(u), \lambda_k(v))$. Then we give the following definition, first introduced in 1971 by Graham and Pollak.

Definition 1. [26] *A graph G is a* partial cube *if G can be isometrically embedded into a hypercube Q_k, for some k.*

Note that all the isometric subgraphs of the hypercubes are partial cubes by definition.

Remark 2. If G is a partial cube, then the isometric embedding of G in Q_k causes G to inherit from Q_k a labeling of its vertices by binary strings of length k (which makes the distance of the vertices in G equal to the Hamming distance between their corresponding labels). Thus, finding an isometric embedding β of G in Q_k is equivalent to determining a set of binary strings B_k of length k that are "compatible" with the graph according to the Hamming distance.

Example 3. Consider the graph $G = (V(G), E(G))$ in Fig. 1 (a). The graph G is a partial cube. In fact, there exists an isometric embedding β of G in Q_4 where $\beta(A) = 0000$, $\beta(B) = 0010$, $\beta(C) = 0100$, $\beta(D) = 0001$ and $\beta(E) = 1001$ (see Fig. 1 (b)).

Remark 4. Note that if a graph has a maximal vertex degree equal to m, then it cannot be embedded into Q_k with $k < m$.

Let us recall the definition of isometric dimension of a graph that was first studied by Djoković [21] and then explicitly named by Cabello *et al.*

Definition 5. [15] *Let G be a graph. The* isometric dimension of *G, denoted by* $\text{idim}(G)$*, is the smallest integer k such that G admits an isometric embedding into Q_k. If no such k exists, then* $\text{idim}(G) = \infty$.

For example, the isometric dimension of the partial cube G in Fig. 1 (a), is 4. In fact, there exists an isometric embedding of G into Q_4 (see Example 3) and it can be shown that there is no isometric embedding of G into Q_k if $k < 4$.

Many questions about partial cubes can be posed from a dual perspective: the algorithmic one and that of combinatorics on words. From the algorithmic side, we are given a graph G with n vertices and ask what kind of structural properties should the graph satisfy to be a partial cube. Moreover, following Remark 2, a major question could be: If G is a partial cube, how can we get a

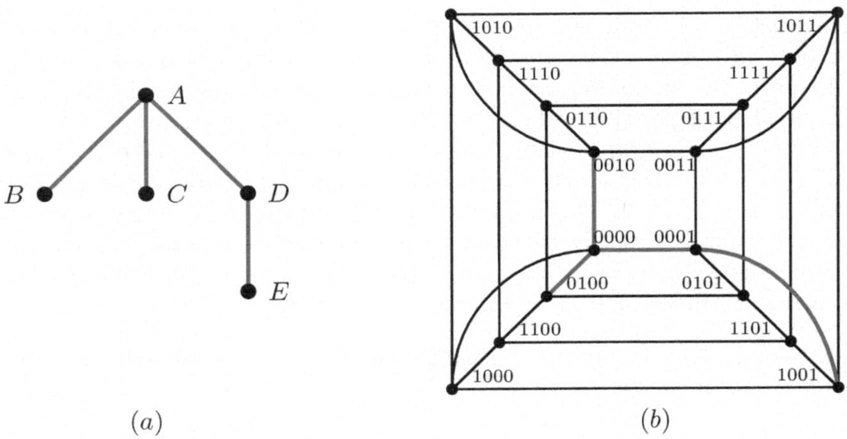

Fig. 1. (a) The graph G of Example 3; (b) The hypercube Q_4 with the isometric embedding β of G into Q_4 marked by red edges, with $\beta(A) = 0000$, $\beta(B) = 0010$, $\beta(C) = 0100$, $\beta(D) = 0001$ and $\beta(E) = 1001$.

set B_k of n strings to label the vertices in a way that the Hamming distance of the labels describes the minimal path lengths between the vertices?

On the other hand, a given set B_k of strings of length k always corresponds to a set of vertices of Q_k and therefore defines a graph G that is a subgraph of Q_k. Hence, it is natural to ask the following question: Are there special combinatorial properties for the string set B_k to correspond to a partial cube?

In the next two sections, we explore both perspectives and report some classical results.

3 Is the Given Graph a Partial Cube?

A structural characterization of partial cubes was first established by Djoković in 1973 [21], who studied distance-preserving subgraphs of hypercubes using the following definition. Given a graph G, a subset of vertices $U \subset V(G)$ is *closed* if, for all $u, v \in U$ and $x \in V(G)$: $\text{dist}_G(u,x) + \text{dist}_G(x,v) = \text{dist}_G(u,v)$ implies $x \in U$. In other words, a closed set includes all the vertices in the minimum path between each u and v in the set. Given a graph G, for each edge (u, v), define the set:

$$G(u,v) = \{x \in V(G) \mid \text{dist}_G(x,u) < \text{dist}_G(x,v)\}$$

Note that the two sets $G(u, v)$ and $G(v, u)$ are disjoint. Then the following theorem of characterization for partial cubes is proved.

Theorem 6. [21] *For a graph G, the following two assertions are equivalent:*

(i) G is isomorphic to an isometric subgraph of a hypercube.

(ii) G is connected and bipartite and if $u, v \in V(G)$ are adjacent, then $G(u,v)$ is a closed subset for G.

Later, in 1984, Winkler [39], by looking deeply in the Djokovi'c's proof, explicitly stated a characterization based on a relation on the edges of the graph. The relation denoted Θ is defined as follows. Given two edges $e = (x,y)$ and $f = (u,v)$ in G, we say that:

$$e \,\Theta\, f \iff \mathrm{dist}_G(x,u) + \mathrm{dist}_G(y,v) \neq \mathrm{dist}_G(x,v) + \mathrm{dist}_G(y,u).$$

This relation is reflexive and symmetric, but not necessarily transitive in general. Then, Winkler proved the following theorem.

Theorem 7. [39] *A connected graph is a partial cube if and only if it is bipartite and the relation Θ? is transitive.*

In that case, Θ becomes an equivalence relation, referred to as the Djoković-Winkler relation. Each class corresponds to a coordinate i in the vertex labeling, separating the graph into two connected components, the first with 0 and the second with 1 in the position i of the label. This implies that the isometric dimension of a partial cube coincides with the number of Djoković-Winkler equivalence classes.

Example 8. Consider the graph G of Example 3. It is a partial cube and, hence, the relation Θ is an equivalence relation. No two edges belong to the same equivalence class. Therefore, there are exactly 4 equivalence classes for Θ and, as expected, $\mathrm{idim}(G) = 4$.

The following definition is introduced in [15] to study the isometric dimension of partial cubes. Let $\beta : V(G) \to V(Q_k)$ be an isometric embedding of a graph G into the hypercube Q_k. The embedding β is *redundant* if there exists $i \in \{1, 2, \ldots, k\}$ such that the labels of all vertices in $\beta(V(G))$ have the same i-th coordinate. It is *irredundant* otherwise. If an embedding is redundant, another embedding can be found into a lower-dimensional hypercube by omitting the redundant coordinate.

The following result is proved by linking the equivalence classes of the Djoković-Winkler relation to the isometric dimension of the partial cube. This will be referred to in the next sections.

Proposition 9. [15] *Let G be a graph. An isometric embedding $\beta : V(G) \to V(Q_k)$ is irredundant if and only if $k = \mathrm{idim}(G)$.*

These foundational characterizations led to efficient algorithms for various problems on partial cubes, such as shortest paths and graph drawing.

In particular, Eppstein [24,25] proposed a quadratic time algorithm to recognize partial cubes by computing the Winkler relation classes. If the input graph is a partial cube, the algorithm outputs a minimal-length binary labeling, where the minimal length coincides with the number of Winkler equivalence classes, i.e., the isometric dimension.

4 Is the Given String Set a Labeling of a Partial Cube?

In this section, we address the problem of investigating whether there exist combinatorial properties of sets of strings that ensure that they can serve as the labels of a partial cube. A positive answer can be given by the sets $L_k(f)$ of all binary strings of length k that do not contain f as a factor, in the case f is a so-called *isometric word* as introduced in 2012 in [31].

We now briefly report the definition of isometric words (originally called good words) and some related recent results [29,31,33,36,37].

Let $u, v \in \Sigma^*$ be words of equal length and $\text{dist}_H(u, v) = d$. A *transformation* τ from u to v is a sequence of $d + 1$ words $(w_0, w_1, \ldots, w_d)$ such that $w_0 = u$, $w_d = v$, and for any $i = 0, 1, \ldots, d-1$, $\text{dist}_H(w_i, w_{i+1}) = 1$. Moreover, τ is *f-free* if for any $i = 0, 1, \ldots, d$, the word w_i is f-free, i.e. w_i does not have f as a factor. This is the basis for the following definition.

Definition 10. [31] *A word f is isometric if for any pair of f-free words u and v, with $|u| = |v|$, there exists a f free transformation from u to v. It is non-isometric otherwise.*

A pair of f-free words (u, v) such that any transformation from u to v is not f-free proves that f is non-isometric and is called a *pair of witnesses* for f. The minimal length of a word u in a pair of witnesses for f is called the *index* of f, and is denoted by $i(f)$.

Isometric words can be characterized in terms of 2-error overlap. A word w has a *2-error overlap* (*2-eo*, for short) if, for some ℓ with $1 \leq \ell \leq n-1$, the prefix of w of length ℓ and the suffix of the same length differ in two positions, namely, they have a Hamming distance equal to 2. It is possible to show that if a word f is non-isometric, then it has a 2-error overlap and, vice versa, from a 2-error overlap of a word f a pair of witnesses for f can be constructed. Therefore, the following result holds.

Proposition 11. [33] *A word f is isometric if and only if f has no 2-error overlap.*

Example 12. The word $f = 1000$ is isometric. It follows from Proposition 11, noting that f has no 2-error overlap. In contrast, $f = 1001$ is non-isometric. In fact, 1001 has 2-error overlaps. For example, the prefix and the suffix of length three differ in two positions. Starting from this 2-error overlap, a pair of witnesses can be obtained for 1001, namely, the pair $(10001, 11011)$.

There exists a strict connection between some particular isometric subgraphs of Q_k and isometric words. Let us first recall a definition.

Definition 13. [28] *The* k-hypercube avoiding a word f, *denoted $Q_k(f)$, is the subgraph of Q_k obtained by removing those vertices whose labels contain f as a factor.*

Moreover, we assume implicitly that $Q_k(f)$ is still a labeled graph keeping the original labels of the hypercube Q_k from which it is derived.

Then, one immediately realizes that $Q_k(f)$ is an isometric subgraph of Q_k if and only if f is isometric. Moreover, in [38], the non-obvious result is proved that if f is not isometric then for all $k \geq i(f)$, $Q_k(f)$ is not even a partial cube, i.e., $Q_k(f)$ does not isometrically embed into any $Q_{k'}$ with $k' \geq i(f)$. These considerations lead to the following proposition.

Proposition 14. [38] *The following statements are equivalent:*

1. *f is isometric*
2. *for any $k \geq |f|$, $Q_k(f)$ is an isometric subgraph of Q_k*
3. *for any $k \geq |f|$, $Q_k(f)$ isometrically embeds into Q_k.*

This implies the following.

Corollary 15. *A word f is isometric if and only if for any $k \geq |f|$, $Q_k(f)$ is a partial cube.*

Furthermore, note that if f is a non-isometric word and $i(f)$ is its index, then for any $k < i(f)$, $Q_k(f)$ isometrically embeds into Q_k, while for any $k \geq i(f)$, $Q_k(f)$ does not isometrically embed into Q_k. Finally, remark that if f is isometric, then its complement $\overline{f}$ and its mirror f^R are also isometric. Consequently, also the graphs $Q_k(\overline{f})$ and $Q_k(f^R)$, which can also be obtained from $Q_k(f)$ by, respectively, complementing or reversing all labels, are isometric subgraphs of Q_k.

By Proposition 14, the hypercube that avoids an isometric word f, that is, $Q_k(f)$, is isometrically embedded in Q_k, for any $k \geq |f|$, but, in principle, its isometric dimension could be smaller than k. Nevertheless, the following proposition shows that this is never the case.

Proposition 16. [5] *Let f be a word with $|f| \geq 3$. If f is an isometric word, then, for any $k \geq 0$, $Q_k(f)$ is a partial cube with* $\mathrm{idim}(Q_k(f)) = k$.

Remark 17. Note that if f is a non-isometric word, with $|f| \geq 3$ and index $i(f)$, then for any $k < i(f)$ the equality $idim(Q_k(f)) = k$, still holds.

5 Fibonacci Cubes and Fibonacci Dimension

In 1993, Hsu [27] introduced the *Fibonacci cube* Γ_k, a special subgraph of the k-cube Q_k obtained by deleting a selected set of vertices.

Definition 18. [27] *The* Fibonacci cube of order k, *denoted Γ_k, is the subgraph of Q_k obtained by selecting only the vertices whose labels avoid* 11 *as a factor.*

Note that the vertices of Γ_k, the Fibonacci cube of order k, are the so-called Fibonacci strings of length k. Hence, the number of vertices of Γ_k is equal to the $(k+2)$-th Fibonacci number. Moreover, one of the main properties of Fibonacci cubes is their recursive structure that has been extensively studied [27,30]. In particular, Γ_k is a partial cube, so the Hamming distance remains valid as a graph distance metric.

In 2011, Cabello, Eppstein, and Klavžar introduced the *Fibonacci dimension* of a graph G to explore the embeddability of other networks in the Fibonacci cubes.

Definition 19. [15] *Let G be a graph. The* Fibonacci dimension *of G, denoted* $\mathrm{fdim}(G)$, *is the smallest integer d such that G admits an isometric embedding into the Fibonacci cube Γ_d. The $fdim(G) = \infty$ if there is no such d.*

This is strictly related to the isometric dimension, since a graph is shown to be isometrically embedded in some hypercube Q_k iff it isometrically embedded in some Fibonacci cube Γ_d. Note that in general $d \geq k$. The upper and lower bounds for the fdim of a graph G were established as stated in the following proposition.

Proposition 20. [15] *Let G be a graph. Then $fdim(G)$ is finite if and only if $idim(G)$ is finite. Moreover, $idim(G) \leq fdim(G) \leq 2\,idim(G) - 1$.*

It is interesting to ask which partial cubes have extremal Fibonacci dimension. There is a neat characterization for the partial cubes whose Fibonacci dimension is as large as possible compared to their isometric dimension. It is given in terms of complete crossing graphs (see [15]). However, it is difficult to characterize the partial cubes whose Fibonacci dimension is as small as possible. Indeed, although the embedding algorithm in a Fibonacci cube is linear [32] and the algorithm for computing the isometric dimension of a graph is quadratic [25], it is shown instead that deciding for a given graph G whether or not $\mathrm{idim}(G) = \mathrm{fdim}(G)$ is a NP-complete problem. In general, it is difficult to find graphs with the "small" Fibonacci dimension. An important family of graphs with this property is the set of trees for which the following result holds.

Theorem 21. [15] *For any tree T of n vertices,* $\mathrm{fdim}(T) = \mathrm{idim}(T) = n - 1$.

Note that trees are very special partial cubes, since they have the maximal possible isometric dimension, namely each Djoković-Winkler class contains only one edge.

It is worth remarking that, also for Fibonacci cubes, the equality $\mathrm{fdim}(\Gamma_k) = \mathrm{idim}(\Gamma_k)$ holds. In fact, the vertex with label 0^k and all its k neighboring vertices are in Γ_k. Therefore, the labeling of Γ_k inherited from Q_k is irredundant and, hence, from Proposition 9, $\mathrm{idim}(\Gamma_k) = k$. Moreover, Γ_k trivially embeds into itself and, from $\mathrm{idim}(\Gamma_k) \leq \mathrm{fdim}(\Gamma_k)$, $\mathrm{fdim}(\Gamma_k) = k$ follows.

Interestingly, the same holds in for all the partial cubes $Q_k(f)$ with f of length 2.

Proposition 22. [5] *Let $f \in \Sigma^*$, $k \geq |f|$. If $|f| = 2$, then $Q_k(f)$ is a partial cube with* $\mathrm{idim}(Q_k(f)) = \mathrm{fdim}(Q_k(f)) = k$.

6 A New Family of Partial Cubes with Minimal Fibonacci Dimension

In this section, we exhibit a new class of graphs with minimal Fibonacci dimension, that is, graphs for which the isometric dimension equals the Fibonacci one. Our main result specifies a family of graphs G_k defined starting from sets $S = \{11, f\}$, and provides sufficient conditions on f such that G_k satisfies $\mathrm{idim}(G_k) = \mathrm{fdim}(G_k) = k$. In the last part, we show that this family is incomparable under inclusion with the previously known class of trees, although they share a nonempty intersection. All the results presented here were announced at conference [5].

The idea is to look for graphs with minimal Fibonacci dimension among those of the form $Q_k(f)$ for some word f. Unfortunately, in general, $\mathrm{fdim}(Q_k(f))$ can be strictly greater than $\mathrm{idim}(Q_k(f))$, even if f is isometric, as shown in the next example.

Example 23. Let $f = 1000$. Since f is isometric then $\mathrm{idim}(Q_4(1000)) = 4$ by Proposition 16. Note that $Q_4(1000)$ has 15 vertices; and Γ_4 only has 8 (the 6th Fibonacci number) vertices, so embedding into Γ_4 is not possible. More precisely, the smallest Fibonacci cube to be considered is Γ_6 that has 21 vertices, then $\mathrm{fdim}(Q_4(1000)) \geq 6$.

In order to overcome this problem, we extend the concept of a subgraph of Q_k avoiding a word with a subgraph avoiding a *set of words*. In analogy to the definition of $Q_k(f)$, the *k-hypercube avoiding a set S*, denoted $Q_k(S)$, is the subgraph of Q_k obtained by keeping only the nodes of Q_k labeled with S-free words (that is, f-free words, for all $f \in S$). In particular, if $S = \{f_1, \ldots, f_m\}$ then $Q_k(S) = Q_k(f_1) \cap \ldots \cap Q_k(f_m)$.

In addition, the notion of isometricity, given for a single word, can be extended to sets of words. Let max be the length of a longest word in S. Then S is *isometric* if for any pair (u, v) of S-free words, with $|u| = |v| \geq \max$, there exists a transformation from u to v such that all its intermediate words are S-free. S is *non-isometric* if it is not isometric, and in this case a *pair of witnesses* for S is a pair (u, v) with $|u| = |v| \geq \max$, such that each transformation from u to v contains a word that is not S-free.

The following proposition shows that isometric sets define a family of isometric subgraphs of Q_k.

Proposition 24. *Let $S \subseteq \Sigma^*$ and let* max *be the length of a longest word in S. The set S an isometric iff for any $k \geq \max$, $Q_k(S)$ is an isometric subgraph of Q_k.*

Recall that a graph admits an isometric embedding into Q_k if and only if it also admits an isometric embedding into some Fibonacci cube Γ_d. Since we are interested in identifying isometric subgraphs of Q_k that embed isometrically into Γ_d, we focus on subgraphs of the form $Q_k(S)$, where the set S includes the word 11. Indeed, by definition, we have:

$$Q_k(\{11, f_1, \ldots, f_m\}) = Q_k(11) \cap Q_k(f_1, \ldots, f_m) = \Gamma_k(f_1, \ldots, f_m).$$

In the next example the word $f = 1000$, from Example 23, is now considered together with 11.

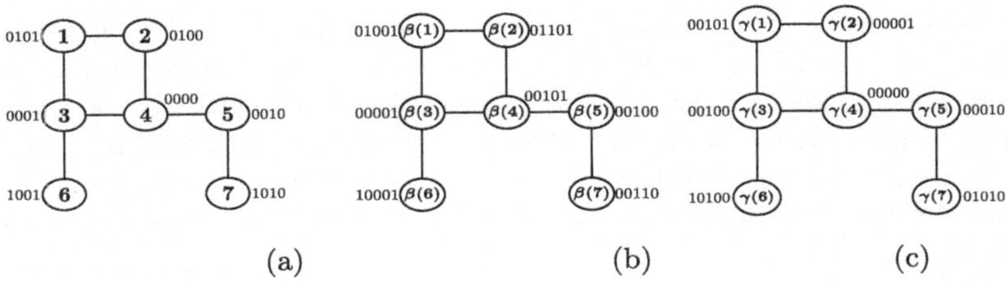

Fig. 2. $Q_4(\{11, 1000\})$ with (a) its labeling λ_4, (b) an isometric embedding β in Q_5 and (c) an isometric embedding γ in Γ_5.

Example 25. Let $S = \{11, 1000\}$. The graph $G_4 = Q_4(S)$ (see Fig. 2(a)) cannot be isometrically embedded into Q_4. In fact, S is a non-isometric set, since (u, v) with $u = 1001$ and $v = 1010$ is a pair of witnesses for S because any transformation from u to v meets either 11 or 1000. According to Proposition 24, G_4 is not an isometric subgraph of Q_4. Let x and y be the vertices in G_4 labeled 1001 and 1010, respectively. One can see that $\text{dist}_{G_4}(x, y) = 4$ (Fig. 2(a)), while $\text{dist}_H(1001, 1010) = 2$. Nevertheless, Eppstein's algorithm in [25] produces the isometric embedding β of G_4 into Q_5 shown in Fig. 2(b), yielding $\text{idim}(Q_4(S)) = 5$ (i.e. it is a partial cube). Moreover, also $\text{fdim}(G_4) = 5$ since the map γ shown in Fig. 2(c) is an isometric embedding of G_4 into Γ_5. Then, $\text{idim}(G_4) = \text{fdim}(G_4) = 5$.

The previous example is emblematic. The set $S = \{11, 1000\}$ is non-isometric; then the graph $Q_4(S)$ is not isometrically embedded into Q_4, but it isometrically embeds into Q_5. This proves that "sets" and "single words" behave differently with respect to the non-isometricity. (Recall that, by Corollary 15 a word f is isometric if and only if $Q_k(f)$ is a partial cube). Another interesting fact is that $Q_4(S)$ has minimal Fibonacci dimension.

We now investigate the isometric and Fibonacci dimensions of hypercubes avoiding a set of words S in the special case $11 \in S$. First, we introduce some notation. For a set S of words, let $L_k(S)$ be the set of words of length k that avoid S, that is, $L_k(S) = \Sigma^k \setminus \Sigma^* S \Sigma^*$. Moreover, coherently with the definition for redundant embedding, we say that the set $L_k(S)$ is *redundant* if there exists $i \in \{1, 2, \ldots, k\}$ such that all strings in $L_k(S)$ have the same symbol at position i. It is *irredundant* otherwise.

Example 26. Let $S = \{11, 010\}$. For any $k \geq 3$, the set that avoids S is $L_k(S) = \{0^k, 10^{k-1}, 0^{k-1}1, 10^{k-2}1\}$. This set is rendundant, since for all $i = 2, \ldots k-1$ all strings in $L_k(S)$ have the same symbol at position i.

Theorem 27. *Let* $S \subseteq \Sigma^*$ *with* $11 \in S$. *Let* max *be the maximal length of a word in* S *and let* $k \geq \max$. *If* S *is an isometric set, then* $Q_k(S)$ *is a partial cube with* $\mathrm{idim}(Q_k(S)) \leq \mathrm{fdim}(Q_k(S)) \leq k$; *moreover,* $\mathrm{idim}(Q_k(S)) = k$ *iff* $L_k(S)$ *is irredundant.*

Corollary 28. *Let* $S \subset \Sigma^*$ *with* $11 \in S$. *If* S *is isometric and* $\mathrm{idim}(Q_k(S)) = k$, *then* $\mathrm{fdim}(Q_k(S)) = k$.

The following theorem characterize sets $S = \{11, f\}$ which are isometric.

Theorem 29. *A set* $S = \{11, f\}$ *is isometric if and only if*

1. 11 *is a factor of* f *or*
2. $f \notin 000\Sigma^* \cup \Sigma^*000 \cup \Sigma^*0000\Sigma^*$ *and* f *is isometric or*
3. $f \notin 000\Sigma^* \cup \Sigma^*000 \cup \Sigma^*0000\Sigma^*$, f *is non-isometric and any pair* (u, v) *of witnesses of minimal distance for* f *is such that* 11 *is a factor of* u *or* v.

The following, instead, characterize the sets $S = \{11, f\}$ for which $L_k(S)$ is redundant.

Theorem 30. *Let* $S = \{11, f\}$, *with* $|f| \geq 3$, *and* $k \geq 3$. *Then* $L_k(S)$ *is redundant if and only if* $f = 010$.

Example 31. Let $S = \{11, 010\}$. For any $k \geq 3$, the subgraph $Q_k(S)$ has 4 vertices whose labels are in $L_k(S) = \{0^k, 10^{k-1}, 0^{k-1}1, 10^{k-2}1\}$ and form a cycle. Hence, for any $k \geq 3$, $Q_k(S)$ is a partial cube with $\mathrm{idim}(Q_k(S)) = 2$ and $\mathrm{fdim}(Q_k(S)) = 3$.

The characterizations shown in Theorems 29 and 30 allow us to specify Theorem 27 and provide a new family $\mathcal{F}$ of partial cubes with minimal Fibonacci dimension.

Theorem 32. *Let* f *be a word such that* $f \neq 010$ *and* f *satisfies conditions 1., 2., or 3. of Theorem 29. Then, for any* $k \geq |f|$, $Q_k(\{11, f\})$ *is a partial cube with* $\mathrm{idim}(Q_k(\{11, f\})) = \mathrm{fdim}(Q_k(\{11, f\})) = k$.

Note that the graph $G_4 = Q_4(\{11, 1000\})$ in Example 25 does not satisfy the hypotheses of Theorem 32 and is a partial cube with $\mathrm{idim}(G_4) = \mathrm{fdim}(G_4) \neq 4$.

Conversely, if f satisfies the conditions of Theorem 32, $G_k = Q_k(\{11, f\})$ is a partial cube with $\mathrm{idim}(G_k) = \mathrm{fdim}(G_k) = k$, as shown in the following example.

Example 33. The set $S = \{11, 1001\}$ is isometric. In fact, $f = 1001 \notin 000\Sigma^* \cup \Sigma^*000 \cup \Sigma^*0000\Sigma^*$ and f is non-isometric. Moreover, the pairs of witnesses for f of minimal distance are $(10001, 11011)$ and $(100001, 101101)$, and both pairs contain a word that is not 11-free, then S satisfies condition 3. in Theorem 29 holds.

Furthermore, let $G_4 = Q_4(\{11, 1001\})$, $f \neq 010$ and then, according to Theorem 30, $L_4(S) = \{0101, 0100, 0001, 0000, 0010, 1000, 1010\}$ is irredundant. According to Theorem 32, G_4 is a partial cube with $\mathrm{idim}(G_4) = \mathrm{fdim}(G_4) = 4$ (See Figura 3).

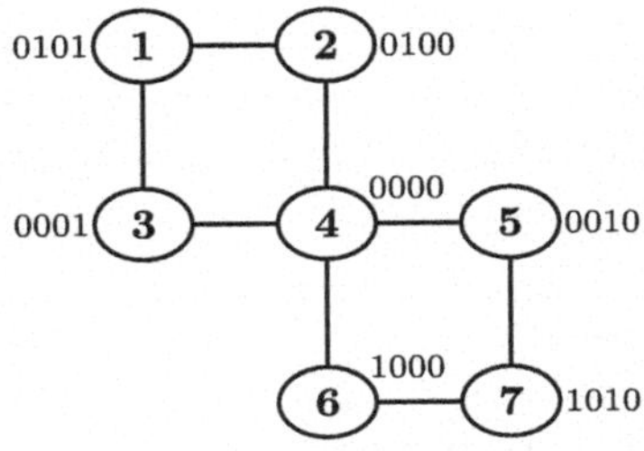

Fig. 3. $Q_4(\{11, 1001\})$

It is natural to ask whether the family of graphs $\mathcal{F}$ is related by inclusion to the family $\mathcal{T}$ of trees, as both share the property of having minimal Fibonacci dimension. The following remark shows that this is not the case.

Remark 34. Denote by T_k the tree with one vertex with k children. Consider the family of partial cubes $\mathcal{F}$ as defined in Theorem 32 and the family $\mathcal{T}$ of trees. Neither of these two families is included in the other one. In fact, $Q_4(\{11, 1001\})$ in Example 33 has a cycle, so it is not a tree. Moreover $\text{idim}(T_4) = \text{fdim}(T_4) = 4$, but T_4 cannot be described as $Q_4(\{11, f\})$. In fact, one can prove that there is no f such that T_4 coincides with $\Gamma_4(f)$. Finally, the tree T_3 with one vertex with three children is an example of graph belonging to both families, because it coincides with $Q_3(\{11, 101\})$.

The following proposition provides a family of graphs $Q_k(S)$, with $card(S) > 2$, such that $\text{idim}(Q_k(S)) = \text{fdim}(Q_k(S))$.

Proposition 35. *For* $k \geq 0$, *let* $D_k = \{11, 101, \ldots, 10^k1\}$. *Then* $Q_k(D_{k-2}) = T_k$ *and* $\text{idim}(Q_k(D_{k-2})) = \text{fdim}(Q_k(D_{k-2})) = k$.

7 Related Works and Further Research

The study of isometric words combines distance-preserving properties with the concept of *forbidden* or *absent* factors of a word. This last is a key topic in both string combinatorics and computational biology (particularly in the analysis of DNA sequences) [14,16,18–20,23]. Research on isometric words remains active [10,13,22,34,35], and has been extended to richer alphabets, various distance metrics, and two-dimensional words [1–4,6–9,11,12,17].

The characterization of isometric words given in Proposition 16 enables the construction of a family of string sets: for each k and each isometric word f, we define the set $L_k(f)$ consisting of all binary strings of length k that avoid f. These sets correspond to the labeling of isometric subgraphs of the hypercube Q_k, and therefore give rise to partial cubes with isometric dimension k. Moreover, isometric words are the only ones for which this property holds.

Extending the concept of isometry to sets of words [3], we found that isometric sets of words S also define sets of words $L_k(S)$ that induce isometric subgraphs of Q_k. Surprisingly, however, there exist non-isometric sets S for which the graph $Q_k(S)$ is not an isometric subgraph of Q_k, but still constitutes a partial cube—though with isometric dimension greater than k, realized via a different labeling. This phenomenon introduces a new class of structures that has not yet been thoroughly studied.

Future research directions include the identification of classes of word sets defined by alternative combinatorial constraints, beyond forbidden factor avoidance, that induce isometric subgraphs of hypercubes or provide new characterizations of partial cubes.

References

1. Anselmo, M., Castiglione, G., Flores, M., Giammarresi, D., Madonia, M., Mantaci, S.: Hypercubes and isometric words based on swap and mismatch distance. In: DCFS 2023. LNCS, vol. 13918, pp. 21–35. Springer, Cham (2023)
2. Anselmo, M., Castiglione, G., Flores, M., Giammarresi, D., Madonia, M., Mantaci, S.: Isometric words based on swap and mismatch distance. In: Developments in Language Theory. DLT 2023. LNCS, vol. 13911, pp. 23–35. Springer (2023)
3. Anselmo, M., Castiglione, G., Flores, M., Giammarresi, D., Madonia, M., Mantaci, S.: Isometric sets of words and generalizations of the Fibonacci cubes. In: Computability in Europe. CIE 2024. LNCS, vol. 14773, pp. 1–14. Springer (2024)
4. Anselmo, M., Castiglione, G., Flores, M., Giammarresi, D., Madonia, M., Mantaci, S.: Characterization of isometric words based on swap and mismatch distance. Int. J. Found. Comput. Sci. **36**(03), 221–245 (2025). https://doi.org/10.1142/S0129054125430051
5. Anselmo, M., Castiglione, G., Flores, M., Giammarresi, D., Madonia, M., Mantaci, S.: A family of partial cubes with minimal Fibonacci dimension. In: 36th Annual Symposium on Combinatorial Pattern Matching, CPM 2025, June 17–19, 2025, Milano, Italy. LIPIcs, vol. 331. Schloss Dagstuhl - Leibniz-Zentrum für Informatik (2025)
6. Anselmo, M., Flores, M., Madonia, M.: Quaternary n-cubes and isometric words. In: Combinatorics on Words. LNCS, vol. 12842, pp. 27–39 (2021)
7. Anselmo, M., Flores, M., Madonia, M.: Fun slot machines and transformations of words avoiding factors. In: Fun with Algorithms. LIPIcs, vol. 226, pp. 4:1–4:15 (2022)
8. Anselmo, M., Flores, M., Madonia, M.: On k-ary n-cubes and isometric words. Theor. Comput. Sci. **938**, 50–64 (2022)
9. Anselmo, M., Flores, M., Madonia, M.: Density of Ham- and Lee- non-isometric k-ary words. In: ICTCS 2023 Italian Conference on Theoretical Computer Science CEUR Workshop Proceedings, vol. 3587, pp. 116–128 (2023)
10. Anselmo, M., Flores, M., Madonia, M.: Computing the index of non-isometric k-ary words with Hamming and Lee distance. Computability **13**(3–4), 199–222 (2024). IOS Press
11. Anselmo, M., Flores, M., Madonia, M.: Density of k-ary words with 0, 1, 2 - error overlaps. Theor. Comput. Sci. **1025**, 114958 (2025)

12. Anselmo, M., Giammarresi, D., Madonia, M., Selmi, C.: Bad pictures: some structural properties related to overlaps. In: Jirásková, G., Pighizzini, G. (eds.) DCFS 2020. LNCS, vol. 12442, pp. 13–25. Springer, Cham (2020). https://doi.org/10.1007/978-3-030-62536-8_2
13. Azarija, J., Klavžar, S., Lee, J., Pantone, J., Rho, Y.: On isomorphism classes of generalized Fibonacci cubes. Eur. J. Comb. **51**, 372–379 (2016)
14. Béal, M.-P., Mignosi, F., Restivo, A.: Minimal forbidden words and symbolic dynamics. In: Puech, C., Reischuk, R. (eds.) STACS 1996. LNCS, vol. 1046, pp. 555–566. Springer, Heidelberg (1996). https://doi.org/10.1007/3-540-60922-9_45
15. Cabello, S., Eppstein, D., Klavzar, S.: The Fibonacci dimension of a graph. Electron. J. Comb. **18**(1) (2011)
16. Manuel, C., von Haeseler, A.: Structure of the space of taboo-free sequences. J. Math. Biol. 1029–1057 (2020). https://doi.org/10.1007/s00285-020-01535-5
17. Castiglione, G., Flores, M., Giammarresi, D.: Isometric words and edit distance: main notions and new variations. In: Cellular Automata and Discrete Complex Systems, AUTOMATA 2023. LNCS, vol. 14152, pp. 6–13. Springer (2023)
18. Castiglione, G., Mantaci, S., Pizzuto, S.L., Restivo, A.: A comparison between similarity measures based on minimal absent words: an experimental approach. In: ICTCS 2024 Italian Conference on Theoretical Computer Science CEUR Workshop Proceedings, vol. To appear (2024)
19. Castiglione, G., Mantaci, S., Restivo, A.: Some investigations on similarity measures based on absent words. Fundam. Informaticae **171**(1–4), 97–112 (2020)
20. Charalampopoulos, P., Crochemore, M., Fici, G., Mercas, R., Pissis, S.P.: Alignment-free sequence comparison using absent words. Inf. Comput. **262**, 57–68 (2018)
21. Djoković, D.: Distance-preserving subgraphs of hypercubes. J. Comb. Theor. Ser. B **14**(3), 263–267 (1973)
22. Epifanio, C., Forlizzi, L., Marzi, F., Mignosi, F., Placidi, G., Spezialetti, M.: On the k-hamming and k-edit distances. In: ICTCS 203 Italian Conference on Theoretical Computer Science CEUR Workshop Proceedings, vol. 3587, pp. 143–156 (2023)
23. Epifanio, C., Gabriele, A., Mignosi, F., Restivo, A., Sciortino, M.: Languages with mismatches. Theor. Comput. Sci. **385**(1), 152–166 (2007)
24. Eppstein, D.: Recognizing partial cubes in quadratic time. In: Teng, S. (ed.) Proceedings SODA 2008, pp. 1258–1266. SIAM (2008)
25. Eppstein, D.: Recognizing partial cubes in quadratic time. J. Graph Algorithms Appl. **15**(2), 269–293 (2011)
26. Graham, R.L., Pollak, H.: On addressing problem for loop switching. Bell Syst. Tech. J. **55**, 2495–2519 (1971)
27. Hsu, W.J.: Fibonacci cubes-a new interconnection topology. IEEE Trans. Parallel Distrib. Syst. **4**(1), 3–12 (1993)
28. Ilić, A., Klavžar, S., Rho, Y.: Generalized Fibonacci cubes. Discrete Math. **312**(1), 2–11 (2012)
29. Ilić, A., Klavžar, S., Rho, Y.: The index of a binary word. Theor. Comput. Sci. **452**, 100–106 (2012)
30. Klavžar, S.: Structure of Fibonacci cubes: a survey. J. Comb. Optim. **25**(4), 505–522 (2013)
31. Klavžar, S., Shpectorov, S.V.: Asymptotic number of isometric generalized Fibonacci cubes. Eur. J. Comb. **33**(2), 220–226 (2012)
32. Vesel, A.: Linear recognition and embedding of Fibonacci cubes. Algorithmica **71**(4), 1021–1034 (2015)

33. Wei, J.: The structures of bad words. Eur. J. Comb. **59**, 204–214 (2017)
34. Wei, J.: Proof of a conjecture on 2-isometric words. Theor. Comput. Sci. **855**, 68–73 (2021)
35. Wei, J., Yang, Y., Wang, G.: Circular embeddability of isometric words. Discret. Math. **343**(10), 112024 (2020)
36. Wei, J., Yang, Y., Zhu, X.: A characterization of non-isometric binary words. Eur. J. Comb. **78**, 121–133 (2019)
37. Wei, J., Zhang, H.: Proofs of two conjectures on generalized Fibonacci cubes. Eur. J. Comb. **51**, 419–432 (2016)
38. Wei, J., Zhang, H.: A negative answer to a problem on generalized Fibonacci cubes. Discret. Math. **340**(2), 81–86 (2017)
39. Winkler, P.M.: Isometric embedding in products of complete graphs. Discret. Appl. Math. **7**(2), 221–225 (1984)

A Comparative Analysis of Deletion Closure Operations and Their Properties

Da-Jung Cho[1] and Tikhon Pshenitsyn[2](✉)

[1] Department of Software and Computer Engineering, Ajou University, Suwon 16499, Republic of Korea
dajungcho@ajou.ac.kr

[2] Steklov Mathematical Institute of Russian Academy of Sciences, Moscow, Russia
tpshenitsyn@mi-ras.ru

Abstract. Deletion operations play a fundamental role in the formal language theory, with applications spanning both computational and biological systems. Deletions with specific constraints, in particular, site-directed deletion, are of special interest for applications. In this paper, we examine the computational and structural differences between deletion closure and site-directed deletion closure operations, addressing some open questions. First, we prove that any recursively enumerable language can be obtained as the site-directed deletion closure of a context-free language. We observe several differences between deletion and site-directed deletion applied to context-free languages. Secondly, we establish that regular languages are closed under deletion closure, resolving an open question from [Ito, Kari, Thierrin 1997]; the closure of regular languages under site-directed deletion closure remains an open problem. In contrast, we show that closures of regular languages under deletion systems are computationally universal. Finally, we show that the membership problem for site-directed deletion closure is NP-hard. These findings contribute to a deeper understanding of complexity of iterated deletion operations.

Keywords: Deletion · Site-directed deletion · Closure properties · Deletion closures

1 Introduction

The deletion operation on strings, a fundamental concept in the formal language theory, has been extensively studied due to its relevance in both theoretical computation and biological processes [4,5,7–9,11,12]. In its weakest form, the deletion operation removes any substring from a string, without restricting the choice to a specific substring. Such an operation with no restrictions is too uncontrolled, so various conditions can be imposed on its application. Depending on these conditions, different variants of deletion arise. In the works [5,11], deletion of a string that belongs to a given language L_1 from a string that belongs

S.-K. Ko and F. Manea (Eds.): DLT 2025, LNCS 16036, pp. 30–44, 2026.
https://doi.org/10.1007/978-3-032-01475-7_3

to another language L_2 is allowed. In contrast, in site-directed deletion, strings from L_1 specify a context ("guide") in which deletion is possible. Site-directed deletion operation plays a critical role in modeling phenomena such as RNA editing and DNA recombination [2,3,16].

Several studies have explored deletion operations from both computational and formal language perspectives. We focus on closures under deletion operations rather than on their one-step applications, because iterated operations exhibit rich and complicated dynamic behaviour. Kari [11] showed that regular and context-free languages behave differently under iterative deletion. Domaratzki and Okhotin [5] studied closure properties and decidability of the deletion closure operation and showed that every recursively enumerable language can be represented as the deletion closure of a linear context-free language. More recently, Cho et al. [4] introduced site-directed deletion, inspired by site-directed mutagenesis, and analyzed its closure properties and computational complexity.

We examine and compare properties of the deletion closure and the site-directed deletion closure operations in order to understand how they affect language structure. The deletion closure of a language $L \subseteq \Sigma^*$, denoted as $\mathrm{D}^*(L)$, is the smallest language containing L that is closed under deletions of the form $x \to y = \{x_1x_2 \mid x_1yx_2 = x, \text{ and } x, y \in L\}$. Similarly, the site-directed deletion closure of L, denoted as $\mathrm{SDD}^*(L)$, is the smallest language that includes L and is closed under site-directed deletion $x \overset{\text{sdd}}{\leftarrow} y = \{x_1uvx_2 \mid x = x_1uwvx_2, y = uv, \text{ and } u, v \neq \varepsilon\}$. We also study deletion systems, which represent a particular case of insertion-deletion systems with deletion rules only. A deletion rule is of the form (u, w, v), and it is applied as follows: $x_1uwvx_2 \Rightarrow x_1uvx_2$. A deletion system is a finite set D of deletion rules, and the closure of a language L under D is the smallest language containing L closed under applications of rules from D. Although deletion rules are very similar to site-directed deletion, there is a fundamental difference between the two. A deletion system explicitly lists finitely many strings which can be deleted along with the contexts required to perform deletion. The definition of a deletion system is external to a language L to which deletions are applied. In contrast, in site-directed deletion, a deleted string and its contexts are defined internally, by the language L itself; if L is infinite, then it defines infinitely many deletion rules.

Table 1 presents the closure properties of variants of deletion and site-directed operations such as iterated deletion, iterated site-directed deletion, deletion closure, site-directed deletion closure, and deletion system closure. The results include known ones and new ones. In this paper, we show that regular languages are closed under deletion closure, which answers an open question from [8, p. 9]. The proof uses the fact that regular languages are closed under iterated deletion with an arbitrary language. A similar problem for site-directed deletion closure remains open. An interesting question arises whether regular languages are too weak so that the closure of a regular language under any reasonable deletion operation cannot result in an undecidable language. This turns out to be false. We show (Theorem 4) that closures of regular languages under deletion systems include some undecidable languages.

Table 1. Closure properties of iterated deletion, iterated site-directed deletion, deletion closure, site-directed deletion closure, and deletion system closure for regular and context-free languages. 'RE' stands for the fact that closures of languages under consideration include all recursively enumerable languages under a certain encoding.

Operation	Regular	Context-free
Iterated deletion	REG [11]	RE [5]
Iterated site-directed deletion	Open	Not closed [4]
Deletion closure	REG (Theorem 3)	RE [5]
Site-directed deletion closure	Open	RE (Theorem 2)
Deletion system closure	RE (Theorem 4)	RE

Speaking of context-free languages, we prove that any recursively enumerable language can be expressed as the iterated site-directed deletion of a linear context-free language (Theorem 2). This result is similar to the one obtained for deletion closure [5]. Finally, we consider site-directed deletion closures of finite languages. There is little sense to describe them in language-theoretic terms because the closure of a finite language under any deletion operation is a finite language. Nevertheless, one can study the algorithmic problem of determining whether a given string w belongs to $C(L)$ where L is a given finite language and C represents deletion closure, site-directed deletion closure, or a deletion system closure. We show that determining whether $w \in \mathrm{SDD}^*(L)$ is NP-hard (Theorem 5) and discuss difficulties that arise when dealing with finite languages.

2 Preliminaries

Let Σ be a finite alphabet of symbols and let Σ^* be the set of all words (strings) over Σ. The symbol ε denotes the empty string. Given a string x, $|x|$ denotes the length of x and x^{R} denotes the reversal of x. Given a string $x = uv$, where $u, v \in \Sigma^*$, we say that u is a *prefix* of x and v is a *suffix* of x.

A language is an arbitrary subset of Σ^*. It is recursively enumerable (RE) if equals the domain of some computable function. A language is RE-complete if any RE language can be many-one reduced to it.

The *sequential deletion* (*deletion* in short) [11] of strings $x, y \in \Sigma^*$, denoted by $x \rightarrow y$, is defined as $x \rightarrow y = \{x_1x_2 \mid x = x_1yx_2\}$. The *site-directed deletion* [4] from x by y, denoted by $x \overset{\mathrm{sdd}}{\leftarrow} y$, is defined as $x \overset{\mathrm{sdd}}{\leftarrow} y = \{x_1uvx_2 \mid x = x_1uwvx_2, y = uv,\ \text{and}\ u, v \neq \varepsilon\}$. If L_1 and L_2 are languages over Σ, then let

$$L_1 \rightarrow L_2 := \bigcup_{x \in L_1, y \in L_2} x \rightarrow y, \text{ and let } L_1 \overset{\mathrm{sdd}}{\leftarrow} L_2 := \bigcup_{x \in L_1, y \in L_2} x \overset{\mathrm{sdd}}{\leftarrow} y.$$

The iterated deletion $L_1 \overset{*}{\rightarrow} L_2$ of L_2 from L_1 is defined as follows:

$$L_1 \overset{0}{\rightarrow} L_2 := L_1,\ L_1 \overset{n}{\rightarrow} L_2 := (L_1 \overset{n-1}{\rightarrow} L_2) \rightarrow L_2, \text{ and } L_1 \overset{*}{\rightarrow} L_2 := \bigcup_{n \geq 0} L_1 \overset{n}{\rightarrow} L_2.$$

The iterated site-directed deletion $\overset{\text{sdd}^*}{\longleftarrow}$ from L_1 by L_2 is defined similarly (replace $\to$ by $\overset{\text{sdd}^*}{\longleftarrow}$).

A language $L \in \Sigma^*$ is *del-closed* if and only if $L \to L \subseteq L$, and *sdd-closed* if and only if $L \overset{\text{sdd}}{\leftarrow} L \subseteq L$. The del-closure $\mathrm{D}^*(L)$ of L is the smallest del-closed language containing L, defined recursively as follows [5]:

$$\mathrm{D}^0(L) = L;\ \ \mathrm{D}^{i+1}(L) = \mathrm{D}^i(L) \to (\mathrm{D}^i(L) \cup \{\varepsilon\}) \text{ for } i \geq 0;\ \ \mathrm{D}^*(L) = \bigcup_{n \in \mathbb{N}} \mathrm{D}^n(L).$$

Similarly, the sdd-closure of L is defined as

$$\mathrm{SDD}^0(L) = L; \quad \mathrm{SDD}^{i+1}(L) = \mathrm{SDD}^i(L) \overset{\text{sdd}}{\longleftarrow} \mathrm{SDD}^i(L), \text{ for } i \geq 0;$$
$$\mathrm{SDD}^*(L) = \bigcup_{n \in \mathbb{N}} \mathrm{SDD}^n(L).$$

(Note that $\mathrm{SDD}^{i+1}(L) \supseteq \mathrm{SDD}^i(L)$ because site-directed deletion of a string from itself contains this string.)

We shall use the following two properties of site-directed deletion in proofs.

Proposition 1. *Let $x, y \in \Sigma^*$ be two strings such that $y = ay'b$ for $a, b \in \Sigma$.*

1. *If ab is not a subsequence of x, then $x \overset{\text{sdd}}{\leftarrow} y = \emptyset$.*
2. *If $x = ax'b$, and both a and b are absent in x', then $x \overset{\text{sdd}}{\leftarrow} y \subseteq \{y\}$.*

Proof. Let $x = x_1y_1zy_2x_2$ and let $y = y_1y_2$ where y_1 starts with a and y_2 ends with b. Then, clearly, ab is a subsequence of x.

Now, if $x = ax'b$ where x' does not contain a and b, then it must be the case that $x_1 = x_2 = \varepsilon$. Consequently, the string $x_1y_1y_2x_2$, which is the result of site-directed deletion, equals $y_1y_2 = y$. □

3 Language-Theoretic Properties of Closure Operations

We start comparing deletion closure and site-directed deletion closure, as applied to languages from well-known classes. Although the two operations are both deletions, they in fact behave very differently. We focus on regular and context-free languages because (site-directed) deletion closures of context-free languages already contain all recursively enumerable languages.

3.1 Closures of Context-Free Languages

Deletion closures of context-free languages were studied in [5] where the following result was proved.

Theorem 1 **([5, Theorem 1]).** *For any recursively enumerable language $L \subseteq \Sigma^*$, there is a linear context-free language L_0 over a larger alphabet Ω and a symbol $\natural \in \Omega \setminus \Sigma$ such that $L = \{w \mid \natural w \natural \in \mathrm{D}^*(L_0)\}$.*

This theorem implies that deletion closure of context-free languages is computationally universal. We would like to investigate whether the same holds for site-directed deletion closure and also to study differences between the latter and the former. Previously, it was proved in [4] that context-free languages are not closed under site-directed deletion closure: a context-free language L was presented such that $\mathrm{SDD}^*(L) \cap \#a^+\$b^+\$c^+\# = \{\#a^n\$b^n\$c^n\# \mid n \geq 1\}$.

It turns out that an analogue to Theorem 1 holds for site-directed deletion as well, which improves the abovementioned result from [4].

Theorem 2. *For any RE language $L \subseteq \Sigma^*$, there is a linear context-free language L_0 over a larger alphabet Ω and words $d_1, d_2 \in \Omega^*$ such that*

$$L = \{w \mid d_1 w d_2 \in \mathrm{SDD}^*(L_0)\}. \tag{1}$$

Proof. Let $\mathtt{M} = (Q, \Sigma, \Gamma, \delta, q_0, B, F)$ be a Turing machine that recognizes L and that always halts after an even number of steps (here $\Gamma \supseteq \Sigma$ is the alphabet of tape symbols, and $B \in \Gamma \setminus \Sigma$ is the blank symbol; $q_0 \in Q$ is the start state and $F \subseteq Q$ is the set of accepting states). By $x \vdash_{\mathtt{M}} y$ one means that $\mathtt{M}$ transforms an instantaneous description x into y in one step. Then, let

$$\begin{aligned} L_1 := \{&\$x_1\#x_3\#\ldots\#x_{2k-1}\#x_{2k}^{\mathrm{R}}\#\ldots\#x_2^{\mathrm{R}}\#x_0^{\mathrm{R}}\heartsuit\$ \mid \\ &x_{2i} \vdash_{\mathtt{M}} x_{2i+1} \text{ for all } i \in \{0,\ldots,k-1\}, x_{2k} \in \Gamma^* F \Gamma^*\}, \\ L_2 := \{&\natural w\$x_1\#x_3\#\ldots\#x_{2k-1}\#x_{2k}^{\mathrm{R}}\#\ldots\#x_2^{\mathrm{R}}\#x_0^{\mathrm{R}}\heartsuit\diamondsuit\$\natural \mid \\ &x_{2i-1} \vdash_{\mathtt{M}} x_{2i} \text{ for all } i \in \{1,\ldots,k\}, x_0 = q_0 w, w \in \Sigma^*\}. \end{aligned}$$

Finally, let $L_0 := L_1 \cup L_2 \cup \{\$\heartsuit\$\natural\}$. The following equalities hold:

- $L_1 \overset{\mathrm{sdd}}{\leftarrow} L_1 = L_1$ (use the second statement of Proposition 1; mind the \$ borders in words of L_1);
- $L_2 \overset{\mathrm{sdd}}{\leftarrow} L_2 = L_2$ (again, by Proposition 1; mind the $\natural$ borders in words of L_2);
- $L_1 \overset{\mathrm{sdd}}{\leftarrow} (L_2 \cup \{\$\heartsuit\$\natural\}) = \emptyset$ (words of L_1 do not contain $\natural$);
- $L_2 \overset{\mathrm{sdd}}{\leftarrow} \$\heartsuit\$\natural = \emptyset$ (because one cannot find non-empty u, v such that $uv = \$\heartsuit\$\natural$ and such that u and v are substrings of a string from L_2);
- $L' := L_2 \overset{\mathrm{sdd}}{\leftarrow} L_1 = \{\natural w\$x_1\#x_3\#\ldots\#x_{2k-1}\#x_{2k}^{\mathrm{R}}\#\ldots\#x_2^{\mathrm{R}}\#x_0^{\mathrm{R}}\heartsuit\$\natural \mid x_0 \vdash_{\mathtt{M}} x_1 \vdash_{\mathtt{M}} \ldots \vdash_{\mathtt{M}} x_{2k}, x_0 = q_0 w, w \in \Sigma^*, x_{2k} \in \Gamma^* F \Gamma^*\}$.

Therefore, $L_0' := L_0 \overset{\mathrm{sdd}}{\leftarrow} L_0 = L_0 \cup L'$. Now, let us look at what happens during the next iteration of site-directed deletion.

- $L_0' \overset{\mathrm{sdd}}{\leftarrow} (L_2 \cup L') \subseteq L_2 \cup L'$ (mind the $\natural$ borders in words of $L_2 \cup L'$);
- $L' \overset{\mathrm{sdd}}{\leftarrow} \$\heartsuit\$\natural = \{\natural v\$\heartsuit\$\natural \mid v \in L\}$ (recall that L is the target RE language);
- $L' \overset{\mathrm{sdd}}{\leftarrow} L_1 \subseteq \natural L\$(\Gamma \cup Q \cup \#)^*\heartsuit\$\natural =: M$.

In what follows, $L_0'' := L_0' \overset{\mathrm{sdd}}{\leftarrow} L_0' \subseteq L_0' \cup M$. Note also that $L_0'' \supseteq \natural L\$\heartsuit\$\natural$. Now, let us consider possible site-directed deletions applied to words in $L_0' \cup M$.

- $(L_0' \cup M) \overset{\text{sdd}}{\leftarrow} M \subseteq M$ (because of the $\natural$ borders in words of M);
- $M \overset{\text{sdd}}{\leftarrow} L_1 \subseteq M$ (because a site-directed deletion from a string of M by a string of L_1 can affect only a substring between \$ and $\heartsuit$ \$, hence the resulting string is also from M);
- $M \overset{\text{sdd}}{\leftarrow} (L_2 \cup L') \subseteq L_2 \cup L'$ (because of the $\natural$ borders);
- $M \overset{\text{sdd}}{\leftarrow} \$\heartsuit\$\natural = \natural L\$\heartsuit\$\natural \subseteq M$.

Therefore, $(L_0' \cup M) \overset{\text{sdd}}{\leftarrow} (L_0' \cup M) \subseteq (L_0' \cup M)$, i.e. this language is sdd-closed. Summing up, the following inclusions hold:

$$\natural L\$\heartsuit\$\natural \subseteq \text{SDD}^*(L_0) \subseteq L_0 \cup \natural L\$(\Gamma \cup Q \cup \#)^*\heartsuit\$\natural$$

Concluding, if one takes $d_1 = \natural$ and $d_2 := \$\heartsuit\$\natural$, then $\{w \mid d_1 w d_2 \in \text{SDD}^*(L_0)\} = L$. Note that there are no words of the form $d_1 w d_2$ in L_0. □

The proof of Theorem 2 as well as that of Theorem 1 are both based on the fact that (site-directed) deletion combines features of intersection and erasing. E.g. when one deletes a string v from a string u, they first check that v is a substring of u (this is a kind of intersection) and then erases v in u. Likewise, for site-directed deletion, one checks that there are decompositions $u = u_1 v_1 z v_2 u_2$, $v = v_1 v_2$ (an "intersection" part) and then erases z in u (an "erasing" part). Each RE language can be obtained in two steps: one intersects two context-free languages resulting in a language of terminating Turing machine protocols (cf. the language L' from the proof of Theorem 2); then, one erases protocol parts in words retaining only the input word. These two steps are reflected in the proof of Theorem 2.

Despite general similarity between Theorems 1 and 2, there are certain differences between deletion and site-directed deletion of context-free languages. Namely, in the proof of Theorem 1, given a RE language L, two linear context-free languages L_1, L_2 are constructed such that $\text{D}^*(L_1 \cup L_2) = \{\varepsilon\} \cup L_1 \cup L_2 \cup (L_2 \rightarrow L_1)$ where $L_2 \rightarrow L_1 = \natural L \natural$. The latter equality is of interest, because it implies that one-step deletion applied to context-free languages already produces all RE languages (modulo having $\natural$ borders). This is not the case for site-directed deletion as the following proposition shows.

Proposition 2. *For any two context-free languages $L_1, L_2 \subseteq \Sigma^*$, the language $L_1 \overset{\text{sdd}}{\leftarrow} L_2$ is context-sensitive.*

Proof. If L_1, L_2 are context-free, then the following languages are context-free as well:

$$L_1' := \{x_1\$u\$v\$x_2 \mid \exists z\, x_1 u z v x_2 \in L_1\};$$

$$L_2' := \{\alpha\$u\$v\$\beta \mid \alpha, \beta \in \Sigma^*, uv \in L_2, u, v \neq \varepsilon\}.$$

Then, their intersection $L_1' \cap L_2'$ is context-sensitive, since context-sensitive languages are closed under intersection. In what follows, if ψ is the homomorphism that erases \$ (i.e. $\psi(\$) = \varepsilon$ and $\psi(a) = a$ for $a \neq \$$), then $\psi(L_1' \cap L_2')$ is context sensitive too since context-sensitive languages are closed under linear erasing. Finally, observe that $\psi(L_1' \cap L_2') = L_1 \overset{\text{sdd}}{\leftarrow} L_2$. □

Thus, one-step site-directed deletion of context-free languages is not computationally universal unlike deletion. This difference arises because site-directed deletion does not specify a part of a string to be deleted, but a context of deletion instead. Nevertheless, two iterations of site-directed deletion are enough to obtain any RE language, as the proof of Theorem 2 shows.

Another difference is related to borders of words that occur in both Theorems 1 and 2. In the former, words are surrounded by a special symbol $\natural$; in the latter, the borders are $d_1 = \natural$ and $d_2 = \$\heartsuit\$\natural$. (Note that, without borders, none of the theorems would be true because there are many RE languages which are not closed under deletion or site-directed deletion.) It turns out that, if one requires borders d_1, d_2 in Theorem 2 be of length 1, as in Theorem 1, then site-directed deletion closure becomes less expressive. To show this, let us introduce a new notion.

Definition 1. *A word $w \in L$ is* symbol-deduplicating in L *($L \subseteq \Sigma^*$) if $w = w_1xw_2$ for some $w_1, w_2 \in \Sigma^*$ and $x \in \Sigma$ such that one of the following holds:*

1. *for each $u, v \in \Sigma^*$, if $uxvw_1xw_2 \in L$, then $uxw_2 \in L$;*
2. *for each $u, v \in \Sigma^*$, if $w_1xw_2uxv \in L$, then $w_1xv \in L$.*

Remark 1. If there is a symbol-deduplicating word in an infinite language $L \subseteq a^*$ over a one-letter alphabet, then $a^* \setminus L$ is finite. Indeed, let $w = a^N = w_1aw_2$ be a symbol-deduplicating word in L; let it satisfy the first of the above properties (the other one is dealt with similarly). It holds that $w_i = a^{k_i}$ for $i = 1, 2$ such that $k_1 + k_2 + 1 = N$. Since L is infinite, for each $m > N$, there is $n > m + N$ such that $a^n \in L$. Take $u = a^{m-(k_2+1)}$, $v = a^{n-(m+k_1+1)}$. Observe that $uavw_1aw_2 = a^{m-(k_2+1)+1+n-(m+k_1+1)+N} = a^{m-N+n-m+N} = a^n$. By the deduplicating property, it holds that $uxw_2 = a^m \in L$. Summing up, for each $m \geq N$, a^m is a member of L.

Proposition 3. *Let $L_0 \subseteq \Sigma^*$ with $\natural_1, \natural_2 \in \Sigma$. Then, for each word w in $L = \{w \mid \natural_1 w \natural_2 \in \mathrm{SDD}^*(L_0)\}$, either $\natural_1 w \natural_2 \in L_0$ or w is symbol-deduplicating in L.*

Proof. Let $\natural_1 w \natural_2 \in \mathrm{SDD}^*(L_0) \setminus L_0$. This means that this word is obtained by non-trivial site-directed deletion, i.e. $\natural_1 w \natural_2 = t_1y_1y_2t_2$ where $y = y_1y_2 \in \mathrm{SDD}^*(L_0)$, $t_1y_1zy_2t_2 \in \mathrm{SDD}^*(L_0)$ and either t_1 or t_2 is nonempty. Without loss of generality, assume that t_1 is nonempty. Let w_1 be the prefix of w such that $\natural_1 w_1 = t_1$; let x be the first symbol of y_1 (it is not $\natural_2$ because y_2 is nonempty); finally, let w_2 be defined by the equality $w_1xw_2 = w$. Let w_2' be a word such that $xw_2 = yw_2'$; let y' be a word such that $y = xy'$.

Assume that $uxvw_1xw_2 \in L$, i.e. $\natural_1 uxvw_1xy'w_2' \natural_2 \in \mathrm{SDD}^*(L_0)$. Since $y = xy'$ belongs to $\mathrm{SDD}^*(L_0)$ and this language is sdd-closed, the word $\natural_1 uxy'w_2' \natural_2$ also belongs to $\mathrm{SDD}^*(L_0)$. Indeed, it is obtained by site-directed deletion from $\natural_1 uxvw_1xy'w_2' \natural_2$ and xy'. Thus, $uxy'w_2' = uxw_2 \in L$, which shows that w is symbol-deduplicating. □

Corollary 1. *Let $L_0 \subseteq \Sigma^*$ be a context-free language, let $\natural_1, \natural_2 \in \Sigma$ be distinguished symbols, and let $L = \{w \mid \natural_1 w \natural_2 \in \mathrm{SDD}^*(L_0)\} \subseteq a^*$ be a one-letter language. Then, L is semilinear (and hence regular).*

Proof. If L is finite, then there is nothing to prove. If L is infinite and there is a symbol-deduplicating word in L, then L contains all but finitely many words from a^*, hence it is semilinear. Otherwise, Proposition 3 implies that $L = \{w \mid \natural_1 w \natural_2 \in L_0\}$. Since L_0 is context-free, the language $\{w \mid \natural_1 w \natural_2 \in L_0\}$ is semilinear by Parikh's theorem. Therefore, L is semilinear. □

Concluding, Theorem 2 is false if $|d_1| = |d_2| = 1$, namely, one-letter word borders are not enough to obtain some non-regular languages as site-directed deletion closures of context-free languages.

3.2 Closures of Regular Languages

Theorem 2, together with the result of [5], shows that linear context-free languages suffice to generate all recursively enumerable languages by (site-directed) deletion closures. What if one starts with a regular language instead of a context-free one? Unlike context-free languages, one cannot generate every recursively enumerable language by applying intersection and erasing to a regular language. One might conjecture that regular languages are closed under (site-directed) deletion closure, but direct proofs of this conjecture remain elusive. Indeed, given a regular language L, an ability to construct a finite automaton for $\mathrm{D}^*(L)$ or for $\mathrm{SDD}^*(L)$ is unclear due to the branching nature of (site-directed) deletion closure. The closure of regular languages under deletion was left open in [8], and a similar question for site-directed deletion remains unresolved since [4]. To our knowledge, these questions remain unanswered.

It turns out that it is not hard to answer the first question by reducing deletion closure to iterated deletion. Below, we provide a positive answer.

Theorem 3. *If L is regular, then $\mathrm{D}^*(L)$ is regular too.*

Proof. In Kari's thesis, it is proved [11, Theorem 3.8] that, if L_1 is regular (and L_2 is arbitrary), then $L_1 \xrightarrow{*} L_2$ is regular too. It suffices to notice that $\mathrm{D}^*(L) = L \xrightarrow{*} \mathrm{D}^*(L)$. The '$\supseteq$' inclusion is trivial. The '$\subseteq$' inclusion is shown by induction using the inductive definition of $\mathrm{D}^*(L)$. For the induction step, assume that $\mathrm{D}^i(L) \subseteq L \xrightarrow{*} \mathrm{D}^*(L)$. If $w \in \mathrm{D}^{i+1}(L) \setminus \mathrm{D}^i(L)$, then $w = w_1 w_2$ such that, for some $u \in \mathrm{D}^i(L)$, it holds that $w_1 u w_2 \in \mathrm{D}^i(L)$. By the induction hypothesis, $w_1 u w_2 \in L \xrightarrow{*} \mathrm{D}^*(L)$. Therefore,

$$w \in (L \xrightarrow{*} \mathrm{D}^*(L)) \to \mathrm{D}^*(L) \subseteq L \xrightarrow{*} \mathrm{D}^*(L)$$

since w is obtained from $w_1 u w_2$ by deleting $u \in \mathrm{D}^*(L)$.

Summing up, $\mathrm{D}^*(L) = L \xrightarrow{*} \mathrm{D}^*(L)$, hence it is regular. □

The above proof fails for site-directed deletion. Indeed, if L_1 is regular, then, in general, it is not the case that $L_1 \xleftarrow{\mathrm{sdd}^*} L_2$ is regular. For example, let Σ be an alphabet and let $\natural, \heartsuit \notin \Sigma$; let $L_1 = \natural \heartsuit \Sigma^* \natural$ and let $L_2 = \natural L \natural$ for an arbitrary language L. Then, it is not hard to prove that $L_1 \xleftarrow{\mathrm{sdd}^*} L_2 = L_1 \cup L_2$. The resulting language is not regular unless L is regular. The question whether regular languages are closed under SDD* still remains an interesting open problem.

Theorem 3 shows that deletion closure is unable to produce complex languages from regular ones. This result can be compared with that by Hofbauer and Waldmann [7], which says that finite deleting string-rewriting systems preserve regularity. Thus, one might be curious whether there is a natural and simple kind of deletion such that closure under it applied to regular languages produces some complex languages, especially some undecidable ones.

It turns out, in a sharp contrast with the abovementioned results, that deletion system closures of regular languages do not preserve regularity and in fact are computationally universal in a certain sense. Deletion systems are insertion-deletion systems [16, Chapter 6] which have only deletion rules; let us present their definition below.

Definition 2. *A* deletion system *is a finite set of word triples* $D \subseteq (\Sigma^*)^3$. *A deletion rule* $(x, y, z) \in D$ *is applied as follows:* $\alpha xyz\beta \Rightarrow_D \alpha xz\beta$.
The closure of L under D is the least set that contains L and that is closed under applications of deletion rules from D; it is denoted by $\mathrm{del}_D^*(L)$. *In other words,* $\mathrm{del}_D^*(L) = \{\beta \mid \exists \alpha \in L\, \alpha \Rightarrow_D^* \beta\}$.

Deletion systems have received little attention in the literature, unlike the dual formalism—*insertion systems* (cf. [16, Sections 6.2, 6.4] for properties of insertion systems). Supposedly, this is due to the fact that insertion-deletion systems usually have a finite set of words as the input. For deletion systems, if L is finite, then $\mathrm{del}_D^*(L)$ is finite too, thus no interesting language-theoretic properties can be investigated. Nevertheless, if the initial language L is at least regular, then deletion systems are computationally universal, as the following theorem shows.

Theorem 4. *For each RE set* $A \subseteq \mathbb{N}$, *there exist a regular language* $R \subseteq \Sigma^*$, *words* $x, y_1, y_2, z \in \Sigma^*$, *and a deletion system D such that* $n \in A$ *if and only if* $xy_1^{2^n} y_2^{2^n} z \in \mathrm{del}_D^*(R)$.

Proof (of Theorem 4). It is well known that, for any RE set $A \subseteq \mathbb{N}$, there is a 2-counter Minsky machine that, given $(2^n, 0)$ as an input, halts iff $n \in A$. Let us fix such a Minsky machine `M` for the set A from the theorem. We assume that a Minsky machine has a finite number of states and that its instructions are of one of the two forms: "being in the state q, increase the i-th counter by 1 and change the state to r" or "being in the state q, check whether the i-th counter equals 0; if so, then change the state to r_1; otherwise, decrease the i-th counter by 1 and change the state to r_2." A machine begins in a state s_0 and accepts by reaching a state s_a.

To prove the theorem, we shall use the results from [14]. In [14, Theorem 4.9], it is shown how 2-counter Minsky machines can be simulated by insertion systems. An insertion system I is defined by a finite set of insertion rules of the form $r = (x, y, z)$. Such a rule is applied as follows: $uxzv \Rightarrow uxyzw$. We write $\alpha \Rightarrow_I^* \beta$ if β can be obtained from α by iteratively applying insertion rules from

I. The paper [14] presents a construction of an insertion system that, starting with a word $\alpha_0 := \triangleright\texttt{=}_2 s_0 \texttt{=}_1 \triangleleft$, simulates computation of a given 2-counter Minsky machine. Such a system produces a word that contains the symbol s_a if and only if the machine halts on the input $(0, 0)$.

In fact, the construction from [14, Theorem 4.9] works not only for the input $(0, 0)$ but for an arbitrary input. Indeed, one can simply increment counter values a given number of times at the beginning of computation. The increment operation is implemented in [14] by insertion; namely, the increment of the first counter in $(0, 0)$ by k, i.e. the pair $(k, 0)$, being simulated by insertions starting with α_0, results in the string $\alpha_k := \triangleright\texttt{=}_2 s_0 (\texttt{=}_2 s_0)^k (\texttt{>}\overline{s}_0\texttt{>}1)^k \texttt{=}_1 \triangleleft$. (We use the construction from the cited article as a black box because this is enough to prove our theorem.) So, the following is in fact a corollary of the result from [14]: given the 2-counter Minsky machine $\mathtt{M}$, one can construct an insertion system $I \subseteq (\Sigma^*)^3$ such that $\alpha_k \Rightarrow_I^* \beta s_a \gamma$ for some β, γ if and only if $\mathtt{M}$ halts on $(k, 0)$. Let us fix such an insertion system. Let $D = I$ be the deletion system with the same rules. It is quite clear that $w_1 \Rightarrow_I^* w_2$ iff $w_2 \Rightarrow_D^* w_1$; indeed, a deletion rule is simply the inverse of the very same insertion rule.

Let Σ be the alphabet of the deletion system D; Σ includes states of the Minsky machine, in particular, $s_a \in \Sigma$. Let $R = \Sigma^* s_a \Sigma^*$; let $x = \triangleright\texttt{=}_2 s_0$, $y_1 = \texttt{=}_2 s_0$, $y_2 = \texttt{>}\overline{s}_0\texttt{>}1$, $z = \texttt{=}_1\triangleleft$. Then, the following facts are equivalent: $\alpha_{2^n} = x y_1^{2^n} y_2^{2^n} z$ belongs to $\mathrm{del}_D^*(R) \iff$ there are $\beta, \gamma \in \Sigma^*$ such that $\beta s_a \gamma \Rightarrow_D^* \alpha_{2^n}$ $\iff$ there are $\beta, \gamma \in \Sigma^*$ such that $\alpha_{2^n} \Rightarrow_I^* \beta s_a \gamma \iff \mathtt{M}$ halts on $(2^n, 0) \iff$ $n \in A$. This completes the proof. □

The following is a corollary of this theorem.

Corollary 2. *There are a regular language R and a deletion system D such that the language $\mathrm{del}_D^*(R)$ is RE-complete.*

Proof. Take a RE-complete set $A \subseteq \mathbb{N}$ and apply Theorem 4 to it. Clearly, A is m-reducible to $\mathrm{del}_D^*(R)$ where R, D are from the statement of that theorem. □

4 On Complexity of Closures of Finite Languages

All conceivable variants of deletion, including the ones studied in the current paper, share the following property: deletion from a string results in a shorter string. Consequently, if one applies any kind of deletion to a finite language iteratively, then the resulting language is always finite. Therefore, there is little sense to search for a description of the class of language closures under a given deletion. Instead, one might be interested in the complexity of the membership problem, namely, whether a given string w_0 is contained in the closure of a given finite language $L = \{w_1, \ldots, w_n\}$ under a given deletion operation. This has biological applications in motif discovery and genetic engineering, where deletion models how genetic sequences lose material while preserving functional motifs [10]. In synthetic biology, understanding deletion complexity aids in designing robust

DNA/RNA templates, such as in CRISPR and RNA origami [6,15]. The computational challenge of predicting deletion closure parallels the difficulty of reconstructing ancestral sequences and verifying genetic circuit stability [1].

Let C be either D^*, SDD^*, or del_D^*. Consider the following problem.

Closure membership problem.
Input: a finite language $L = \{w_1, \ldots, w_n\}$; a word w_0.
Question: does w_0 belong to $C(L)$?

For deletion systems, this problem is in NP, because each deletion rule decreases the length of a word; so, if $w_0 \in \mathrm{del}_D(\{w_i\})$, then this is justified by a linear size derivation w.r.t. $|w_i|$. Besides, it is proved in [17, Theorem 4.8.1] that checking whether, given a deletion system D and a word w, it holds that $\varepsilon \in \mathrm{del}_D^*(\{w\})$ is NP-complete. In the cited thesis, the formalism used is called *clearing restarting automata*, but, essentially, the latter are equivalent to insertion systems.

The situation is more involved for (site-directed) deletion closure membership. A trivial upper bound for the closure membership problem both for D^* and SDD^* is EXPTIME. Indeed, since all words from $C(L)$ are subsequences of words from L, the size of $C(L)$ does not exceed $n2^l$ where l is the length of the largest string in L; thus, it can be constructed in exponential time.

One can view the set $C(L)$ as the set of theorems where axioms are members of L and where the inference rule is deletion if $C = \mathrm{D}^*$ or site-directed deletion if $C = \mathrm{SDD}^*$. In particular, if $w \in C(L)$, then one can consider a derivation tree of w. For example, if $L = \{abbc, b, aacb\}$, then $w = c$ belongs to $\mathrm{D}^*(L)$ with the following derivation tree:

$$\dfrac{\dfrac{abbc \quad b}{abc} \qquad \dfrac{aacb \qquad \dfrac{\dfrac{abbc \quad b}{abc} \qquad b}{ac}}{ab}}{c}$$

Clearly, one could optimize the above tree by merging two identical subtrees with the root *abc*; so, in general, one can consider a derivation DAG justifying that a string w belongs to $C(L)$. If one shows that, whenever $w \in C(L)$, there is a derivation tree of w of size (depth) polynomially bounded by the sizes of w and L, then the closure membership problem is in NP (PSPACE resp.). However, it is not clear even whether this problem is in PSPACE. Indeed, imagine that one tries to check whether a word w belongs to $\mathrm{SDD}^*(L)$. They guess the last application of site-directed deletion, which can be represented by the scheme

$$\frac{x_1uzvx_2 \quad uv}{x_1uvx_2}$$

The second premise of this rule application is shorter than its conclusion. The former, in turn, can be obtained from another string by means of sdd:

$$\dfrac{x_1uzvx_2 \qquad \dfrac{x'_1u'z'v'x'_2 \quad u'v'}{uv = x'_1u'v'x'_2}}{x_1uvx_2}$$

Thus, there are no obvious reasons why the size or the depth of a derivation tree is polynomially bounded, hence we have only the EXPTIME upper bound. As for lower bounds, though, we can prove only the following result.

Theorem 5. *Site-directed deletion closure membership problem is NP-hard.*

Proof. We reduce the longest subsequence problem to the sdd-closure membership problem. To recall, the former one is stated as follows [13]: given strings $u_1, \dots, u_M$ and an integer $p \in \mathbb{N}$, check whether there is a common subsequence of $u_1, \dots, u_M$ of length at least p. (Without loss of generality, we may assume that p is given in unary.) If a string u' is a subsequence of a string u, then let us denote this by $u' \preceq u$. For example, $bd \preceq abcde$ and $abcde \preceq abcde$.

Let us introduce the following strings.

1. Let $u_i = a_1 \dots a_k$ (for $1 \le i < M$). Then $\xi_i(u_i) := \heartsuit_i\clubsuit_i \dots \heartsuit_1\clubsuit_1\#\%a_1\#\%a_2\#\% \dots \#\%a_k\#\%\diamondsuit_1\spadesuit_1 \dots \diamondsuit_{M-i}\spadesuit_{M-i}$.
2. Let $u_M = b_1 \dots b_l$. Then $\xi_M(u_M) :=$ $\natural\heartsuit_{M-1}\clubsuit_{M-1} \dots \heartsuit_1\clubsuit_1\#\%b_1\#\%b_2\#\% \dots \#\%b_l\#\%\diamondsuit_1\$\spadesuit_1 \dots \diamondsuit_M\$\spadesuit_M\natural$.
3. Let $\delta_i := (\#\%)^i\diamondsuit_1\spadesuit_1 \dots \diamondsuit_M\spadesuit_M\natural$ for $1 \le i \le p+1$.
4. Let $w_0 := \natural\heartsuit_{M-1}\clubsuit_{M-1} \dots \heartsuit_1\clubsuit_1(\#\%)^{p+1}\diamondsuit_1\spadesuit_1 \dots \diamondsuit_M\spadesuit_M\natural$.

Consider the sdd-closure membership problem with the instance

$$L = \{\xi_1(u_1), \dots, \xi_M(u_M), \#\%, \delta_{p+1}\};\ w_0.$$

We claim that the answer to this problem is "YES" if and only if so is the answer to the longest subsequence problem. To prove this, let us first find $\mathrm{SDD}^*(\{\xi_1(u_1), \dots, \xi_M(u_M), \#\%\})$. We shall use the following notation: if $v = v_1 \dots v_n$ is a string ($v_i \in \Sigma$), then

$$\xi^j_M(v) := \natural\heartsuit_{M-1}\clubsuit_{M-1} \dots \heartsuit_1\clubsuit_1 \\ \#\%v_1\#\% \dots \#\%v_n\#\%\diamondsuit_1\spadesuit_1 \dots \diamondsuit_{j-1}\spadesuit_{j-1}\diamondsuit_j\$\spadesuit_j \dots \diamondsuit_M\$\spadesuit_M\natural.$$

Lemma 1. $\mathrm{SDD}^*(\{\xi_1(u_1), \dots, \xi_M(u_M), \#\%\}) =$

$$\bigcup_{i=1}^{M-1}\{\xi_i(u) \mid u \preceq u_i\} \cup \bigcup_{j=1}^{M}\{\xi^j_M(u) \mid \forall k > M-j\ u \preceq u_k\} \cup \{\#\%\}. \qquad (2)$$

Proof. Let us denote the set in the right-hand side of (2) by S. Clearly, S contains $\{\xi_1(u_1), \dots, \xi_M(u_M), \#\%\}$ (observe that $\xi_M(u_M) = \xi^1_M(u_M)$). Let us check that S is sdd-closed. In the below reasonings, we use Proposition 1 (often implicitly).

- For $i, j < M$, $\xi_i(u) \overset{\text{sdd}}{\leftarrow} \xi_j(v)$ equals $\emptyset$ if $i \neq j$ and $\emptyset$ or $\{\xi_j(v)\}$ otherwise. Indeed, if $i < j$, then $\xi_j(v)$ contains the symbol $\heartsuit_j$, which is not present in $\xi_i(v)$; if $i > j$, then $\xi_j(v)$ contains $\spadesuit_{M-j}$. If $i = j$, then observe that the first and the last symbols of $\xi_j(v)$ are unique in this word, so the only possible site-directed deletion of this word applied to itself is trivial.
- $\xi^i_M(u) \overset{\text{sdd}}{\leftarrow} \xi^j_M(v)$ equals either $\emptyset$ or $\{\xi^j_M(v)\}$ (by Proposition 1).
- $\xi_i(u) \overset{\text{sdd}}{\leftarrow} \xi^j_M(v) = \emptyset$ (there is no $\natural$ in $\xi_i(u)$).
- $\xi^j_M(u) \overset{\text{sdd}}{\leftarrow} \xi_i(v) = \emptyset$ if $M - i > j$. Indeed, if $M - i > j$, then $\diamondsuit_j\spadesuit_j\diamondsuit_{j+1}\spadesuit_{j+1}$ is a substring of $\xi_i(v)$. If site-directed deletion was possible, then at least one of the strings $\diamondsuit_j\spadesuit_j$ or $\diamondsuit_{j+1}\spadesuit_{j+1}$ would be a substring of $\xi^j_M(u)$. However, in $\xi^j_M(u)$, $\diamondsuit_j$ and $\spadesuit_j$ are separated by \$ as well as $\diamondsuit_{j+1}$ and $\spadesuit_{j+1}$.
- If $M - i < j$, then $\xi^j_M(u) \overset{\text{sdd}}{\leftarrow} \xi_i(v)$ results in $\xi^j_M(u')$ for $u' \preceq u$.
- If $M - i = j$, then $\xi^j_M(u) \overset{\text{sdd}}{\leftarrow} \xi_i(v)$ results in removing \$ between $\diamondsuit_j$ and $\spadesuit_j$ iff $u = v$. Note that $u \preceq u_k$ for $M - j < k \leq M$ and $v \preceq u_i = u_{M-j}$. We thus obtain $\xi^{j+1}_M(u)$ where $u \preceq u_{M-j}, \ldots, u_M$.
- $\xi_i(u) \overset{\text{sdd}}{\leftarrow} \#\% \subseteq \{\xi_i(u') \mid u' \preceq u\}$ ($\#\%$ deletes a substring in u).
- $\xi^j_M(u) \overset{\text{sdd}}{\leftarrow} \#\% \subseteq \{\xi^j_M(u') \mid u' \preceq u\}$.
- Site-directed deletion from the string $\#\%$ is not possible.

Thus, we have exhausted all possibilities of site-directed deletion of two strings belonging to S and checked that the result is from S too. Moreover, the above reasonings show that one can obtain all words of S from $\{\xi_1(u_1), \ldots, \xi_M(u_M), \#\%\}$ by site-directed deletions. Namely, let $\xi^j_M(u) \in S$ where, for all $k > M - j$, $u \preceq u_k$ and let v be a common subsequence of u and u_{M-j}. Then, first, one can obtain $\xi^j_M(v)$ from $\xi^j_M(u)$ by site-directed deletions involving $\#\%$; similarly, one can obtain $\xi_{M-j}(v)$ from $\xi_{M-j}(u_{M-j}) \in L$. Secondly, one can obtain $\xi^{j+1}_M(v)$ as the result of $\xi^j_M(v) \overset{\text{sdd}}{\leftarrow} \xi_{M-j}(v)$. □

Given $u = u_1 \ldots u_n$ ($u_i \in \Sigma$) and $0 \leq q \leq n$, we define the word $\theta_q(u)$ as follows:

$$\theta_q(u) := \natural\heartsuit_{M-1}\clubsuit_{M-1}\ldots\heartsuit_1\clubsuit_1\#\%u_1\ldots\#\%u_{n-q}(\#\%)^{q+1}\diamondsuit_1\spadesuit_1\ldots\diamondsuit_M\spadesuit_M\natural.$$

Let $\Theta := \{\theta_q(u) \mid 0 \leq q \leq \min(p, |u|), \forall k = 1, \ldots, M\ u \preceq u_k\}$. Let also $\Delta := \{\delta_i \mid 1 \leq i \leq p+1\}$. Finally, let $T := S \cup \Delta \cup \Theta$.

We claim that $\mathrm{SDD}^*(L) = T$. Clearly, $L \subseteq T$ since $\{\xi_1(u_1), \ldots, \xi_M(u_M), \#\%\}$ is contained in S and δ_{p+1} is contained in Δ. Let us check that T is sdd-closed.

- If $u \in S$ and $v \in \Delta$, then $u \overset{\text{sdd}}{\leftarrow} v$ is empty except, possibly, the case where $u = \xi^M_M(w)$ for some common subsequence $w = w_1 \ldots w_n$ of $u_1, \ldots, u_M$ and for $v = \delta_1$. In this case, the only way to apply site-directed deletion is as follows: $\xi^M_M(w) = x_1y_1zy_2x_2$ where $x_1 = \natural\heartsuit_{M-1}\clubsuit_{M-1}\ldots\heartsuit_1\clubsuit_1\#\%w_1\#\%\ldots\#\%w_n$, $y_1 = \#\%\diamondsuit_1\spadesuit_1\ldots\diamondsuit_{M-1}\spadesuit_{M-1}\diamondsuit_M$, $z = \$$, $y_2 = \spadesuit_M\natural$, $x_2 = \varepsilon$. Then, $v = \delta_1 = y_1y_2$, and $x_1y_1y_2x_2$ is obtained from u by deleting \$, i.e. the resulting string is $\theta_0(w) \in \Theta$.

- Site-directed deletion of one string from Δ from another string from Δ, clearly, results in a string from Δ.
- The only way to obtain a string from $\Delta \overset{\text{sdd}}{\leftarrow} S$ is by site-directed deletion $\delta_i \overset{\text{sdd}}{\leftarrow} \#\%$; this is because no string of Δ contains $\heartsuit_1$. Clearly, $\delta_i \overset{\text{sdd}}{\leftarrow} \#\% = \{\delta_j \mid j \leq i\}$ because $\#\%$ can erase a substring between $\#$ and $\%$ in δ_i.
- $T \overset{\text{sdd}}{\leftarrow} \Theta \subseteq \Theta$ (Proposition 1; mind the $\natural$ borders).
- $\Theta \overset{\text{sdd}}{\leftarrow} S \subseteq \Theta \cup S$. This is justified by Proposition 1 except for two cases. The first one is $\theta_q(u) \overset{\text{sdd}}{\leftarrow} \xi_i(v)$. The result of site-directed deletion is $\theta_0(v)$ because this is how $\xi_i(v)$ "overwrites" the part in $\theta_q(u)$ between $\heartsuit_i$ and $\spadesuit_{M-i}$. Moreover, in the site-directed deletion $\theta_q(u) \overset{\text{sdd}}{\leftarrow} \xi_i(v)$, a substring of u is removed, so v is a subsequence of u. Thus, if $u \preceq u_k$ for each $k = 1, \ldots, M$, then $v \preceq u_k$ as well. Thus, the result of site-directed deletion, which is $\theta_0(v)$, belongs to Θ.

  The second interesting case is $\theta_q(u) \overset{\text{sdd}}{\leftarrow} \#\%$. The result is $\theta_r(u')$ for $u' \preceq u$ and $r \leq q$. Clearly, $\theta_r(u') \in \Theta$.
- It remains to find $\Theta \overset{\text{sdd}}{\leftarrow} \Delta$. The site-directed deletion $\theta_q(u) \overset{\text{sdd}}{\leftarrow} \delta_i$ is non-empty if and only if $i \leq q + 2$. In this case, some part of $\theta_q(u)$ is removed between an occurrence of the string $\#\%$ and the substring δ_{i-1} of $\theta_q(u)$. This results in a string of the form $\theta_r(u')$ where $r \leq q + 1$ and $u' \preceq u$. Thus, the resulting string belongs to Θ.

The above reasonings also prove that $T \subseteq \text{SDD}^*(L)$ because they show how to obtain all words of T by site-directed deletions from words of S and from δ_{p+1}. Thus, $T = \text{SDD}^*(L)$. Note that the target word w_0 equals $\theta_p(u)$ for any u of length p. Therefore, $w_0 \in \text{SDD}^*(L)$ iff $w_0 \in \Theta$ iff there is a word u of length p such that $u \preceq u_k$ for $k = 1, \ldots, M$ (see the definition of Θ). The latter means exactly that there is a common subsequence of $u_1, \ldots, u_M$ of length p. □

5 Conclusion

We investigated deletion and site-directed deletion operations, emphasizing the differences in their computational power. It is known that deletion closure applied to context-free languages produces all RE languages, and we showed the same property for sdd-closure. Regular languages, however, present a more nuanced case: although they are closed under deletion closure via a reduction to iterated deletion, a similar approach fails for site-directed deletion due to its branching nature. Consequently, it remains an open question whether regular languages are closed under sdd-closure.

Since all deletion variants produce shorter strings, iterated deletions from a finite language always yield a finite language. This limits the study of closure classes and shifts the focus to the closure membership problem, which has biological relevance and is theoretically interesting. We showed that the sdd-closure membership problem is NP-hard but the exact complexity remains open.

Funding. Cho was supported by Institute of Information & Communications Technology Planning & Evaluation (IITP) under the Artificial Intelligence Convergence Innovation Human Resources Development (IITP-2025-RS-2023-00255968) grant funded by the Korea government (MSIT).
The work of Pshenitsyn was performed at the Steklov International Mathematical Center and supported by the Ministry of Science and Higher Education of the Russian Federation (agreement no. 075-15-2025-303). Pshenitsyn was also supported by the Foundation for the Advancement of Theoretical Physics and Mathematics "BASIS".

Disclosure of Interests. The authors have no competing interests to declare that are relevant to the content of this article.

References

1. Abroshan, M., Venkataramanan, R., Dolecek, L., Guillén i Fabregas, A.: Coding for deletion channels with multiple traces. In: 2019 IEEE International Symposium on Information Theory (ISIT), pp. 1372–1376. IEEE (2019)
2. Adli, M.: The CRISPR tool kit for genome editing and beyond. Nat. Commun. **9**(1), 1911 (2018)
3. Benne, R.: RNA Editing: the Alteration of Protein Coding Sequences of RNA. Ellis Horwood Series in Molecular Biology. Ellis Horwood, New York (1993)
4. Cho, D.-J., Han, Y.-S., Kim, H., Salomaa, K.: Site-directed deletion. In: Developments in Language Theory, pp. 219–230 (2018)
5. Domaratzki, M., Okhotin, A.: Representing recursively enumerable languages by iterated deletion. Theoret. Comput. Sci. **314**(3), 451–457 (2004)
6. Doudna, J.A., Charpentier, E.: The new frontier of genome engineering with CRISPR-Cas9. Science **346**(6213), 1258096 (2014)
7. Hofbauer, D., Waldmann, J.: Deleting string rewriting systems preserve regularity. Theoret. Comput. Sci. **327**(3), 301–317 (2004)
8. Ito, M., Kari, L., Thierrin, G.: Insertion and deletion closure of languages. Theoret. Comput. Sci. **183**(1), 3–19 (1997)
9. Ito, M., Silva, P.V.: Remark on deletions, scattered deletions and related operations on languages. In: Semigroups and Applications, pp. 97–105 (1998)
10. Jinek, M., Chylinski, K., Fonfara, I., Hauer, M., Doudna, J.A., Charpentier, E.: A programmable dual-RNA–guided DNA endonuclease in adaptive bacterial immunity. Science **337**(6096), 816–821 (2012)
11. Kari, L.: On Insertion and Deletion in Formal Languages. PhD thesis, University of Turku (1991)
12. Kari, L., Thierrin, G.: K-insertion and k-deletion closure of languages. Soochow J. Math. **21**(4), 479–495 (1995)
13. Maier, D.: The complexity of some problems on subsequences and supersequences. J. ACM **25**(2), 322–336 (1978)
14. Malcher, A.: Decidability questions for insertion systems and related models. Fund. Inform. **180**(1–2), 53–76 (2021)
15. McRae, E.K.S., et al.: Structure, folding and flexibility of co-transcriptional RNA origami. Nature Nanotechnol. **18**(7), 808–817 (2023)
16. Păun, G., Rozenberg, G., Salomaa, A.: DNA Computing: New Computing Paradigms. Springer, Heidelberg (1998)
17. Černo, P.: Restricted Restarting Automata. PhD thesis, Charles University in Prague (2015)

Universally Wheeler Languages

Ruben Becker[1], Giuseppa Castiglione[2], Giovanna D'Agostino[3], Alberto Policriti[3], Nicola Prezza[1], Antonio Restivo[2], and Brian Riccardi[4(✉)]

[1] Department of Environmental Sciences, Informatics and Statistics, Ca' Foscari University of Venice, Venice, Italy
[2] Department of Mathematics and Computer Science, University of Palermo, Palermo, Italy
[3] Department of Mathematics, Computer Science and Physics, University of Udine, Udine, Italy
[4] Department of Informatics, Systems and Communication, University of Milano-Bicocca, Milan, Italy
brian.riccardi@unimib.it

Abstract. The notion of Wheeler languages is rooted in the Burrows-Wheeler transform (BWT), one of the most central concepts in data compression and indexing. The BWT has been generalized to finite automata, the so-called Wheeler automata, by Gagie, Manzini, and Sirén. Wheeler languages have subsequently been defined as the class of regular languages for which there exists a Wheeler automaton accepting them. Besides their advantages in data indexing, these Wheeler languages also satisfy many interesting properties from a language theoretic point of view. A characteristic yet unsatisfying feature of Wheeler languages however is that their definition depends on a *fixed* order of the alphabet. In this paper we introduce the *Universally Wheeler languages* UW, i.e., the regular languages that are Wheeler with respect to *all* orders of a given alphabet. Our first main contribution is to relate UW to some very well known regular language classes. We first show that the *Strictly Locally Testable* languages are strictly included in UW. After noticing that UW is not closed under taking the complement, we prove that the class of languages for which both the language and its complement are in UW exactly coincides with those languages that are *Definite* or *Reverse Definite*. Secondly, we prove that deciding if a regular language given by a DFA is in UW can be done in quadratic time. We also show that this is optimal unless the Strong Exponential Time Hypothesis fails.

Keywords: String Matching · Deterministic Finite Automata · Wheeler languages · Graph Indexing · Co-lexicographical Sorting

1 Introduction

Probably the most powerful single algorithmic primitive for indexing data is sorting. The Burrows-Wheeler transform (BWT) is a string transformation that is based on sorting all rotations of a given string. It is at the base of the most efficient text indexes developed in the last three decades such as, for example, the FM-index [11] and the r-index [13]. This technique of indexing compressed data

S.-K. Ko and F. Manea (Eds.): DLT 2025, LNCS 16036, pp. 45–60, 2026.
https://doi.org/10.1007/978-3-032-01475-7_4

based on sorting has been extended to sets of strings that can be described by finite automata, i.e., regular languages, by Gagie et al. [12]. The class of automata that permits sorting its states so that the underlying language can be indexed was termed *Wheeler automata*. Subsequently, there has been an interesting line of works [1,3,8] that aims at understanding the class of regular languages accepted by such automata, the so-called *Wheeler languages*.

The definition of Wheeler languages is, however, characterized by the peculiarity that it depends on a fixed total order of the alphabet. This peculiarity is inherited from the definition of Wheeler automata as follows. Given a totally ordered alphabet, a language is called Wheeler if there exists an automaton accepting that language, which is Wheeler with respect to this given order. This dependency of an alphabet order is unsatisfying for several reasons. Firstly, from a practical point of view, in terms of the original indexing problem, this alphabet order is not very meaningful as any order of the alphabet is equally interesting, e.g. after a suitable alphabet remapping. Secondly, from a more theoretical perspective, if we want to understand the characteristics of Wheeler languages by relating them to other classes of regular languages, those classes typically do not depend on alphabet orders and thus such a relation is not easily possible.

There are two canonical ways of removing this dependency of an alphabet order: to consider the languages for which there *exists* an alphabet order that makes the language Wheeler or to consider those languages that are Wheeler with respect to *all* alphabet orders. The first approach was followed by D'Agostino et al. [9] leading to the class of *Existentially Wheeler* languages EW (named *Generalized Wheeler* in [9]). However, their approach is unsatisfying from a computational viewpoint. While the Wheeler language recognition problem, i.e., the problem of checking if a language $\mathcal{L}$ is Wheeler given a DFA accepting it, can be solved in quadratic time in the size of the DFA [3], the problem of deciding whether $\mathcal{L}$ is in EW is NP-complete [9], even when the language is given by a DFA.

Our Contribution. In this article we follow the second approach by defining the class of *Universally Wheeler languages* UW, that is, those languages that are Wheeler with respect to every order of the alphabet. We present two main contributions in this area. (1) We manage to relate UW to some very well known classes of regular languages, namely the *Definite* (DEF), *Reverse Definite* (RDEF), and the *Strictly Locally Testable* (SLT) languages (Sect. 3). More precisely, we first show that SLT is strictly included in UW. As UW is not closed under taking the complement, we then consider the class of languages $\mathcal{L}$ such that both $\mathcal{L}$ and $\overline{\mathcal{L}}$ belong to UW and prove that this class coincides with $\mathsf{DEF} \cup \mathsf{RDEF}$. (2) From a computational point of view, we prove that UW behaves better than EW: checking whether the language accepted by a given DFA is in UW can be done in quadratic time in the size of the DFA (Sect. 4). We complete this algorithmic result with a fine-grained complexity lower bound showing that no truly subquadratic time algorithm is possible unless the Strong Exponential Time Hypothesis (SETH) fails.

Due to space limits, some proofs are omitted but can be found in [2].

2 Preliminaries

Automata. A *deterministic finite automaton* (DFA) is a tuple $\mathcal{D} = (Q, s, \Sigma, \delta, F)$, where Q is a finite non-empty set of *states*, s is the *initial state* (or *source*), Σ is the finite *alphabet*, $\delta \colon Q \times \Sigma \to Q$ is the (possibly partial) *transition function*, and $F \subseteq Q$ is the set of *final states*.

If δ is not defined on (q, σ) we write $\delta(q, \sigma) = \bot$. If δ is a total function we say that the DFA is *complete*. We sometimes omit the alphabet Σ and simply write a DFA as a tuple $\mathcal{D} = (Q, s, \delta, F)$. If we are dealing with a single DFA, the symbols Q, s, δ, F always refer to the set of $\mathcal{D}$-states, the $\mathcal{D}$-initial state, etc. Sometimes we shall describe the transition function δ using triples (edges), where (q, a, q') stands for $q' = \delta(q, a)$.

We denote with $|\delta|$ and $|Q|$ the number of edges and states of an automaton $\mathcal{D} = (Q, s, \delta, F)$, and with $|\mathcal{D}| = |Q| + |\delta|$ its total size.

As customary, we extend δ to strings as follows: for all $q \in Q$, $a \in \Sigma$ and $\alpha \in \Sigma^*$, we define $\delta(q, \varepsilon) = q$ and $\delta(q, \alpha a) = \delta(\delta(q, \alpha), a)$, where $\delta(\bot, a) = \bot$.

Given a DFA $\mathcal{D} = (Q, s, \Sigma, \delta, F)$ and any of its states $q \in Q$, we denote with I_q the set of all strings labeling paths starting from the source of $\mathcal{D}$ and ending in q, i.e., $I_q = \{\, \alpha \in \Sigma^* : \delta(s, \alpha) = q \,\}$. The language *accepted* (or *recognized*) by $\mathcal{D}$ is then defined as $\mathcal{L}(\mathcal{D}) = \bigcup_{q \in F} I_q$. Moreover, for any $q \in Q$, we denote by $\lambda(q)$ the set of all letters $a \in \Sigma$ that label transitions of the form $(\cdot, a, q)$, i.e., $\lambda(q) :- \{\, a \in \Sigma : (\exists p \in Q)(\delta(p, a) = q) \,\}$.

We mostly consider *trimmed* DFAs, where every state is reachable from the initial state and can reach at least one final state. This is not restrictive: every automaton can be turned into a trimmed and equivalent one by simply deleting unreachable states as well as states not reaching at least one final state. For trimmed automata the following hold: there can be at most one state without incoming transitions, namely s, and every string that can be read starting from s belongs to the set of prefixes, $\mathrm{Pref}(\mathcal{L})$, of the language $\mathcal{L}$.

The languages accepted by automata form the class of *regular languages*. Given a regular language $\mathcal{L}$, there exists a unique (up to isomorphism) state-wise *minimum* complete DFA recognizing $\mathcal{L}$. The states of this minimum DFA correspond to the classes of the Myhill-Nerode [5,14] equivalence relation $\equiv_{\mathcal{L}}$ on Σ^*, defined as follows:

$$\alpha \equiv_{\mathcal{L}} \beta \iff \{\, \gamma \in \Sigma^* : \alpha\gamma \in \mathcal{L} \,\} = \{\, \gamma \in \Sigma^* : \beta\gamma \in \mathcal{L} \,\}.$$

We denote by $\mathcal{D}^c_{\mathcal{L}}$ the *complete* DFA having as set of states the classes $[\alpha]_{\mathcal{L}}$ of the Myhill-Nerode equivalence, $s = [\varepsilon]_{\mathcal{L}}$, $\delta([\alpha]_{\mathcal{L}}, a) = [\alpha \cdot a]_{\mathcal{L}}$, and $F = \{\, [\alpha]_{\mathcal{L}} : \alpha \in \mathcal{L} \,\}$. Then, it can be proved that $\mathcal{D}^c_{\mathcal{L}}$ is a DFA with minimum number of states among all complete DFAs recognizing $\mathcal{L}$ and it is unique up to isomorphism [14]. Let $\mathcal{D}_{\mathcal{L}}$ be the DFA obtained from $\mathcal{D}^c_{\mathcal{L}}$ by considering only the classes $[\alpha]_{\mathcal{L}}$ with $\alpha \in \mathrm{Pref}(\mathcal{L})$ and the transitions among them. Then, one can easily check that $\mathcal{D}_{\mathcal{L}}$

recognizes $\mathcal{L}$, it is trimmed, and can differ from $\mathcal{D}^c_{\mathcal{L}}$ for (at most) one non-final *absorbing* state—that is, a state $q \in Q$ such that $\delta(q, a) = q$, for all $a \in \Sigma$. The automaton $\mathcal{D}_{\mathcal{L}}$ is the DFA with minimum number of states among (complete or not) DFAs recognizing $\mathcal{L}$.

Orders. If $(Z, \leq)$ is a partial order, we denote by $(Z, <)$ its corresponding strict partial order. If $\preceq$ is a total order on the alphabet Σ, we extend it to strings in Σ^* *co-lexicographically*, that is, for $\alpha, \beta \in \Sigma^*$, we have $\alpha \preceq \beta$ if and only if either α is a suffix of β (denoted by $\alpha \dashv \beta$), or there exist $\alpha', \beta', \gamma \in \Sigma^*$ and $a, b \in \Sigma$, such that $\alpha = \alpha' a \gamma$ and $\beta = \beta' b \gamma$ and $a \prec b$.

Wheeler Automata and Wheeler Languages. Wheeler Automata are a special class of DFAs that leverage an a priori fixed order of the alphabet in order to achieve, among other things, efficient compression and indexing (see [12]). Specifically, the co-lexicographic order is lifted from strings to states of the automaton in such a way that p precedes q if and only if, for every α reaching p and β reaching q, α precedes β co-lexicographically. Axioms (W1) and (W2) below do the job.

Definition 1 (Wheeler Automaton). *Let $\preceq$ be a fixed total order of the alphabet Σ. A Wheeler DFA (WDFA for brevity) is a trimmed DFA endowed with a total order $(Q, \leq)$ on the set of states such that the initial state s has no incoming transitions, it is minimum for $\leq$, and such that the following two* Wheeler axioms *are satisfied. Let $p' = \delta(p, a)$ and $q' = \delta(q, b)$:*

(W1) if $a \prec b$, then $p' < q'$;
(W2) if $a = b$, $p < q$, and $p' \neq q'$, then $p' < q'$.

Notice that (W1) implies that a WDFA is *input consistent*, that is, $|\lambda(p)| = 1$ for all states $p \neq s$. Any DFA can be transformed into an equivalent input consistent DFA in $\mathcal{O}(|Q| \cdot |\Sigma|)$ time by simply creating, for each state $q \in Q$, at most $|\Sigma|$ copies of q, one for each different incoming label of q.

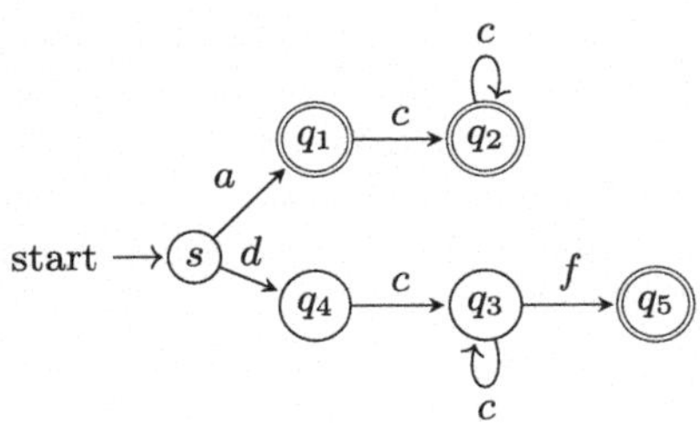

Fig. 1. A WDFA $\mathcal{D}$ recognizing the language $\mathcal{L}_d = ac^* \cup dc^+f$. The only order that makes $\mathcal{D}$ Wheeler is $s < q_1 < q_2 < q_3 < q_4 < q_5$.

It can be proved (see [1,8]) that, if such a total order satisfying Definition 1 exists, then it is unique and allows the lifting of the co-lexicographic order from strings to states (Fig. 1). More precisely, given an order $\preceq$ on Σ and a trimmed, input consistent DFA $\mathcal{D}$ in which the initial state has no incoming edges, we can define a partial order $\leq_{\mathcal{D}}$ on Q by $q \leq_{\mathcal{D}} q' \Leftrightarrow (q = q') \vee (\forall \alpha \in I_q)(\forall \beta \in I_{q'})(\alpha \prec \beta)$. Then, it can be proved that $\leq_{\mathcal{D}}$ always satisfies properties (W1) and (W2) and that the DFA is Wheeler precisely when the partial order $\leq_{\mathcal{D}}$ is total. In this case, we can decide whether $q \leq_{\mathcal{D}} q'$ by simply checking the relative co-lexicographical order among a pair (α, β) with $\alpha \in I_q$ and $\beta \in I_{q'}$.

Switching from automata to languages we define Wheeler languages [1].

Definition 2. *A language is* Wheeler *if there exists a WDFA recognizing it.*

One can easily generalize Definition 1 to WNFA and prove that WDFA and WNFA have the same expressive power in terms of accepted languages (see [1, Theorem 3.16]). We denote by $\mathsf{W}(\preceq)$ the class of all regular languages that are Wheeler w.r.t. the order $\preceq$. In [3] a characterization of languages which are not in $\mathsf{W}(\preceq)$ was based on the existence of particular cycles in the product automaton $\mathcal{D}^2_{\mathcal{L}}$. Here we shall use a variant of this result based on the following definition:

Definition 3. *Let $\mathcal{D} = (Q, \Sigma, \delta, s, F)$ be a DFA and let $\preceq$ be an order of the alphabet. If $p, q \in Q$ we define:*

$$p \bowtie_{\preceq} q \iff (\exists \alpha, \alpha' \in I_p)(\exists \beta, \beta' \in I_q)(\alpha \prec \beta \wedge \beta' \prec \alpha').$$
$$p \bowtie q \iff (\exists \preceq)(p \bowtie_{\preceq} q).$$

Recall that the automaton $\mathcal{D}^2_{\mathcal{L}}$ is defined by $\mathcal{D}^2_{\mathcal{L}} = (Q \times Q, \Sigma, \delta', (s, s), F \times F)$ where $\delta'((p, q), a) = (\delta(p, a), \delta(q, a))$. Then we have:

Lemma 1 ([3, Theorem 3]). *Let $\mathcal{L}$ be a regular language, let $\preceq$ be an order of the alphabet, and let $\mathcal{D}_{\mathcal{L}}$ be the minimum trimmed DFA accepting $\mathcal{L}$. Then, $\mathcal{L} \notin \mathsf{W}(\preceq)$ if and only if $\mathcal{D}^2_{\mathcal{L}}$ contains a cycle $(p_1, q_1) \to (p_2, q_2) \to \cdots \to (p_k, q_k) \to (p_1, q_1)$ such that the following hold: (i) $p_1 \neq q_1$, and (ii) $p_1 \bowtie_{\preceq} q_1$.*

It is not difficult to prove that the class $\mathsf{W}(\preceq)$ is closed under intersection. However, it is neither closed under complement nor union: indeed, there are examples of languages $\mathcal{L} \in \mathsf{W}(\preceq)$ for which no *complete* WDFA recognizing $\mathcal{L}$ exists (see [1]) and this prevent the possibility of finding a WDFA for the complement by simply switching final and non-final states [7].

One peculiarity of the Wheeler notion is that it depends on a fixed order of the alphabet. In order to compare Wheeler languages with other sub-classes of regular languages, which do not depend on a fixed order of the alphabet, it is natural to ask for a definition that is independent of such an order. Two natural options are the following.

Definition 4. *A regular language $\mathcal{L}$ is:*

- Existentially Wheeler *if there exists an order $\preceq$ such that $\mathcal{L} \in \mathsf{W}(\preceq)$;*
- Universally Wheeler *if $\mathcal{L} \in \mathsf{W}(\preceq)$ for all orders $\preceq$ of the alphabet.*

The class of Existentially (Universally) Wheeler languages is denoted by EW *(*UW*).*

In [10] it is proved that deciding EW is NP-complete. In Sect. 4 we shall prove that deciding UW is in P, but first in the next lemma we give a necessary condition that languages in EW must satisfy.

Lemma 2. *If the minimum trimmed automaton $\mathcal{D}_{\mathcal{L}}$ of a language $\mathcal{L}$ contains three equally labeled cycles starting from three different states, then $\mathcal{L} \notin$ EW.*

Note that the condition of Lemma 2 can easily be checked in cubic time.

3 Definite and Reverse Definite Languages Meet Wheelerness

We first notice that the class UW is not closed under complement (see Example 1). In this section we give a characterization of the class of languages $\mathcal{L}$ for which both $\mathcal{L}$ and $\overline{\mathcal{L}}$ are in the class UW. This characterization involves two well known subclasses of star-free languages, the definite and reverse definite ones [4]. Remember that, given two languages $\mathcal{L}, \mathcal{L}'$, we denote by $\mathcal{L}\mathcal{L}'$ the set containing all concatenations of a word in $\mathcal{L}$ with a word in $\mathcal{L}'$ (with $\mathcal{L}\mathcal{L}'$ equal to the empty language if either $\mathcal{L}$ or $\mathcal{L}'$ is).

Definition 5. *A regular language $\mathcal{L}$ is:* definite *(*DEF*) if $\mathcal{L} = F \cup \Sigma^* G$, for finite $F, G \subseteq \Sigma^*$;* reverse definite *(*RDEF*) if $\mathcal{L} = F \cup G\Sigma^*$, for finite $F, G \subseteq \Sigma^*$;* strictly locally testable *(*SLT*, see [6]) if there are finite sets H, K, W, F s.t. $\mathcal{L} = F \cup (H\Sigma^* \cap \Sigma^* K) \setminus \Sigma^* W \Sigma^*$.*

Our aim is to prove that the class DEF ∪ RDEF consists in exactly the UW-languages for which the complement is in UW as well. First, we recall a characterization of SLT due to Caron and then we compare SLT and UW languages.

Lemma 3 ([6, Proposition 3.2 and 3.3]). *Let $\mathcal{L}$ be a regular language and let $\mathcal{D}_{\mathcal{L}}$ be its minimum trimmed DFA. Then $\mathcal{L} \notin$ SLT if and only if there exists a pair of equally labeled cycles C_p, C_q in $\mathcal{D}_{\mathcal{L}}$ starting respectively from states $p \neq q$.*

Since a pair of equally labeled cycles in $\mathcal{D}_{\mathcal{L}}$ corresponds to a single cycle in $\mathcal{D}^2_{\mathcal{L}}$, combining the previous lemma with Lemma 1 we obtain:

Lemma 4. SLT ⊆ UW. *In particular, all languages in* DEF ∪ RDEF *are in* UW.

In order to prove that the class of regular languages $\mathcal{L}$ for which both $\mathcal{L}$ and $\overline{\mathcal{L}}$ are in UW coincides with the class DEF ∪ RDEF, we need a careful inspection, for a regular language $\mathcal{L}$, on the differences between the minimum complete DFA for $\mathcal{L}$, namely $D^c_{\mathcal{L}}$, its trimmed version $D_{\mathcal{L}}$, the minimum complete DFA for $\overline{\mathcal{L}}$, namely $D^c_{\overline{\mathcal{L}}}$, and its trimmed version $D_{\overline{\mathcal{L}}}$, as in the following remark.

Remark 1. Since the Myhill-Nerode equivalence on Σ^* is the same for $\mathcal{L}$ and $\overline{\mathcal{L}}$, we have that $D^{c}_{\mathcal{L}}$ and $D^{c}_{\overline{\mathcal{L}}}$ only differ for the final states, which in $D^{c}_{\overline{\mathcal{L}}}$ are the complement of the ones in $D^{c}_{\mathcal{L}}$. Let Q be the set of $D^{c}_{\mathcal{L}}$-states. Consider the subsets $Q_1, Q_2, Q_3 \subseteq Q$ defined as follows:

$$Q_1 = \{\, q \in Q : I_q \cap (\mathrm{Pref}(\mathcal{L}) \setminus \mathrm{Pref}(\overline{\mathcal{L}})) \neq \emptyset \,\},$$
$$Q_2 = \{\, q \in Q : I_q \cap (\mathrm{Pref}(\overline{\mathcal{L}}) \setminus \mathrm{Pref}(\mathcal{L})) \neq \emptyset \,\},$$
$$Q_3 = \{\, q \in Q : I_q \cap (\mathrm{Pref}(\overline{\mathcal{L}}) \cap \mathrm{Pref}(\mathcal{L})) \neq \emptyset \,\}.$$

States in Q_1 are final for $\mathcal{D}^{c}_{\mathcal{L}}$ and can reach only final states, hence, by minimality, $|Q_1| \leq 1$. Moreover, $|Q_1| = 1$ iff $\mathrm{Pref}(\mathcal{L}) \setminus \mathrm{Pref}(\overline{\mathcal{L}}) \neq \emptyset$, and if $Q_1 = \{q\}$ then q is an absorbing state (that is, $\delta(q, a) = q$, for all $a \in \Sigma$) and it is final for $\mathcal{D}^{c}_{\mathcal{L}}$. The analogous holds for Q_2 with respect to $\mathcal{D}^{c}_{\overline{\mathcal{L}}}$: $|Q_2| = 1$ iff $\mathrm{Pref}(\overline{\mathcal{L}}) \setminus \mathrm{Pref}(\mathcal{L}) \neq \emptyset$, and if $Q_2 = \{\overline{q}\}$ then $\overline{q}$ is an absorbing state which is final in $\mathcal{D}^{c}_{\overline{\mathcal{L}}}$. Finally, $D_{\mathcal{L}}$ is obtained from $D^{c}_{\mathcal{L}}$ by erasing the absorbing state $\overline{q}$ and all transition arriving in $\overline{q}$, while $D_{\overline{\mathcal{L}}}$ is obtained from $D^{c}_{\overline{\mathcal{L}}}$ by erasing the absorbing state q and all transition arriving in q (see Fig. 2).

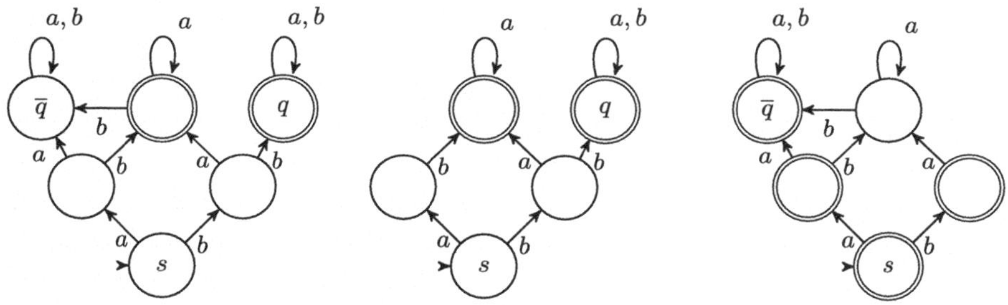

Fig. 2. *Left:* The minimum complete DFA $\mathcal{D}^{c}_{\mathcal{L}}$ for a language $\mathcal{L}$, with both the absorbing state $\overline{q}$ for $\overline{\mathcal{L}}$ and the absorbing state q for $\mathcal{L}$. *Center:* The minimum trimmed DFA $\mathcal{D}_{\mathcal{L}}$ for $\mathcal{L}$. *Right:* The minimum trimmed DFA $\mathcal{D}_{\overline{\mathcal{L}}}$ for $\overline{\mathcal{L}}$.

Remark 2. For every order $(\Sigma, \preceq)$, if $\mathrm{Pref}(\mathcal{L}) = \Sigma^*$ and $\mathcal{L} \in \mathsf{W}(\preceq)$ then $\overline{\mathcal{L}} \in \mathsf{W}(\preceq)$. This follows by the fact that if $\mathrm{Pref}(\mathcal{L}) = \Sigma^*$, then any deterministic automaton accepting $\mathcal{L}$ is complete and, thus, a Wheeler DFA for $\mathcal{L}$ becomes a Wheeler DFA for $\overline{\mathcal{L}}$ by just exchanging final and not final states. In particular, if $\mathcal{L} \in \mathsf{UW}$ and $\mathrm{Pref}(\mathcal{L}) = \Sigma^*$, then $\overline{\mathcal{L}} \in \mathsf{UW}$. Notice that the converse is not true, because, as the following example shows, there are languages $\mathcal{L}$ with $\mathrm{Pref}(\mathcal{L}) = \Sigma^*$, $\overline{\mathcal{L}} \in \mathsf{UW}$ but $\mathcal{L} \notin \mathsf{UW}$ (hence UW is not closed under complementation).

Example 1. We give an example of a language $\mathcal{L}$ such that $\mathrm{Pref}(\mathcal{L}) = \Sigma^*$, the minimum trimmed automaton for $\mathcal{L}$ (on the left in Fig. 3) satisfies the condition of Lemma 1 w.r.t. the usual order of the alphabet (hence $\mathcal{L} \notin \mathsf{UW}$), while the minimum trimmed automaton for $\overline{\mathcal{L}}$ (on the right in Fig. 3) does not contain two equally labeled cycles, proving that $\overline{\mathcal{L}} \in \mathsf{UW}$.

Using the previous remarks we can characterize definite languages using UW (for a proof, see [2]).

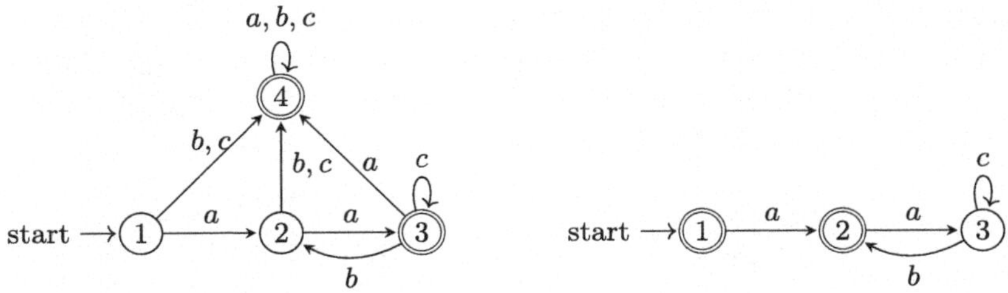

Fig. 3. A language $\mathcal{L} \notin \mathsf{UW}$ (minimum DFA on the left) with $\mathrm{Pref}(\mathcal{L}) = \Sigma^*$, $\overline{\mathcal{L}} \in \mathsf{UW}$ (minimum DFA on the right)

Lemma 5. *$\mathcal{L} \in \mathsf{DEF}$ iff $\mathcal{L}$ is finite, or $\mathcal{L} \in \mathsf{UW}$ and $\mathrm{Pref}(\mathcal{L}) = \Sigma^*$.*

We now consider the case $\mathrm{Pref}(\mathcal{L}) \neq \Sigma^*$. If $|\Sigma| = 1$ then $\mathsf{UW} = \mathsf{EW}$ coincide with the class of finite and cofinite languages which, for a unary alphabet, is equal to $\mathsf{DEF} \cup \mathsf{RDEF}$. Hence, for unary languages, $\mathcal{L}, \overline{\mathcal{L}} \in \mathsf{UW}$ iff $\mathcal{L} \in \mathsf{DEF} \cup \mathsf{RDEF}$ holds. To handle the case $|\Sigma| \geq 2$ we prove:

Lemma 6. *Let $|\Sigma| \geq 2$, and $\mathcal{L} \in \mathsf{UW}$. If $\mathrm{Pref}(\mathcal{L}) \neq \Sigma^*$, then following are equivalent:*

1. *there exists an order $\preceq$ such that $\overline{\mathcal{L}} \notin \mathsf{W}(\preceq)$;*
2. *$\mathrm{Pref}(\mathcal{L}) \cap \mathrm{Pref}(\overline{\mathcal{L}})$ is an infinite set.*

Proof. ($1 \Rightarrow 2$) If there exists an order $\preceq$ such that $\overline{\mathcal{L}} \notin \mathsf{W}(\preceq)$ but $\mathcal{L} \in \mathsf{UW}$, then, in particular, $\mathcal{L} \in \mathsf{W}(\preceq)$. Since $\overline{\mathcal{L}} \notin \mathsf{W}(\preceq)$, then $\overline{\mathcal{L}} \notin \mathsf{SLT}$. By Lemma 3, $\mathcal{D}_{\overline{\mathcal{L}}}$ must contain two equally labeled cycles (say, with label γ) starting from two different states p, q. One of them, say p, must be different from the absorbing state $\overline{q}$ for $\overline{\mathcal{L}}$ (see the discussion on minimum complete or trimmed DFA for a language and its complement). Since all words arriving in p are in $\mathrm{Pref}(\mathcal{L}) \cap \mathrm{Pref}(\overline{\mathcal{L}})$, we have $\alpha\gamma^* \in \mathrm{Pref}(\mathcal{L}) \cap \mathrm{Pref}(\overline{\mathcal{L}})$ and $\mathrm{Pref}(\mathcal{L}) \cap \mathrm{Pref}(\overline{\mathcal{L}})$ is infinite.

($2 \Rightarrow 1$) Suppose $\mathrm{Pref}(\mathcal{L}) \cap \mathrm{Pref}(\overline{\mathcal{L}})$ is infinite. Since $\mathrm{Pref}(\mathcal{L}) \neq \Sigma^*$ implies $\mathrm{Pref}(\overline{\mathcal{L}}) \setminus \mathrm{Pref}(\mathcal{L}) \neq \emptyset$, in $\mathcal{D}_{\overline{\mathcal{L}}}$ there is an absorbing final state $\overline{q}$. Let β be a word arriving in $\overline{q}$ in $\mathcal{D}_{\overline{\mathcal{L}}}$. From the hypothesis we get that $\beta \neq \epsilon$ (otherwise $\overline{\mathcal{L}} = \Sigma^*$ would contradict $\mathrm{Pref}(\mathcal{L}) \cap \mathrm{Pref}(\overline{\mathcal{L}}) \neq \emptyset$), and, since $\mathrm{Pref}(\mathcal{L}) \cap \mathrm{Pref}(\overline{\mathcal{L}})$ is infinite, in $\mathcal{D}_{\overline{\mathcal{L}}}$ there must be a cycle not passing through $\overline{q}$. Let $p \neq \overline{q}$ be on this cycle, α be a word arriving in p, and $\gamma \neq \epsilon$ be a label of a cycle starting from p and avoiding $\overline{q}$. We may suppose w.l.o.g. that γ is not a suffix of α (if so, just consider a power of γ instead of γ). Notice that there is a cycle labeled γ both from p and from $\overline{q}$ ($\overline{q}$ is an absorbing state, hence there are cycles starting from $\overline{q}$ labeled by any word).

We claim that there exists an alphabet order $\preceq$ such that $p \bowtie_{\preceq} \overline{q}$. Consider two cases:

1. $\alpha \dashv \gamma$. Let g be the last letter of γ and let $x \in \Sigma$ with $x \neq g$. Fix an order $\preceq$ of the alphabet such that $x \prec g$. Since α arrives in p and βx arrives in $\overline{q}$ we

have either $\alpha \prec \beta x$ or $\alpha \succ \beta x$. In the first case, we may consider the words $\alpha \prec \beta x \prec \alpha\gamma$, while if $\alpha \succ \beta x$ we may consider the words $\beta x \prec \alpha \prec \beta x\gamma$. In both cases the words witness that $p \bowtie_{\preceq} \overline{q}$.
2. $\alpha \not\dashv \gamma$. Then there is an order $\preceq$ of the alphabet such that $\gamma \prec \alpha$ holds. If $\preceq$ is such an order, then, since we already supposed $\gamma \not\dashv \alpha$, we have $\beta\gamma \prec \alpha \prec \beta\alpha$, witnessing $p \bowtie_{\preceq} \overline{q}$.

This proves the claim. Since a cycle labeled γ starts from both p and $\overline{q}$, in $\mathcal{D}^2_{\overline{\mathcal{L}}}$ there is a cycle starting from $(p, \overline{q})$, with $p \neq \overline{q}$ and $p \bowtie_{\preceq} \overline{q}$, and Lemma 1 applied to $\overline{\mathcal{L}}$ yields $\overline{\mathcal{L}} \notin \mathsf{W}(\preceq)$. □

Using Lemma 6 we can finally prove:

Corollary 1. *Let* $|\Sigma| \geq 2$ *and let* $\mathcal{L}$ *be a regular language such that* $\text{Pref}(\mathcal{L}) \neq \Sigma^*$. *Then* $\mathcal{L}, \overline{\mathcal{L}} \in \mathsf{UW}$ *iff* $\mathcal{L} \in \mathsf{RDEF}$.

Proof. If $\mathcal{L} \in \mathsf{RDEF}$ then $\mathcal{L}, \overline{\mathcal{L}} \in \mathsf{SLT}$ and hence $\mathcal{L}, \overline{\mathcal{L}} \in \mathsf{UW}$ by Lemma 4.

Suppose that $\mathcal{L}, \overline{\mathcal{L}} \in \mathsf{UW}$ and $\text{Pref}(\mathcal{L}) \neq \Sigma^*$. Then from Lemma 6 we obtain that $\text{Pref}(\mathcal{L}) \cap \text{Pref}(\overline{\mathcal{L}}) = F$ where F is a finite set.
If $G = \{\, \alpha x \in \text{Pref}(\mathcal{L}) \setminus \text{Pref}(\overline{\mathcal{L}}) : \alpha \in F,\ x \in \Sigma \,\}$, we claim that $\mathcal{L} = (F \cap \mathcal{L}) \cup G\Sigma^*$. If $\beta \in G\Sigma^*$ then $\beta = \alpha x \gamma$ with $\alpha x \in \text{Pref}(\mathcal{L}) \setminus \text{Pref}(\overline{\mathcal{L}})$ and $\alpha \in F$. Since $\alpha x \notin \text{Pref}(\overline{\mathcal{L}})$, for every $\delta \in \Sigma^*$ we have $\alpha x \delta \in \mathcal{L}$. In particular, $\beta = \alpha x \gamma \in \mathcal{L}$, proving $(F \cap \mathcal{L}) \cup G\Sigma^* \subseteq \mathcal{L}$.

Conversely, we prove that if $\beta \in \mathcal{L}$, then $\beta \in (F \cap \mathcal{L}) \cup G\Sigma^*$. If $\beta \in \text{Pref}(\overline{\mathcal{L}})$ then $\beta \in \text{Pref}(\mathcal{L}) \cap \text{Pref}(\overline{\mathcal{L}}) = F$. Hence $\beta \in F \cap \mathcal{L}$.

If $\beta \notin \text{Pref}(\overline{\mathcal{L}})$, let α be its longest prefix belonging to $\text{Pref}(\mathcal{L}) \cap \text{Pref}(\overline{\mathcal{L}}) = F$. Since $\alpha \neq \beta$, there exists $x \in \Sigma$ such that αx is still a β-prefix and hence $\alpha x \in \text{Pref}(\mathcal{L})$. From the definition of α it follows that $\alpha x \in \text{Pref}(\mathcal{L}) \setminus \text{Pref}(\overline{\mathcal{L}})$. Hence $\alpha x \in G$ and $\beta \in G\Sigma^*$.

Since we proved that $\mathcal{L} = (F \cap L) \cup G\Sigma^*$ with G, F finite sets, we obtain $\mathcal{L} \in \mathsf{RDEF}$. □

Putting all previous results together we finally obtain:

Theorem 1. *Let* $\mathcal{L}$ *be a regular language. Then* $\mathcal{L}, \overline{\mathcal{L}} \in \mathsf{UW} \Leftrightarrow \mathcal{L} \in \mathsf{DEF} \cup \mathsf{RDEF}$.

Proof. Since the class $\mathsf{DEF} \cup \mathsf{RDEF}$ is closed for complements, implication from right to left is Lemma 4. The implication from left to right is proved just before Lemma 6 for $|\Sigma| = 1$ and, if $\text{Pref}(\mathcal{L}) = \Sigma^*$ then the result follows from Lemma 5; if $\text{Pref}(\mathcal{L}) \neq \Sigma^*$ and $|\Sigma| \geq 2$ and then the result follows from Corollary 1. □

4 Deciding UW

In this section we consider the problem of deciding, given a DFA $\mathcal{D}$, whether $\mathcal{L}(\mathcal{D}) \in \mathsf{UW}$. First, we propose an algorithm that solves this problem in quadratic time, more precisely in time $\mathcal{O}(nm)$, where n and m are respectively the number of states and transitions of the minimal DFA $\mathcal{D}_{\mathcal{L}}$ accepting the language $\mathcal{L}$ (note that n and m are upper bounded by the number of states and transitions of the input DFA $\mathcal{D}$). Finally, we prove that this is optimal unless the Strong Exponential Time Hypothesis (SETH) [15,16] fails.

The Algorithm. A characterization of UW can be given by looking at the structure of the minimum trimmed automaton. Indeed, since the existence of a cycle in $\mathcal{D}_{\mathcal{L}}^2$ (see Sect. 2) does not depend on the alphabet's order, using the relation $\bowtie$ of Def. 3, from Lemma 1 we obtain:

Corollary 2. *Let $\mathcal{D}_{\mathcal{L}}$ be the minimum trimmed DFA accepting $\mathcal{L}$. Then, $\mathcal{L} \notin$ UW if and only if $\mathcal{D}_{\mathcal{L}}^2$ contains a cycle $(p_1, q_1) \to (p_2, q_2) \to \cdots \to (p_k, q_k) \to (p_1, q_1)$ such that the following hold: (i) $p_1 \neq q_1$, and (ii) $p_1 \bowtie q_1$.*

This corollary provides us with a strategy to decide, for a given regular language $\mathcal{L}$, whether $\mathcal{L} \in$ UW without testing every possible order $\preceq$ of the alphabet, by checking the existence of a cycle in $\mathcal{D}_{\mathcal{L}}^2$ with the above properties. It thus motivates the following definition of a directed (unlabeled) graph. Here and in the rest of this section Q, δ always refer to the set of states and the transition function of $\mathcal{D}_{\mathcal{L}}$, the minimum trimmed DFA for $\mathcal{L}$.

Definition 6. *Let (V, E) be such that $V := \{ (p, q) \in Q^2 : p \neq q \wedge p \bowtie q \}$ and $E := \{ \langle (p, q), (p', q') \rangle : (\exists a \in \Sigma)\big(\delta(p, a) = p' \wedge \delta(q, a) = q'\big) \}$.*

Using the graph (V, E), we can rephrase Corollary 2 as follows.

Corollary 3. *$\mathcal{L} \in$ UW if and only if graph $G = (V, E)$ of Definition 6 is acyclic.*

More Tractable Definition of V. Given the above corollary, in order to decide UW it is sufficient to test the directed graph (V, E) for acyclicity. The way V is currently defined does not directly translates into an efficient way to compute it and, thus, we give an alternative way (see Lemma 7 below) to determine whether $p \bowtie q$ holds for states p and q.

First, we define a table P, indexed by pairs $(p, q) \in Q^2$, such that $P[p, q]$ collects witnesses of the existence of words $\alpha \in I_p$ and $\beta \in I_q$ such that either $\alpha \dashv \beta$, $\beta \dashv \alpha$, or $\alpha \prec \beta$ (with α being not a suffix of β) for some co-lex order $\prec$. Specifically, in the former two cases $P[p, q]$ contains pairs $(\dashv, \dashv)$ and $(\vdash, \vdash)$, respectively, whether in the latter it contains all pairs $(a, b) \in \Sigma^2$ (with $a \neq b$) for which $\alpha = \alpha' a \gamma$ and $\beta = \beta' b \gamma$.

Definition 7. *Let $p, q \in Q$ with $p \neq q$. Let $p \dashv q$ stands for $(\exists \alpha \in I_p)(\exists \beta \in I_q)(\alpha \dashv \beta)$. Furthermore,*

$$P[p, q] := \{ (a, b) \in \Sigma^2 : a \neq b \wedge (\exists \alpha, \beta, \gamma \in \Sigma^*)(\alpha a \gamma \in I_p \wedge \beta b \gamma \in I_q) \} \cup \{ (\dashv, \dashv) : p \dashv q \} \cup \{ (\vdash, \vdash) : q \dashv p \}.$$

Notice that, since $\mathcal{D}_{\mathcal{L}}$ is trimmed we have, for $(a, b) \in \Sigma^2$ with $a \neq b$:

$$\begin{aligned} &(a, b) \in P[p, q] \iff \\ &(\exists p', q' \in Q)(\exists \gamma \in \Sigma^*)\,(a \in \lambda(p') \wedge b \in \lambda(q') \wedge p = \delta(p', \gamma) \wedge q = \delta(q', \gamma)). \end{aligned} \tag{1}$$

Lemma 7. *Let $p, q \in Q$ with $p \neq q$. Then, $p \bowtie q$ if and only if $|P[p,q]| \geq 2$.*

Proof. ($\Rightarrow$) If $p \bowtie q$, then there exists an order $\preceq$ and strings $\alpha, \alpha' \in I_p$ and $\beta, \beta' \in I_q$ such that $\alpha \prec \beta$ and $\beta' \prec \alpha'$. If α is a suffix of β, then $(\dashv, \dashv) \in P[p,q]$. Otherwise, there exist $a, b \in \Sigma$, with $a \prec b$, and a string $\gamma \in \Sigma^*$ such that $a\gamma \dashv \alpha$ and $b\gamma \dashv \beta$, and thus $(a,b) \in P[p,q]$. Similarly, for $\beta' \prec \alpha'$ we have either $(\vdash, \vdash) \in P[p,q]$ or there exists $a', b' \in \Sigma$, with $b' \prec a'$ and (a', b') in $P[p,q]$. Furthermore, if such a, b, a', b' exists, we must have $(a,b) \neq (a',b')$ because $a \prec b$ and $b' \prec a'$. Hence, $|P[p,q]| \geq 2$.

($\Leftarrow$) Let $(a,b) \neq (a',b')$ be two elements in $P[p,q]$. Suppose $\{(a,b),(a',b')\} \cap \{(\dashv,\dashv)\} = \emptyset$. One of the pairs, say (a,b), must be different from $(\vdash,\vdash)$.

If $(a',b') = (\vdash,\vdash)$, then any total order $\preceq$ on Σ for which $a \prec b$ is such that $p \bowtie_{\preceq} q$. Otherwise, any total order $\preceq$ on Σ for which $a \prec b$ and $b' \prec a'$ is such that $p \bowtie_{\preceq} q$. Similarly, if $\{(a,b),(a',b')\} \cap \{(\vdash,\vdash)\} = \emptyset$ then there exists $\preceq$ such that $p \bowtie_{\preceq} q$. Finally, if $\{(a,b),(a',b')\} = \{(\dashv,\dashv),(\vdash,\vdash)\}$ then $p \bowtie_{\preceq} q$ holds for any order $\preceq$ of the alphabet Σ. □

In order to decide whether $p \bowtie q$ it will be sufficient to compute just two elements (if they exist) from $P[p,q]$. Since we are only interested in deciding whether $P[p,q]$ contains at least two elements, it seems plausible that it is sufficient to examine only a few labels of $\lambda(q)$ for each $q \in Q$. Indeed, in this spirit suppose we have computed a maximal set $\lambda'(p) \subseteq \lambda(p)$ such that $|\lambda'(p)| \leq 2$; in other words, $\lambda' \colon Q \to 2^{\Sigma}$ is a function such that $\lambda'(p) = \lambda(p)$, if $|\lambda(p)| \leq 1$, while $\lambda'(p)$ is a subset of $\lambda(p)$ with two elements, if $|\lambda(p)| > 1$. For example, we may define $\lambda'(p)$ as the set containing the two smallest elements in $\lambda(p)$ (or $\lambda(p)$ itself, in case there are not enough elements). Given such a function λ', we define:

$$
\begin{aligned}
P_{\lambda'}[p,q] :=\{\, (a,b) \;:\; & a \neq b \wedge (\exists p', q' \in Q)(\exists \gamma \in \Sigma^*) \\
& (a \in \lambda'(p') \wedge b \in \lambda'(q') \wedge p = \delta(p',\gamma) \wedge q = \delta(q',\gamma)) \,\} \\
\cup \{\, (\dashv,\dashv) : p \dashv q \,\} & \cup \{\, (\vdash,\vdash) : q \dashv p \,\}\,.
\end{aligned}
$$

We note that, due to the characterization of $P[p,q]$ in (1), the definition of $P_{\lambda'}[p,q]$ is equivalent to the one of $P[p,q]$ up to replacing λ by λ'. Moreover, since $\lambda'(\cdot) \subseteq \lambda(\cdot)$, it holds that $P_{\lambda'}[p,q] \subseteq P[p,q]$ for every $p, q \in Q$.

The following lemma ensures that we can use $\lambda'(\cdot)$ instead of $\lambda(\cdot)$.

Lemma 8. *Let $p, q \in Q$, with $p \neq q$. Then $|P[p,q]| \geq 2$ iff $|P_{\lambda'}[p,q]| \geq 2$.*

Proof. ($\Rightarrow$) We prove the statement by contraposition. Suppose $|P_{\lambda'}[p,q]| < 2$, hence $(P_{\lambda'}[p,q] \setminus \{(\dashv,\dashv),(\vdash,\vdash)\}) \subseteq \{(a,b)\}$, for some $a, b \in \Sigma$. Then, for every pair $p', q' \in Q$ such that there exists $\gamma \in \Sigma^*$ for which $p = \delta(p',\gamma)$ and $q = \delta(q',\gamma)$ it holds $\lambda'(p') \subseteq \{a\}$ and $\lambda'(q') \subseteq \{b\}$. Now recall that $\lambda'(p')$ and $\lambda'(q')$ where chosen as a maximal subset of $\lambda(p')$ and $\lambda(q')$ with at most two elements and thus also $\lambda(p') \subseteq \{a\}$ and $\lambda(q') \subseteq \{b\}$ must hold, from which it follows that $(P[p,q] \setminus \{(\dashv,\dashv),(\vdash,\vdash)\}) \subseteq \{(a,b)\}$. Then $|P[p,q]| < 2$ since $(\dashv,\dashv) \in P[p,q]$ implies $(\dashv,\dashv) \in P_{\lambda'}[p,q]$ and the same holds for $(\vdash,\vdash)$.

($\Leftarrow$) Since $P_{\lambda'}[p,q] \subseteq P[p,q]$, the result is immediate. □

Algorithm. We are now ready to describe the algorithm that tests whether a language $\mathcal{L}$ is in UW using Corollary 3. We start by constructing the vertices V of the graph (V, E) as follows. We compute sets $\lambda'(p)$ for each state p in $\mathcal{D}_\mathcal{L}$ as above, i.e., $\lambda'(p)$ is maximal with $\lambda'(p) \subseteq \lambda(p)$ and $|\lambda'(p)| \leq 2$. Then, according to Lemma 7 and Lemma 8, it follows that the vertex set V can be equivalently defined as $V = \{ (p, q) : |P_{\lambda'}[p, q]| \geq 2 \}$. The pseudocode is given in Algorithm 1.

First, in lines 2 to 4 we compute the sets $\lambda'(q)$, for all $q \in Q$. Then, from line 6 to 11 the algorithm computes a queue A containing all quadruples $(s, q, \dashv, \dashv)$, $(p, s, \vdash, \vdash)$, where s is the initial state of $\mathcal{D}_\mathcal{L}$ and $s \neq p, q$. Notice that, since $\mathcal{D}_\mathcal{L}$ is a trimmed automaton, both $s \dashv q$ and $s \dashv p$ hold so that $(\vdash, \vdash) \in P_{\lambda'}[s, q]$ and $(\dashv, \dashv) \in P_{\lambda'}[p, s]$, for all $s \neq p, q$. Then, from line 12 to 14, all quadruples (p, q, a, b) such that $p \neq q$, $a \neq b$, $a \in \lambda'(p)$, $b \in \lambda'(q)$ are added to A. These quadruples represent the "base case" in which $(a, b) \in P_{\lambda'}[p, q]$, i.e., where in the definition of $P_{\lambda'}[p, q]$ we consider $\gamma = \epsilon$, $p' = p$, and $q' = q$.

Starting from these quadruples, from line 16 to 23 we "propagate", generating new pairs $(a, b) \in P_{\lambda'}[p, q]$, as long as $|P_{\lambda'}[p, q]| < 2$. This is correct because, by definition, for all (p', q') we have $(a, b) \in P_{\lambda'}[p', q']$ iff there exists $\gamma \in \Sigma^*$ and (p, q) such that (p, q, a, b) belongs to A and $\delta(p, \gamma) = p', \delta(q, \gamma) = q'$. At the same time we compute a table T such that, for each $p \neq q$, $T[p, q]$ contains two pairs in $P_{\lambda'}[p, q]$, if there are such.

Finally, from line 25 to 26, we select the vertices V using T (i.e., those pairs (p, q) with $|T[p, q]| \geq 2$), build the edge set E, and test acyclicity of the subgraph $G = (V, E)$ of $\mathcal{D}^2_\mathcal{L}$.

Running Time. For an automaton $\mathcal{D}_\mathcal{L}$ with n states and m transitions (whose squared automaton $\mathcal{D}^2_\mathcal{L}$ has n^2 states and at most nm transitions), the time required by the respective steps of the above algorithm is as follows:

1. The computation of λ' requires $\mathcal{O}(m)$ time,
2. Computing the table T can be done in $\mathcal{O}(nm)$ time,
3. Computing the graph $G = (V, E)$ takes $\mathcal{O}(nm)$ time,
4. Finally, testing acyclicity of G takes $\mathcal{O}(nm)$ as well.

Altogether, Algorithm 1 runs in $\mathcal{O}(n^2 + nm) \subseteq \mathcal{O}(nm)$ time.

Conditional Lower Bound. First, we define the Orthogonal Vectors problem.

Definition 8 (Orthogonal Vectors problem (OV)). *Given two sets A and B, each containing N vectors from $\{0, 1\}^d$, decide whether there exist $a \in A$ and $b \in B$ such that $a^\intercal b = 0$.*

Through a classical reduction [17], for $d \in \omega(\log N)$, it is known that OV cannot be solved in time $\mathcal{O}(N^{2-\eta}\mathsf{poly}(d))$ for any constant $\eta > 0$ unless the Strong Exponential Time Hypothesis [15,16] (SETH) fails. Our goal now is to prove the following theorem.

Theorem 2. *Let $\mathcal{L}$ be a regular language, let $\mathcal{D}_\mathcal{L}$ be the minimum DFA accepting $\mathcal{L}$ and let m be the number of transitions in $\mathcal{D}_\mathcal{L}$. If the problem of checking*

Algorithm 1 Test if $\mathcal{L}(D_{\mathcal{L}}) \in \mathsf{UW}$.

```
function UW-Tester(D_L = (Q, Σ, δ, F))
    for all (p, a, q) ∈ δ do                          ▷ Compute λ'(q) for every q ∈ Q.
        if |λ'(q)| < 2 ∧ a ∉ λ'(q) then
            λ'(q).Push(a)

    A ← EmptyQueue                                    ▷ Base cases.
    for all p, q ∈ Q s.t. p ≠ q do
        if p = s then
            A.Push((s, q, ⊣, ⊣))
        if q = s then
            A.Push((p, s, ⊢, ⊢))
        for all (a, b) ∈ λ'(p) × λ'(q) do
            if a ≠ b then
                A.Push((p, q, a, b))

    T ← Empty(Q × Q)                                  ▷ Propagation phase.
    while A ≠ ∅ do
        (p, q, a, b) ← A.Pop()
        if |T[p, q]| < 2 ∧ (a, b) ∉ T[p, q] then
            T[p, q].Push((a, b))
            for all (p', q') outgoing from (p, q) in D² do
                if p' ≠ q' then
                    A.Push((p', q', a, b))

    V ← { (p, q) ∈ Q × Q : |T[p, q]| ≥ 2 }            ▷ Subgraph of D² of interest.
    E ← { (u, v) ∈ V × V : v is outgoing from u in D² }
    return IsAcyclic(V, E)
```

whether $\mathcal{L} \in \mathsf{UW}$ *can be solved in time* $\mathcal{O}(m^{2-\eta})$ *for some constant* $\eta > 0$, *then the Orthogonal Vectors problem with* $d \in \Omega(\log N)$ *can be solved in time* $\mathcal{O}(N^{2-\eta}\mathsf{poly}(d))$, *thus contradicting* SETH.

We prove this theorem using the following lemma, which reduces an instance of the OV problem with two sets of N vectors in d dimensions each into a minimum DFA (and thus a regular language $\mathcal{L}$ for which we can check if $\mathcal{L} \in \mathsf{UW}$) of size $\Theta(Nd)$.

Lemma 9. *For an instance of* OV, *we can in* $\mathcal{O}(N(d+\log N))$ *time construct a DFA* $\mathcal{D}$ *with* $m = \Theta(N(d+\log N))$ *edges that is minimum for its language such that the* OV *instance is a* YES-*instance if and only if* $\mathcal{D}$ *contains two distinct nodes* u, v *with* $\delta(u, \gamma) = u$ *and* $\delta(v, \gamma) = v$ *for some non-empty string* γ *and* $u \bowtie v$.

Once this lemma is established, we can take an OV instance with sets of size N containing vectors of dimension $d \in \omega(\log N)$ and construct the DFA $\mathcal{D}$ of size $m = \Theta(N(d+\log N)) = \Theta(Nd)$. Now assume that we can check if $\mathcal{L} \in \mathsf{UW}$

for the language recognized by $\mathcal{D}$ in $\mathcal{O}(m^{2-\eta})$. Using Corollary 2 and Lemma 9, we can thus solve the OV instance in $\mathcal{O}((Nd)^{2-\eta}) = \mathcal{O}(N^{2-\eta}\mathsf{poly}(d))$ time, as the OV instance is a YES-instance if and only if $\mathcal{L} \notin \mathsf{UW}$. This shows Theorem 2. In the rest of this section we argue why Lemma 9 holds. Our reduction is indeed identical to the one of Becker et al. [3], however we need a few additional observations in order to adjust the proof of their Proposition 1 to a proof of Lemma 9. The construction given in Proposition 1 of [3] gives a minimum DFA $\mathcal{D}$ over the alphabet $\{\#, 0, 1\}$ from an instance $A = \{a_1, ..., a_N\}, B = \{b_1, ..., b_N\}$ of OV such that:

1. $\mathcal{D}$ contains pairwise distinct nodes $\hat{a}_1, ..., \hat{a}_N, \hat{b}_1, ..., \hat{b}_N$, such that, for all states $u \neq v$ in $\mathcal{D}$, there exists γ with $\delta(u, \gamma) = u$ and $\delta(v, \gamma) = v$ for some string γ iff $u = \hat{a}_i, v = \hat{b}_j$, and $a_i^\intercal b_j = 0$ holds.
2. every node $\hat{a}_i$ is reached by a word ending in 000 and by a word ending in 110; every node $\hat{b}_j$ is reached by a word ending in 010.

These properties are sufficient to prove Lemma 9. On the one hand, if $A = \{a_1, ..., a_N\}, B = \{b_1, ..., b_N\}$ are such that $a_i^\intercal b_j = 0$ for some i, j, then, by 1 above, in $\mathcal{D}$ we can find $u = \hat{a}_i, v = \hat{b}_j$ and a string γ with $\delta(u, \gamma) = u$ and $\delta(v, \gamma) = v$. Moreover, by 2 above, there are $\alpha, \alpha' \in I_u$ and $\beta \in I_v$ such that $000 \dashv \alpha, 110 \dashv \alpha'$ and $010 \dashv \beta$. Then $u \bowtie v$ holds, because we have $\alpha \prec \beta$ iff $\beta \prec \alpha'$, for any order $\preceq$ of the alphabet $\{\#, 0, 1\}$.

On the other hand, if $\mathcal{D}$ contains two distinct nodes u, v with $\delta(u, \gamma) = u$ and $\delta(v, \gamma) = v$ for some string γ and $u \bowtie v$, then, by 1 above, there are i, j such that $u = \hat{a}_i, v = \hat{b}_j$ and $a_i^\intercal b_j = 0$ holds.

5 Conclusions

In this paper, we established a link between the class of Wheeler languages and—well studied—(sub-)classes of *locally testable* languages. Our link is provided on both theoretical as well as algorithmic grounds. On the one hand, we proved that the class of languages that, together with their complement, are Wheeler with respect to *any* order of the alphabet corresponds exactly to the union of Definite and Reverse Definite languages—that is, the collections of words characterized either by a finite set of prefixes or suffixes. On the other hand we provided an optimal algorithm for characterizing the *universally* (with respect to the alphabet ordering) Wheeler languages.

The next steps along the path we started to follow here are, for example, to characterize languages that, together with their complement, are Wheeler with respect to a fixed order of the alphabet. Moreover, somehow along the style proposed in [8], it would be interesting to reformulate and study analogues of the problems studied here for general—i.e. co-lexicographically *partially* ordered—languages/automata.

Funding Information. Ruben Becker and Nicola Prezza are funded by the European Union (ERC, REGINDEX, 101039208). Views and opinions expressed are however those of the authors only and do not necessarily reflect those of the European Union or the European Research Council Executive Agency. Neither the European Union nor the granting authority can be held responsible for them. Giuseppa Castiglione is funded by the project "ACoMPA" (CUP B73C24001050001), NextGeneration EU programme PNRR ECS00000017 Tuscany Health Ecosystem (Spoke 6). Brian Riccardi received grants from the European Union's Horizon 2020 Research and Innovation Programme under the Marie Skłodowska-Curie grant agreement PANGAIA No. 872539, and from MUR 2022YRB97K, PINC, Pangenome INformatiCs: from Theory to Applications. Alberto Policriti is founded by Project funded under the National Recovery and Resilience Plan (NRRP), Mission 4 Component 2 Investment 1.4 âĂŞ Call for tender No. 3138 of 16 December 2021, rectified by Decree n.3175 of 18 December 2021 of Italian Ministry of University and Research funded by The European Union âĂŞ NextGenerationEU. Award Number: Project code CN_00000033, Concession Decree No. 1034 of 17 June 2022 adopted by the Italian Ministry of University and Research, CUP G23C22001110007, Project title: "National Biodiversity Future Center âĂŞ NBFC".

References

1. Alanko, J., D'Agostino, G., Policriti, A., Prezza, N.: Wheeler languages. Inf. Comput. **281**, 104820 (2021). https://doi.org/10.1016/J.IC.2021.104820
2. Becker, R., et al.: Universally wheeler languages. arXiv:2504.19537 (2025)
3. Becker, R., Cenzato, D., Kim, S.H., Kodric, B., Policriti, A., Prezza, N.: Optimal wheeler language recognition. In: International Symposium on String Processing and Information Retrieval, pp. 62–74. Springer (2023)
4. Brzozowski, J.: Characterizations of locally testable events. Discrete Math. **4**(3), 243–271 (1973). https://doi.org/10.1016/S0012-365X(73)80005-6
5. Brzozowski, J.A., Szykuła, M., Ye, Y.: Syntactic complexity of regular ideals. Theory Comput. Syst. **62**, 1175–1202 (2018)
6. Caron, P.: Families of locally testable languages. Theor. Comput. Sci. **242**(1-2), 361–376 (2000). https://doi.org/10.1016/S0304-3975(98)00332-6
7. Castiglione, G., Restivo, A.: Completing wheeler automata. In: de'Liguoro, U., Palazzo, M., Roversi, L. (eds.) Proceedings of the 25th Italian Conference on Theoretical Computer Science, Torino, Italy, September 11–13, 2024. CEUR Workshop Proceedings, vol. 3811, pp. 120–132. CEUR-WS.org (2024). https://ceur-ws.org/Vol-3811/paper060.pdf
8. Cotumaccio, N., D'Agostino, G., Policriti, A., Prezza, N.: Co-lexicographically ordering automata and regular languages - part I. J. ACM **70**(4), 27:1–27:73 (2023). https://doi.org/10.1145/3607471
9. D'Agostino, G., Martincigh, D., Policriti, A.: Ordering regular languages: a danger zone. In: Coen, C.S., Salvo, I. (eds.) Proceedings of the 22nd Italian Conference on Theoretical Computer Science, Bologna, Italy, September 13–15, 2021. CEUR Workshop Proceedings, vol. 3072, pp. 46–69. CEUR-WS.org (2021). https://ceur-ws.org/Vol-3072/paper5.pdf
10. D'Agostino, G., Martincigh, D., Policriti, A.: Ordering regular languages and automata: complexity. Theor. Comput. Sci. **949**, 113709 (2023). https://doi.org/10.1016/J.TCS.2023.113709
11. Ferragina, P., Manzini, G.: Indexing compressed text. J. ACM **52**(4), 552–581 (2005). https://doi.org/10.1145/1082036.1082039

12. Gagie, T., Manzini, G., Sirén, J.: Wheeler graphs: a framework for BWT-based data structures. Theor. Comput. Sci. **698**, 67 – 78 (2017). Algorithms, Strings and Theoretical Approaches in the Big Data Era (In Honor of the 60th Birthday of Professor Raffaele Giancarlo). https://doi.org/10.1016/j.tcs.2017.06.016
13. Gagie, T., Navarro, G., Prezza, N.: Fully functional suffix trees and optimal text searching in BWT-runs bounded space. J. ACM **67**(1), 2:1–2:54 (2020). https://doi.org/10.1145/3375890
14. Hopcroft, J.E., Motwani, R., Ullman, J.D.: Introduction to automata theory, languages, and computation. ACM SIGACT News **32**(1), 60–65 (2001)
15. Impagliazzo, R., Paturi, R.: On the complexity of k-SAT. J. Comput. Syst. Sci. **62**(2), 367–375 (2001). https://doi.org/10.1006/jcss.2000.1727
16. Vassilevska Williams, V.: Hardness of easy problems: basing hardness on popular conjectures such as the strong exponential time hypothesis (invited talk). In: 10th International Symposium on Parameterized and Exact Computation (IPEC 2015), pp. 17–29. Schloss Dagstuhl–Leibniz-Zentrum fuer Informatik (2015)
17. Williams, R.: A new algorithm for optimal 2-constraint satisfaction and its implications. Theor. Comput. Sci. **348**(2–3), 357–365 (2005). https://doi.org/10.1016/j.tcs.2005.09.023

Relativized Codes, Finite Decodability, and Bounded Languages

Oscar H. Ibarra[1] and Ian McQuillan[2(✉)]

[1] Department of Computer Science, University of California, Santa Barbara, CA 93106, USA
ibarra@cs.ucsb.edu

[2] Department of Computer Science, University of Saskatchewan, Saskatoon, SK S7N 5A9, Canada
mcquillan@cs.usask.ca

Abstract. A language C is a *code relative to* L if every word in L has a unique factorization into words of C; this is a generalization of a code. We extend this notion to d-decodability (respectively, finite-decodability) for $d \geq 1$, which means that every word in L has at most d (respectively, a finite number of) factorizations into words of C. We study decidability of testing this property on languages accepted (respectively, generated) by different machine (respectively, grammar) models. Then, we study applications of finite decodability towards a new notion regarding bounded languages called C-boundedness for a language C, leading to several new and general decidability results. In particular, we show that in any family with a decidable finiteness problem that is effectively closed under homomorphism, inverse homomorphism, and intersection with regular languages, it is decidable, given a language L in the family and a set $\Sigma^{\leq l}$ of all strings of length at most l over Σ, whether there exist words $w_1, \ldots, w_n$ in $\Sigma^{\leq l}$ such that $L \subseteq w_1^* \cdots w_n^*$. This can be considered as a finite analog of the boundedness problem. This also implies that the letter-boundedness problem is always decidable in these families.

1 Introduction

Codes are languages C over some alphabet Σ whereby each word in C^+ has a unique factorization over C. Codes are a well-studied concept as they model the communication of information from a source to a destination over a channel using a set of codewords C such that the set of messages that can be sent C^+ can be uniquely decomposed into concatenated words of C. Different classes of codes have been studied in the literature such as prefix-, infix-, and suffix-codes [25], and codes have been generalized in multiple ways. First, what if instead of having a message having a unique decomposition, there is $d \geq 1$ such that each word in C^+ has at most d decompositions. In this case, C is called d-decodable and it

The research of I. McQuillan was supported, in part, by Natural Sciences and Engineering Research Council of Canada.

S.-K. Ko and F. Manea (Eds.): DLT 2025, LNCS 16036, pp. 61–76, 2026.
https://doi.org/10.1007/978-3-032-01475-7_5

is finitely-decodable if it is d-decodable for some d. This concept was studied in [26] for certain types of codes and decidability questions were addressed. Next, in [7], another generalization was studied where instead of considering all possible messages in C^+ and whether each has a unique decomposition, a set L of possible messages was considered. Then, we say C is a *code relative to* L if each word in L has a unique factorization into words of C. Certainly, one could also define d-decodability and finite-decodability of codes relative to a language as well, and we study this notion in this paper.

Separately, in formal language and automata theory, one of the most commonly studied restrictions on languages are bounded languages. A language $L \subseteq \Sigma^*$ is *bounded* if there exists $w_1, \ldots, w_n \in \Sigma^+$ such that $L \subseteq w_1^* \cdots w_n^*$. Furthermore, a language is *letter-bounded* if $w_1, \ldots, w_n$ are letters. The following two problems have been studied:

> The *boundedness problem* for a machine model (or grammar system) $\mathcal{M}$: for $M \in \mathcal{M}$ over Σ, is $L(M)$ bounded? And if so, what are words $w_1, \ldots, w_n \in \Sigma^+$ such that $L(M) \subseteq w_1^* \cdots w_n^*$?

The *letter-boundedness problem* is similar except the words are restricted to be letters. It is known that the boundedness and letter-boundedness problems are decidable for context-free languages [14,15], and even more general classes such as one-way nondeterministic pushdown automata augmented with reversal-bounded counters ([11] for letter-boundedness, and [2] for boundedness). It has also recently been shown that boundedness is decidable for vector addition system with states (equivalent to partially-blind multicounter machines) [5], and certain restrictions of Turing machines [24].

In this paper, we study the property of a language being d-decodable (finitely-decodable) relative to L. We show that if a language family $\mathcal{L}$ is effectively closed under inverse homomorphism and intersection with regular languages and it has a decidable emptiness problem ("given $L \in \mathcal{L}$, is $L = \emptyset$?"), then it is decidable, given a regular language C, $L \in \mathcal{L}$, and $d \geq 1$, whether C is d-decodable relative to L. This implies decidability of being a code relative to L.

Next a generalization of boundedness is studied. A language L is *C-bounded* if $L \subseteq w_1^* \cdots w_n^*$ where $w_1, \ldots, w_n \in C$. Hence, the normal concept of boundedness coincides with Σ^+-boundedness. Connections are made to coding theory, and decidability problems are studied. For example, if $\mathcal{L}$ is a full trio (with all properties including an effective construction) with a decidable finiteness problem ("given $L \in \mathcal{L}$, is L finite?"), then 1) it is decidable, given finite C, and $L \in \mathcal{L}$ such that C is a code relative to L, whether L is C-bounded; and 2) it is decidable, given $L \in \mathcal{L}$ and $l \geq 1$, whether L is $\Sigma^{\leq l}$-bounded. The latter result can be considered as a finite analog of the boundedness problem. It is an interesting open question whether decidability of the boundedness problem implies decidability of the finite analog problem, and vice-versa. Thus, the problems are of independent interest. Our result also implies that the letter-boundedness problem is always decidable in these families, which includes families such as the matrix languages [8,19] and the OI hierarchy [9]. In the past, the letter-boundedness problem has always been proven using ad hoc techniques for each model. But with this new theorem, the result and algorithm follow for free using closure properties and the finiteness problem. Finally, it is proven that for general families $\mathcal{L}$ such as the matrix languages, it is decidable, given $L \in \mathcal{L}$ and C finite such that C is finitely-decodable relative to L, whether L is C-bounded.

2 Preliminaries

Let $\mathbb{N}_0$ be the set of non-negative integers. A representation S of a linear set in $\mathbb{N}_0^k$, for some $k \geq 1$, consists of a vector $\vec{v_0} \in \mathbb{N}_0^k$ called *the constant*, and a finite set of vectors $\vec{v_1}, \ldots, \vec{v_n} \in \mathbb{N}_0^k$, for some $n \geq 0$ called *the periods*. Given a representation of a linear set S, denote by $\widehat{S} = \{\vec{v_0} + i_1\vec{v_1} + \cdots + i_n\vec{v_n} \mid i_1, \ldots, i_n \in \mathbb{N}_0\} \subseteq \mathbb{N}_0^k$, which is called a *linear set*. A representation S of a semilinear set is a finite set $\{S_1, \ldots, S_l\}$ of representations of linear sets where $\widehat{S_1}, \ldots, \widehat{S_l} \subseteq \mathbb{N}_0^k$, for some $k \geq 1$; and $\widehat{S} = \widehat{S_1} \cup \cdots \cup \widehat{S_l}$ is called a *semilinear set*.

An alphabet Σ is a finite set of symbols. A word over Σ is a finite sequence of symbols from Σ obtained by juxtaposing the letters. We denote the set of all words over Σ by Σ^*, which includes the empty word λ, and Σ^+ is the set of all non-empty words over Σ. Given $k \in \mathbb{N}$, $\Sigma^{\leq k}$ is the set of all non-empty words of length at most k, and $\Sigma^{\geq k}$ is the set of all words of length at least k. A language L over Σ is any $L \subseteq \Sigma^*$. The complement of L with respect to Σ is $\Sigma^* - L$.

Given $w \in \Sigma^*$, $|w|$ is the length of w. For $a \in \Sigma$, $|w|_a$ is the number of a's in w. If $\Sigma = \{a_1, \ldots, a_k\}$, then $\psi(w) = (|w|_{a_1}, \ldots, |w|_{a_k})$, and given $L \subseteq \Sigma^*$, $\psi(L) = \{\psi(w) \mid w \in L\}$, which is called the *Parikh image* of L. A language is *semilinear* if its Parikh image is a semilinear set. The *length image* of L is $\{n \mid w \in L, n = |w|\}$. A language is *length-semilinear* if its length image is a semilinear set. This means that after mapping all its letters onto one letter, the Parikh image is semilinear. A language family is semilinear (respectively length-semilinear) if all languages in it are semilinear (respectively length-semilinear). We say that a machine model or grammar system is effectively semilinear (respectively effectively length-semilinear), if given a machine or grammar of that type, a representation S of a semilinear set can be built such that $\widehat{S}$ is the Parikh image (respectively length image) of the language accepted or generated.

Given $w = a_1 \cdots a_n, a_i \in \Sigma$, a *period* of w is an integer k such that $a_{k+i} = a_i$ for i from $1 \leq i \leq n - k$. Given w, v is a *subsequence* of w (denoted by $v \leq_{\mathrm{e}} w$), if $v = a_{i_1} a_{i_2} \cdots a_{i_m}$ for some $1 \leq i_1 < i_2 < \cdots < i_m \leq n$. Given $w \in \Sigma^*$, if $w = xyz, x, y, z \in \Sigma^*$, then x is a *prefix* of w (denoted by xy), y is an *infix* of w (denoted by $y \leq_{\mathrm{i}} w$), and z is a *suffix* of w (denoted by zw).

A *nondeterministic generalized sequential machine with accepting states* (gsm, also called a-transducer or finite transducer) is a tuple $M = (Q, \Sigma, \Delta, \delta, q_0, F)$, where Q is the finite state set, Σ is the input alphabet, Δ is the output alphabet, δ is a finite subset of $Q \times \Sigma^* \times \Delta^* \times Q$, $q_0 \in Q$ is the initial state and $F \subseteq Q$ is the set of final states. Let $\vdash$ be the relation on $Q \times \Sigma^* \times \Delta^*$ defined by letting $(q, xw, z_1) \vdash (p, w, z_2)$ for each $w \in \Sigma^*$ if $(q, x, y, p) \in \delta$ and $z_2 = z_1 y$. A triple (q, w, z) represents that M is in state q, w is the input still to be read, and z the accumulated output. Let $\vdash^*$ be the reflexive, transitive closure of $\vdash$. Let $M = (Q, \Sigma, \Delta, \delta, q_0, F)$ be such a gsm. For each word $w \in \Sigma^*$, let $M(w) = \{z \mid (q_0, w, \lambda) \vdash^* (q, \lambda, z)$ for some $q \in F\}$. For every set $L \subseteq \Sigma^*$, let $M(L) = \bigcup_{w \in L} M(w)$.

The emptiness problem for a family $\mathcal{L}$ is, "given $L \in \mathcal{L}$ is $L = \emptyset$?". The finiteness problem is, "given $L \in \mathcal{L}$, is L finite?". And, the membership problem is, "given $L \in \mathcal{L}$ and $w \in \Sigma^*$, is $w \in L$?". For these and other decision problems here, it is assumed that there is some finite representation such as a machine model or grammar system that describes the languages in $\mathcal{L}$, and that is the input. Many of the results in this paper are general and apply to any family closed under certain operations, rather than proving them for specific machine or grammar models. Furthermore, for decision properties, they often require closure under certain operations with effective procedures. Towards this, we will employ some theory from the Abstract Families of Languages [13]. A language family $\mathcal{L}$ is a *full trio* if $\mathcal{L}$ is closed under inverse homomorphism, homomorphism, and intersection with regular languages. It is known that the full trios are exactly the families closed under nondeterministic gsms with accepting states. A full trio is a *full AFL* if it is closed under union, concatenation, and Kleene-*. When discussing a language family that has certain properties, such as a full trio, we say that the family has *all properties effective* if all of these closure properties provide effective constructions. For semilinearity, this means that the model is effectively semilinear, and similarly for length-semilinearity.

We mention the following, which is commonly known and can be easily shown.

Proposition 1. *Let $\mathcal{L}$ be a full trio with all properties effective and a decidable finiteness problem. Then $\mathcal{L}$ has a decidable emptiness problem, and also we can effectively determine, given $L \in \mathcal{L}$, an integer n if it exists such that n is the length of the longest word in L.*

We will define counter machines. A one-way nondeterministic machine with k counters is a tuple $M = (Q, \Sigma, \delta, q_0, F)$ where Q is a finite set of states, Σ is the input alphabet, $q_0 \in Q$ is the initial state, $F \subseteq Q$ is the set of final states, δ is a partial function with transitions of the form $(p, d_1, \ldots, d_k) \in \delta(q, a, c_1, \ldots, c_k)$ where $c_1, \ldots, c_k \in \{0, 1\}$ and $d_1, \ldots, d_k \in \{-1, 0, +1\}$, $q, p \in Q$ and $a \in \Sigma \cup \{\lambda, \lhd\}$ where $\lhd$ is the right input end marker. Such a transition can be applied to a configuration $(q, aw, x_1, \ldots, x_k)$ either if $x_i > 0$ and $c_i = 1$ or $x_i = 0$ and $c_i = 0$ to produce $(p, w, x_1 + d_1, \ldots, x_k + d_k)$. An accepting computation is $(q_0, w\lhd, 0, \ldots, 0) \vdash^* \cdots \vdash^* (q_f, \lambda, 0, \ldots, 0), q_f \in F$ (we require all counters to end in 0). Also M is deterministic if $|\delta(q, a, c_1, \ldots, c_k) \cup \delta(q, \lambda, c_1, \ldots, c_k)| \leq 1$ for all $q \in Q, a \in \Sigma \cup \{\lhd\}, c_1, \ldots, c_k \in \{0, 1\}$. We say M is r-reversal-bounded if, for each accepting computation, each counter switches between increasing or decreasing (or vice versa) at most r times. Also, M is 1-*reversal-bounded blind* if M is 1-reversal-bounded and $\delta(q, a, c_1, \ldots, c_k) = \delta(q, a, c_1', \ldots, c_k')$ for all $c_i, c_i' \in \{0, 1\}$; i.e. the status component is not relevant and can be left off. While in general, blind counter machines as defined in the literature are allowed to store negative numbers, if they are 1-reversal-bounded, there is not any difference as they would need a second reversal to go below zero and then finish with zero. Let $\mathsf{NCM}(k, r)$ be the set of one-way nondeterministic r-reversal-bounded k counter machines, and $\mathsf{DCM}(k, r)$ is the deterministic restriction, and NCM is all $\mathsf{NCM}(k, r)$ for any k, r.

Remark 2. The following are examples of models that accept language families that are semilinear full trios, where all properties are effective:

- deterministic finite automata (DFA), nondeterministic finite automata (NFA), and one-way nondeterministic pushdown automata (NPDA) [21],
- one-way nondeterministic pushdown automata with some number of reversal-bounded counters, NPCM [22],
- finite-index ET0L languages (ET0L systems where the number of non-active symbols in each derivation is bounded by a constant) [27],
- k-flip NPDA (NPDA that can "flip" their pushdown up to k times) [20],
- nondeterministic Turing machines with a one-way read-only input tape and a finite-crossing worktape [18],
- uncontrolled finite-index indexed grammars (indexed grammars where every accepting derivation has a bounded number of nonterminals), [6].
- multi-push-down machines (a machine with multiple pushdowns where the machine can simultaneously push to all pushdowns, but can only pop from the first non-empty pushdown) [3].

Because semilinearity is effective, this means that all of those models have a decidable emptiness problem (check if any of the linear representations are non-empty), have a decidable finiteness problem (as we can check if all of the linear sets have no periods), and have a decidable membership problem (check if $L \cap \{w\} = \emptyset$).

Remark 3. The following are examples of models that generate/accept full trios where all properties are effective, which have a decidable emptiness, finiteness, and membership problem, but that are not semilinear:

- context-free matrix grammars, which are also closed under intersection with NCM languages [8,19]; of note, this family is length-semilinear despite not being semilinear,
- stack automata [16], ET0L systems [27], indexed grammars [1], and the OI hierarchy (languages accepted by one-way machines with a pushdown of pushdowns of ... of pushdowns for some k number of times) [10].
- a family yR-TBY$^+$ that contains yields of tree languages of ranges of a large class of tree transducers (in [9], this large family was shown to be a full AFL with a decidable finiteness problem that contains the OI Hierarchy).

It is worth noting that an undecidable emptiness problem implies an undecidable finiteness problem by Proposition 1. And in [9], Drewes and Engelfriet point out that to their knowledge, all full AFLs with a known decidable emptiness problem are contained in the full AFL yR-TBY$^+$; i.e., for all known full AFLs, the emptiness problem is decidable exactly when the finiteness problem is decidable. But it is open whether there is a full AFL with a decidable emptiness problem but an undecidable finiteness problem.

3 Decidability of Properties of Relativized Codes

We study the notions of codes and relativized codes from the area of coding theory. Let $C \subseteq \Sigma^+$ be a language. A word $w \in \Sigma^+$ is *C-decodable* if there are words $u_1, \ldots, u_n \in C$ such that $w = u_1 \cdots u_n$; and $(u_1, \ldots, u_n)$ is called a *C-decoding of w*. The language C is said to be *uniquely decodable* if every word in Σ^+ has at most one C-decoding. Therefore, C is uniquely decomposable if and only if each word in C^+ has exactly one C-decoding. A language $C \subseteq \Sigma^+$ is a *code* if C is a non-empty uniquely decodable language.

A set $C \subseteq \Sigma^+$ is a *code relative to* $L \subseteq \Sigma^+$ if every word $w \in L$ has at most one C-decoding [7]. Therefore, a language $C \subseteq \Sigma^+$ is a code if and only if C is a code relative to C^+. We generalize this notion with a new definition as follows. Let $d \in \mathbb{N}$. A word $w \in \Sigma^+$ is d-C-decodable (or simply d-decodable if C is understood), if there are at most d different C-decodings of w. A set $C \subseteq \Sigma^+$ is d-decodable relative to $L \subseteq \Sigma^+$ if every word $w \in L$ is d-C-decodable. Further, we say that C is finitely-decodable relative to $L \subseteq \Sigma^+$ if C is d-decodable relative to L for some d. In these definitions, we call L the *relativized language*.

Remark 4. Certainly, a language $C \subseteq \Sigma^+$ is a code relative to $L \subseteq \Sigma^+$ if and only if C is 1-decodable relative to $L \subseteq \Sigma^+$. Also, we see next that there are non-trivial examples of d-decodable languages.

If we have a set of words $C \subseteq \Sigma^+$ that is not a code, then there is some word $w \in C^+$ which has more than one C-decoding. In this case, by examining words in w^+, there are words in C^+ with arbitrarily many C-decodings (and thus C would not be d-decodable relative to Σ^+ or relative to C^+ for any d). However, with relativized codes, this does not need to be the case. For example, consider $C = \{\$, \$0, 0\}$. Clearly, C is not a code since $\$0$ has two C-decodings. Thus, by considering words in $(\$0)^+$, we can produce words with arbitrarily many C-decodings. However, consider $L = \$0^+$. Then every word in L has exactly two C-decodings; $\$0^m = (\$0)(0^{m-1})$ and $\$0^m = (\$)(0^m)$. Thus, C is 2-decodable relative to L.

Next, we address some decidability properties. It is known that it is undecidable if a linear language (or a context-free language) is a code [12] (this is related to the undecidability of the ambiguity problem for minimal linear grammars [17]). In fact, this is even undecidable for one-way deterministic 1-reversal-bounded BLIND 1-counter automata. Furthermore, in that proof, deterministic machines were defined to not use the input end-marker. Let $\mathsf{DBLIND}(1,1)$ be this class of machines (which accept a subset of the languages accepted by $\mathsf{DCM}(1,1)$). This is one of the most restrictive language families that has been studied that is defined by a standard class of automata that contains non-regular languages. Hence, this result carries over to relativized codes as well.

Proposition 5. *It is undecidable, given a* $\mathsf{DBLIND}(1,1)$ *machine M (respectively* $\mathsf{DCM}(1,1)$*) accepting $C \subseteq \Sigma^+$, whether C is 1-decodable relative to Σ^+.*

This is equivalent to testing whether C is 1-decodable relative to C^+.

The next question to study is what happens when C is regular, while possibly allowing the relativized language to be more general.

First, we will define a concept relevant to finite-decodability, and prove an interesting property. Given $C \subseteq \Sigma^+$ and $d \geq 1$, let $V_d = \{\$_v \mid v$ is a non-empty subsequence of $12 \cdots d\}$ be an alphabet of new symbols and let \$ be a new symbol. Define the following homomorphisms. For $d \geq 1$ and $1 \leq i \leq d$, let $h_{i,d}$ from $(\Sigma \cup V_d)^*$ to $(\Sigma \cup \{\$\})^*$ defined by

$$h_{i,d}(\$_v) = \begin{cases} \lambda, & \text{if } i \not\leq_{\mathrm{e}} v \\ \$ & \text{if } i \leq_{\mathrm{e}} v, \end{cases}$$

and $h_{i,d}(a) = a$ for all $a \in \Sigma$. Let h_d be a homomorphism from $(\Sigma \cup V_d)^*$ to Σ^* that erases all letters of V_d and fixes the others. Define the following language $C^{(\geq d)}$ over $(\Sigma \cup V_d)^*$ equal to:

$$\{w \mid \forall i, 1 \leq i \leq d, h_{i,d}(w) = u_{i,1}\$u_{i,2}\$ \cdots u_{i,m_i}\$, u_{i,l} \in C \text{ for } l, 1 \leq l \leq m_i, \\ \text{and } \forall i, j, i \neq j, (m_i, (u_{i,1}, \ldots, u_{i,m_i})) \neq (m_j, (u_{j,1}, \ldots, u_{j,m_j}))\},$$

and

$$C^{(d)} = \{w \mid w \in C^{(\geq d)}, h_{d+1}^{-1}(h_d(w)) \cap C^{(\geq d+1)} = \emptyset\}.$$

Thus, $C^{(\geq d)}$ contains all words that have at least d distinct decodings of C relative to C^+, where each different decoding i is annotated with i as a subscript on a \$ symbol; and $C^{(d)}$ contains all words that have d distinct decodings annotated, and it ensures that there are no additional decodings (and so there are exactly d decodings).

Proposition 6. *Let C be a regular language with $C \subseteq \Sigma^+$, and let $d \geq 1$. Then $C^{(\geq d)}$ and $C^{(d)}$ are regular. Also, given a* DFA *(respectively* NFA*) M accepting C where d is fixed, another* DFA *(respectively* NFA*) M' can be constructed accepting $C^{(\geq d)}$ in polynomial time.*

Proof. Consider first a DFA accepting C (the case is similar for NFA). Consider

$$\bigcap_{i, 1 \leq i \leq d} X_i \cap \bigcap_{i,j, 1 \leq i < j \leq d} Y_{i,j}, \tag{1}$$

where $X_i = h_{i,d}^{-1}((C\$)^+) = \{x \mid h_{i,d}(x) \in (C\$)^+\}, 1 \leq i \leq d$ and $Y_{i,j} = \{x \mid x \in (\Sigma \cup V_d)^*$, there is some occurrence of $\$_v$ in x where $i \leq_{\mathrm{e}} v$ or $j \leq_{\mathrm{e}} v$ but not both$\}$. Thus, each X_i makes sure that each i is properly placed on subscript of \$ between words of C, and $Y_{i,j}$ enforces that each i in $\{1, \ldots, d\}$ represents a different decoding. This is exactly $C^{(\geq d)}$.

For fixed d, there is a fixed number of intersections in Equation (1) and it is clear that we can construct a DFA accepting $C^{(\geq d)}$ in polynomial time.

Next, consider $C_1 = h_d(C^{(\geq d)}) \cap \overline{h_{d+1}(C^{(\geq d+1)})}$. This consists of all words with at least d decodings but do not have at least $d+1$. By constructing $h_d^{-1}(C_1) \cap \bigcap_{i,1 \leq i \leq d} X_i \cap \bigcap_{i,j,1 \leq i < j \leq d} Y_{i,j}$ as above, this is exactly $C^{(d)}$. $\square$

We are able to use $C^{(\geq d)}$ in a general way to see if C is d-decodable relative to L, for L in any family $\mathcal{L}$ satisfying simple properties.

Theorem 7. *Let $\mathcal{L}$ be a language family effectively closed under inverse homomorphism and intersection with regular languages with a decidable emptiness problem. It is decidable, given a regular language C, $L \in \mathcal{L}$, and $d \geq 1$, whether C is d-decodable relative to L. Hence, it is decidable whether C is a code relative to L. In particular, these results hold for any of the full trios from Remarks 2 and 3.*

Proof. The second statement follows from the first by setting $d = 1$.

Given C, there is a DFA M accepting $C^{(\geq d+1)}$. Given $L \in \mathcal{L}$, let $L' = h_{d+1}^{-1}(L) \cap L(M)$, which is empty if and only if C is d-decodable. □

This holds for almost every deterministic or nondeterministic machine model defined in the literature that has a decidable emptiness problem.

An interesting open question arises from Theorem 7. For a given language family $\mathcal{L}$ (perhaps satisfying some closure properties, or a specific family), is it decidable, given a regular language C and $L \in \mathcal{L}$, whether C is finitely-decodable relative to L? The result above implies that this is decidable for a given d, but it is open when we do not know d in advance. This is even open when C is finite.

4 C-Boundedness

Let $C \subseteq \Sigma^+$ be a set of words, where Σ is an alphabet of symbols. A language L is C-bounded if $L \subseteq w_1^* \cdots w_n^*$ for some not-necessarily distinct words $w_1, \ldots, w_n$ in C. Note that Σ-bounded is the same as letter-bounded.

First, we report the following observations regarding decodability:

Remark 8. The following are true:

1. Let C and L be languages with C finite. If L is C-bounded, then C is not necessarily finitely-decodable relative to L. This can be seen with the following example. Let $C = \{01, 0101\}$ and $L = \{(01)^{2n} \mid n > 0\}$. Then $L \subseteq (01)^*$, so L is C-bounded but C is not finitely-decodable relative to L.
2. Let C and L be languages in Σ^+ with C finite. If C is finitely-decodable relative to L, then it is not necessarily true that L is C-bounded. This can be seen using the following example. Let $C = \{0, 1\}$ and $L = \{0, 1\}^+$. Then C is 1-decodable relative to L, but L is not C-bounded, i.e., not letter-bounded (nor bounded [15]).

Thus, the notions of C-bounded, and C being finite-decodable relative to L are independent.

For the next propositions, we require the following definitions.

Definition 9. *Let $C \subseteq \Sigma^+$. For $w \in \Sigma^*$,*

$$\begin{aligned}\text{count}_C(w) =& \{n \mid w = w_1^{i_1} \cdots w_n^{i_n}, w_1, \ldots, w_n \in C, i_1, \ldots, i_n \geq 1 \\ & \text{with } w_i \neq w_{i+1} \text{ for } 1 \leq i < n\}.\\ \text{mincount}_C(w) =& \min\{i \mid i \in \text{count}_C(w)\},\\ \text{maxcount}_C(w) =& \max\{i \mid i \in \text{count}_C(w)\}.\end{aligned}$$

For a language $L \subseteq \Sigma^*$, let $\text{count}_C(L) = \{\text{count}_C(w) \mid w \in L\}$, $\text{mincount}_C(L) = \{\text{mincount}_C(w) \mid w \in L\}$ *and* $\text{maxcount}_C(L) = \{\text{maxcount}_C(w) \mid w \in L\}$.

In the following, the proof is omitted due to space constraints, but 1) is proven using properties of bounded languages, 2) follows immediately from 1), and 3) follows from the definitions.

Proposition 10. *Consider a language $C \subseteq \Sigma^+$ and a language L. The following are true:*

1. *L is C-bounded if and only if there exists $D \subseteq C$, D finite, $L \subseteq D^*$, and $\text{mincount}_D(L)$ is finite;*
2. *Assume that C is finite. Then L is C-bounded if and only if $L \subseteq C^*$ and $\text{mincount}_C(L)$ is finite;*
3. *$\text{count}_C(L)$ is finite implies $\text{mincount}_C(L)$ is finite.*

The rest of the paper will only consider the case where C is finite. When C is finite, a nondeterministic gsm with accepting states can, on input $w \in \Sigma^*$, compute $\{1^i \mid i \in \text{count}_C(w)\}$ by, at each step, guessing a $w_i \in C$ and verifying that the next part of the input is w_i, while outputting a 1 at the beginning and every time a word w_i is guessed that is different from the previous word guessed. If L is a language in a full trio $\mathcal{L}$, then $\text{count}_C(L)$ is also in $\mathcal{L}$, since full trios are closed under nondeterministic gsms with accepting states [13]. Notice that the gsms need to be nondeterministic with accepting states since on a given input, it guesses what words in C to match with the input. If at the end of the input the guesses match all of the symbols in the input string, then it accepts, otherwise it rejects.

Proposition 11. *Let $\mathcal{L}$ be a full trio with all properties effective with a decidable finiteness problem. It is decidable, given a finite language $C \subseteq \Sigma^+$ and a language $L \in \mathcal{L}$ with $L \subseteq C^*$, whether $\text{count}_C(L)$ is finite. Furthermore, if $\text{count}_C(L)$ is finite, then L is C-bounded and words $w_1, \ldots, w_n \in C$ can be effectively determined such that $L \subseteq w_1^* \cdots w_n^*$.*

Proof. We have already observed that we can construct $\text{count}_C(L) \in \mathcal{L}$ since every full trio is closed under nondeterministic gsms with accepting states, and use the decidability of the finiteness problem to test whether it is finite. Proposition 10 implies that L is C-bounded.

If $\text{count}_C(L)$ is finite and $L \subseteq C^*$, then because $\text{count}_C(L) \in \mathcal{L}$, Proposition 1 implies we can determine the length of the longest word in it, m say. This implies every word $w \in L$ is in $w_1^* \cdots w_m^*$ for some $w_1, \ldots, w_m$, but they can be different depending on each w. But for each $w_1, \ldots, w_m \in C$, let $D = w_1^* \cdots w_m^*$, and we can concatenate all such D's. □

Even though in general we need to compute $\text{mincount}_C(L)$ to determine C-boundedness, the next proposition shows that for finite C where C is a code relative to L, we can decide C-boundedness of L with $\text{count}_C(L)$.

Proposition 12. *Consider $C \subseteq \Sigma^+$, C finite and C is a code relative to L. Then L is C-bounded if and only if $L \subseteq C^*$ and* $\text{count}_C(L)$ *is finite.*

Proof. Suppose L is C-bounded, i.e., $L \subseteq w_1^* \cdots w_n^*$ where $w_i \in C$ for $1 \leq i \leq n$ (and $w_j \neq w_{j+1}$ for $1 \leq j < n$). But because C is a code relative to L and is also C-bounded, the unique decoding of each $w \in L$ must be obtained with some number of w_1's, followed by w_2, etc. up to w_n, and hence the longest word in $\text{count}_C(L)$ must be at most n in length and $\text{count}_C(L)$ is finite.

If $\text{count}_C(L)$ is finite and $L \subseteq C^*$, then Proposition 10 parts 3 and 2 shows L is C-bounded. □

Hence, from Proposition 11 and Proposition 12, we obtain:

Proposition 13. *Let $\mathcal{L}$ be a full trio with all properties effective with a decidable finiteness problem. It is decidable, given a finite language $C \subseteq \Sigma^+$ and a language $L \in \mathcal{L}$ such that C is a code relative to L, whether L is C-bounded. If it is, then $w_1, \ldots, w_n \in C$ can effectively be found such that $L \subseteq w_1^* \cdots w_n^*$. In particular, these results hold for any family in Remarks 2 and 3.*

Proposition 13 is an interesting result, keeping in mind that decidability of the boundedness problem is open for all of the families in Remark 3 and some in Remark 2 (multi-push-down machines and finite-index ET0L).

Although $\text{count}_C(L)$ can be computed with a gsm, we cannot obviously do so with mincount. And, when C is not a code relative to L, it is not enough to determine whether $\text{count}_C(L)$ is finite as seen from the next proposition which is weakened only slightly, to 2-decodable languages.

Proposition 14. *There is a finite language C and a regular language $L \subseteq C^+$ such that L is C-bounded and 2-decodable relative to L, but* $\text{count}_C(L)$ *is infinite.*

Proof. Let $C = \{\$0123, \#, 0123, \$0, 12, 30, 3\#\}$ and $L = \$(0123)^+\#$. Notice that L is C-bounded because $L \subseteq (\$0123)^*(0123)^*\#^*$. Also, C is 2-decodable relative to L because each word $w = \$(0123)^i\#, i \geq 1$ has two decodings (which are highlighted with parentheses):

1. $(\$0123)(0123)^{i-1}\#$,
2. $(\$0)((12)(30))^{i-1}(12)(3\#)$,

and no other decodings. But $\text{count}_C(L)$ is infinite, because given any k, consider the word $w = (\$0)((12)(30))^{k+1}(12)(3\#)$. Indeed, $\text{count}_C(w) > k$ because it alternates between (12) and (30) more than k times, and each time the count goes up. □

It is an open problem to determine whether it is decidable, given a finite language $C \subseteq \Sigma^+$ and a language $L \subseteq \Sigma^*$ (in a full trio or a standard family with a decidable finiteness problem), whether L is C-bounded. If $\text{count}_C(L)$ is finite, then we can immediately conclude that it is C-bounded, but when $\text{count}_C(L)$ is infinite, we do not know a procedure. However, we will investigate two special cases where it is decidable. We start with the following definition.

Let $L \subseteq \Sigma^*, C \subseteq \Sigma^+$ and let $e \geq 1$. Then L is e-C-bounded if each word $w \in L$ is in $w_1^* \cdots w_d^*$ for some $w_1, \ldots, w_d \in C, d \leq e$.

Lemma 15. *Let $L \subseteq \Sigma^*$ and let $C \subseteq \Sigma^+$ be a finite language. Then L is e-C-bounded for some $e \geq 1$ if and only if L is C-bounded.*

Proof. If L is e-C-bounded, then each $w \in L$ is in $w_1^* \cdots w_d^* \in C, d \leq e$. Thus, each $w \in L$ is in $w_1^* \cdots w_d^*$ for some $w_1, \ldots, w_d \in C, d \leq e$. By concatenating all these, L is C-bounded.

If L is C-bounded, then L is e-C-bounded for some e. □

If we are given this e value in advance, then we can decide e-C-boundedness.

Lemma 16. *Let $\mathcal{L}$ be a language family effectively closed under intersection with regular languages and with a decidable emptiness problem (or a full trio with all properties effective and a decidable finiteness problem). It is decidable, given $L \in \mathcal{L}$, C finite, and $e \geq 1$, whether L is e-C-bounded.*

Proof. For the case where finiteness is decidable, so is emptiness by Proposition 1. Let

$$X = \bigcup_{\substack{w_1, \ldots, w_d \in C \\ d \leq e}} w_1^* \cdots w_d^*.$$

Clearly this is regular since regular languages are closed under concatenation, Kleene-*, and union. Further, consider $\overline{X}$ which is also regular. Thus, $L \cap \overline{X} \in \mathcal{L}$ is empty if and only if L is e-C-bounded. □

Still though, given $L \in \mathcal{L}$ and C finite, it is not clear how to determine e such that, if L is C-bounded, then L is e-C-bounded. If we were to develop such an algorithm, then we would have an algorithm to determine if $L \in \mathcal{L}$ and C finite has L being C-bounded.

The next special case of interest can be helped with another gsm.

Definition 17. *Given finite C, let z be the length of the longest word in C. We build a deterministic gsm g_C (somewhat similar to the gsm computing* count_C *above) that works as follows: g_C keeps a buffer of the last $z_1 = 3z$ letters read of the input, and keeps refilling it as it reads letters until the end of the input while always processing input from the front of the buffer. Once the buffer is full or it hits the end of the input, g_C switches to state q_1 where g_C does the following:*

- *It checks if the full buffer contains $x^i x', i \geq 3$ for some $x \in C, x'x$. If it is, then for the shortest such x, g_C remembers x and outputs 1. Then it repeatedly removes the frontmost x from the front of the buffer, and refills the buffer without outputting until the buffer no longer has x as prefix. Then it switches back to q_1.*
- *If the buffer does not contain such a repeat, then g_C outputs 1, removes the frontmost letter from the buffer, and fills the buffer with one symbol from the input if there is one, and switches back to q_1.*

Lemma 18. *Given $L \subseteq \Sigma^*, C \subseteq \Sigma^+$ with C finite and let z be the length of the longest word in C. If L is C-bounded with $L \subseteq w_1^* \cdots w_n^*, w_1, \ldots, w_n \in C$, then $g_C(L)$ is finite. Furthermore, all words in $g_C(L)$ are of length at most $3zn$.*

Proof. Assume L is C-bounded. Then $L \subseteq w_1^* \cdots w_n^*, w_1, \ldots, w_n \in C$. Let $w \in L$. Then $w = w_1^{i_1} \cdots w_n^{i_n}, i_1, \ldots, i_n \geq 0$. As g_C reads $w_1^{i_1}$, either 1) $|w_1^{i_1}| \leq z_1$ is not big enough to fill up the buffer with only copies of w_1 and therefore g_C outputs at most $z_1 = 3z$ 1's corresponding to these symbols from the buffer; or 2) $|w_1^{i_1}| \geq z_1$ and therefore g_C fills up the buffer and the buffer contains $w_1^i w_1'$ for some $i \geq 3$ and $w_1' w_1$. Assume first that w_1 is the shortest such word. Then it outputs only 1 as it reads all of $w_1^{i_1}$. Otherwise, assume there is some word $x \in C$ minimal such that $|x| < |w_1|$ and the buffer also contains $x^j x'$ for some $j, x'x$. Then $w_1^i w_1' = x^j x'$. Then both $|w_1|$ and $|x|$ are periods of this word. According to Corollary 6.1 of [4], if a word has periods p and q and is of length at least $p+q-\gcd(p,q)$, then it also has period $\gcd(p,q)$. Let y be the prefix of w_1 and x of length $\gcd(p,q)$, and so $w_1^j w_1' = x^j x' = y^k y'$ where $y'y$. Hence both w_1 and x are powers of y. Thus g_C outputs 1 as it reads x^l, for l maximal. But $w_1^{i_1} = x^j y^m$ for some j and m such that $|y^m| < |w_1|$. Hence g_C will output 1 as it reads x^j and less than $|w_1|$ 1's as it finishes scanning $w_1^{i_1}$. For every l from 2 to n, it is similar except the buffer can match $x^j x', x'x \in C, j \geq 3$, where less than z characters from $w_1^{i_1} \cdots w_{l-1}^{i_{l-1}}$ can be used as part of this. Thus, less than z letters of the buffer will be taken up with a word before w_l starts, leaving at least $2z$ letters of $w_l^{i_l}$ of the buffer. Let $x = x_1 x_2$ be such that x_2 is where $w_l^{i_l}$ starts, and x_1 is the part before that within x. Thus, the buffer contains $x^l x_1 (x_2 x_1)^m x_2 x'$ where $|x^l x_1| < z$. The same analysis of periods above can be used to see $x_2 x_1$ and w_l are powers of a common word and the remaining proof is similar. □

The next series of results involves the special case of sets C where $\Sigma \subseteq C$.

Lemma 19. *Given $L \subseteq \Sigma^*, C \subseteq \Sigma^+$ with C finite and $\Sigma \subseteq C$. If $g_C(L)$ is finite with 1^e being the longest word output, then L is e-C-bounded.*

Proof. We know that on an input $w \in L$, g_C continually fills the buffer and checks if the input contains multiple copies of some $w_i \in C$ and if so, for the smallest such $|w_i|$, it outputs 1 for w_i^j with j maximal. If not, then it outputs 1 for each symbol read. Let $w_1, \ldots, w_n$ be such that from left to right, $1 \leq j \leq n$, while reading the input, w_j is either the repeat identified by the gsm, or the single letter if no repeat was identified. Then n is the length of the output (and hence $n \leq e$). In addition, because $\Sigma \subseteq C$, $w_1, \ldots, w_n \in C$. Hence, $w \in w_1^* \cdots w_n^*$, and therefore L is e-C-bounded. □

Proposition 20. *Given $L \subseteq \Sigma^*, C \subseteq \Sigma^+$ with C finite and $\Sigma \subseteq C$. Then L is C-bounded if and only if $g_C(L)$ is finite and L is e-C-bounded where 1^e is the length of the longest word in $g_C(L)$.*

Proof. If L is C-bounded, then Lemma 18 implies $g_C(L)$ is finite. Let 1^e be the length of the longest word output by $g_C(L)$. Thus, L is e-C-bounded by Lemma 19.

If $g_C(L)$ is finite and L is e-C-bounded where 1^e is the length of the longest word in $g_C(L)$, then L is C-bounded by Lemma 15. □

Theorem 21. *Let $\mathcal{L}$ be a full trio with all properties effective and a decidable finiteness problem. It is decidable, given $L \in \mathcal{L}$ and C finite, $L \subseteq \Sigma^*, C \subseteq \Sigma^+$ with $\Sigma \subseteq C$, whether L is C-bounded. In particular, this result holds for any family in Remarks 2 and 3.*

Proof. First, every full trio is closed under gsm mappings, and therefore it is possible to construct $g_C(L) \in \mathcal{L}$.

By Proposition 1, we can determine, given L, whether $g_C(L)$ is finite and if so, to determine the length of the longest word 1^e, and test whether L is e-C-bounded by Lemma 16. Hence, it is decidable whether L is C-bounded by Proposition 20. □

Since for every $l \geq 1$, $\Sigma^{\leq l}$ is a finite language such that $\Sigma \subseteq \Sigma^{\leq l}$, the following is true.

Corollary 22. *Let $\mathcal{L}$ be a full trio with all properties effective and a decidable finiteness problem. It is decidable, given $L \in \mathcal{L}$ and $l \geq 1$ with $L \subseteq \Sigma^*$, whether L is $\Sigma^{\leq l}$-bounded. Hence, letter-boundedness is decidable. In particular, this result holds for any family in Remarks 2 and 3.*

This can be thought of as the finite version of the boundedness problem; that is, the boundedness problem is equivalent to deciding whether a language is Σ^+-bounded, and here we have all words up to a given length. The normal version of the boundedness problem is decidable for context-free languages and is unknown for most incomparable families, but for the finite version, it is decidable for every full trio with a decidable finiteness problem (and as mentioned, to our knowledge, every full trio that has been studied with a decidable emptiness problem). Therefore, this property is quite general. And, this finite-version of the boundedness problem is general enough to always show decidability of the letter-boundedness problem for full trios with a decidable finiteness problem.

The second special case is where C is finitely-decodable relative to L. Recall the languages $C^{(d)}$ and $C^{(\geq d)}$, and the homomorphisms h_d and $h_{i,d}$ from Sect. 3. For $C^{(d)}$, this only contains d-decodable words, and each decoding is annotated with a distinct symbol i in the subscript of symbols of V_d. For $w \in (\Sigma \cup V_d)^*$, we can extend the functions $\text{count}_C(w)$, $\text{mincount}_C(w)$, and $\text{maxcount}_C(w)$ so that they ignore the symbols of V_d. By Propositions 6 and 10, if we could determine if $\text{mincount}_{C^{(d)}}(L)$ is finite, we could determine $C^{(d)}$-boundedness. Although we cannot obviously compute $\text{mincount}_{C^{(d)}}(L)$ with a gsm, we can use reversal-bounded counters to compute the minimum, and it is known that for a semilinear full trio, there is another one closed under intersection with NCM languages [23], and we can compute finiteness of $\text{mincount}_{C^{(d)}}(L)$ within the larger family. Altogether, we obtain the following general result (proof omitted due to space constraints).

Proposition 23. *Let $\mathcal{L}$ be a full trio which is either*

- *semilinear, or*
- *length-semilinear, and closed under intersection with* NCM*,*

with all properties effective. It is decidable, given $L \in \mathcal{L}$, C finite such that C is finitely-decodable relative to L, whether L is C-bounded. In particular, this result holds for any family in Remarks 2 and for context-free matrix grammars.

All families in Remark 2 are semilinear full trios with a decidable finiteness problem. And as mentioned in Remark 3, the matrix languages are a length-semilinear full trio with a decidable finiteness problem that is closed under intersection with NCM languages.

However, the problem is still open for C finite where C is not finitely-decodable relative to L (where L is in various important families).

5 Conclusions and Future Directions

In this paper, we studied generalizations of codes, and the notion of a language C being d-decodable relative to a language L (each word in L has at most d factorizations in terms of words of C). While it is undecidable to decide the property of whether very simple classes of non-regular languages are d-decodable (or are codes) relative to regular languages, it is often decidable to decide whether regular languages are codes relatives to languages in quite general language families. This is indeed the case when they are in any full trio (with closure properties providing effective constructions) with a decidable emptiness problem.

Next, we studied generalizations of bounded languages, where a language L is C-bounded if $L \subseteq w_1^* \cdots w_n^*$ with $w_1, \ldots, w_n \in C$. It was shown that for any full trio $\mathcal{L}$ with a decidable finiteness problem, then it is decidable, given a finite language C and $L \in \mathcal{L}$ where C is a code relative to L, whether L is C-bounded. The condition of being a code relative to L can be weakened to finitely-decodable relative to L and remain decidable for certain general families including for semilinear full trios or even for families such as the matrix languages. Finally, it was shown that if $\mathcal{L}$ is any full trio with a decidable finiteness problem, then it is decidable, given $L \in \mathcal{L}$ and $l \geq 1$, whether L is $\Sigma^{\leq l}$ bounded. Hence, letter-boundedness is always decidable for these families without any need for ad hoc constructions.

Many open problems remain. For various non-regular language families $\mathcal{L}$, is it decidable, given a regular language C (or even finite C), and $L \in \mathcal{L}$, whether

C is finitely-decodable relative to L. It is also open to determine whether it is decidable, given a regular language (or even finite language) $C \subseteq \Sigma^+$ and a language L from a full trio or a standard family with a decidable finiteness problem, whether L is C-bounded. Also for families that are not length-semilinear, but are full trios with a decidable finiteness problem, is it decidable, given C finite and finitely-decodable relative to L, whether L is C-bounded. Finally, decidability of the boundedness problem is open for all of the families in Remark 3 and some in Remark 2 (multi-push-down machines and finite-index ET0L).

References

1. Aho, A.: Indexed grammars – an extension of context-free grammars. J. ACM **15**, 647–671 (1968)
2. Baumann, P., et al.: Unboundedness problems for machines with reversal-bounded counters. In: Lecture Notes in Computer Science. Proceedings of the 26th International Conference on Foundations of Software Science and Computation Structures, FoSSaCS 2023, vol. 13992, pp. 240–264 (2023)
3. Breveglieri, L., Cherubini, A., Citrini, C., Reghizzi, S.: Multi-push-down languages and grammars. Int. J. Found. Comput. Sci. **7**(3), 253–291 (1996)
4. Handbook of Formal Languages. Springer, Heidelberg (1997). https://doi.org/10.1007/978-3-642-59136-5_9
5. Czerwinski, W., Hofman, P., Zetzsche, G.: Unboundedness problems for languages of vector addition systems. In: Chatzigiannakis, I., Kaklamanis, C., Marx, D., Sannella, D. (eds.) 45th International Colloquium on Automata, Languages, and Programming (ICALP 2018). Leibniz International Proceedings in Informatics (LIPIcs), vol. 107, p. 119. Schloss Dagstuhl–Leibniz-Zentrum fuer Informatik, Dagstuhl, Germany (2018)
6. D'Alessandro, F., Ibarra, O.H., McQuillan, I.: On finite-index indexed grammars and their restrictions. Inf. Comput. **279**, 104613 (2021)
7. Daley, M., Jürgensen, H., Kari, L., Mahalingam, K.: Relativized codes. Theoret. Comput. Sci. **429**, 54–64 (2012)
8. Dassow, J., Păun, G., Salomaa, A.: Handbook of Formal Languages, vol. 2, chap. Grammars with Controlled Derivations. Springer-Verlag, Berlin (1997)
9. Drewes, F., Engelfriet, J.: Decidability of the finiteness of ranges of tree transductions. Inf. Comput. **145**, 1–50 (1998)
10. Engelfriet, J.: Iterated stack automata and complexity classes. Inf. Comput. **95**, 21–75 (1991)
11. Eremondi, J., Ibarra, O.H., McQuillan, I.: On the density of context-free and counter languages. In: Potapov, I. (ed.) DLT 2015. LNCS, vol. 9168, pp. 228–239. Springer, Cham (2015). https://doi.org/10.1007/978-3-319-21500-6_18
12. Fernau, H., Reinhardt, K., Staiger, L.: Decidability of code properties. RAIRO — Theor. Inform. Appl. **41**, pp. 243–259 (2007)
13. Ginsburg, S.: Algebraic and Automata-Theoretic Properties of Formal Languages. North-Holland Publishing Company, Amsterdam (1975)
14. Ginsburg, S., Spanier, E.: Bounded Algol-like languages. Trans. Am. Math. Soc. **113**(2), 333–368 (1964)
15. Ginsburg, S.: The Mathematical Theory of Context-Free Languages. McGraw-Hill Inc, New York, NY, USA (1966)

16. Greibach, S.: Checking automata and one-way stack languages. J. Comput. Syst. Sci. **3**(2), 196–217 (1969)
17. Greibach, S.A.: Minimal linear grammars. Inf. Control **6**, 119–125 (1963)
18. Greibach, S.A.: One way finite visit automata. Theoret. Comput. Sci. **6**, 175–221 (1978)
19. Hauschildt, D., Jantzen, M.: Petri net algorithms in the theory of matrix grammars. Acta Informatica **31**, 719–728 (1994)
20. Holzer, M., Kutrib, M.: Flip-pushdown automata: $k + 1$ pushdown reversals are better than k. In: Baeten, J.C.M., Lenstra, J.K., Parrow, J., Woeginger, G.J. (eds.) Automata, Languages and Programming: 30th International Colloquium, ICALP 2003 Eindhoven, The Netherlands, June 30 - July 4, 2003 Proceedings, pp. 490–501. Springer, Berlin Heidelberg (2003)
21. Hopcroft, J.E., Ullman, J.D.: Introduction to Automata Theory, Languages, and Computation. Addison-Wesley, Reading, MA (1979)
22. Ibarra, O.H.: Reversal-bounded multicounter machines and their decision problems. J. ACM **25**(1), 116–133 (1978)
23. Ibarra, O.H., McQuillan, I.: Semilinearity of families of languages. Int. J. Found. Comput. Sci. **31**(8), 1179–1198 (2020)
24. Ibarra, O.H., McQuillan, I.: Techniques for showing the decidability of the boundedness problem of language acceptors. In: Lecture Notes in Computer Science. Proceedings of the 28th International Conference on Developments in Language Theory, DLT 2024, vol. 14791, pp. 156–172 (2024)
25. Handbook of Formal Languages. Lecture Notes in Computer Science, Springer, Heidelberg (1997). https://doi.org/10.1007/978-3-642-59136-5_9
26. Ko, S.K., Han, Y.S., Salomaa, K.: Generalizations of code languages with marginal errors. Int. J. Found. Comput. Sci. **32**(05), 509–529 (2021)
27. Rozenberg, G., Vermeir, D.: On ET0L systems of finite index. Inf. Control **38**, 103–133 (1978)

Turn Complexity of Context-Free Languages, Pushdown and One-Counter Automata

Giovanni Pighizzini(✉)

Dipartimento di Informatica, Università degli Studi di Milano, Milan, Italy
pighizzini@di.unimi.it

Abstract. A turn in a computation of a pushdown automaton is a switch from a phase in which the height of the pushdown store increases to a phase in which it decreases. Given a pushdown or one-counter automaton, we consider for each string in its language the *minimum number of turns made in accepting computations*. We prove that it cannot be decided if this number is bounded by any constants. Furthermore, we obtain a non-recursive trade-off between pushdown and one-counter automata accepting in a finite number of turns and finite-turn pushdown automata, that are defined requiring that the constant bound is satisfied by *each accepting computation*. We prove that there are languages accepted in a sublinear but not constant number of turns, with respect to the input length. Furthermore, there exists an infinite proper hierarchy of complexity classes, with the number of turns bounded by different sublinear functions. In addition, there is a language requiring a number of turns which is not constant but grows slower than each of the functions defining the above hierarchy.

1 Introduction

Recently, we investigated two complexity measures related to context-free languages, pushdown automata, and one-counter automata. The *height of the pushdown* measures the space used in the pushdown store [13]. In the restricted case of one-counter automata, in which the pushdown is used only to represent a counter written in unary, this measure corresponds to the maximum value reached by the counter [12]. The *push complexity* takes into account the number of push operations used in computations [11]. It should be clear that when one of these measures is bounded by a constant, namely it does not depend on the input length, then the number of possible configurations of the pushdown store is finite. Hence, the store can be encoded in the finite control. This implies the regularity of the accepted language.

The non-constant case is more interesting. First of all, there are fundamental differences if the costs of all computations are taken into account, using the

G. Pighizzini: The author is a member of the Gruppo Nazionale Calcolo Scientifico-Istituto Nazionale di Alta Matematica (GNCS-INdAM).

S.-K. Ko and F. Manea (Eds.): DLT 2025, LNCS 16036, pp. 77–91, 2026.
https://doi.org/10.1007/978-3-032-01475-7_6

so-called strong measure, or only the less expensive accepting computation is considered for each accepted input (weak measure). In the strong measure, if one of the above-mentioned resources is not bounded by any constant, then an amount that linearly grows with respect to the length of the input is required [11, 12]. Similar results hold if the costs of *all accepting* computations (instead of those of all computations) are taken into account (accept measure) [12].

The situation for the weak measure is more complicated, but more challenging. We point out that this measure is strictly related to the idea on nondeterminism in a broad sense: a nondeterministic machine is not only able to make choices that lead to acceptance, if an accepting strategy does exist, but it is also able of taking choices that lead to the *less expensive* accepting strategy. For the two resources mentioned at the beginning, it has been proved that if a machine uses an amount which is not bounded by any constants, then this amount should grow at least as a double logarithmic function in the length of the input. This lower bound is optimal for input alphabets of at least two symbols, while different and higher bounds (between a logarithmic and a linear amount, depending on the resource) have been obtained in the unary case.

In this paper, we continue this line of investigation by considering a further resource, namely the number of *turns* made in computations. Roughly, a pushdown automaton makes a turn during a computation when a sequence of moves in which the height of the store increases is followed by a sequence in which the height decreases. In a pioneering paper on formal languages and pushdown automata, Ginsburg and Spanier started the investigation of the role of turns in pushdown automata computations [2]. In particular, they introduced *finite-turn pushdown automata*, namely pushdown automata for which there exists a constant limiting the number of turns made in *each accepting computation*. The class of languages accepted by these devices is a proper subclass of context-free languages and properly contains the class of regular languages. Even restricting to one-counter automata that makes only one turn, the resulting class of languages is wider than regular languages. In particular, 1-turn pushdown automata characterize the class of *linear context-free languages*.

These facts already reveal a main difference between the number of turns and resources as pushdown height and push complexity: one turn is enough to have more power than only with a finite control. In the same paper, Ginsburg and Spanier presented several interesting results. Among them, they proved that it is decidable if a pushdown automaton is finite turn and if it is k-turn, for a given k. They also obtained a characterization of languages accepted by finite-turn pushdown automata in terms of a special kind of context-free grammars, called *ultralinear grammars*. For this reason, the class of languages accepted by finite-turn automata is called class of *ultralinear languages*. We point out that *all accepting* computations of the machine are taken into account, namely the accept measure is considered.

It is quite natural to ask what happens by considering the weak measure, i.e., if for each accepted input, *only* an accepting computation using the lowest number of turns is taken into account. In the following, when we say that a machine *accepts in* a certain number of turns, we always refer to the weak measure. Ginsburg and Rice observed that when a machine accepts in a constant number k of turns, namely for each accepted input there is an accepting computation containing at most k turns, then the machine can be modified in order to stop when the number of turns exceeds k. The resulting *k-turn pushdown automaton* is equivalent to the original one. So, even under the weak measure, the class of languages accepted in a finite number of turns coincides with the class of ultralinear languages. However, in contrast with the fact that it can be decided whether a pushdown automaton is finite turn (accept measure), Ginsburg and Rice also proved that it is not decidable whether a pushdown automaton accepts an ultralinear language [2].

Going deeper into this investigation, in the first part of this paper, we prove that it is also undecidable whether a pushdown automaton accepts in a finite number of turns (weak measure).[1] Actually we prove that the property is undecidable already in the restricted case of one-counter automata. We also prove that it cannot be decided if a one-counter automaton accepts in k turns, for a given $k > 0$. Only acceptance in 0 turns is decidable for these devices as well as for pushdown automata.

We mentioned that each pushdown automaton accepting in k turns can be modified to obtain an equivalent k-turn pushdown automaton. However, as we prove in the paper, the value k cannot be bounded by any recursive function in the size of the given machine. As a consequence, we are able to prove that the size trade-off between pushdown automata accepting in a constant number of turns and finite-turn pushdown automata is nonrecursive.[2] [3]

The second part of the paper is devoted to study how the number of turns can grow with respect to the length of the input, still considering the weak measure. It is easy to give examples of languages accepted in a constant number of turns and of languages accepted in a linear number of turns. Here, we show the existence of languages accepted in a sublinear number of turns, for several sublinear functions. First, we prove that there is a language accepted by a one-counter automaton in $O(\sqrt{n})$ turns, where n is the length of the input. Furthermore, the same language cannot be accepted in $o(\sqrt[3]{n})$ turns, even using a pushdown automaton. We then prove that there are machines and languages using even a lower number of turns. In particular, for each integer k, we prove the existence of a language L_k accepted by a one-counter automaton in $\lg^{(k)} n$ turns but that cannot be accepted by any pushdown automaton in a number of turns growing less than $\lg^{(k)} n$.[4] This result reveals a big difference with pushdown height and push

---

1 We point out that this problem is different from the previous one. Indeed, the fact that a pushdown automaton accepts a ultralinear language does not imply that it accepts in a finite number of turns.

2 For a discussion on nonrecursive trade-offs we address the reader to [6].

3 Further results concerning the role of turns in the computations of pushdown automata, in both nondeterministic and deterministic cases, have been obtained by Malcher [8] and further expanded in [9].

4 $\lg^{(k)}$ denotes the composition of the base 2 logarithm with itself k times.

complexity, for which there are double logarithmic lower bounds when they are not constant, and defines a proper infinite hierarchy of complexity classes with respect to the number of turns.

It is possible to go further below: we show the existence of a language U_* which is accepted by a one-counter automaton in $\lg^* n$ turns (the *iterated logarithm* of n, a function growing slower than $\lg^{(k)} n$, for any k), but that cannot be accepted in a constant number of turns.

For brevity reasons, some of the proofs are omitted or just outlined in this version of the paper.

2 Preliminaries

We assume that the reader is familiar with the standard notions of automata and formal language theory. Given an alphabet Σ, the set of strings over Σ is denoted by Σ^*, with the empty string denoted by ε. The length of a string $x \in \Sigma^*$ is denoted as $|x|$. If $\Delta \subseteq \Sigma$, then the number of occurrences of symbols from Δ in x is indicated as $|x|_\Delta$. Given a set S, its cardinality is denoted as $\#S$ and the family of its subsets as 2^S.

Let $\log_2 n$ denote the logarithm in base 2 of $n > 0$. We consider the following modification of the logarithm, denoted as $\lg n$, where $n \geq 0$: $\lg n = \log_2 n$, when $n \geq 1$; $\lg n = 0$, otherwise.

Given an integer $k \geq 0$, $\lg^{(k)}$ denotes the function obtained by composing k times the function lg with itself, namely $\lg^{(k)} n = \underbrace{\lg \lg \cdots \lg}_{k \text{ times}} n$. We point out that $\lg^{(0)}$ is the identity function.

The *iterated logarithm* is the function $\lg^* : \mathbb{N} \to \mathbb{N}$ defined as $\lg^* n = \min\{k \mid \lg^{(k)} n \leq 1\}$. This function grows very slowly. For instance $\lg^* n = 4$ for $16 < n \leq 65536$, while $\lg^* n = 5$ for $65536 < n \leq 2^{65536}$.

We shortly remember the notion of pushdown automata as presented in [2,5]. A *pushdown automaton* (PDA, for short) is defined as $\mathcal{M} = (Q, \Sigma, \Gamma, \delta, q_0, Z_0)$, where Q, Σ, and Γ are finite sets: the set of states, the input alphabet, the pushdown alphabet, respectively; $q_0 \in Q$ is the initial state; $Z_0 \in \Gamma$ is the start symbol on the pushdown store; δ is the *transition function* from $Q \times (\Sigma \cup \{\varepsilon\}) \times \Gamma$ to the finite subsets of $Q \times \Gamma^*$, where $(p, \alpha) \in \delta(q, \sigma, A)$, with $p, q \in Q$, $\sigma \in \Sigma \cup \{\varepsilon\}$, $A \in \Gamma$, $\alpha \in \Gamma^*$, means that the automaton in the state q, with A on the top of the pushdown store, reading the symbol $\sigma \in \Sigma$ from the tape, or without reading any symbol when $\sigma = \varepsilon$, can move in the state p after replacing A on the top of the stack with the string α. We assume *acceptance by empty store*, namely an input w is accepted when there exists a computation starting from the initial configuration (input head on the first symbol of w, finite control in the state q_0, pushdown store containing only Z_0) that, according to the transition function, reaches a configuration in which all the input has been scanned (i.e., the input head is to the right of the rightmost symbol of w) and the pushdown store is empty. It is well known that the class of language accepted by these devices coincides with the class of *context-free languages*.

A *turn* in a computation $\mathcal{C}$ of a PDA is a sequence of moves in which at the beginning the height of the pushdown increases, then it does not change for some moves, and finally decreases, namely it is defined by $k + 1$ consecutive configurations, with $k > 1$, in which the pushdown contents are $\gamma_0, \gamma_1, \ldots, \gamma_k$, and $|\gamma_0| < |\gamma_1| = |\gamma_2| = \cdots = |\gamma_{k-1}| > |\gamma_k|$.

According to [2], a PDA $\mathcal{M}$ is said to be *k-turn*, for an integer $k \geq 0$, when *each accepting computation* of $\mathcal{M}$ contains at most k turns. A PDA $\mathcal{M}$ is *finite turn* if it is k-turn for some $k \geq 0$. It is easy to see that 0-turns PDAs recognize only regular languages. Furthermore, the class of languages accepted by 1-turn PDAs is the class of *linear languages*. This class is strictly included in the class of languages accepted by finite turns PDAs, which is the class of *ultralinear languages*, a proper subclass of context-free languages.

In the above definitions, the number of turns in *each accepting computation* is taken into account. Since in a nondeterministic machine several computations using different numbers of turns could be possible, we are going to introduce a different notion. In particular, we are going to consider the number of turns which are sufficient to accept input strings. More precisely, we give the following definition:

Definition 1. Let $\mathcal{M}$ be a PDA accepting a language L. We say that *a string* $w \in L$ *is accepted in* k *turns*, for an integer $k \geq 0$, when there exists a computation of $\mathcal{M}$ accepting w and containing at most k turns.

Given a function $t : \mathbb{N} \to \mathbb{N}$, we say that $\mathcal{M}$ accepts L in $t(n)$ turns, when for each $w \in L$, with $|w| = n$, there is an accepting computation containing at most $t(n)$ turns. Furthermore, if the function $t(n)$ is bounded by a constant, then we say that $\mathcal{M}$ accepts in a finite number of turns.

If we know that a PDA accepts in k turns, for some given $k \geq 0$, then we can always obtain an equivalent k-turn PDA just counting in the finite control the number of turns and stopping the computation when such a number exceeds k. However, as we will see, acceptance in k turns cannot be decided, unless $k = 0$.

According to [3], the size of a PDA is the number of symbols used to write down its description and it depends on the cardinalities of its alphabets and set of states, and of the maximum number of pushdown symbols that appear on the right-hand sides of transition rules.

A *one-counter automaton* (OCA, for short) is a pushdown automaton in which the pushdown alphabet consists only of one symbol A, besides Z_0. The symbol Z_0 is used only to mark the bottom. In this way, the pushdown store always contains a string of the form $A^k Z_0$, representing the integer $k \geq 0$, with the exception of accepting configurations, in which the store is empty. It is well known that the class of languages recognized by OCAs is a proper subclass of context-free languages and properly contains the class of regular languages.

In the paper, we use the language $\mathsf{Eq} = \{a^n b^n \mid n > 0\}$, which is a linear context-free language accepted by a trivial deterministic OCA $\mathcal{A}_{\mathsf{Eq}}$ that makes exactly one turn in each accepting computation. By $\mathcal{A}_{\mathsf{Eq}^*}$ we denote the trivial deterministic OCA accepting Eq^* and making exactly k turns to accept each string in Eq^k, $k > 0$. It is folklore that k turns are necessary and sufficient to accept Eq^k, for any fixed $k \geq 0$. Furthermore, each PDA accepting Eq^* should make k turns to accept any string of the form $a^{n_1}b^{n_1}a^{n_2}b^{n_2}\cdots a^{n_k}b^{n_k}$, for sufficiently large $n_1, n_2, \ldots, n_k$. We give a generalization of these lower bounds that will be used several times in the paper.

Lemma 1. *Let $\Sigma = \Delta \cup \{a, b\}$ be an alphabet, with $a, b \notin \Delta$, and let $\mathcal{M}$ be a* PDA *accepting a language $L \subseteq \Sigma^*$. There exists an integer $N > 0$ such that for each integer $k > 0$ and strings $y_0, y_1, \ldots, y_k \in \Delta^*$, if the language $L' = y_0\{a,b\}^*y_1\{a,b\}^*\cdots y_{k-1}\{a,b\}^*y_k \cap L$ is a subset of $y_0\mathsf{Eq}y_1\mathsf{Eq}\cdots y_{k-1}\mathsf{Eq}y_k$ then, for $n_1, n_2, \ldots, n_k \geq N' = N\cdot(|y_0y_1\cdots y_k|+1)$, each accepting computation of $\mathcal{M}$ on input $y_0a^{n_1}b^{n_1}y_1a^{n_2}b^{n_2}\cdots y_{k-1}a^{n_k}b^{n_k}y_k$ contains at least k turns.*

3 Undecidability and Non-Recursive Trade-Offs

Ginsburg and Spanier proved that it is decidable whether a PDA is finite turn or whether it is k-turn for any given integer $k \geq 0$ [2]. We underline that all the accepting computations are considered to count the turns. So, for each accepted input, the bound must be satisfied by the most expensive accepting computation.

Here, we consider the number of turns which are *sufficient* to accept a given input. Hence, the bound must be satisfied, for each accepted input, by the accepting computation using the lowest number of turns. With this measure, we obtain completely different results. In particular, all the above mentioned questions become undecidable, except acceptance in 0 turns. We will also prove a non-recursive trade-off result, by showing that in the case of acceptance in a number of turns bounded by a value k, constant with respect to the length of the input, such a value cannot be bounded by any recursive function in the size of the description of the machine. This will allow to prove that the trade-offs between OCAs and PDAs accepting in a finite number of turns and finite-turn PDAs are not recursive.

Let us start by presenting the undecidability results, that are obtained by adapting a well-known technique introduced by Hartmanis and which is based on encodings of single-tape Turing machine computations [4].

Configurations of a single-tape Turing machine $\mathcal{T}$ with state set Q and tape alphabet Γ are denoted, in a standard way, as strings in $\Gamma^*Q\Gamma^*$. A computation is encoded as a string $\alpha_1\$\alpha_2\$\cdots\alpha_m\$$, where the blocks α_i, $i = 1, \ldots, m$, are encodings of configurations of $\mathcal{T}$ and $\$ \notin Q \cup \Gamma$ is a delimiter. Hence, the (encoding of a) *valid computation* of $\mathcal{T}$ on input w is a string $\alpha_1\$\alpha_2\$\cdots\alpha_m$, for some integer $m \geq 1$, such that:

1. $\alpha_i \in \Gamma^*Q\Gamma^*$, i.e., α_i encodes a configuration of $\mathcal{T}$, $i = 1, \ldots, m$;
2. α_1 encodes the initial configuration on input w by the string q_Iw, where q_I is the initial state of $\mathcal{T}$;
3. α_m encodes a halting configuration of $\mathcal{T}$, namely a configuration from which no move is possible;
4. α_{j+1} is reachable in one step from α_j, $j = 1, \ldots, m-1$.

Let us denote by $\mathsf{Valid}(\mathcal{T}, w)$ the set of valid computations of $\mathcal{T}$ on input w and by $\mathsf{Valid}(\mathcal{T}, w)^c$ its complement.

It can be proved that for fixed $\mathcal{T}$ and w, the set $\mathsf{Valid}(\mathcal{T}, w)^c$ is accepted by a 1-turn OCA. This allows us to obtain:

Theorem 1. *The following properties are undecidable:*

- *whether a* OCA *accepts in a finite number of turns,*
- *whether a* OCA *accepts in k turns, for a given integer $k > 0$.*

Proof. We give a reduction from the blank-tape halting problem, namely the problem of deciding whether a Turing machine halts on the empty tape. Given a Turing machine $\mathcal{T}$, let us consider the language

$$L = \{x\$y \mid x \in \{a,b\}^* \text{ and } (x \in \mathsf{Eq}^* \text{ or } y \notin \mathsf{Valid}(\mathcal{T}, \varepsilon))\}.$$

We build a OCA $\mathcal{M}$ that given an input z of the form $x\$y$, with $x \in \{a,b\}^*$, decides if $z \in L$ by nondeterministically choosing between verifying that $x \in \mathsf{Eq}^*$ (this is done by simulating the OCA $\mathcal{A}_{\mathsf{Eq}^*}$), or verifying that $y \notin \mathsf{Valid}(\mathcal{T}, \varepsilon)$, which is done by simulating a 1-turn OCA accepting $\mathsf{Valid}(\mathcal{T}, \varepsilon)^c$.

If $\mathcal{T}$ halts on input ε, then fix $\hat{y} \in \mathsf{Valid}(\mathcal{T}, \varepsilon)$. For each $k > 0$, the only accepting computation on input $(ab)^k\$\hat{y}$ uses k turns. So the number of turns used by $\mathcal{M}$ to accept cannot be bounded by any constant.

However, if $\mathcal{T}$ does not halt on input ε, then $\mathsf{Valid}(\mathcal{T}, \varepsilon)$ is empty and, hence, each input $z = x\$y$, with $x \in \{a,b\}^*$, is accepted by $\mathcal{M}$ in a computation that verifies that $y \in \mathsf{Valid}(\mathcal{T}, \varepsilon)^c$ using at most one turn.

Hence, we can conclude that $\mathcal{M}$ accepts in a finite number of turns if and only if $\mathcal{T}$ does not halt on input ε, which is undecidable. So, it cannot be decided if a OCA accepts in a finite number of turns.

The second statement follows from the fact that when the above-described machine $\mathcal{M}$ accepts in a finite number of turns, then it accepts in one turn. □

In contrast to the second statement of Theorem 1, for $k = 0$ we prove:

Theorem 2. *It is decidable whether a* PDA *$\mathcal{M}$ accepts in 0 turns.*

Proof. By removing from $\mathcal{M}$ all the transitions making push or pop operations, we obtain a finite automaton $\mathcal{A}$ accepting all the strings that are accepted by $\mathcal{M}$ without using any turn. Then $\mathcal{M}$ accepts in 0 turns if and only if $L(\mathcal{A}) = L(\mathcal{M})$ that, being $L(\mathcal{A}) \subseteq L(\mathcal{M})$, is equivalent to $L(\mathcal{M}) \cap (L(\mathcal{A}))^c = \emptyset$. Since context-free languages are effectively closed by intersection with regular languages and emptiness is decidable for context-free languages, we conclude that the property is decidable. □

The last part of this section is devoted to investigate size trade-offs.

Theorem 3. *There exists a function $T : \mathbb{N} \to \mathbb{N}$ that cannot be bounded by any recursive function, such that for infinitely many integers s there exists a context-free language L_s having a description of size s such that:*

– *There exists a* OCA *accepting* L_s *in* $T(s)$ *turns.*[5]
– *Each* PDA *accepting* L_s *uses at least* $T(s)$ *turns to accept strings of length* n*, for infinitely many integers* n*.*

Proof. The argument is derived from [10, Prop. 7]. Let us start by introducing the following modification of the complement of $\mathsf{Valid}(\mathcal{T}, w)$, in which each occurrence of the delimiter \$ is replaced by a string in $\{a, b\}^+$. For technical reasons and without loss of generality, let us suppose $a, b \notin \Gamma \cup Q \cup \{\$\}$:

$$\begin{aligned}\mathsf{pNotValid}(\mathcal{T}, w) = \{\alpha_1 z_1 \alpha_2 z_2 \cdots \alpha_m z_m \mid\ & m > 0,\\ & \alpha_i \in (\Gamma \cup Q)^+, z_i \in \{a, b\}^+,\ i = 1, \ldots, m, \text{ and:}\\ & \text{(a) } \alpha_1\$\alpha_2\$\cdots\alpha_m\$ \in \mathsf{Valid}(\mathcal{T}, w)^c, \text{ or}\\ & \text{(b) } \forall i, 1 \le i \le m,\ z_i \in \mathsf{Eq}\}.\end{aligned}$$

By adapting the arguments used to prove Theorem 1, we define a OCA $\mathcal{M}$ that accepts $\mathsf{pNotValid}(\mathcal{T}, \varepsilon)$ with the following strategy. At the beginning of the computation, on input $x = \alpha_1 z_1 \alpha_2 z_2 \cdots \alpha_m z_m$, $\mathcal{M}$ nondeterministically decides to verify one of the two conditions (a) and (b). Condition (a) can be tested, as described in the proof of Theorem 1, using at most one turn. To test condition (b), the input is scanned while simulating the OCA $\mathcal{A}_{\mathsf{Eq}}$ on each block z_i. This uses exactly m turns. We point out that the size of the description of $\mathcal{M}$ is polynomial in the number of states of the given machine $\mathcal{T}$.

Let us consider any halting single-tape deterministic Turing machine $\mathcal{T}_s$ with a description of size s. Let $\mathcal{C}_s$ be the encoding of the computation of $\mathcal{T}_s$ on ε. It is well known that the length of $\mathcal{C}_s$ cannot be bounded by any recursive function in the number of states of $\mathcal{T}_s$ and, so, in s.

Let us denote by $\mathcal{M}_s$ the OCA accepting $\mathsf{pNotValid}(\mathcal{T}_s, \varepsilon)$ obtained according to the previous construction.

Suppose that $\mathcal{T}_s$ halts on ε. Being $\mathcal{T}_s$ deterministic, there is only one computation $\mathcal{C}_s$ on ε. This computation is described by a string $\alpha_1\$\alpha_2\$\cdots\alpha_m\$ \in \mathsf{Valid}(\mathcal{T}_s, \varepsilon)$. This implies that each string $x = \alpha_1 z_1 \alpha_2 z_2 \cdots \alpha_m z_m$, with $z_i \in \mathsf{Eq}$, for $i = 1, \ldots, m$, satisfies *only* condition (b). Hence it requires m turns to be accepted, regardless its length. We take $T(s) = m$. Since, the number m of configurations occurring in $\mathcal{C}_s$ cannot be bounded by any recursive function in s, it follows that the function $T(s)$ cannot be bounded by any recursive function in s.

To prove the lower bound on the number of turns for PDAs accepting L_s, we consider again the encoding $\alpha_1\$\alpha_2\$\cdots\alpha_m\$$ of the computation $\mathcal{C}_s$ on ε. The intersection of $\alpha_1\{a, b\}^*\alpha_2\{a, b\}^* \cdots \alpha_m\{a, b\}^*$ with $\mathsf{pNotValid}(\mathcal{T}_s, \varepsilon)$ coincides with $\alpha_1\mathsf{Eq}\alpha_2\mathsf{Eq}\cdots\mathsf{Eq}\alpha_m\mathsf{Eq}$. Hence, according to Lemma 1, for each PDA accepting $\mathsf{pNotValid}(\mathcal{T}_s, \varepsilon)$, m turns are necessary to recognize any input of the form $\alpha_1 z \alpha_2 z \cdots \alpha_m z$, with $z = a^K b^K$, for each sufficiently large K. □

[5] We point out that the parameter s is the size of the description of the language L_s. Hence, $T(s)$ is constant with respect to the input length. This means that for each string in L_s the OCA has an accepting computation making at most $T(s)$ turns.

In contrast to Theorem 3, by considering all accepting computations, we obtain a recursive bound:

Theorem 4. *There exists a recursive function f such that if $\mathcal{M}$ is a k-turn* PDA *then $k \leq f(s)$, where s is the size of $\mathcal{M}$.*

As a consequence of Theorems 3 and 4 we obtain:

Corollary 1. *The trade-offs between* OCA*s and* PDA*s accepting in a constant number of turns and finite-turn* PDA*s are not recursive.*

4 A Sublinear Number of Turns

In this section, we present a language $\mathsf{L_{sq}}$ which is accepted in a sublinear number of turns. This language is a 'padded' version of a language presented in [1], whose strings of length n can be accepted by a OCA using a counter bounded by $O(\sqrt{n})$. We will prove that $\mathsf{L_{sq}}$ is accepted by a OCA which uses $O(\sqrt{n})$ turns. Furthermore, this bound cannot be significantly improved, even using a PDA.

The language $\mathsf{L_{sq}}$ is over the alphabet $\{0, a, b\}$. Roughly speaking, a string belongs to it if the sequence of its maximal factors in 0^+ does not coincide with the sequence of unary representations of all integers from 1 to some $m \geq 1$, or all its maximal factors in $\{a, b\}^+$ belong to the language Eq:

$$\begin{aligned}\mathsf{L_{sq}} = \{x_1 z_1 x_2 z_2 \cdots x_m z_m \mid\ & m > 0, x_i \in \{0\}^+, z_i \in \{a,b\}^+, i = 1, \ldots, m, \text{ and:}\\ & \text{(a) } \exists i, 1 \leq i \leq m, |x_i| \neq i, \text{ or}\\ & \text{(b) } \forall i, 1 \leq i \leq m,\ z_i \in \mathsf{Eq}\,\}\,.\end{aligned}$$

Theorem 5. *The language $\mathsf{L_{sq}}$ is accepted by a* OCA *in $O(\sqrt{n})$ turns. Furthermore, each* PDA *cannot accept $\mathsf{L_{sq}}$ using $o(\sqrt[3]{n})$ turns.*

Proof. Given a string $w = x_1 z_1 x_2 z_2 \cdots x_m z_m$, a OCA $\mathcal{A}$ can decide if $w \in \mathsf{L_{sq}}$ by nondeterministically selecting one between the conditions (a) and (b), and verifying it. (We are supposing that w is in the appropriate format. This can be tested using the finite control while scanning w.)

Condition (a) can be tested by nondeterministically choosing a block x_i, with the aim of verifying that it is the first one with $|x_i| \neq i$. If $i = 1$, then the verification is done by using the finite control. Otherwise, the choice of the block is actually done when $\mathcal{A}$ starts to scan the previous block x_{i-1}: its length is saved in the counter to be compared with that of x_i, after performing one turn.

Condition (b) can verified by simulating on each block z_i the OCA $\mathcal{A}_{\mathsf{Eq}}$ accepting Eq. This uses one turn for each z_i, hence m turns.

Since to accept strings satisfying (a) one turn is enough, to obtain the total number of turns we estimate the number m of blocks with respect to the length n of any string w for which *only* condition (b) is satisfied, namely $w = 0^1 z_1 0^2 z_2 \cdots 0^m z_m$, with $m > 0$ and $z_1, z_2, \ldots, z_m \in \mathsf{Eq}$. This gives $n = |w| > \sum_{i=1}^{m} i = m(m+1)/2$. Hence, the number of turns used to accept w is $O(\sqrt{n})$.

To prove the lower bound, let us consider a PDA $\mathcal{M}$ accepting $\mathsf{L_{sq}}$. Let N be the constant of Lemma 1 for $\mathcal{M}$. Given an integer $m > 0$, according to the definition of $\mathsf{L_{sq}}$, the intersection of $0^1\{a,b\}^*0^2\{a,b\}^* \cdots 0^m\{a,b\}^*$ with L is $0^1\mathsf{Eq}0^2\mathsf{Eq}\cdots 0^m\mathsf{Eq}$. Let us take $N' = N \cdot (m(m+1)/2 + 1)$. By Lemma 1, to be accepted by $\mathcal{M}$, the string $w = 0^1 a^{N'} b^{N'} 0^2 a^{N'} b^{N'} \cdots 0^m a^{N'} b^{N'}$ requires m turns. Furthermore $n = |w| = \frac{m(m+1)}{2} + 2N'm = \Theta(m^3)$. Hence, for each m there is an input length n for which m turns are required, where $m = \Theta(\sqrt[3]{n})$. This allows to conclude that $\mathcal{M}$ cannot accept using $o(\sqrt[3]{n})$ turns. □

5 An Infinite Hierarchy

In Sect. 4, we presented a language $\mathsf{L_{sq}}$ accepted in a sublinear number of turns, not growing less than $\sqrt[3]{n}$. We now show that there are languages for which the number of turns, necessary and sufficient, can be dramatically reduced. In particular, we will prove that for each integer k there is a language L_k accepted in a number of turns of the order of $\lg^{(k)} n$ which is also necessary. This gives an infinite hierarchy with respect to the number of turns.

To obtain such result, let us start by introducing some preliminary tools. Let us denote by $\mathsf{bin}(x)$ the binary representation of any integer $x > 0$. It is well known that $|\mathsf{bin}(x)| = \lfloor \lg x \rfloor + 1$.

We consider the following language over the alphabet $\{0, 1, \$\}$:

$$\mathsf{ListBin} = \{\mathsf{bin}(1)\$\mathsf{bin}(2)\$\cdots\$\mathsf{bin}(m)\$ \mid m > 0\}\,.$$

Strings in $\mathsf{ListBin}$ are lists of binary encoded integers from 1 to some $m \geq 1$, where each encoding is ended by the delimiter \$. Given $y \in \mathsf{ListBin}$, we denote by $\mathsf{last}(y)$ the rightmost binary block in y, namely the binary encoding $\mathsf{bin}(m)$ of the largest integer m represented in y.

Lemma 2. *If* $y = x_1\$x_2\$\cdots\$x_m\$ \in \mathsf{ListBin}$, *with* $x_i \in \{0,1\}^*$, $i = 1, \ldots, m$, *then* $|\mathsf{last}(y)| \leq \lg |y|$.

Using a pumping argument, it can be proved that $\mathsf{ListBin}$ is not a context-free language. However, its complement is context free. More precisely, it is accepted by a 1-turn OCA [1,7].

We now introduce the languages defining the announced hierarchy. Let us consider the alphabets $\Delta = \{0, 1, \$\}$ and $\Sigma = \Delta \cup \{a, b\}$. For each integer $k \geq 0$, we define the following language over Σ:

$$\mathsf{L}_k = \begin{cases} (\mathsf{Eq}\$)^+ & \text{if } k = 0, \\ \{x_1\$x_2\$\cdots\$x_m\$\$x \mid x_i \in \{0,1\}^+, i = 1,\ldots,m, \text{ and at} \\ \qquad \text{least one of the following holds:} \\ \qquad \text{(a) } x_1\$x_2\$\cdots\$x_m\$ \notin \mathsf{ListBin} \\ \qquad \text{(b) } |x|_\Delta > |x_m| \\ \qquad \text{(c) } x \in \mathsf{L}_{k-1}\} & \text{otherwise.} \end{cases}$$

We point out that, in the above definition, if $x_1\$x_2\$\cdots\$x_m\$ \notin \mathsf{ListBin}$, i.e., condition (a) is satisfied, then the suffix x could be any string in Σ^*. Otherwise, if x does not contain more than $|x_m|$ occurrences of symbols in Δ (i.e., condition (b) is not satisfied), then it must belong to L_{k-1} (condition (c)). Strings satisfying only condition (c) will be crucial to obtain the next results.

We prove an upper bound for the number of turns that are sufficient to accept L_k:

Theorem 6. *For each $k \geq 0$ there exists a* OCA $\mathcal{A}_k$ *accepting the language* L_k *in* $\lg^{(k)} n$ *turns.*

Proof. It is enough prove that, for each $k \geq 0$, there exists a OCA $\mathcal{A}_k$ accepting each $w \in \mathsf{L}_k$ in $\lg^{(k)}|w|_\Delta$ turns. Since $|w|_\Delta \leq |w|$, this will allow to conclude that $\mathcal{A}_k$ accepts each $w \in \mathsf{L}_k$ in $\lg^{(k)}|w|_\Delta \leq \lg^{(k)}|w|$ turns.

We proceed by induction on k. For $k = 0$, we take a 'trivial' OCA $\mathcal{A}_0$ that accepts each $w \in (\mathsf{Eq}\$)^t$, $t > 0$, by making *exactly* t turns. Hence, the number of turns is exactly $t = |w|_\Delta = \lg^{(0)}|w|_\Delta$.

For $k > 0$, let us suppose to have a OCA $\mathcal{A}_{k-1}$ accepting each $x \in \mathsf{L}_{k-1}$ in $\lg^{(k-1)}|x|_\Delta$ turns. We define a OCA $\mathcal{A}_k$ for L_k that given an input w in the appropriate form $x_1\$x_2\$\cdots\$x_m\$\$x$, nondeterministically selects one among conditions (a), (b), and (c) in the definition of L_k, and verifies it.

Conditions (a) and (b) can be tested using the counter and at most one turn. Hence, one turn is enough when one of these conditions is satisfied by w. Condition (c) is checked by simulating the given OCA $\mathcal{A}_{k-1}$ on the suffix x, using at most $\lg^{(k-1)}|x|_\Delta$ turns. When *only (c) is satisfied* by w then $|x|_\Delta \leq |x_m|$ and, by Lemma 2, $|x_m| \leq \lg|x_1\$x_2\$\cdots\$x_m\$| \leq \lg|w|_\Delta$. Since $\mathcal{A}_{k-1}$ accepts x in $\lg^{(k-1)}|x|_\Delta$ turns, the number of turns used to accept w is bounded by $\lg^{(k-1)}|x|_\Delta \leq \lg^{(k-1)}|x_m| \leq \lg^{(k-1)}\lg|w|_\Delta = \lg^{(k)}|w|_\Delta$. This completes the proof. □

A number of turns of the order of $\lg^{(k)} n$ is necessary to accept L_k, even when the machine is a PDA:

Theorem 7. *For each* PDA $\mathcal{M}_k$ *accepting the language* L_k, $k > 0$, *and for infinitely many integers n, there is a string of length n requiring at least* $\lg^{(k)} n - k - 1$ *turns to be accepted by* $\mathcal{M}_k$.

*Proof (*Sketch). For each integer $t > 0$, we inductively define the following string, where $k \geq 0$:

$$w_{t,k} = \begin{cases} (ab\$)^t & \text{if } k = 0, \\ \mathsf{bin}(1)\$\mathsf{bin}(2)\$\cdots\$\mathsf{bin}(2^{|w_{t,k-1}|_\Delta - 1})\$\$w_{t,k-1} & \text{otherwise.} \end{cases}$$

Since $w_{t,0} \in (\mathsf{Eq}\$)^+$, an easy induction shows that $w_{t,k} \in \mathsf{L}_k$, for $k \geq 0$. From the definition, we immediately see that $w_{t,k}$ does not satisfy condition (a) in definition of L_k. Furthermore, the binary representation of $2^{|w_{t,k-1}|_\Delta - 1}$, namely the last block before the first occurrence of the factor \$\$, consists of exactly $|w_{t,k-1}|_\Delta$

binary digits, i.e., $|w_{t,k-1}|_\Delta = |\mathsf{bin}(2^{|w_{t,k-1}|_\Delta - 1})|$, thus implying that even condition (b) is not satisfied by $w_{t,k}$. Hence, the string $w_{t,k}$ *satisfies only condition (c).*

For each PDA $\mathcal{M}_k$ accepting L_k, we will prove that t turns are necessary to accept a modified version of $w_{t,k}$. This, together with a lower bound on t with respect the length of $w_{t,k}$, will allow to prove the statement of the theorem.

To state the lower bound on t, let us start to compute the number of occurrences of symbols from Δ in $w_{t,k}$, as follows, where $m = 2^{|w_{t,k-1}|_\Delta - 1}$:

$$|w_{t,k}|_\Delta = \sum_{i=1}^{m} |\mathsf{bin}(i)\$| + |\$w_{t,k-1}|_\Delta = \sum_{i=1}^{m} (\lfloor \lg i \rfloor + 2) + 1 + |w_{t,k-1}|_\Delta$$

From the previous equalities, we can obtain that $|w_{t,k}|_\Delta \leq 2^{2|w_{t,k-1}|_\Delta}$ and, hence, $\lg |w_{t,k}|_\Delta \leq 2|w_{t,k-1}|_\Delta$, that gives $\lg^{(2)} |w_{t,k}|_\Delta \leq 1 + \lg |w_{t,k-1}|_\Delta$. By iterating, we obtain that $\lg^{(k)} |w_{t,k}|_\Delta \leq k + |w_{t,0}|_\Delta$ and, since $|w_{t,0}|_\Delta = t$:

$$\lg^{(k)} |w_{t,k}|_\Delta \leq k + t\,.$$

According to the definition, we observe that $w_{t,k} = y_0 w_{t,0}$, for some $y_0 \in \Delta^*$.

Given a PDA $\mathcal{M}_k$ accepting L_k, let N be the constant of Lemma 1. We are now going to modify the string $w_{t,k}$ in order to obtain a string that needs t turns to be accepted by $\mathcal{M}_k$.

To this aim, let us consider $N' = N \cdot (|w_{t,k}|_\Delta + 1) \geq N \cdot (|y_0| + 1)$. The 'modified' $w_{t,k}$ is the string $w'_{t,k} = y_0(a^{N'} b^{N'} \$)^t$, where we replaced the block $w_{t,0}$ by a longer block in $(\mathsf{Eq}\$)^t$, in order to be able to apply Lemma 1.

We now prove that t turns are necessary to accept $w'_{t,k}$. To this aim, we observe that $w'_{t,k} = y_0 z_1 y_1 z_2 \cdots z_t y_t$ with $z_i = a^{N'} b^{N'}$ and $y_i = \$$, $i = 1, \ldots, t$. We remind the reader that the string $w_{t,k}$ has been defined in such a way that only condition (c) in the definition of L_k is satisfied, hence, also $w'_{t,k}$ satisfies only condition (c). This implies that if in $w'_{t,k}$ we replace some of the blocks z_i with some string in $\{a,b\}^* \setminus \mathsf{Eq}$, then we would obtain a string not in L_k, while if the replacements are made only using strings in Eq, then the resulting string belongs to L_k. In other words, $y_0\{a,b\}^* y_1 \{a,b\}^* \cdots \{a,b\}^* y_t \cap \mathsf{L}_k = y_0 \mathsf{Eq} y_1 \mathsf{Eq} \cdots \mathsf{Eq} y_t$, Hence, from Lemma 1, it follows that t turns are necessary to accept $w'_{t,k}$.

To complete the proof, we now state a relationship between the number t of turns necessary to accept $w'_{t,k}$ and the length of $w'_{t,k}$.

Let us start to compute $n = |w'_{t,k}|$. According to the definition of $w'_{t,k}$, we obtain that:

$$n = |w_{t,k}|_\Delta + 2N't = |w_{t,k}|_\Delta + 2\,N \cdot (|w_{t,k}|_\Delta + 1)t \leq 3Nt \cdot |w_{t,k}|_\Delta\,.$$

Hence $\lg^{(k)} n \leq \lg^{(k)}(3Nt \cdot |w_{t,k}|_\Delta)$. Since N is fixed, for t sufficiently large and $k > 0$, we have that $3Nt < |w_{t,k}|_\Delta$. This allows to obtain $\lg^{(k)} n \leq 1 + \lg^{(k)} |w_{t,k}|_\Delta \leq 1 + k + t$, and hence:

$$t \geq \lg^{(k)} n - k - 1$$

In this way, we conclude that the number of turns necessary to accept strings of length n is at least $\lg^{(k)} n - k - 1$, for infinitely many integers n. □

6 Even Much Less Turns

In this section, we present a language U_* that can be accepted with less turns than each language in the hierarchy discussed in Sect. 5. Indeed, we prove that $\lg^* n$ turns are sufficient to recognize U_*. However, U_* cannot be recognized using a constant number of turns.

To introduce the language U_*, we preliminarily consider, for each $k \geq 0$, the following modification of the language L_k:

$$\begin{aligned}\mathsf{U}_k = \{y_k z_k y_{k-1} \cdots z_2 y_1 z_1 y_0 \mid\ & y_i \in (\{0,1\}^+\$)^+, z_i \in \{a,b\}^+, i = 1,\ldots,k,\\ & y_0 = 0, \text{ and at least one of the following holds:}\\ & \text{(a) } \exists i, 1 \leq i \leq k, y_i \notin \mathsf{ListBin}\\ & \text{(b) } \exists i, 1 \leq i \leq k, |y_{i-1} \cdots y_1 y_0| > |\mathsf{last}(y_i)|\\ & \text{(c) } \forall i, 1 \leq i \leq k, z_i \in \mathsf{Eq}\,\}.\end{aligned}$$

We observe that each string $w \in \mathsf{L}_k$ satisfying *only condition (c)* in the definition of L_k (e.g., the strings used in the proof of Theorem 7), can be written as $w = y_k\$y_{k-1}\$\cdots\$y_1\$y_0$, where $y_i \in (\{0,1\}^+\$)^+$, for $i = 1,\ldots,k$, while $y_0 \in (\{a,b\}^+\$)^+$. Strings in U_k have a similar structure, with the difference that the block y_0 consists of the symbol 0, while factors from $\{a,b\}^*$ are 'distributed' along the input, replacing the delimiters between consecutive blocks y_i's. The conditions (a), (b), and (c) are similar to those defining L_k. Hence, the recognition of U_k can be done in a similar way, using k turns.

We now define the language we are interested in as $\mathsf{U}_* = \bigcup_{k\geq 0} \mathsf{U}_k$.

Theorem 8. *The language* U_* *is accepted by a* OCA *in* $\lg^* n$ *turns.*

Proof. Given an input string $w = y_k z_k y_{k-1} \cdots z_2 y_1 z_1 y_0$, $k \geq 0$, in the appropriate form (this can be verified while scanning the input, using the finite control), in order to check if $w \in \mathsf{U}_*$, a OCA $\mathcal{B}$ can guess one condition among (a), (b), and (c). Conditions (a) and (b) are verified nondeterministically choosing one block y_i and using one turn, as in the proof of Theorem 6. To test condition (c), $\mathcal{B}$ simulates the OCA $\mathcal{A}_{\mathsf{Eq}}$ on each block z_i, so making exactly k turns. This allows us to conclude that the maximum number of turns k is reached when w satisfies *only* condition (c).

We are now going to estimate how large can be k, with respect to the length n of an input w satisfying only (c).[6] For $i = 1,\ldots,k$, using Lemma 2 and the fact that condition (a) is not satisfied, from $y_i \in \mathsf{ListBin}$ we get that $|\mathsf{last}(y_i)| \leq \lg |y_i|$. Furthermore, also condition (b) is not satified. Hence, $|y_{i-1} \cdots y_1 y_0| \leq |\mathsf{last}(y_i)| \leq \lg |y_i| \leq \lg |y_i \cdots y_1 y_0|$, for $i = 1\ldots,k$. Using the

[6] We give some example of strings satisfying *only* condition (c). For $k = 0$, we have only $y_0 = 0$. For $k = 1$, we can consider $1\$ab0 = y_1 ab y_0$, where $y_1 = 1\$$. Since $|y_0| = |\mathsf{last}(y_1)|$ this is the *shortest string* in U_1 satisfying only condition (c). In a similary way, for $k = 2$, the shortest string satisfying only (c) is $1\$10\$11\$100\$ab1\$ab0 = y_2 ab y_1 ab y_0$, where $y_2 = 1\$10\$11\$100\$$. Indeed, $|y_1 y_0| = 3 = |\mathsf{last}(y_2)|$. For $k = 3$, we take $y_3 ab y_2 ab y_1 ab y_0$, where $y_3 \in \mathsf{ListBin}$ is the list of the binary encodings of all integers from 1 to 2^{14}. In this way, $|\mathsf{last}(y_3)| = 15 = |y_2 y_1 y_0|$.

previous inequalities, we now verify by induction on i that $|y_0| \le \lg^{(i)}|y_i \cdots y_1 y_0|$. The basis $i = 0$ is trivial. Supposing the inequality true for $i-1$, we then obtain:

$$|y_0| \le \lg^{(i-1)}|y_{i-1} \cdots y_1 y_0| \le \lg^{(i-1)}\lg|y_i \cdots y_1 y_0| = \lg^{(i)}|y_i \cdots y_1 y_0|,$$

that completes the induction.

For $i = k$, we obtain $|y_0| \le \lg^{(k)}|y_k \cdots y_1 y_0| \le \lg^{(k)}n$, where $n = |w|$. Since $|y_0| = 1$, we obtain $\lg^{(k)}n \ge 1$ and, remembering that $\lg^* n = \min\{j \mid \lg^{(j)}n \le 1\}$, we conclude that $k \le \lg^* n$.

Hence $\lg^* n$ turns are enough to recognize strings of length n in U_*. □

We conclude this section by proving that the language U_* cannot be accepted using a constant number of turns:

Theorem 9. *Each* PDA *accepting the language* U_* *should make an unbounded number of turns.*

Proof. For each $k > 0$, we take a string $w = y_k z_k y_{k-1} \cdots z_2 y_1 z_1 y_0 \in \mathsf{U}_*$ satisfying only condition (c). By replacing some of the z_i's with strings in $\{a,b\}^*$, we can obtain strings in U_*, only if strings in Eq are used in the replacements. From Lemma 1, we derive that k turns are necessary. □

7 Concluding Remarks

In some recent papers on pushdown and one-counter automata, we considered two complexity measures: the *height of the pushdown* and the *push complexity* [11–13]. In this paper, we continued that line of research, investigating *turn complexity*.

While there are immediate relationships between the first two measures, it is not clear if turn complexity can be in some extent related to the other two measures. The examples of PDAs using minimal non-constant $O(\log\log n)$ height or push complexity presented in the above-cited papers make at most one turn in each accepting computation. It would be interesting to know if there exist machines using a 'small' but unbounded number of turns and, at the same time, $O(\log\log n)$ height or push complexity, or if the use of an unbounded number of turns implies a larger use of height and push operations. We point out that the OCA $\mathcal{A}_k$ accepting the language L_k and described in the proof of Theorem 6, can use an unbounded height to recognize strings satisfying only condition (c). It would be interesting to investigate a complexity measure that keeps into account at the same time the height (or the number of push operations) and the number to turns.

Another question concerns the possibility of reformulating our results in terms of grammars. It is know that finite-turn pushdown automata corresponds to ultralinear and to non-terminal bounded grammars [2]. Suppose to have a PDA $\mathcal{M}$ accepting in a number of turns bounded by a non-constant function $f(n)$. Can we relate $f(n)$ to some measure on a context-free grammar obtained from $\mathcal{M}$ using a 'standard' transformation?

References

1. Gabarró, J.: Pushdown space complexity and related full-AFLs. In: Fontet, M., Mehlhorn, K. (eds.) STACS 84, Symposium of Theoretical Aspects of Computer Science, Paris, France, 11-13 April, 1984, Proceedings. LNCS, vol. 166, pp. 250–259. Springer (1984)
2. Ginsburg, S., Spanier, E.H.: Finite-turn pushdown automata. SIAM J. Control **4**(3), 429–453 (1966)
3. Harrison, M.A.: Introduction to Formal Language Theory. Addison-Wesley Longman Publishing Co., Inc, Boston, MA, USA (1978)
4. Hartmanis, J.: Context-free languages and Turing machine computations. In: Mathematical Aspects of Computer Science. Proceedings of Symposia in Applied Mathematics, vol. 19, pp. 42–51. American Mathematical Society (1967)
5. Hopcroft, J.E., Ullman, J.D.: Introduction to Automata Theory. Addison-Wesley, Languages and Computation (1979)
6. Kutrib, M.: The phenomenon of non-recursive trade-offs. Int. J. Found. Comput. Sci. **16**(5), 957–973 (2005)
7. Lewis II, P.M., Stearns, R.E., Hartmanis, J.: Memory bounds for recognition of context-free and context-sensitive languages. In: 6th Annual Symposium on Switching Circuit Theory and Logical Design, Ann Arbor, Michigan, USA, October 6-8, 1965, pp. 191–202. IEEE Computer Society (1965)
8. Malcher, A.: On recursive and non-recursive trade-offs between finite-turn pushdown automata. J. Autom. Lang. Comb. **12**(1–2), 265–277 (2007)
9. Malcher, A., Pighizzini, G.: Descriptional complexity of bounded context-free languages. Inf. Comput. **227**, 1–20 (2013)
10. Meyer, A.R., Fischer, M.J.: Economy of description by automata, grammars, and formal systems. In: 12th Annual Symposium on Switching and Automata Theory, East Lansing, Michigan, USA, October 13-15, 1971, pp. 188–191. IEEE Computer Society (1971)
11. Pighizzini, G.: Push complexity: optimal bounds and unary inputs. In: Fazekas, S.Z. (ed.) Implementation and Application of Automata - 28th International Conference, CIAA 2024, Akita, Japan, September 3-6, 2024, Proceedings. LNCS, vol. 15015, pp. 289–301. Springer (2024)
12. Pighizzini, G., Prigioniero, L.: Pushdown and one-counter automata: constant and non-constant memory usage. Inf. Comput. **306**, 105329 (2025)
13. Pighizzini, G., Prigioniero, L.: Pushdown automata and constant height: decidability and bounds. Acta Informatica **60**(2), 123–144 (2023)

Positive Varieties of Lattice Languages

Yusuke Inoue(✉) and Yuji Komatsu

Nagoya University, Nagoya, Japan
y-inoue@sqlab.jp

Abstract. While a language assigns a value of either 'yes' or 'no' to each word, a lattice language assigns an element of a given lattice to each word. An advantage of lattice languages is that joins and meets of languages can be defined as generalizations of unions and intersections. This fact also allows for the definition of positive varieties—classes closed under joins, meets, quotients, and inverse homomorphisms—of lattice languages. In this paper, we extend Pin's positive variety theorem, proving a one-to-one correspondence between positive varieties of regular lattice languages and pseudo-varieties of finite ordered monoids. Additionally, we briefly explore algebraic approaches to finite-state Markov chains as an application of our framework.

1 Introduction

Multi-valued logic, which includes options such as 'unknown' in addition to 'yes' and 'no', has been studied for a long time. There have also been several studies on *multi-valued automata* and *multi-valued languages* as natural counterparts in formal language theory [4,11]. In multi-valued automata with a set of values V, each state is assigned a value (or color) in V instead of 'accept' or 'reject'. Therefore, the language accepted by such an automaton is a mapping $L : \Sigma^* \to V$ rather than a subset $L : \Sigma^* \to \{\text{yes}, \text{no}\}$.

As mentioned in [4,11], the significant case is that of *lattice languages*, where the set of values V forms a lattice $\Lambda = (V, \vee, \wedge)$. Note that the usual languages can be viewed as lattice languages over the Boolean lattice $\mathbb{B} = (\{0,1\}, \vee, \wedge)$. In addition to this, we can consider the lattice $\Lambda_{\text{ter}} = \{0 < 1/2 < 1\}$ that includes $1/2$='unknown', the lattice $\Lambda_{\text{fuz}} = [0,1]$ corresponding to fuzzy logic, and so on (see [4] for more examples). One reason lattice languages are significant is that we can define the join and meet operations for lattice languages, which are natural extensions of the union and intersection. In particular, if Λ is distributive, then lattice languages can be considered as weighted languages by regarding Λ as a semiring [6]. Also, any multi-valued language defined on V can be treated as a lattice language by replacing its values with the lattice of the powerset $(\mathcal{P}(V), \cup, \cap)$. Therefore, properties of lattice languages are worth investigating.

Monoids are very powerful tools for studying regular languages. In particular, the Eilenberg's variety theorem [7], which establishes a one-to-one correspondence between *varieties* of regular languages and *pseudo-varieties* of finite

S.-K. Ko and F. Manea (Eds.): DLT 2025, LNCS 16036, pp. 92–106, 2026.
https://doi.org/10.1007/978-3-032-01475-7_7

monoids, is one of the most outstanding results in such studies. A variety of languages is a class of languages closed under Boolean combinations (unions, intersections, and complements), left and right quotients, and inverse homomorphisms. Notably, Boolean combinations play important roles in the proof of the theorem. Let us now consider whether the variety theorem can be extended to lattice languages. As we mentioned in the previous paragraph, we can define joins and meets for lattice languages; however, in contrast, we cannot define an operation corresponding to the complement. Thus, the concept of *positive variety* introduced by Pin [12] can be helpful for our problem. Positive varieties, which do not require closure under complements, serve as alternatives to varieties, and they correspond to pseudo-varieties of ordered monoids. Following the idea of positive varieties, we can prove the variety theorem for lattice languages. However, due to the generalization to lattices, the proof is more complex and technical compared to the original proof in [12].

As the main part of this paper, we prove a one-to-one correspondence between positive varieties of lattice languages and pseudo-varieties of ordered monoids. The definitions and proofs follow the same approach as in [12], but some technical adjustments are required. First, in [12], the concept of *order ideals* is introduced as downward-closed subsets that behave like accepting states. However, when considering a lattice, subsets are not sufficient to characterize lattice languages. Therefore, we need to extend order ideals to *order-preserving colorings*. This is a key idea of our framework. Also, we require that the definition of a positive variety be closed under certain homomorphisms, which are called Λ-*morphisms*. As we will see in Sect. 3, without this modification, it is impossible to establish a one-to-one correspondence between positive varieties and pseudo-varieties.

In this paper, in addition to proving the variety theorem, we briefly discuss its potential applications. Specifically, we claim that lattice languages and ordered monoids can contribute to the study of finite-state Markov chains. Historically, algebraic approaches to Markov chains have been explored such as random walks on groups [16] and semigroup-theoretic analysis of stationary distributions [15]. Building on this context, we outline how lattice languages can provide meaningful semantics for Markov chains and help describe specific classes of Markov chains.

This paper is structured as follows. In Sect. 2, we introduce basic concepts such as ordered monoids, order-preserving colorings, and lattice languages. In Sect. 3, we introduce syntactic ordered monoids and prove the variety theorem. In addition, we explain the relationship with the usual (positive) varieties using some examples. Finally, in Sect. 4, we explain how lattice languages and ordered monoids can provide useful algebraic approaches to Markov chains. Some proofs have been moved to the appendix due to space limitations.

2 Preliminaries

A (partial) order is a relation $\leq$ on a set that is reflexive, antisymmetric, and transitive. For two ordered sets $(S_1, \leq_1)$ and $(S_2, \leq_2)$, a mapping $\varphi : S_1 \to S_2$ is *order-preserving* if $s \leq_1 s'$ implies $\varphi(s) \leq_2 \varphi(s')$ for each $s, s' \in S_1$. Clearly, the composition of two order-preserving mappings is also order-preserving.

A *lattice* is an ordered set in which any two elements have a unique least upper bound and a unique greatest lower bound. A lattice Λ can be considered as an algebra with two binary operations the *join* $\vee$ and the *meet* $\wedge$ where $\lambda_1 \vee \lambda_2$ (resp. $\lambda_1 \wedge \lambda_2$) is the least upper bound (resp. greatest lower bound) of λ_1 and λ_2 in Λ. A lattice Λ is *complete* if every subset $I \subseteq \Lambda$ has both the least upper bound $\bigvee I$ and the greatest lower bound $\bigwedge I$. For a complete lattice Λ, let $\mathbb{1} = \bigvee \Lambda$ and $\mathbb{0} = \bigwedge \Lambda$. For instance, for any set V, the powerset $(\mathcal{P}(V), \subseteq)$ forms the complete lattice $(\mathcal{P}(V), \cup, \cap)$ with $\mathbb{1} = V$ and $\mathbb{0} = \emptyset$. In particular, the Boolean lattice $\mathbb{B} = (\{0,1\}, \vee, \wedge)$ is isomorphic to the case that V is a singleton set. Note that every finite lattice is complete. In this paper, we mainly consider the case that Λ is complete. Additionally, we assume that Λ is non-trivial, i.e., $\mathbb{0} \neq \mathbb{1}$.

For a lattice Λ, an order-preserving mapping $\alpha : \Lambda \to \Lambda$ is called a Λ-*morphism*. For example, for any element $\lambda \in \Lambda$, the constant mapping $\mathrm{cons}(\lambda) : \Lambda \to \Lambda$ defined as $\mathrm{cons}(\lambda)(x) = \lambda$ for each $x \in \Lambda$ is a Λ-morphism.

2.1 Ordered Monoids

An *ordered monoid* is a pair $(M, \leq)$ of a monoid M and an order $\leq$ on M satisfying for every $u, v, m_1, m_2 \in M$, $m_1 \leq m_2$ implies $um_1v \leq um_2v$. We simply write an ordered monoid M instead of $(M, \leq)$ when the order $\leq$ is clear from the context. A *morphism of ordered monoids* is an order-preserving monoid morphism. A morphism $\varphi : (M_1, \leq_1) \to (M_2, \leq_2)$ is an *isomorphism* if there is a morphism $\psi : (M_2, \leq_2) \to (M_1, \leq_1)$ such that both of $\varphi \circ \psi$ and $\psi \circ \varphi$ are identity. As is mentioned in [12], a bijective morphism is not necessarily an isomorphism.

An ordered monoid $(M_1, \leq_1)$ is a *submonoid* of $(M_2, \leq_2)$ if M_1 is an (unordered) submonoid of M_2, and $\leq_1$ is the restriction of $\leq_2$ to M_1. An ordered monoid $(M_1, \leq_1)$ is a *quotient* of $(M_2, \leq_2)$ if there is a surjective morphism $\varphi : (M_2, \leq_2) \to (M_1, \leq_1)$. A quotient of a submonoid is also called a *divisor*. For a family of ordered monoids $\{(M_i, \leq_i)\}_i$, the *direct product* $\prod_i (M_i, \leq_i)$ is the ordered monoid on the product of monoids $\prod_i M_i$ with the order $\leq$ defined as $(u_i)_i \leq (v_i)_i \iff \forall i.(u_i \leq_i v_i)$.

Let Σ be a finite alphabet, and let Σ^* be the free monoid on Σ. The following properties of the free (ordered) monoid are shown in [12].

Proposition 1 (Proposition 2.5, *[12]***).** *For any mapping $\varphi : \Sigma \to M$ with an ordered monoid $(M, \leq)$, there is a unique morphism $\overline{\varphi} : (\Sigma^*, =) \to (M, \leq)$ such that $\varphi(a) = \overline{\varphi}(a)$ for each $a \in \Sigma$.* ■

Proposition 2 (Corollary 2.6, *[12]***).** *Let $\eta : (\Sigma^*, =) \to (M_1, \leq_1)$ be a morphism and $\varphi : (M_2, \leq_2) \to (M_1, \leq_1)$ be a surjective morphism. There is a morphism $\psi : (\Sigma^*, =) \to (M_2, \leq_2)$ such that $\eta = \phi \circ \psi$.* ■

2.2 Order-Preserving Colorings

Let Λ be a complete lattice. For an ordered monoid $(M, \leq)$, an order-preserving mapping $P : M \to \Lambda$ is called an *order-preserving coloring* (or *op-coloring*) on $(M, \leq)$.

Remark 1 A subset I of an ordered monoid $(M, \leq)$ is called an *order ideal* in [12] if $x \leq y$ and $y \in I$ implies $x \in I$ for all $x, y \in M$. It is clear that $I \subseteq M$ is an order ideal iff $I = P^{-1}(0)$ with an op-coloring $P : M \to \mathbb{B}$. In this sense, the concept of op-colorings is a natural extension of order ideals. ∎

For a family $\{P_i\}_i : M \to \Lambda$ of op-colorings on an ordered monoid $(M, \leq)$, define the *join* $\bigvee_i P_i : M \to \Lambda$ and the *meet* $\bigwedge_i P_i : M \to \Lambda$ as the pointwise extensions, that is,

$$(\bigvee_i P_i)(m) = \bigvee_i (P_i(m)), \quad (\bigwedge_i P_i)(m) = \bigwedge_i (P_i(m))$$

for each $m \in M$. We can easily show that both of $\bigvee_i P_i$ and $\bigwedge_i P_i$ are order-preserving. In addition, let $\{(M_i, \leq_i)\}_i$ be a family of ordered monoids, and let $P_i : M_i \to \Lambda$ be op-colorings for each i. Define the *product join* $\bigoplus_i P_i$ and the *product meet* $\bigotimes_i P_i$ as the mappings from $\prod_i (M_i, \leq_i)$ to Λ such that

$$(\bigoplus_i P_i)(m) = \bigvee_i P_i(m_i), \quad (\bigotimes_i P_i)(m) = \bigwedge_i P_i(m_i)$$

for each $m = (m_i)_i \in \prod_i (M_i, \leq_i)$.

Proposition 3. *Let $\{(M_i, \leq_i)\}_i$ be a family of ordered monoids, and let $P_i : M_i \to \Lambda$ be op-colorings. Then, both of $\bigoplus_i P_i$ and $\bigotimes_i P_i$ are op-colorings.* ∎

Let P be an op-coloring on $(M, \leq)$. For an element $u \in M$, the *left quotient* $u \backslash P : M \to \Lambda$ and the *right quotient* $P/u : M \to \Lambda$ with u are defined as $u \backslash P(x) = P(ux)$ and $P/u(x) = P(xu)$ for each $x \in M$.

Proposition 4. *For an op-coloring P on $(M, \leq)$ and an element $u \in M$, both of $u \backslash P$ and P/u are op-colorings.* ∎

Proposition 5. *Let $\eta : M_1 \to M_2$ be a morphism with ordered monoids M_1 and M_2. Also, let P be an op-coloring on M_2. For each element $u \in M_1$, $u \backslash (P \circ \eta) = (\eta(u) \backslash P) \circ \eta$ holds.* ∎

Let P be an op-coloring on $(M_2, \leq_2)$, and let $h : (M_1, \leq_1) \to (M_2, \leq_2)$ be a morphism of ordered monoids. We say that $P \circ h$ is the *inverse homomorphism* with h. This is because if we regard the condition $P(x) = \mathbb{0}$ as "$x \in M_2$ is in the subset $\hat{P} \subseteq M_2$"[1], then,

$$Ph(x) = \mathbb{0} \iff h(x) \in \hat{P} \iff x \in h^{-1}(\hat{P})$$

holds for each $x \in M_1$. It is clear that Ph is order preserving.

[1] Following [12], we interpret $\mathbb{1}$ as 'rejected' and $\mathbb{0}$ as 'accepted'. However, by considering the dual of each lattice, this interpretation can be reversed without affecting the substance of the argument.

2.3 Lattice Languages

Let Σ be a finite alphabet, and let Λ be a complete lattice. A Λ-*language* over Σ is a mapping $L : \Sigma^* \to \Lambda$. Note that a Λ-language is also an op-coloring $L : (\Sigma^*, =) \to \Lambda$. Therefore, joins, meets, left and right quotients, and inverse homomorphisms are defined in the same way as the previous subsection.

A Λ-language L is *recognized* by an ordered monoid $(M, \leq)$ if there is a morphism $\eta : (\Sigma^*, =) \to (M, \leq)$ and an op-coloring $P : (M, \leq) \to \Lambda$ such that the diagram

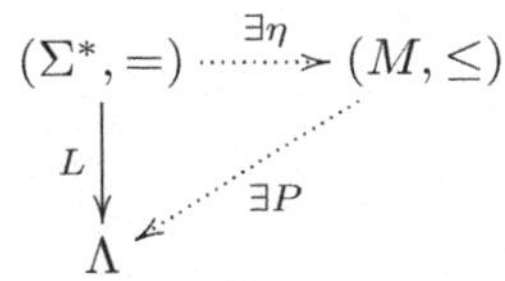

commutes, that is, $L = P \circ \eta$. In this case, we also say that L is recognized by the triple $(\eta, (M, \leq), P)$. This definition is an extension of the definition of recognition in [12] (see also Remark 1). A Λ-language L is *regular* if it is recognized by a finite ordered monoid. The following are basic properties of recognition.

Theorem 1. *Let Λ be a complete lattice, and let L_1 and L_2 be lattice languages over Σ recognized by $(M_1, \leq_1)$ and $(M_2, \leq_2)$, respectively. Then,*

(i) the join $L_1 \vee L_2$ and the meet $L_1 \wedge L_2$ are recognized by $(M_1, \leq_1) \times (M_2, \leq_2)$,
(ii) for a word $u \in \Sigma^*$, the quotients $u \backslash L_1$ and L_1/u are recognized by $(M_1, \leq_1)$,
(iii) for a language homomorphism $h : (\Sigma')^* \to \Sigma^*$, the inverse homomorphism $L_1 h$ is recognized by $(M_1, \leq_1)$, and
(iv) for a Λ-morphism $\alpha : \Lambda \to \Lambda$, αL_1 is recognized by $(M_1, \leq_1)$.

Proof. **(i)** Assume that L_1 and L_2 are recognized by $(\eta_1, (M_1, \leq_1), P_1)$ and $(\eta_2, (M_2, \leq_2), P_2)$, respectively. Let $\langle \eta_1, \eta_2 \rangle : \Sigma^* \to (M_1, \leq_1) \times (M_2, \leq_2)$ be the homomorphism defined as $\langle \eta_1, \eta_2 \rangle (w) = (\eta_1(w), \eta_2(w))$ for each $w \in \Sigma^*$. Then, we can easily show that $L_1 \vee L_2$ is recognized by $(\langle \eta_1, \eta_2 \rangle, (M_1, \leq_1) \times (M_2, \leq_2), P_1 \oplus P_2)$, and $L_1 \wedge L_2$ is recognized by $(\langle \eta_1, \eta_2 \rangle, (M_1, \leq_1) \times (M_2, \leq_2), P_1 \otimes P_2)$.

(ii) $u \backslash L_1$ is recognized by $(\eta_1, (M_1, \leq_1), \eta_1(u) \backslash P_1)$ because $u \backslash L_1 = u \backslash (P_1 \circ \eta_1) = (\eta_1(u) \backslash P_1) \circ \eta_1$ by Proposition 5. The same holds for L_1/u.

(iii and iv) The equation $\alpha \circ L_1 \circ h = (\alpha P_1) \circ (\eta_1 h)$ holds as in the following commutative diagram:

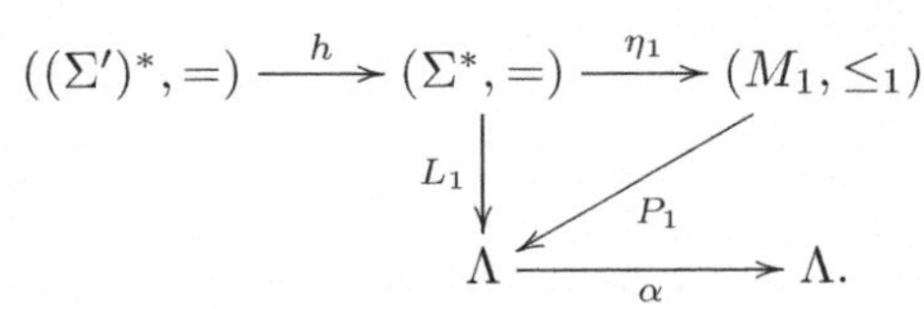

Therefore, $\alpha L_1 h$ is recognized by $(\eta_1 \circ h, (M_1, \leq_1), \alpha \circ P_1)$. □

Next, let us discuss recognition by divisors. The case of quotients is almost the same as in [12], but the case of submonoids requires some additional discussion.

Theorem 2. *Let $(M_1, \leq_1)$ and $(M_2, \leq_2)$ be ordered monoids. For a complete lattice Λ, let L be a Λ-language over Σ recognized by $(M_1, \leq_1)$. Then,*

(i) if $(M_1, \leq_1)$ is a quotient of $(M_2, \leq_2)$, then $(M_2, \leq_2)$ also recognizes L,
(ii) if $(M_1, \leq_1)$ is a submonoid of $(M_2, \leq_2)$, then $(M_2, \leq_2)$ also recognizes L, and
(iii) if $(M_1, \leq_1)$ is a divisor of $(M_2, \leq_2)$, then $(M_2, \leq_2)$ also recognizes L.

Proof. Assume that L is recognized by $(\eta, (M_1, \leq_1), P)$ with a morphism $\eta : (\Sigma^*, =) \to (M_1, \leq_1)$ and an op-coloring $P : M_1 \to \Lambda$. **(i)** If $(M_1, \leq_1)$ is a quotient of $(M_2, \leq_2)$, there is a surjective morphism $\varphi : (M_2, \leq_2) \to (M_1, \leq_1)$. Then, by Proposition 2, there is a morphism $\psi : (\Sigma^*, =) \to (M_2, \leq_2)$ such that $\eta = \phi \circ \psi$. Thus, we have that L is recognized by $(\psi, (M_2, \leq_2), P \circ \phi)$.

(ii) If $(M_1, \leq_1)$ is a submonoid of $(M_2, \leq_2)$, then there is an embedding morphism $\phi : (M_1, \leq_1) \to (M_2, \leq_2)$ such that $\phi(m) = m$ for each $m \in M_1$. Define $P' : M_2 \to \Lambda$ as

$$P'(m) = \bigvee\{P(x) \mid x \in M_1, x \leq_2 m\}.$$

Then, P' is order-preserving because if $m \leq_2 m'$ holds with some elements $m, m' \in M_2$, we have that $\{P(x) \mid x \in M_1, x \leq_2 m\} \subseteq \{P(x) \mid x \in M_1, x \leq_2 m'\}$. Therefore,

$$P'(m) = \bigvee\{P(x) \mid x \in M_1, x \leq_2 m\} \leq \bigvee\{P(x) \mid x \in M_1, x \leq_2 m'\} = P'(m')$$

holds. Also, because $P'(m) = P(m)$ holds if $m \in M_1$, we obtain $P' \circ \phi = P$. Consequently, $L = P' \circ \phi \circ \eta$ is recognized by $(\phi \circ \eta, (M_2, \leq_2), P')$.

(iii) The statement follows from **(i)** and **(ii)**. □

3 Eilenberg's Correspondence for Lattice Languages

For a complete lattice Λ, we say that a class $\mathbf{L}$ of Λ-languages is closed under Λ-morphisms if for any language $L \in \mathbf{L}$ and any Λ-morphism $\alpha : \Lambda \to \Lambda$, it holds that $\alpha L \in \mathbf{L}$. A class of regular Λ-languages is called a *positive variety* if it is closed under joins, meets, left and right quotients, inverse homomorphisms, and Λ-morphisms. Also a class of finite ordered monoids is called a *pseudo-variety* if it is closed under taking divisors and finite direct products. The goal of this section is to prove the Eilenberg's correspondence:

Claim. Let Λ be a complete lattice. There is a natural correspondence between positive varieties of Λ-languages and pseudo-varieties of finite ordered monoids.

The standard definition of a positive variety requires closure under joins, meets, left and right quotients, and inverse homomorphisms. However, our definition additionally requires closure under Λ-morphisms. This is based on the following technical reason: Without Λ-morphisms, a one-to-one correspondence cannot be established. Let us consider the simplest example. For a lattice Λ with three or more elements, the class of languages **Cons** $= \{\mathrm{cons}(\lambda) : \Sigma^* \to \Lambda \mid \lambda \in \Lambda, \Sigma \text{ is a finite alphabet}\}$ consisting of all constant colorings is clearly closed under joins, meets, left and right quotients, and inverse homomorphisms. Also, its subclass $\mathbf{B} = \{\mathrm{cons}(\mathbb{0}), \mathrm{cons}(\mathbb{1})\}$ is closed under these operations. (This class corresponds to $\{\emptyset, \Sigma^*\}$ in the usual sense.) However, both **Cons** and **B** correspond to the same pseudo-variety of ordered monoids, which consists only of the trivial monoid. This leads to a many-to-one correspondence between standard positive varieties and pseudo-varieties. Intuitively, **B** is inadequate as a class of Λ-language because it is inherently a class of $\mathbb{B}$-languages. The solution to this problem is taking closure of Λ-morphisms. Because all constant coloring can be constructed using Λ-morphisms, **B** does not form a positive variety in our framework. Note that the positive variety of $\mathbb{B}$-languages corresponds to the positive variety of languages in the usual sense. This is because a $\mathbb{B}$-morphism is either cons(0), cons(1), or the identity mapping. Therefore, our result is a valid generalization of [12].

3.1 Syntactic Ordered Monoids

We first define the syntactic ordered monoids of Λ-languages. Let Λ be a complete lattice with an order $\leq$, and let L be a Λ-language over Σ. Define the preorder $\preceq_L$ on Σ^* as

$$w_1 \preceq_L w_2 \iff \forall u, v \in \Sigma^*. \, (L(uw_1v) \leq L(uw_2v)) ,$$

and define the equivalence relation $\simeq_L$ on Σ^* as $w_1 \simeq_L w_2$ iff ($w_1 \preceq_L w_2$ and $w_2 \preceq_L w_1$). We call $\simeq_L$ the *syntactic congruence* of L. This is indeed a congruence, and thus, the quotient $M_L = \Sigma^*/\simeq_L$ forms a monoid. The ordered monoid $(M_L, \leq_L)$ with the order $\leq_L$ induced by $\preceq_L$ is called the *syntactic ordered monoid* of L. In addition, the natural surjective morphism $\eta_L : (\Sigma^*, =) \to (M_L, \leq_L)$ where $\eta_L(w)$ is the equivalence class of w is called the *syntactic morphism* of L.

The key property of the syntactic ordered monoid M_L is that it is the 'smallest' monoid recognizing L. The following theorem, which states this, can be shown in almost the same way as in [12].

Theorem 3. *Let L be a Λ-language over an alphabet Σ. If an ordered monoid $(M, \leq)$ recognizes L, then the syntactic ordered monoid $(M_L, \leq_L)$ is a divisor of $(M, \leq)$.* ■

3.2 Ideal Colorings

Let Λ be a lattice, and let $m \in M$ be an element of an ordered monoid $(M, \leq)$. The *m-ideal coloring* is the op-coloring $\iota[m] : M \to \Lambda$ such that

$$\iota[m](x) = \begin{cases} \mathbb{0} & \text{if } x \leq m, \\ \mathbb{1} & \text{otherwise.} \end{cases}$$

Ideal colorings play an important role in the following fact.

Proposition 6. *Let Λ be a complete lattice, and let $(M, \leq)$ be an ordered monoid. For an op-coloring $P : M \to \Lambda$, it holds that $P = \bigwedge_{m \in M}(\iota[m] \vee \mathrm{cons}(P(m)))$.* ■

The following lemma is an essential part of the variety theorem.

Lemma 1. *Let $L : \Sigma^* \to \Lambda$ be a regular Λ-language, and let M_L and η_L be the syntactic ordered monoid and the syntactic morphism of L. Then, for every element $m \in M_L$, the language $\iota[m] \circ \eta_L$ can be represented by using joins, meets, quotients, and Λ-morphisms with L.*

Proof. If $m \in M_L$ is the greatest element, then the statement is clear because $\iota[m] \circ \eta_L = \mathrm{cons}(\mathbb{0}) \circ L$ holds with the Λ-morphism $\mathrm{cons}(\mathbb{0}) : \Lambda \to \Lambda$. We assume that $m \in M_L$ is not the greatest element. Let $P_L : M_L \to \Lambda$ denote the op-coloring such that $L = P_L \circ \eta_L$. Pick $y \in M_L$ such that $y \not\leq m$. By the definition of the syntactic ordered monoid, there are words $w, w' \in M_L$ such that $P_L(uyu') \not\leq P_L(umu')$ with $u = \eta_L(w)$ and $u' = \eta_L(w')$. Then, $(u \backslash P_L / u')(y) = P_L(uyu') \not\leq P_L(umu') = (u \backslash P_L / u')(m)$. Here, let $\alpha_y : \Lambda \to \Lambda$ be the Λ-morphism defined as

$$\alpha_y(\lambda) = \begin{cases} \mathbb{0} & \text{if } \lambda \leq (u \backslash P_L / u')(m), \\ \mathbb{1} & \text{otherwise.} \end{cases}$$

Let $Q_y = \alpha_y \circ (u \backslash P_L / u')$. Then, we have $Q_y(x) = \mathbb{0}$ if $x \leq m$, and $Q_y(x) = \mathbb{1}$ if $x = y$ by the following reasons: The first part follows from that $x \leq m$ implies $(u \backslash P_L / u')(x) \leq (u \backslash P_L / u')(m)$ because $(u \backslash P_L / u')$ is order-preserving. The second part follows because $(u \backslash P_L / u')(y) \not\leq (u \backslash P_L / u')(m)$ as we have seen before. Therefore, $\forall y \not\leq m. (Q_y(x) = \mathbb{0})$ if $x \leq m$, and $Q_x(x) = \mathbb{1}$ if $x \not\leq m$. Consequently, the ideal coloring $\iota[m]$ can be constructed as $\iota[m] = \bigvee_{y \not\leq m} Q_y$. Then, we have that

$$\begin{aligned} \iota[m] \circ \eta_L &= (\bigvee_{y \not\leq m} Q_y) \circ \eta_L && [\iota[m] = \bigvee_{y \not\leq m} Q_y] \\ &= \bigvee_{y \not\leq m} (\alpha_y \circ (u \backslash P_L / v') \circ \eta_L) && [Q_y = \alpha \circ (u \backslash P_L / u')] \\ &= \bigvee_{y \not\leq m} (\alpha_y \circ (\eta_L(w) \backslash P_L / \eta_L(w')) \circ \eta_L) && [u = \eta_L(w), u' = \eta_L(w')] \\ &= \bigvee_{y \not\leq m} (\alpha_y \circ (w \backslash L / w')) && [\text{by Proposition 5 }] \end{aligned}$$

holds. Thus, $\iota[m] \circ \eta_L$ is represented by using joins, meets, quotients, and Λ-morphisms with L. □

We have the following lemma based on this fact.

Lemma 2. *Let* $\mathbf{L}$ *be a positive variety of* Λ*-languages. Let* $L_i \in \mathbf{L}$ *be a* Λ*-language over* Σ_i *and let* M_{L_i} *be its syntactic ordered monoid with* $1 \le i \le k$*. Then, every* Λ*-language recognized by* $M = \prod_{1 \le i \le k} M_{L_i}$ *is contained in* $\mathbf{L}$*.*

Proof. Let L be a Λ-language over Σ recognized by (η, M, P). We show that L is in $\mathbf{L}$. Let $\pi_i : M \to M_i$ denote the i-the projection for each $1 \le i \le k$, that is, π_i is the monoid morphism such that $\pi_i(m) = m_i$ for each $m = (m_1, \ldots, m_k) \in M$. By Proposition 6, $P : M \to \Lambda$ can be represented by $P = \bigwedge_{m \in M}(\iota[m] \vee \mathrm{cons}(P(m)))$. Thus, we have that

$$
\begin{aligned}
L &= P \circ \eta \\
&= \bigwedge_{m \in M} (\iota[m] \vee \mathrm{cons}(P(m))) \circ \eta \\
&= \bigwedge_{m \in M} (\iota[m] \circ \eta \vee \mathrm{cons}(P(m)))
\end{aligned}
$$

holds. Note that every constant mapping is a Λ-morphism. In addition, the op-coloring $\iota[m] \circ \eta : \Sigma^* \to \Lambda$ is equal to $\bigvee_{1 \le i \le k}(\iota[m_i] \circ \pi_i \circ \eta)$ with $m = (m_1, \ldots, m_k)$ by the following reasons: First, it is clear that both op-colorings take only the values $\mathbb{0}$ or $\mathbb{1}$. Then,

$$
\begin{aligned}
(\iota[m] \circ \eta)(w) = \mathbb{0} &\iff \eta(w) \le (m_1, \ldots, m_k) \\
&\iff \pi_i(\eta(w)) \le m_i \text{ for each } 1 \le i \le k \\
&\iff \iota[m_i](\pi_i \circ \eta(w)) = \mathbb{0} \text{ for each } 1 \le i \le k \\
&\iff \bigvee_{1 \le i \le k} ((\iota[m_i] \circ \pi_i \circ \eta)(w)) = \mathbb{0}
\end{aligned}
$$

holds for each $w \in \Sigma^*$. This implies $\iota[m] \circ \eta = \bigvee_i (\iota[m_i] \circ \pi_i \circ \eta)$. Therefore, L can be represented by using meets, joins, and Λ-morphisms of languages $\iota[m_i] \circ \pi_i \circ \eta : \Sigma^* \to \Lambda$ with $m = (m_1, \ldots, m_k) \in M$ and $1 \le i \le k$.

We conclude the proof by showing that the language $\iota[m_i] \circ \pi_i \circ \eta : \Sigma^* \to \Lambda$ is in $\mathbf{L}$ for each $m = (m_1, \ldots, m_k) \in M$ and $1 \le i \le k$. Let $\eta_{L_i} : \Sigma_i^* \to M_i$ be the syntactic morphism of L_i. Because η_{L_i} is surjective, there is a morphism $\psi : \Sigma^* \to \Sigma_i^*$ such that the diagram

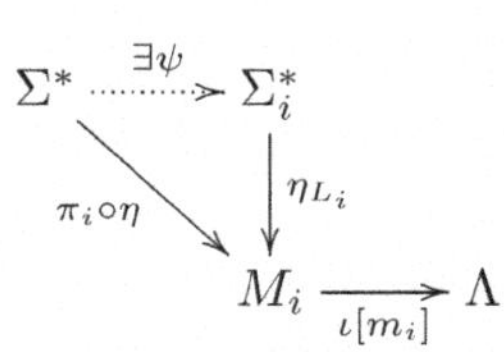

commutes by Proposition 2. Now, by Lemma 1, the language $\iota[m_i] \circ \eta_{L_i}$ can be represented by using joins, meets, quotients, and Λ-morphisms with L_i. This implies $\iota[m_i] \circ \eta_{L_i} \in \mathbf{L}$. Therefore, its inverse image $(\iota[m_i] \circ \eta_{L_i}) \circ \psi = \iota[m_i] \circ \pi_i \circ \eta$ under ψ is also in $\mathbf{L}$. □

3.3 The Variety Theorem

Let us prove the variety theorem. Let Λ be a complete lattice. For a positive variety $\mathbf{L}$ of Λ-languages, define the class of ordered monoids $\mathcal{V}(\mathbf{L})$ as the smallest variety containing $\{M \mid M$ is the syntactic ordered monoid of $L \in \mathbf{L}\}$. In addition, for a pseudo-variety of ordered monoids $\mathcal{M}$, define the class of Λ-languages $\mathbf{V}(\mathcal{M})$ as $\{L \mid L$ is recognized by $M \in \mathcal{M}\}$. Note that $\mathbf{V}(\mathcal{M})$ forms a positive variety by Theorem 1.

In [12], the following is proved.

Proposition 7 (Proposition 5.3, *[12]***).** *Let $\mathcal{M}$ be a pseudo-variety of ordered monoids, and let $(M, \leq) \in \mathcal{M}$. Then there are languages $L_1, \ldots, L_k \in \mathbf{V}(\mathcal{M})$ such that $(M, \leq)$ is a submonoid of $M_{L_1} \times \cdots \times M_{L_k}$ where M_{L_i} is the syntactic ordered monoid of L_i with $1 \leq i \leq k$.* ■

Note that a $\mathbb{B}$-language can be considered as a Λ-language. Therefore, we have the following corollary.

Corollary 1. *Let $\mathcal{M}$ be a pseudo-variety of ordered monoids, and let $(M, \leq) \in \mathcal{M}$. Then there are Λ-languages $L_1, \ldots, L_k \in \mathbf{V}(\mathcal{M})$ such that $(M, \leq)$ is a submonoid of $M_{L_1} \times \cdots \times M_{L_k}$.* ■

Then, we can prove the main theorem.

Theorem 4 (The Variety Theorem). *For a complete lattice Λ, $\mathcal{V}$ and $\mathbf{V}$ are mutually inverse. That is, $\mathbf{V} \circ \mathcal{V}(\mathbf{L}) = \mathbf{L}$ and $\mathcal{V} \circ \mathbf{V}(\mathcal{M}) = \mathcal{M}$ for a positive variety $\mathbf{L}$ of Λ-languages and a pseudo-variety $\mathcal{M}$ of ordered monoids.*

Proof. We first show that $\mathbf{V} \circ \mathcal{V}(\mathbf{L}) = \mathbf{L}$. The direction $\mathbf{L} \subseteq \mathbf{V} \circ \mathcal{V}(\mathbf{L})$ is clear because the syntactic ordered monoid of L recognizes L. Let $M \in \mathcal{V}(\mathbf{L})$ be an ordered monoid. Then, M is a divisor of the direct product N of syntactic monoids of languages in $\mathbf{L}$. By Lemma 2, every language recognized by N is in $\mathbf{L}$. In addition, if a language L is recognized by M, then N also recognizes L by Theorem 2. Thus, L is in $\mathbf{L}$. This implies $\mathbf{V} \circ \mathcal{V}(\mathbf{L}) \subseteq \mathbf{L}$.

Next, we show that $\mathcal{V} \circ \mathbf{V}(\mathcal{M}) = \mathcal{M}$. The direction $\mathcal{M} \subseteq \mathcal{V} \circ \mathbf{V}(\mathcal{M})$ is implied by Corollary 1. Let $M_L \in \mathcal{V} \circ \mathbf{V}(\mathcal{M})$ be the syntactic ordered monoid of a Λ-language $L \in \mathbf{V}(\mathcal{M})$. By the definition of $\mathbf{V}(\mathcal{M})$, L is recognized by an ordered monoid $M \in \mathcal{M}$. By Theorem 3, M_L is a divisor of $M \in \mathcal{M}$, and therefore, $M_L \in \mathcal{M}$. This implies $\mathcal{V} \circ \mathbf{V}(\mathcal{M}) \subseteq \mathcal{M}$. □

We have considered a *local* version of the variety theorem for lattices, where a lattice Λ is fixed in advance. However, since the definition of pseudo-varieties of ordered monoids does not depend on the choice of Λ, we can conclude that for any two complete lattices Λ and Λ', there is a one-to-one correspondence between positive varieties of Λ-languages and those of Λ'-languages. Based on this observation, it is plausible that a *global* version of the variety theorem, where Λ is not fixed, can also be established. This is similar to the usual variety theorem, where it does not matter which alphabet Σ is used. Intuitively, as inverse homomorphisms allow us to freely change the alphabet from Σ to Σ', Λ-morphisms allow us to freely change the lattice from Λ to Λ'.

As noted in [12], a variety of languages is also a positive variety. Although we considered ordered monoids for technical reasons, familiar varieties, such as groups or aperiodic monoids, are also covered by our result.

3.4 Connection with Languages as Subsets

In this subsection, we examine how positive varieties of Λ-languages relate to positive varieties of ordinary (i.e., non-lattice) languages when Λ is finite. To distinguish languages in the sense of subsets from lattice languages, we will use the hat symbol such as $\hat{L} \subseteq \Sigma^*$.

For a Λ-language $L : \Sigma^* \to \Lambda$, define $L_{\leq\lambda} \subseteq \Sigma^*$ as $\hat{L}_{\leq\lambda} = \{w \mid L(w) \leq \lambda\}$ for each $\lambda \in \Lambda$. Then, we show that an ordered monoid recognizing $\hat{L}_{\leq\lambda}$ for each $\lambda \in \Lambda$ has sufficient power to recognize L.

Theorem 5. *Let Λ be a finite lattice, and $L : \Sigma^* \to \Lambda$ be a Λ-language. Also, let M_λ be an ordered monoid recognizing $\hat{L}_{\leq\lambda} = \{w \mid L(w) \leq \lambda\} \subseteq \Sigma^*$ for each $\lambda \in \Lambda$. Then, L is recognized by $\prod_{\lambda\in\Lambda} M_\lambda$.* ■

In [12], shuffle ideals are introduced as an example of a class of languages characterized by ordered monoids. Here, let us discussed the lattice version of shuffle ideals and its characterization. For a word $w = a_1 \cdots a_n$ with $a_1, \ldots, a_n \in \Sigma$, w is a *subword* of a word v if $v = v_0 a_1 v_1 \cdots v_{n-1} a_n v_n$ for some words $v_0, \ldots, v_n \in \Sigma^*$. We write $w \sqsubseteq v$ if $w \in \Sigma^*$ is a subword of $v \in \Sigma^*$. For a lattice Λ, a Λ-language $L : \Sigma^* \to \Lambda$ is a *shuffle ideal* if $w \sqsubseteq v$ implies $L(v) \leq L(w)$ for each $w, v \in \Sigma^*$. By Theorem 5, we have the following as an extension of Theorem 6.4 in [12].

Proposition 8. *Let Λ be a finite lattice. A Λ-language $L : \Sigma^* \to \Lambda$ is a shuffle ideal if and only if it is recognized by a finite ordered monoid M such that the identity element $1_M \in M$ is the greatest element.*

Proof. For each $\lambda \in \Lambda$, define $\hat{L}_{\leq\lambda} \subseteq \Sigma^*$ as $\hat{L}_{\leq\lambda} = \{w \mid L(w) \leq \lambda\}$. Then, $\hat{L}_{\leq\lambda}$ is the shuffle ideal in the sense of [12] because if $w \sqsubseteq v$ and $L_{\leq\lambda}(w) \leq \lambda$, then $L_{\leq\lambda}(v) \leq L_{\leq\lambda}(w) \leq \lambda$ and $v \in \hat{L}_{\leq\lambda}$. Thus, by Theorem 6.4 of [12], each $\hat{L}_{\leq\lambda}$ is recognized by an ordered monoid M_λ such that $1_{M_\lambda} \in M_\lambda$ is the greatest element. By Theorem 5, L is recognized by $M = \prod_\lambda M_\lambda$, which is an ordered monoid such that $1_M = (1_{M_\lambda})_\lambda \in M$ is the greatest element.

The other direction can be proved in a similar way as [12] (see the appendix for details). □

Of course, the class of shuffle ideals discussed here is just an example of positive varieties of lattice languages. As in this example, (positive) varieties of usual languages and their characterization can be extended to the lattice version in an appropriate way.

4 Applications to Markov Chains

In this section, we briefly discuss an applied aspect of our framework. Specifically, we demonstrate that lattice languages and ordered monoids can provide a helpful method to analyze finite-state Markov chains. Let us consider a Markov chain with a state space $\Omega = \{p_1, \ldots, p_n\}$ and a transition matrix $\Pi \in [0,1]^{\Omega \times \Omega}$. A subset C of Ω is called a *communicating class* if for any two states $s_1, s_2 \in C$, there exists a path from s_1 to s_2 with a positive probability. Also, a subset $C \subseteq \Omega$ is called an *ergodic class* if it is communicating and closed, i.e., $\Pi(s,t) = 0$ for each $s \in C$ and $t \in \Omega \setminus C$ (see Chapter II of [10] in detail). For example, the Markov chain $\mathcal{C}$ shown in Fig. 1 has two ergodic classes $C_1 = \{s_{11}, s_{12}\}$ and $C_2 = \{s_{21}, s_{22}\}$.

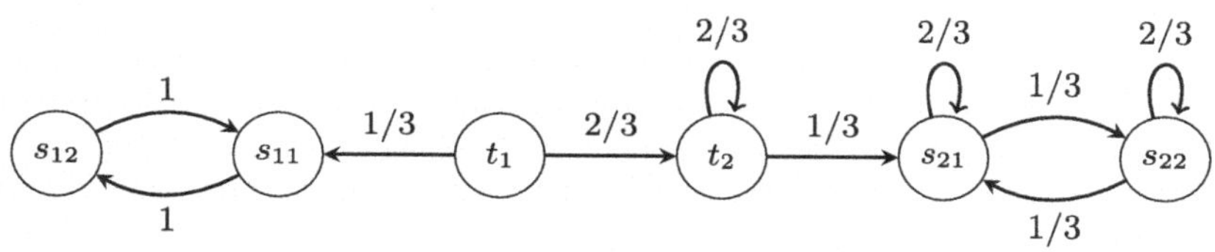

Fig. 1. Markov chain $\mathcal{C}$.

A Markov chain with a state space Ω is *irreducible* if Ω itself is ergodic. Because ergodic classes are convenient for analyzing Markov chains, reducible Markov chains are usually divided into ergodic classes and other components. For example, the Markov chain $\mathcal{C}$ is divided into the class of transient states $\{t_1, t_2\}$ and ergodic classes C_1, C_2.

There have been studies that mention the relationship between Markov chains and formal languages such as [2,8] and [9]. In particular, it is shown in [15] that every finite-state Markov chain has an action of a finite monoid on its state space that simulates the chain. Because an action of a finite monoid can be viewed as a DFA, every Markov chain can be converted into a DFA. For example, $\mathcal{C}$ can be converted into the following DFA $\mathcal{A}$ with the alphabet $\Sigma = \{a, b, c\}$.

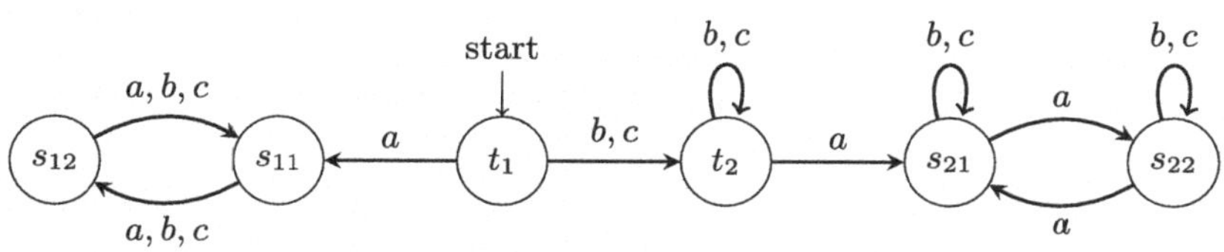

Fig. 2. DFA $\mathcal{A}$ simulating $\mathcal{C}$.

In this context, it is natural to ask the *problem of ergodic classes*: Which ergodic class does each word reach? To consider this, let us use lattice languages.

First, let $\Lambda = \mathcal{P}(\{1,2\})$ be the lattice of the powerset of $\{1,2\}$. That is, $\emptyset = \mathbb{0}$ is the least element, $\{1,2\} = \mathbb{1}$ is the greatest element, and $\{1\}$ and $\{2\}$ are incomparable. Next, let $F : \Omega \to \Lambda$ be the coloring on the DFA $\mathcal{A}$ defined as

$$F(s) = \begin{cases} \{1\} & \text{if } s \in C_1 = \{s_{11}, s_{12}\}, \\ \{2\} & \text{if } s \in C_2 = \{s_{21}, s_{22}\}, \\ \{1,2\} & \text{otherwise.} \end{cases}$$

The Λ-language $L : \Sigma^* \to \Lambda$ is defined in the usual way, e.g., $L(ab) = \{1\}$ and $L(bbc) = \{1,2\}$. Then, of course, $L^{-1}(\{i\}) = \{w \in \Sigma^* \mid w \text{ reaches a state in } C_i\}$ holds for each $i \in \{0,1\}$. In addition, the probability that a randomly chosen word reaches an ergodic class C_i can be computed as the *probability* of the language $L^{-1}(\{i\})$ (see e.g. Chapter III of [17] in detail).

More importantly, it is easily shown that this Λ-language L is a shuffle ideal (see Sect. 3.4). Therefore, we immediately obtain that the syntactic ordered monoid M_L of L is an ordered monoid in which the identity element is the greatest element by Proposition 8. In general, any language calculating the problem of ergodic classes is a shuffle ideal. Consequently, various results from language theory can be applied to these languages and ordered monoids. For example, because shuffle ideals are star-free (e.g., Chapter VII of [13]), syntactic ordered monoids are always aperiodic by Schützenberger's theorem [18]. This is also consistent with the fact that a random walk on a group cannot be equivalent to any reducible Markov chain.[2]

In this section, for simplicity, we have only discussed ergodic classes. However, more generally, transient states and its hierarchy can be effectively described using shuffle ideals. For instance, in $\mathcal{C}$, even if we modify the lattice Λ and coloring F so that $F(t_2) \lneq F(t_1) \in \Lambda$ by reflecting the fact that t_2 cannot return to t_1, L is still a shuffle ideal. Further discussion is omitted due to space limitations, but the authors believe that various other discussions in Markov chains can be effectively described using lattice languages and ordered monoids.

5 Conclusion

In this paper, we proved a natural one-to-one correspondence between positive varieties of lattice languages and pseudo-varieties of ordered monoids, as an extension of the result in [12]. Additionally, we outlined how our framework can be applied to the algebraic approach to Markov chains. Here, we explain two directions for future work.

One of the future directions is to engage in a deeper discussion regarding the application to Markov chains. The description of reducibility and the relationship with probabilities in language theory were briefly discussed in Sect. 4. Additionally, there is an important unsolved problem known as the *cutoff phenomenon* [1,5]. The cutoff phenomenon refers to the situation where a Markov

[2] Groups and aperiodic monoids are orthogonal: only the trivial monoid is both invertible and aperiodic.

chain quickly approaches its stationary distribution after a specific point in time. This is empirically known to occur in Markov chains with certain symmetries, such as random walks on groups. Therefore, it is possible that our approach, based on pseudo-varieties, could provide insights into this problem.

In addition, there are subsequent studies of [12] such as Polák's correspondence between conjunctive varieties of languages and pseudo-varieties of idempotent semirings [14], and Ballester-Bolinches's correspondence between formations of languages and formations of monoids [3]. It is an interesting question whether these variants can also be extended to lattice versions.

References

1. Aldous, D., Diaconis, P.: Shuffling cards and stopping times. Am. Math. Mon. **93**(5), 333–348 (1986)
2. Baier, C., Kiefer, S., Klein, J., Müller, D., Worrell, J.: Markov chains and unambiguous automata. J. Comput. Syst. Sci. **136**, 113–134 (2023)
3. Ballester-Bolinches, A., Pin, J.E., Soler-Escrivà, X.: Formations of finite monoids and formal languages: Eilenberg's variety theorem revisited. Forum Mathematicum, 26(6):1737–1761, (2014)
4. Bruns, G., Godefroid, P.: Model checking with multi-valued logics. In: International Colloquium on Automata, Languages, and Programming, pages 281–293. Springer, (2004)
5. Diaconis, P., Shahshahani, M.: Generating a random permutation with random transpositions. Zeitschrift für Wahrscheinlichkeitstheorie und Verwandte Gebiete **57**(2), 159–179 (1981). https://doi.org/10.1007/BF00535487
6. Droste, M., Kuich, W., Vogler, H.: Handbook of weighted Automata. Springer Science and Business Media, (2009)
7. Eilenberg, S.: Automata, languages, and machines. Academic press, (1974)
8. Etessami, K., Yannakakis, M.: Recursive Markov chains, stochastic grammars, and monotone systems of nonlinear equations. J. ACM (JACM) **56**(1), 1–66 (2009)
9. Inoue, Y., Hashimoto, K., Seki, H.: Semidirect product decompositions for periodic regular languages. In: International Conference on Developments in Language Theory, pages 173–188. Springer, (2024)
10. Kemeny, J.G., Snell, J.L.: Finite Markov Chains. Van Nostrand, (1960)
11. Kupferman, O., Lustig, Y., Lattice automata. In: International Workshop on Verification, Model Checking, and Abstract Interpretation, pages 199–213. Springer, (2007)
12. Pin, J.: A variety theorem without complementation. Russian Mathematics (Izvestija vuzov. Matematika), 39:80–90, (1995)
13. Pin, J.É.: Mathematical foundations of automata theory. Lect. Notes LIAFA, Université Paris **7**, 73 (2010)
14. Polák, L.: Syntactic semiring of a language. In: International Symposium on Mathematical Foundations of Computer Science, pages 611–620. Springer, (2001)
15. Rhodes, J., Schilling, A.: Unified theory for finite Markov chains. Adv. Math. **347**, 739–779 (2019)
16. Saloff-Coste, L.: Random walks on finite groups. In: Probability on Discrete Structures, pages 263–346. Springer, (2004)

17. Salomaa, A., Soittola, M.: Automata-theoretic aspects of formal power series. Springer Science and Business Media (2012)
18. Marcel Paul Schützenberger: On finite monoids having only trivial subgroups. Inf. Control **8**(2), 190–194 (1965)

On the Transformation of Two-Way Nondeterministic Finite Automata to Unambiguous Finite Automata

Semyon Petrov and Alexander Okhotin(✉)

Department of Mathematics and Computer Science, St. Petersburg State University, 7/9 Universitetskaya nab., Saint Petersburg 199034, Russia
semenuska2010@yandex.ru , alexander.okhotin@spbu.ru

Abstract. This paper establishes a lower bound on the number of states necessary in the worst case to simulate an n-state two-way nondeterministic finite automaton (2NFA) by a one-way unambiguous finite automaton (UFA). It is proved that for every n, there is a language recognized by an n-state 2NFA that requires a UFA with at least $\sum_{k=1}^{n}(k-1)!k!\left\{{n \atop k}\right\}\left\{{n+1 \atop k}\right\} = \Omega\left(n^{2n+2}/e^{2n}\right)$ states, where $\left\{{n \atop k}\right\}$ denotes Stirling's numbers of the second kind. This result is proved by estimating the rank of a certain matrix, which describes every possible behaviour of n-state 2NFAs during their computation.

Keywords: Two-way finite automata · unambiguous finite automata · state complexity

1 Introduction

A recurring question in finite automata theory is the number of states in automata of one kind needed to simulate an n-state finite automaton of another kind. It remains unknown whether nondeterministic two-way finite automata (2NFA) with n states can be simulated by deterministic two-way finite automata (2DFA) with polynomially many states, and this question is known to be connected to the *L vs. NL* and *L/poly vs. NL* problems in the complexity theory, see Kapoutsis [14] and Kapoutsis and Pighizzini [15]. Much progress has been made on the complexity of simulating two-way automata by one-way automata. The possibility of such a transformation is one of the earliest results of automata theory [27], and the exact number of states sufficient and necessary to simulate 2NFA and 2DFA by NFA and DFA was later determined by Kapoutsis [12,13]. Other related results include the complexity of transforming two-way automata to one-way automata recognizing the complement of the original language, first investigated by Vardi [28], followed by Birget [1], and later extended to alternating automata by Geffert and Okhotin [7]. Attention was given to transforming two-way automata to one-way in the case of alphabets of limited size: in

Research supported by Russian Science Foundation, project 23-11-00133.

S.-K. Ko and F. Manea (Eds.): DLT 2025, LNCS 16036, pp. 107–122, 2026.
https://doi.org/10.1007/978-3-032-01475-7_8

the unary case, the complexity was studied by Chrobak [2], Mereghetti and Pighizzini [19], Geffert et al. [4,5] and Kunc and Okhotin [16], and later Geffert and Okhotin [6,8] improved the bounds for alphabets of subexponential size. Recently, the complexity of transforming sweeping permutation automata to one-way permutation automata was determined by Radionova and Okhotin [24].

Among the results of Kapoutsis [12,13], let us note the precise complexity of transforming 2NFA to one-way DFA and NFA. For an n-state 2NFA, the worst-case size of an equivalent DFA is $2^{n^2-O(n)}$ states [13], whereas an equivalent NFA requires $\binom{2n}{n+1}$ states in the worst case [12]. Between these two perfectly conclusive results, there is an open question involving an intermediate model between DFA and NFA: the *unambiguous finite automata* (UFA), which can use nondeterminism, yet are bound to accept each string in at most one computation.

State complexity of UFA is an active research topic. Leung [17,18] studied relative succinctness of NFA with different degrees of ambiguity, and, in particular, proved that transforming an n-state UFA to a DFA requires 2^n states in the worst case [18]. For the case of a unary alphabet, Okhotin [20] showed that the state complexity of transforming an n-state UFA to DFA is of the order $e^{\Theta(n^{1/3}(\ln n)^{2/3})}$, which was recently sharpened to $e^{(1+o(1))4^{-1/3}n^{1/3}(\ln n)^{2/3}}$ by F. Petrov [21]. A notable fact is that the complement of every n-state UFA can be recognized by a UFA with much fewer than 2^n states: this was first proved by Jirásek Jr. et al. [11], who demonstrated that at most $2^{0.79n}$ states are sufficient, and soon thereafter Indzhev and Kiefer [10] improved the upper bound to $\sqrt{n+1} \cdot 2^{n/2}$. The most recent work on the state complexity of complementing UFA is by Göös, Kiefer and Yuan [9] and by Czerwiński et al. [3].

The complexity of transforming 2DFA to UFA was investigated recently by Petrov and Okhotin [22]. Since UFA is an intermediate model between NFA and DFA, this complexity has to lie between the two bounds of Kapoutsis [12]: $\binom{2n}{n+1}$, which is of the order $\Theta(\frac{4^n}{\sqrt{n}})$, and $n(n^n - (n-1)^n)$. Both bounds were improved. An improved upper bound of $2^n \cdot n!$ states was obtained by a new 2DFA-to-UFA transformation [22], which augments the NFA constructed by Kapoutsis [12] to ensure the uniqueness of its accepting computation. A lower bound of $\Omega(5.656^n)$ states was proved by reducing the task to estimating the rank of a certain matrix [22]. The key element in the estimation is the rank of a particular submatrix, which was earlier determined by Raz and Spieker [25]. Using it in the calculation leads to an improved lower bound of the order $9^n \cdot n^{-3/2}$ on the complexity of the 2DFA-to-UFA transformation [23].

The complexity of the 2NFA-to-UFA transformation has not been investigated up to date. It has to lie between the lower bound of the order $9^n \cdot n^{-3/2}$ on the 2DFA to UFA transformation, and Kapoutsis' bound of the 2NFA to DFA transformation, which is of the order $2^{n^2-O(n)}$. It is natural to ask what is the exact function in the case of the 2NFA to UFA transformation; these known bounds are even farther apart than in the 2DFA-to-UFA case. This gap is narrowed in this paper by improving the lower bound to $\sum_{k=1}^{n}(k-1)!k!\left\{{n \atop k}\right\}\left\{{n+1 \atop k}\right\}$, where $\left\{{n \atop k}\right\}$ denotes Stirling's numbers of the second kind. The new lower bound is asymptotically estimated as $\Omega(\frac{n^{2n+2}}{e^{2n}})$: this is greater than the upper bound

$2^n \cdot n!$ for the transformation of 2DFA to UFA, showing that these two functions are definitely distinct.

2 Definitions

For a set X, the set $\mathcal{P}(X)$ is defined as $\mathcal{P}(X) = \{Y \mid Y \subseteq X\}$.

Definition 1. *A* nondeterministic finite automaton *(NFA) is a quintuple $\mathcal{B} = (\Sigma, Q, Q_0, \delta, F)$, in which Σ is a finite alphabet; Q is a finite set of states; $Q_0 \subseteq Q$ is the set of initial states; $\delta\colon Q \times \Sigma \to \mathcal{P}(Q)$ is the transition function; $F \subseteq Q$ is the set of accepting states.*

An accepting computation on an input string $a_1 \ldots a_m$ is a sequence of states $p_0, p_1, \ldots, p_m$, with $p_0 \in Q_0$, $p_{i+1} \in \delta(p_i, a_{i+1})$ for all i, and $p_\ell \in F$. The automaton recognizes the set of strings with at least one accepting computation.

An NFA is said to be *unambiguous* (UFA), if there is at most one accepting computation on each string.

Definition 2. *A* two-way nondeterministic finite automaton *(2NFA) is a quintuple $\mathcal{C} = (\Sigma, Q, Q_0, \delta, F)$, in which Σ is a finite alphabet, which does not contain a left end-marker $\vdash \notin \Sigma$ and a right end-marker $\dashv \notin \Sigma$; Q is a finite set of states; $Q_0 \subseteq Q$ is the set of initial states; $\delta\colon Q \times (\Sigma \cup \{\vdash, \dashv\}) \to \mathcal{P}(Q \times \{-1, +1\})$ is the transition function; $F \subseteq Q$ is the set of accepting states, applied at the right end-marker $\dashv$, with $\delta(q, \dashv) = \varnothing$ for all $q \in F$.*

An accepting computation on a string $w = a_1 \ldots a_\ell$ augmented with end-markers $a_0 = \vdash$ and $a_{\ell+1} = \dashv$ is a finite sequence of configurations $(p_0, k_0), (p_1, k_1), \ldots, (p_m, k_m)$, with $(p_i, k_i) \in Q \times \{0, \ldots, \ell + 1\}$, in which $(p_0, k_0) \in Q_0 \times \{0\}$, $(p_{i+1}, k_{i+1} - k_i) \in \delta(p_i, a_{k_i})$ for all i, and $(p_m, k_m) \in F \times \{\ell + 1\}$. A string is accepted if such a computation exists.

3 The Witness Language and Its 2NFA

Let n be a positive integer. The goal is to construct a witness language L_n, which is recognized by a 2NFA using n states, but any equivalent UFA would require a substantial number of states.

Consider computations of a 2NFA on inputs of the form $\vdash xy \dashv$, where $\vdash x$ is a *prefix* and $y \dashv$ is a *suffix*. If a computation is accepting, it would start in the prefix, end in the suffix, and cross the boundary one or more times—those moves form a *crossing sequence*. The crossing sequence can be fairly large, in which case a DFA simulating that 2NFA on the string xy would need to remember much information about the computations of the 2NFA on the prefix. As shown by Kapoutsis [13, Ch. 4], this would require the DFA to have $2^{n^2 - O(n)}$ states in the worst case. An NFA can use non-determinism to remember less information, bringing the required number of states down to $\binom{2n}{n+1}$ [12]. However, in the latter construction, more than one sequence of guesses may lead to the acceptance, which complicates checking this unambiguously. Intuitively, a UFA would not be able to use its nondeterminism to the full extent, and will end up using substantially more states.

The proposed witness language L_n consists of all strings of the form uv, where u encodes a possible behaviour of an arbitrary n-state 2NFA on some prefix, v encodes its behaviour on some suffix, and these two behaviours concatenated together allow at least one accepting computation. This is in some sense a universal language for all n-state 2NFAs, and it is recognized by a fixed n-state 2NFA.

Consider a computation of a 2NFA on an input consisting of a prefix $\vdash x$ and a suffix $y \dashv$. It can be split into the computations solely on the prefix and the computations solely on the suffix, connected by the crossing sequence.

For each prefix, the relevant computations on this prefix include the computations from the initial configuration to the end of the prefix, as well as computations beginning at the last symbol of the prefix and eventually exiting to the right from that symbol. The outcomes of all these computations on a prefix can be represented as a function, see Kapoutsis [13, §4.1]; this paper uses the following variant of these functions.

Denote $[n] = \{1, 2, \dots, n\}$.

Definition 3. *A* prefix table *is a function $f\colon [n] \to \mathcal{P}([n]) \setminus \varnothing$, such that there exists at least one state $i \in [n]$ with $f(i) \subseteq f(j)$ for all $j \in [n]$. Every such i is called a* starting state *of f.*

A prefix table can be naturally represented as a bipartite directed graph $G(f)$ on vertices $\{\mathtt{l}, \mathtt{r}\} \times [n]$, with $E(G(f))$ containing all edges $((\mathtt{l}, i), (\mathtt{r}, j))$ with $j \in f(i)$.

The representation of computations by a prefix table is illustrated in Fig. 1(a). Possible transitions of the underlying 2NFA are depicted with solid arrows. In the resulting graph $G(f)$, vertices of the form $(\mathtt{l}, q)$ correspond to the last symbol of the prefix, and vertices $(\mathtt{r}, q)$ correspond to the next symbol to come. All states at the last symbol of the prefix reachable from the initial configuration form the set of starting states, which are encircled in the figure. Furthermore, for every transition from any of the starting states, one can pretend that the same transition is also possible from all other states; adding these transitions does not affect the acceptance of any strings. These extra transitions are depicted in the figure as dotted arrows.

Note that if i and j are both starting states of f, then $f(i) \subseteq f(j)$ and $f(j) \subseteq f(i)$; hence, $f(i) = f(j)$.

Denote by $s(f)$ the minimal integer that is a starting state of f.

For a prefix table f, its *size* is defined as $|f| = \sum_{i=1}^{n} |f(i)|$.

The information about all computations on a suffix is stored in a *suffix table*, which, for every state i, represents all possible outcomes of the computations beginning at the first symbol of a suffix in the state i: the outcomes include crossing to the prefix in a certain state, as well as acceptance. If acceptance is among the possible outcomes, then one can pretend that all other outcomes are

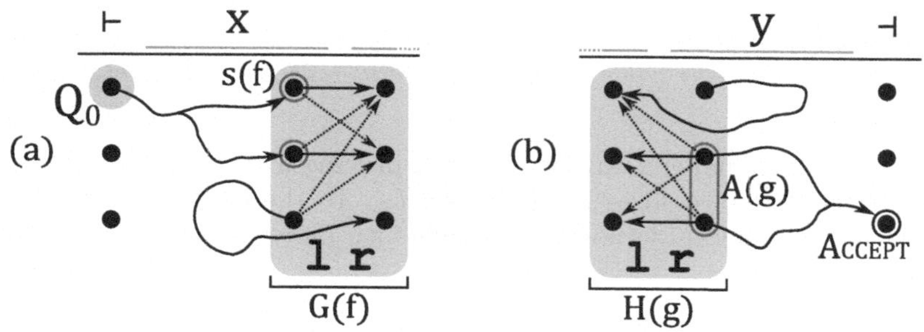

Fig. 1. (a) A prefix table on a prefix x; (b) A suffix table on a suffix y.

possible as well: indeed, if a string can already be accepted, then no additional transitions could make it otherwise. Naturally, acceptance must be a possible outcome for at least one state, otherwise the string cannot be accepted at all.

Definition 4. *A* suffix table *is a function* $g\colon [n] \to \mathcal{P}([n] \cup \{\textsc{Accept}\})$*, such that* $\textsc{Accept} \in g(i)$ *for some* i*, and for every state* i*, if* $\textsc{Accept} \in g(i)$*, then* $g(i) = [n] \cup \{\textsc{Accept}\}$*.*

For a suffix table g, denote as $H(g)$ a bipartite directed graph on the vertices $\{\mathtt{l}, \mathtt{r}\} \times [n]$, with $((\mathtt{r}, i), (\mathtt{l}, j)) \in E(H(g))$ for all $j \in g(i)$. Also, define $A(g) = \{i \mid \textsc{Accept} \in g(i)\}$; this is the set of states leading to acceptance.

Computations on a suffix are represented by a suffix table as illustrated in Fig. 1(b). Possible transitions of the underlying 2NFA are depicted with solid arrows, and the edges that were added are dotted. Vertices $(\mathtt{l}, q)$ of $H(g)$ correspond to the symbol left of the suffix, and vertices $(\mathtt{r}, q)$ are at the first symbol of the suffix.

Let $\Gamma_n = \{(\mathtt{l}, i) \mid i \in [n]\} \cup \{(\mathtt{l}, f) \mid f \text{ is a prefix table}\} \cup \{(\mathtt{r}, g) \mid g \text{ is a suffix table}\}$ be the alphabet of L_n, the universal language for n-state 2NFA. This language is constructed as the language of a 2NFA $\mathcal{A}_n$ with the following properties: the input alphabet is Γ_n; the set of states is $[n]$; the set of starting states is $\{1\}$; the set of accepting states is $[n]$. It uses the transition function δ, which is defined on $[n] \times \Gamma_n$ as follows:

- $\delta(q, (\mathtt{l}, i)) = \{(i, +1)\}$;
- $\delta(q, (\mathtt{l}, f)) = f(q) \times \{+1\}$;
- $\delta(q, (\mathtt{r}, g)) = g(q) \times \{-1\}$ if $\textsc{Accept} \notin g(q)$, and $\{(q, +1)\}$ otherwise.

If the input string is of a form $(\mathtt{l}, s(f))(\mathtt{l}, f)(\mathtt{r}, g)$, for some prefix table f and suffix table g, then the automaton $\mathcal{A}_n$ takes the starting state of f from the first symbol, and then follows the transitions in prefix and suffix tables, checking whether the accepting state is reachable. An equivalent representation of the accepting computation is a path in the graph $G(f) \cup H(g)$ that starts in $(\mathtt{l}, s(f))$ and ends in $(\mathtt{r}, q)$ for some $q \in A(g)$; the vertices $(\mathtt{l}, q)$ represent $\mathcal{A}_n$ being in state q on the second symbol $(\mathtt{l}, f)$, and the vertices $(\mathtt{r}, q)$ represent $\mathcal{A}_n$ being in state q on the third symbol $(\mathtt{r}, g)$.

Lemma 1. *The automaton $\mathcal{A}_n$ accepts the string $(\mathtt{l}, s(f))(\mathtt{l}, f)(\mathtt{r}, g)$ if and only if the graph $G(f) \cup H(g)$ has a path that starts in $(\mathtt{l}, s(f))$ and ends in $(\mathtt{r}, q)$ for some $q \in A(g)$.*

The rest of this paper is concerned with proving a lower bound on the size of every UFA that recognizes the language L_n. To the best of our knowledge, the only known general method for proving such lower bounds is the following theorem.

Theorem A (Schmidt [26], see also Leung [18]). *Let L be a regular language, and let X and Y be sets of strings. Let M be an integer matrix with rows labeled by elements of X and columns labeled by elements of Y, defined by $M_{x,y} = 1$, if $xy \in L$, and $M_{x,y} = 0$ otherwise. Then, every UFA for L has at least* rank M *states.*

A string $(\mathtt{l}, s(f))(\mathtt{l}, f)(\mathtt{r}, g)$ can be naturally divided into a prefix $(\mathtt{l}, s(f))(\mathtt{l}, f)$ that depends only on a prefix table, and a suffix $(\mathtt{r}, g)$ that depends only on a suffix table. The set of all such prefixes shall be used as X in Schmidt's theorem, and all such suffixes form the set Y. Accordingly, the following matrix can be defined.

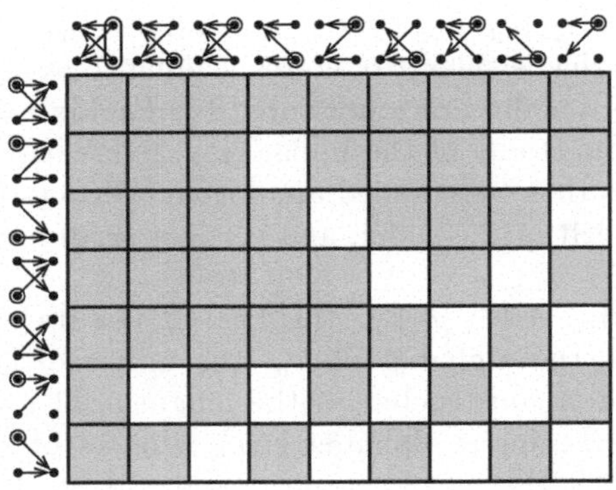

Fig. 2. The matrix $M^{(2)}$.

Definition 5. *The matrix $M^{(n)}$ is a matrix with rows indexed by prefix tables and columns indexed by suffix tables, with $M^{(n)}_{f,g} = 1$ if the automaton $\mathcal{A}_n$ accepts the string $(\mathtt{l}, s(f))(\mathtt{l}, f)(\mathtt{r}, g)$, and 0 otherwise.*

The matrix $M^{(2)}$ is illustrated in Fig. 2. The rows are labeled by graphs corresponding to prefix tables f, with circled vertices corresponding to $s(f)$. The columns are labeled by graphs corresponding to suffix tables g, with circled vertices representing the ones that are in $A(g)$. Filled cells are those containing 1, and empty cells contain zeroes.

One can see that $M^{(2)}$ is full-rank. For larger n, the matrix $M^{(n)}$ is no longer full-rank. In order to obtain a lower bound on the size of a UFA recognizing the language L_n via Schmidt's theorem, the rank of this matrix needs to be calculated.

4 Rank-Preserving Matrix Reduction

To calculate the rank of $M^{(n)}$, we shall simplify it first. This can be done by identifying some prefixes of the form $(\mathtt{l}, s(f))(\mathtt{l}, f)$, for which the corresponding row of the matrix is a linear combination of other rows. Then, those prefix tables can be excluded without affecting the rank of the resulting matrix.

For a bipartite directed graph $G(U, V, E)$ together with a vertex $s \in U$ and a set $A \subseteq V$, define the function $\mathrm{HasPath}(G, s, A)$, which is equal to 1 if there is a path in G that starts in s and ends in A, and 0 otherwise. Note that, by Lemma 1, the automaton $\mathcal{A}_n$ accepts the string $(\mathtt{l}, s(f))(\mathtt{l}, f)(\mathtt{r}, g)$ if and only if $\mathrm{HasPath}\big(G(f) \cup H(g), (\mathtt{l}, s(f)), \{\mathtt{r}\} \times A(g)\big) = 1$.

For each graph $G = (U, V, E)$ and for an edge $e \in (U \times V) \setminus E$, denote by $G + e$ the graph $(U, V, E \cup \{e\})$ obtained from G by adding the edge e.

The redundancy of some rows in the matrix $M^{(n)}$ is ultimately inferred from the following lemma.

Lemma 2. *Let $G(U, V, E)$ be a directed bipartite graph. Let $s \in U$ be a vertex, and let $A \subseteq V$ be a set of vertices. Let $u_1, u_2 \in U$ and $v_1, v_2 \in V$ be such vertices that $(u_1, v_1), (u_2, v_2) \in E$, but the edges $e = (u_1, v_2), e' = (u_2, v_1)$ are not in E.*

Then, $\mathrm{HasPath}(G, s, A) + \mathrm{HasPath}(G + e + e', s, A) = \mathrm{HasPath}(G + e, s, A) + \mathrm{HasPath}(G + e', s, A)$.

Both the left-hand side and the right-hand side are numbers between 0 and 2. To see that they are equal, note that if $\mathrm{HasPath}(G, s, A) = 1$ (that is, there is a path without any of the two extra edges), then all four graphs have a path, and the equality holds as $2 = 2$. Similarly, if $\mathrm{HasPath}(G + e + e', s, A) = 0$ (that is, there is no path in the largest graph), then there are no paths in any graphs $(0 = 0)$. In the remaining case, there is $1 = 0 + 1$ on the left-hand side, and it can be proved that exactly one of the graphs on the right-hand side has a path.

The same equality shall now be restated in terms of prefix tables.

Lemma 3. *Let f be a prefix table. Let $u_1 \neq u_2 \in [n]$ and $v_1 \neq v_2 \in [n]$ be such elements that $v_1 \in f(u_1), v_2 \in f(u_2)$, but $v_2 \notin f(u_1), v_1 \notin f(u_2)$. Let f_e be a function equal to f except $f_e(u_1) = f(u_1) \cup \{v_2\}$; let $f_{e'}$ be a function equal to f except $f_{e'}(u_2) = f(u_2) \cup \{v_1\}$; and let $f_{e+e'}$ be a function equal to f except $f_{e+e'}(u_1) = f(u_1) \cup \{v_2\}$ and $f_{e+e'}(u_2) = f(u_2) \cup \{v_1\}$.*

Then, the functions f_e, $f_{e'}$ and $f_{e+e'}$ are prefix tables, and $M^{(n)}_{f,g} + M^{(n)}_{f_{e+e'},g} = M^{(n)}_{f_e,g} + M^{(n)}_{f_{e'},g}$ for every suffix table g.

Then, for any f that satisfies the conditions in Lemma 3, the corresponding row $M^{(n)}_f$ in $M^{(n)}$ is a linear combination $M^{(n)}_{f_e} + M^{(n)}_{f_{e'}} - M^{(n)}_{f_{e+e'}}$ of rows for prefix tables of greater size, and can therefore be removed without affecting the rank. By iterating over all prefix tables f in order of increasing size, some rows shall be removed, and all remaining ones shall correspond to prefix tables for which the condition in Lemma 3 does not hold. The remaining prefix tables are called *ordered* (the meaning of this term will be revealed in the next Sect. 5).

Definition 6. *A prefix table f is called* ordered *if there are no such elements $u_1 \neq u_2 \in [n]$ and $v_1 \neq v_2 \in [n]$ that $v_1 \in f(u_1), v_2 \in f(u_2)$, but $v_2 \notin f(u_1), v_1 \notin f(u_2)$.*

A reduced matrix can then be defined.

Definition 7. *The matrix $K^{(n)}$ is a matrix with rows indexed by ordered prefix tables and columns indexed by suffix tables, with $K^{(n)}_{f,g} = 1$ if the automaton $\mathcal{A}_n$ accepts the string $(\mathtt{l}, s(f))(\mathtt{l}, f)(\mathtt{r}, g)$, and 0 otherwise.*

Hence, $K^{(n)}_{f,g} = M^{(n)}_{f,g}$ for any ordered prefix table f and any suffix table g.

Theorem 1. *The matrices $K^{(n)}$ and $M^{(n)}$ are of equal rank:* $\operatorname{rank} K^{(n)} = \operatorname{rank} M^{(n)}$.

Thus, the task is now to determine the rank of $K^{(n)}$. Later it will turn out that this matrix is full-rank, and hence it will be sufficient to calculate the number of ordered prefix tables.

5 Ordered Prefix Tables

There is a nicer characterization of a prefix table being ordered, which explain the choice of a term used.

Lemma 4. *A prefix table f is ordered if and only if for every two states $i, j \in [n]$ either $f(i) \subseteq f(j)$ or $f(j) \subseteq f(i)$.*

Proof. Assume f is not ordered, but for every $i, j \in [n]$ either $f(i) \subseteq f(j)$ or $f(j) \subseteq f(i)$. Then, there are states $u_1 \neq u_2 \in [n]$ and $v_1 \neq v_2 \in [n]$ that $v_1 \in f(u_1), v_2 \in f(u_2)$, but $v_2 \notin f(u_1), v_1 \notin f(u_2)$. Therefore, $f(u_1) \not\subseteq f(u_2)$, since $f(u_1)$ contains v_1, and $f(u_2)$ does not. Similarly, $f(u_2) \not\subseteq f(u_1)$, since $f(u_2)$ contains v_2, but $f(u_1)$ does not. This is a contradiction.

Conversely, assume that f is ordered, but there are states $i, j \in [n]$ such that $f(i) \not\subseteq f(j)$ and $f(j) \not\subseteq f(i)$. Then $i \neq j$. Let $u_1 = i$ and $u_2 = j$. Let v_1 be any state in $f(i) \setminus f(j)$, which is not empty since $f(i) \not\subseteq f(j)$. Let $v_2 \in f(j) \setminus f(i)$; it is not empty because $f(j) \not\subseteq f(i)$. Then, $v_1 \in f(u_1) = f(i)$, $v_2 \in f(u_2) = f(j)$, but $v_2 \notin f(u_1) = f(i)$, $v_1 \notin f(u_2) = f(j)$, which contradicts the table f's being ordered. □

It follows that the values of an ordered prefix table f can be ordered: $f(i_1) \subseteq f(i_2) \subseteq \cdots \subseteq f(i_n) \subseteq [n]$. This chain of inclusions can contain identical sets, so an additional sequence is defined. For an ordered prefix table f, let $S^f_0 \subsetneq S^f_1 \subsetneq S^f_2 \subsetneq \cdots \subsetneq S^f_k = [n]$ be the sequence of unique sets in this ordering. Note that the set $[n]$ is always present, even if there is no state i such that $f(i) = [n]$; this is relevant for the definition below.

Definition 8. *The* rank *of an ordered prefix table f, denoted* $\operatorname{rank} f$, *is equal to the number of unique sets $f(i) \neq [n]$.*

In the terms of a sequence S_i^f above, the rank of f is equal to k.

The rank of f is actually equal to the rank of the $n \times n$ matrix $J - M(f)$, where J is a matrix in which every element is equal to 1, and $M(f)$ is defined by $M(f)_{i,j} = 1$ if $j \in f(i)$, and 0 otherwise.

Definition 9. *For an ordered prefix table f and a state $u \in [n]$, the* prefix layer *of u in f, denoted by* $\mathrm{PL}_f(u)$, *is defined as the index i such that $f(u) = S_i^f$.*

A prefix layer is a characterization of how powerful the given state is if obtained on a prefix, that is, how many states are directly reachable from it. All states on the same prefix layer lead to the same states on a suffix, and a state on a higher prefix layer leads to a strictly larger set of states than any state on a lower prefix layer.

Note that the prefix layer of any starting state including $s(f)$ is equal to 0, and the prefix layer of every u is bounded by the rank of the prefix table ($\mathrm{PL}_f(u) \leqslant \operatorname{rank} f$). All layers from 0 up to $\operatorname{rank} f - 1$ must be non-empty, whereas layer $\operatorname{rank} f$ may be empty (it is empty if $[n]$ is not in the image of f).

Definition 10. *For an ordered prefix table f and a state $v \in [n]$, the* suffix layer *of v in f, denoted by* $\mathrm{SL}_f(v)$, *is defined as the minimal index i such that $v \in S_i^f$.*

In terms of the simulated automaton, the suffix layer of a state indicates how powerful a state on a prefix must be to reach this state on the suffix.

Using this new terminology, one can prove that $K^{(n)}$ is full-rank.

Theorem 2. *Rows of $K^{(n)}$ are linearly independent.*

Proof (a brief sketch). Suppose the contrary, that there is a zero linear combination of some rows with non-zero coefficients. There is a row indexed with an ordered prefix table $f_\circ$ of minimal size among these rows. Then a set of $2^{\operatorname{rank} f_\circ}$ suffix tables is constructed, so that for each suffix layer of $f_\circ$, either all arrows from that suffix layer go to the same prefix layer, or all of them go to the next prefix layer (higher by 1); and getting to the element Accept requires reaching the highest layer.

One of these $2^{\operatorname{rank} f_\circ}$ suffix tables, $g_\bullet$, has all arrows going to the next prefix layer, and accordingly $G(f_\circ) \cup H(g_\bullet)$ has a path to the highest layer, and the matrix has 1 in the corresponding cell; on the other hand, $G(f_\circ) \cup H(g)$ with $g \neq g_\bullet$ from this set has no such path, and hence other cells contain zeroes.

At the same time, for each ordered prefix table f in the linear combination different from $f_\circ$, either for some suffix layer it is not necessary to lead to a higher layer for a path to exist, or no path exists at all, and hence, in the row for f, there are either no 1 s in the columns corresponding to these suffix tables, or at least two 1 s. Using this, it is proved that the coefficient at $f_\circ$ must be zero, which yields a contradiction. □

This theorem gives a way to calculate the rank of the original matrix.

Corollary 1. rank $M^{(n)}$ *is equal to the number of ordered prefix tables on* $[n]$.

Proof. By Theorem 1, $\operatorname{rank} M^{(n)} = \operatorname{rank} K^{(n)}$. By Theorem 2, the rows of $K^{(n)}$ are linearly independent, therefore, the rank of $K^{(n)}$ is equal to the number of rows in $K^{(n)}$. By definition, the rows of $K^{(n)}$ are indexed by ordered prefix tables on $[n]$. □

6 Rank Calculation

Corollary 1 allows the rank of the original matrix to be calculated using combinatorial methods. The question is, how many ordered prefix tables on $[n]$ are there for a given n?

Let f be an ordered prefix table. Then, by definition of a prefix layer, $f(u) = S^f_{\mathrm{PL}_f(u)}$ for every $u \in [n]$. Hence, the *prefix layer function* PL_f mapping each $u \in [n]$ to its prefix layer, and the sets $S^f_i \subseteq [n]$ of prefix layer contents, for all $0 \leqslant i \leqslant \operatorname{rank} f$, together uniquely determine f.

However, there are some limitations on both objects. The condition on the prefix layer function is that every layer (except for the optional last one) actually exists, that is, occurs in the image of PL_f. This condition is expressed in the following abstract definition of a prefix layer function that is independent of f.

Definition 11. *Let k be a non-negative integer. A function $p\colon [n] \to \{0, 1, \ldots, k\}$ is called a* valid prefix layer function *of rank k, if for every value $i \in \{0, 1, \ldots, k-1\}$ there is $u \in [n]$ such that $p(u) = i$. Let $P(n, k)$ be the number of valid prefix layer functions of rank k.*

The conditions on the sets S^f_i is that they are non-empty and strictly nested within each other.

Definition 12. *Let k be a non-negative integer. A sequence of sets $S_0 \subsetneq S_1 \subsetneq \cdots \subsetneq S_k = [n]$ is called a* valid nested set sequence *of rank k, if S_0 is non-empty. Let $S(n, k)$ be the number of valid nested set sequences of rank k.*

It will now be proved that pairs of a valid prefix layer function and a valid nested set sequence are in one-to-one correspondence with ordered prefix tables. First, one can naturally extract such a pair from every ordered prefix table.

Lemma 5. *Let f be an ordered prefix table, with* $\operatorname{rank} f = k$. *Then, PL_f is a valid prefix layer function of rank k, and the sequence $S^f_0, S^f_1, \ldots, S^f_k$ is a valid nested set sequence of rank k.*

This representation of an ordered prefix table as a pair of a valid prefix layer function and a valid nested set sequence uniquely determines it.

Lemma 6. *Let f and f' be two ordered prefix tables such that* $\mathrm{PL}_f = \mathrm{PL}_{f'}$ *and $S^f_i = S^{f'}_i$ for all i. Then, $f = f'$.*

Finally, every possible pair of those objects has a matching ordered prefix table.

Lemma 7. *Let p be a valid prefix layer function of rank k, and let $S_0, S_1, \ldots, S_k$ be a valid nested set sequence of rank k. Define the function $f\colon [n] \to \mathcal{P}([n]) \setminus \varnothing$ by $f(u) = S_{p(u)}$. Then, f is an ordered prefix table,* $\operatorname{rank} f = k$, $\mathrm{PL}_f = p$, *and $S_i^f = S_i$ for all $i \in \{0, 1, \ldots, k\}$.*

This bijection provides the following way to calculate the number of ordered prefix tables of certain rank.

Lemma 8. *The number of ordered prefix tables on $[n]$ of rank k is equal to $P(n, k)S(n, k)$.*

Proof. By Lemmata 5, 6 and 7, there is a bijection between ordered prefix tables of rank k and pairs consisting of a valid prefix layer function of rank k and a valid nested set sequence of rank k. Therefore, the sizes of sets of those objects are equal. In addition, the number of pairs is equal to the product of numbers of possible choices for each object in a pair. Hence, the lemma is proved. □

It remains to determine $P(n, k)$ and $S(n, k)$. Both limitations on valid objects can be expressed in terms of partitions. The prefix layer function of rank k can be seen as an ordered partition of $[n]$ into k non-empty parts and one possibly empty part, all of them pairwise disjoint, one for each layer number. The nested set sequence of rank k can be described as an ordered partition of $[n]$ into $k+1$ pairwise disjoint non-empty sets: S_0 together with k sets of differences $S_i \setminus S_{i-1}$. This provides a way to express $P(n, k)$ and $S(n, k)$ in terms of partitions.

Definition 13. *Let $k \leqslant n$ be integers. A* Stirling number of the second kind, *denoted $\left\{ {n \atop k} \right\}$, is defined as the number of ways to partition the set $[n]$ into k non-empty pairwise disjoint parts.*

It is well-known that the number of surjections from an n-element set to a k-element set is exactly $k!\left\{ {n \atop k} \right\}$.

Lemma 9. *The number of valid prefix layer functions of rank k is $P(n, k) = k!\left\{ {n+1 \atop k+1} \right\}$.*

Lemma 10. *The number of valid nested set sequences of rank k is $S(n, k) = (k+1)!\left\{ {n \atop k+1} \right\}$.*

Those results can be combined into the final one.

Theorem 3. *There are $\sum_{k=0}^{n-1} k!(k+1)!\left\{ {n \atop k+1} \right\}\left\{ {n+1 \atop k+1} \right\} = \sum_{k=1}^{n} (k-1)!k!\left\{ {n \atop k} \right\}\left\{ {n+1 \atop k} \right\}$ different ordered prefix tables on $[n]$.*

Proof. An ordered prefix table on $[n]$ can have a rank ranging from 0 to $n-1$, inclusive (since, for an ordered prefix table f, the size of a final set $|S^f_{\text{rank } f}| = n$, but $|S^f_i| \geqslant i+1$ due to being strictly contained into one another). By Lemma 8, for each rank k, there are $P(n,k)S(n,k)$ ordered prefix tables of this rank. Therefore, there are $\sum_{k=0}^{n-1} P(n,k)S(n,k)$ ordered prefix tables on $[n]$ in total.

By Lemma 9, the value $P(n,k)$ is equal to $k!\left\{{n+1 \atop k+1}\right\}$. By Lemma 10, the value $S(n,k) = (k+1)!\left\{{n \atop k+1}\right\}$. Substituting them into the formula above yields the desired result. □

Theorem 4. *For every n, there is a language recognized by a 2NFA with n states, such that every UFA for the same language requires at least $\sum_{k=1}^{n}(k-1)!k!\left\{{n \atop k}\right\}\left\{{n+1 \atop k}\right\}$ states.*

Proof. By Theorem A, for a 2NFA $\mathcal{A}_n$ and its language L_n, every UFA recognizing the same language has at least rank $M^{(n)}$ states. By Corollary 1, the rank of $M^{(n)}$ is equal to the number of ordered prefix tables on $[n]$. Finally, by Theorem 3, there are $\sum_{k=1}^{n}(k-1)!k!\left\{{n \atop k}\right\}\left\{{n+1 \atop k}\right\}$ ordered prefix tables on $[n]$. Therefore, every UFA recognizing the language L_n requires at least $\sum_{k=1}^{n}(k-1)!k!\left\{{n \atop k}\right\}\left\{{n+1 \atop k}\right\}$ states. □

The growth of rate of this sum is estimated as follows.

Theorem 5. $\sum_{k=1}^{n}(k-1)!k!\left\{{n \atop k}\right\}\left\{{n+1 \atop k}\right\} = \Omega\left(\frac{n^{2n+2}}{e^{2n}}\right)$.

Proof. An asymptotic bound on this sum is obtained by estimating a single term, with $k=n$. Then, $\sum_{k=1}^{n}(k-1)!k!\left\{{n \atop k}\right\}\left\{{n+1 \atop k}\right\} \geqslant (n-1)!n!\left\{{n \atop n}\right\}\left\{{n+1 \atop n}\right\}$.

The Stirling numbers of the second kind in this term can be calculated precisely. There is only one partition of $[n]$ into n parts, since then every part consists of a single element (hence, $\left\{{n \atop n}\right\} = 1$). Also, there are $\binom{n}{2} = \frac{n(n-1)}{2}$ partitions of $[n]$ into $n-1$ parts, since all parts except one must consist of a single element, and there are $\binom{n}{2}$ ways to choose the remaining two-element part. Therefore, $(n-1)!n!\left\{{n \atop n}\right\}\left\{{n+1 \atop n}\right\} = \frac{n(n-1)}{2}(n-1)!n! = \frac{n-1}{2}(n!)^2$.

By Stirling's approximation, $n! \sim \sqrt{2\pi n}\cdot\frac{n^n}{e^n}$. Hence, $\frac{n-1}{2}(n!)^2 \sim \frac{n}{2}\cdot 2\pi n\cdot\frac{n^{2n}}{e^{2n}} = \pi\cdot\frac{n^{2n+2}}{e^{2n}} = \Omega\left(\frac{n^{2n+2}}{e^{2n}}\right)$. □

7 No Better Lower Bound via Schmidt's Theorem

The lower bound on the 2NFA to UFA transformation established above (see Theorem 4) is obtained as a rank of a Boolean matrix constructed for a particular 2NFA and two particular sets of strings. Then, Schmidt's theorem ensures that no UFA for the same language may have fewer states. The question is, could a better lower bound be obtained by choosing another language recognized by some n-state 2NFA, and by using different sets of prefixes and suffixes?

The answer turns out to be negative.

Theorem 6. *Let $\mathcal{A}$ be a 2NFA over an alphabet Σ with n states that recognizes a regular language L. Let $X = \{x_1, x_2, \ldots, x_\ell\}$ and $Y = \{y_1, y_2, \ldots, y_m\}$ be sets of strings over the alphabet Σ. Let M be a $\ell \times m$ matrix defined by $M_{i,j} = 1$ if $x_i y_j \in L$, and $M_{i,j} = 0$ otherwise. Then,* $\operatorname{rank} M \leqslant \sum_{k=1}^{n} (k-1)!k!\left\{{n \atop k}\right\}\left\{{n+1 \atop k}\right\}$.

Proof (a sketch). The proof uses the notions of prefix and suffix tables, as in Sect. 3. Whereas the above lower bound proof begins with constructing strings that implement all possible prefix and suffix tables, the proof of Theorem 6 proceeds in the other direction: for any strings in the sets X and Y, it will be shown that the behaviour of the 2NFA on these strings is described by prefix and suffix tables. This implies that the matrix M, with duplicate rows and columns removed, is a submatrix of the matrix $M^{(n)}$, and hence $\operatorname{rank} M \leqslant \operatorname{rank} M^{(n)}$.

There is one complication in this sketch of a proof, caused by prefixes that the 2NFA can never leave, as well as by suffixes that 2NFA cannot pass through to reach an accepting configuration. Such prefixes and suffixes do not have any corresponding prefix and suffix tables. Hence, the definition of prefix and suffix tables has to be slightly expanded to include these special functions. However, it can be shown that these special functions do not affect the rank, since the corresponding rows and columns are all zeros. □

Table 1. The lower bound established in this paper for small values of n, compared to the known tradeoffs from two-way to one-way finite automata.

n	2DFA → UFA 2UFA → UFA (lower bound) $\sum_{k=1}^{n} \binom{n}{k-1}\binom{n}{k}\binom{2k-2}{k-1}$	2DFA → UFA 2UFA → UFA (upper bound) $\sum_{k=1}^{n} \binom{n}{k-1}\binom{n}{k}k!$	**2NFA → UFA** (lower bound) $\sum_{k=1}^{n} (k-1)!k!\left\{{n \atop k}\right\}\left\{{n+1 \atop k}\right\}$	2NFA → DFA $\sum_{i,j=1}^{n} \binom{n}{i}\binom{n}{j}(2^{n-i}-1)^{n-j}$
1	1	1	1	1
2	6	6	7	7
3	39	39	115	133
4	276	292	3 451	7 891
5	2 055	2 505	164 731	1 613 581
6	15 798	24 306	11 467 387	1 201 168 507
7	124 173	263 431	1 096 832 395	3 360 710 751 133
8	992 232	3 154 824	138 027 417 451	36 005 748 492 454 531

8 Conclusion

The lower bound on the state complexity of transforming 2NFA to UFA established in this paper is of the order $\Omega\left(\frac{n^{2n+2}}{e^{2n}}\right)$, which is well above the upper bound of $O(2^n \cdot n!)$ on the 2UFA-to-UFA transformation, established in the previous paper by the authors [22]. This result shows that the tradeoff functions for the 2NFA-to-UFA and 2UFA-to-UFA transformations must be different. However, the new lower bound is still far from the current upper bound, derived from 2NFA-to-DFA transformation, which is of the order $2^{n^2-O(n)}$ [13]. But since it has been shown that no higher lower bound can be obtained using Schmidt's theorem, new methods would be needed to narrow the gap between the lower and the upper bounds.

The lower bound derived in this paper is compared to the known bounds on the 2NFA-to-UFA tradeoff in Table 1.

All bounds presented in the table rely on using an alphabet of size exponential in the number of states; the bounds might be different in the case of alphabets of subexponential size, as is known to be the case for 2DFA-to-DFA and 2DFA-to-NFA transformations [6,8].

References

1. Birget, J.: State-complexity of finite-state devices, state compressibility and incompressibility. Math. Syst. Theory **26**(3), 237–269 (1993). https://doi.org/10.1007/BF01371727
2. Chrobak, M.: Finite automata and unary languages. Theor. Comput. Sci. **47**(3), 149–158 (1986). https://doi.org/10.1016/0304-3975(86)90142-8
3. Czerwinski, W., et al.: Languages given by finite automata over the unary alphabet. J. Comput. Syst. Sci. **151**, 103634 (2025). https://doi.org/10.1016/J.JCSS.2025.103634
4. Geffert, V., Mereghetti, C., Pighizzini, G.: Converting two-way nondeterministic unary automata into simpler automata. Theor. Comput. Sci. **295**, 189–203 (2003). https://doi.org/10.1016/S0304-3975(02)00403-6
5. Geffert, V., Mereghetti, C., Pighizzini, G.: Complementing two-way finite automata. Inf. Comput. **205**(8), 1173–1187 (2007). https://doi.org/10.1016/J.IC.2007.01.008
6. Geffert, V., Okhotin, A.: One-way simulation of two-way finite automata over small alphabets. In: Bensch, S., Drewes, F., Freund, R., Otto, F. (eds.) Fifth Workshop on Non-Classical Models for Automata and Applications - NCMA 2013, Umeå, Sweden, August 13–14, 2013, Proceedings. books@ocg.at, vol. 294, pp. 151–162. Österreichische Computer Gesellschaft (2013)
7. Geffert, V., Okhotin, A.: Transforming two-way alternating finite automata to one-way nondeterministic automata. In: Csuhaj-Varjú, E., Dietzfelbinger, M., Ésik, Z. (eds.) Mathematical Foundations of Computer Science 2014 - 39th International Symposium, MFCS 2014, Budapest, Hungary, August 25–29, 2014. Proceedings, Part I. LNCS, vol. 8634, pp. 291–302. Springer (2014).https://doi.org/10.1007/978-3-662-44522-8_25
8. Geffert, V., Okhotin, A.: Deterministic one-way simulation of two-way deterministic finite automata over small alphabets. In: Han, Y., Ko, S. (eds.) Descriptional Complexity of Formal Systems - 23rd IFIP WG 1.02 International Conference, DCFS 2021, Virtual Event, September 5, 2021, Proceedings. LNCS, vol. 13037, pp. 26–37. Springer (2021). https://doi.org/10.1007/978-3-030-93489-7_3

9. Göös, M., Kiefer, S., Yuan, W.: Lower bounds for unambiguous automata via communication complexity. In: Bojanczyk, M., Merelli, E., Woodruff, D.P. (eds.) 49th International Colloquium on Automata, Languages, and Programming, ICALP 2022, July 4–8, 2022, Paris, France. LIPIcs, vol. 229, pp. 126:1–126:13. Schloss Dagstuhl - Leibniz-Zentrum für Informatik (2022). https://doi.org/10.4230/LIPICS.ICALP.2022.126
10. Indzhev, E., Kiefer, S.: On complementing unambiguous automata and graphs with many cliques and cocliques. Inf. Process. Lett. **177**, 106270 (2022). https://doi.org/10.1016/J.IPL.2022.106270
11. Jirásek, J., Jr., Jirásková, G., Sebej, J.: Operations on unambiguous finite automata. Int. J. Found. Comput. Sci. **29**(5), 861–876 (2018). https://doi.org/10.1142/S012905411842008X
12. Kapoutsis, C.A.: Removing bidirectionality from nondeterministic finite automata. In: Jedrzejowicz, J., Szepietowski, A. (eds.) Mathematical Foundations of Computer Science 2005, 30th International Symposium, MFCS 2005, Gdansk, Poland, August 29 - September 2, 2005, Proceedings. LNCS, vol. 3618, pp. 544–555. Springer (2005). https://doi.org/10.1007/11549345_47
13. D Kapoutsis, C.A.: Algorithms and lower bounds in finite automata size complexity, Ph.D. thesis, Massachusetts Institute of Technology, Cambridge, MA, USA (2006). https://hdl.handle.net/1721.1/37891
14. Kapoutsis, C.A.: Two-way automata versus logarithmic space. Theory Comput. Syst. **55**(2), 421–447 (2014). https://doi.org/10.1007/S00224-013-9465-0
15. Kapoutsis, C.A., Pighizzini, G.: Two-way automata characterizations of L/poly versus NL. Theory Comput. Syst. **56**(4), 662–685 (2014). https://doi.org/10.1007/s00224-014-9560-x
16. Kunc, M., Okhotin, A.: Describing periodicity in two-way deterministic finite automata using transformation semigroups. In: Mauri, G., Leporati, A. (eds.) Developments in Language Theory - 15th International Conference, DLT 2011, Milan, Italy, July 19–22, 2011. Proceedings. LNCS, vol. 6795, pp. 324–336. Springer (2011). https://doi.org/10.1007/978-3-642-22321-1_28
17. Leung, H.: Separating exponentially ambiguous finite automata from polynomially ambiguous finite automata. SIAM J. Comput. **27**(4), 1073–1082 (1998). https://doi.org/10.1137/S0097539793252092
18. Leung, H.: Descriptional complexity of NFA of different ambiguity. Int. J. Found. Comput. Sci. **16**(5), 975–984 (2005). https://doi.org/10.1142/S0129054105003418
19. Mereghetti, C., Pighizzini, G.: Optimal simulations between unary automata. SIAM J. Comput. **30**(6), 1976–1992 (2001). https://doi.org/10.1137/S009753979935431X
20. Okhotin, A.: Unambiguous finite automata over a unary alphabet. Inf. Comput. **212**, 15–36 (2012). https://doi.org/10.1016/J.IC.2012.01.003
21. Petrov, F.: Logarithmic asymptotics of Landau-Okhotin function. Acta Math. Hungar. **169**(1), 272–276 (2023). https://doi.org/10.1007/s10474-023-01299-6
22. Petrov, S., Okhotin, A.: On the transformation of two-way finite automata to unambiguous finite automata. Inf. Comput. **295**(Part A), 104956 (2023). https://doi.org/10.1016/J.IC.2022.104956
23. Petrov, S., Petrov, F., Okhotin, A.: On the rank of the communication matrix for deterministic two-way finite automata. arXiv preprint arXiv:2312.05909 (2023)
24. Radionova, M., Okhotin, A.: Sweeping permutation automata. In: Nagy, B., Freund, R. (eds.) Proceedings of the 13th International Workshop on Non-Classical Models of Automata and Applications, NCMA 2023, Famagusta, North Cyprus,

18th–19th September, 2023. EPTCS, vol. 388, pp. 110–124 (2023). https://doi.org/10.4204/EPTCS.388.11
25. Raz, R., Spieker, B.: On the log rank-conjecture in communication complexity. Combinatorica **15**(4), 567–588 (1995). https://doi.org/10.1007/BF01192528
26. Schmidt, E.M.: Succinctness of descriptions of context-free, regular and finite languages, Ph.D. thesis, Cornell University, Ithaca, New York (1977)
27. Shepherdson, J.C.: The reduction of two-way automata to one-way automata. IBM J. Res. Dev. **3**(2), 198–200 (1959). https://doi.org/10.1147/RD.32.0198
28. Vardi, M.Y.: A note on the reduction of two-way automata to one-way automata. Inf. Process. Lett. **30**(5), 261–264 (1989). https://doi.org/10.1016/0020-0190(89)90205-6

Left Quotients of Deterministic Context-Free Languages

Brennan Lockinger(✉) and Ian McQuillan

Department of Computer Science, University of Saskatchewan, Saskatoon, SK S7N5A9, Canada
{brennan.lockinger,mcquillan}@cs.usask.ca

Abstract. Left quotients play a crucial role in automata and formal language theory, but their use on the important class of one-way deterministic pushdown automata (which accept the deterministic context-free languages, DCFL) has been understudied. In this paper, we study two novel language families: the family of all DCFL languages L such that the left quotient of L by any regular language is in DCFL; and the smallest family containing DCFL that is closed under left quotient by regular languages. The first family is properly contained in DCFL but is powerful enough to accept the Dyck languages and has quite unusual closure properties. Multiple characterizations are given for the second family, including the surprising result that the smallest family containing DCFL that is closed under left quotient with regular languages is equal to the smallest family containing DCFL that is closed under union, concatenation, and Kleene-$*$. This implies that languages in this family are quite practical as they can all be recognized in linear time.

Keywords: Deterministic Pushdown Automata · Left Quotients · Regular Languages · Closure Properties · Decision Problems

1 Introduction

In the study of formal languages and automata theory, the left and right quotient operations play a key role. Both are binary operations on languages: the right quotient contains words from the first language, where a suffix taken from the second language has been removed; and the left quotient contains words with a prefix removed instead.

Left quotients are important in the history of the literature on finite automata and regular languages. Indeed, the Myhill-Nerode Theorem ([18], see also [3]) is a famous result that states that a language L is regular if and only if the number of distinct left quotients of L is finite; furthermore, a minimal deterministic finite automaton which accepts L can be constructed by using left quotients as states. The left quotient operation is a generalization of the Brzozowski derivative (left quotient by a word), and the derivatives of a regular expression can be used as an algorithm to convert regular expressions to finite automata [2].

S.-K. Ko and F. Manea (Eds.): DLT 2025, LNCS 16036, pp. 123–137, 2026.
https://doi.org/10.1007/978-3-032-01475-7_9

It is known that every language family that is closed under homomorphism, inverse homomorphism, and intersection with regular languages is closed under both quotient operations [7]. Closure under all of these operations is common for languages accepted by types of one-way nondeterministic automata. In contrast, families accepted by deterministic automata are sometimes closed just under right quotient with regular languages, which is the case for deterministic context-free languages [8,12], deterministic stack languages [11], deterministic reversal-bounded multicounter languages [5], and several others [13]. For left quotient however, the operation is more powerful. DCFL is not closed under left quotient with regular (or even finite) languages [8]. The family of languages accepted by deterministic reversal-bounded multicounter machines is also not closed under left quotient by regular languages [5].

This paper focuses on the effects of the left quotient operation on DCFL, a family that is important for parsing [12] as they can all be recognized in linear time. Left quotients are a more limited form of erasing than homomorphisms, as applying arbitrary homomorphisms to DCFL gives all context-free languages (CFL), and arbitrary CFL languages still cannot be recognized in $O(n^2)$ time.

Towards this, we study two new language families which allow for a more nuanced analysis beyond the known non-closure of DCFL under left quotients. The first, denoted by $\lfloor \mathsf{DCFL} \rfloor$, is the subset of DCFL whereby the left quotient of each by any regular language is in DCFL. This concept is similar to a "kernel" of DCFL, as studied in [16]. The second is the smallest family containing DCFL that is closed under left quotient by regular languages, $\lceil \mathsf{DCFL} \rceil$. We provide comparisons of the computational capacity for both to well-studied language families, examine closure properties and decision problems of both families, and also develop several characterizations of $\lceil \mathsf{DCFL} \rceil$ in terms of other closure properties, with DPDA that have certain contents preset on the pushdown, and with nondeterministic pushdown automata with limited nondeterminism. These characterizations imply that the family is properly between DCFL and CFL, contain inherently ambiguous languages, but they can still be recognized in linear time. Hence, left quotients by regular languages provide a form of erasing that can increase the power without sacrificing recognition time.

2 Preliminaries

We assume a basic knowledge of formal language and automata theory, as in [12]. We fix some notation next.

Let Σ be a finite alphabet. A *word* of length n is a sequence $a_1 \cdots a_n$ of elements from Σ; the unique word of length 0 is denoted ε. The set of all words over Σ is denoted Σ^*; the length of a word $w \in \Sigma^*$ is denoted $|w|$. Given $a \in \Sigma$, $|w|_a$ is the number of a's in w. A *language* over Σ is any subset $L \subseteq \Sigma^*$; a language *family* is any set of languages.

Let $u, w \in \Sigma^*$. We say u is a *prefix* of w if $w = uv$ for some $v \in \Sigma^*$, denoted $u \leq w$. If $|v| > 0$, then we say that u is a *proper prefix*, and the relation is denoted $u < w$. If $w = a_1 \cdots a_n$ is a word over Σ, $a_i \in \Sigma, 1 \leq i \leq n$, the *reverse* of w, denoted w^{R}, is $a_n \cdots a_1$ (and $w = \varepsilon$ implies $w^{\mathsf{R}} = \varepsilon$); the reverse is extended to languages L^{R} in the natural way.

Let $L, R \subseteq \Sigma^*$ be languages over a common alphabet Σ. The *left quotient* $R^{-1}L$ and *right quotient* LR^{-1} are defined as

$$R^{-1}L = \{\, x \in \Sigma^* \mid \exists w \in R : wx \in L \,\}, \tag{1}$$

$$LR^{-1} = \{\, w \in \Sigma^* \mid \exists x \in R : wx \in L \,\}. \tag{2}$$

We also use the standard operations of union, intersection, concatenation, Kleene-$*$, homomorphism, and inverse homomorphism as usual [12]. We will also use some lesser known operations that are important for deterministic machines, as defined in [4]. For the operations below, \$ and # are special symbols not in Σ. The *marked concatenation* of $L, R \subseteq \Sigma^*$ is $L\$R$, the *marked-$*$* of L is $(\$L)^*$, and the *marked union* of L and R is $\$L \cup \#R$.

A *one-way nondeterministic pushdown automaton* (abbreviated by NPDA) is a tuple $M = (Q, \Sigma, \Gamma, \delta, q_0, Z_0, F)$, where Q is a finite set of states, Σ is the input alphabet, Γ is the pushdown alphabet, $q_0 \in Q$ is the initial state, $Z_0 \in \Gamma$ is the bottom of stack symbol, $F \subseteq Q$ is the set of final states, and δ is a partial function from $Q \times (\Sigma \cup \{\, \varepsilon \,\}) \times \Gamma$ into finite subsets of $Q \times \Gamma^*$. A configuration of M is an element in $Q \times \Sigma^* \times Z_0(\Gamma \setminus Z_0)^*$. Two configurations change as follows: $(q, aw, xb) \vdash_M (p, w, xy)$ if $\delta(q, a, b)$ contains (p, y) (we use simply $\vdash$ if M is clear). We let $\vdash_M^*$ be the reflexive and transitive closure of $\vdash_M$. A *computation of M on w* is a sequence $(q_0, w_0, x_0) \vdash \cdots \vdash (q_n, w_n, x_n)$ where $w_0 = w, x_0 = Z_0$, and it is an *accepting computation of M on w* if also $q_n \in F, w_n = \varepsilon$. The *language accepted by M*, $L(M)$, is the set of all words $w \in \Sigma^*$ such that there is an accepting computation of M on w. We will also need a lesser known concept called the *store language* of M, $S(M) = \{\, xq \mid (q_0, w_0, Z_0) \vdash^* (q, w, Z_0 x) \vdash^* (q_f, \varepsilon, x_f) \mid q_f \in F \,\}$. It is known that given any NPDA M, $S(M)$ is always a regular language [9,13]. (Although the state is sometimes put at the beginning of the words of the store language, it can clearly also be placed at the end and it will still be regular.)

Such a machine is *deterministic* (i.e., a DPDA) if, 1) for each $q \in Q, b \in \Gamma$, if $\delta(q, \varepsilon, b)$ is not empty, then $\delta(q, a, b)$ is empty for all $a \in \Sigma$, and 2) $\delta(q, a, b)$ always contains at most one element for each $q \in Q, a \in \Sigma \cup \{\, \varepsilon \,\}, b \in \Gamma$. The family of languages accepted by one-way nondeterministic pushdown automata is known as the context-free languages (denoted by CFL), and the family of languages accepted by one-way deterministic pushdown automata is known as the deterministic context-free languages (denoted by DCFL).

It is known that DCFL is closed under inverse homomorphism, intersection with regular languages, right quotient with regular languages, marked union, marked concatenation, union with regular languages, complement, and marked-$*$, but it is not closed under homomorphism, union, concatenation, Kleene-$*$, and left quotient with regular languages [4,8,12].

We also define DFA and NFA to be the class of deterministic finite automata and nondeterministic finite automata, and the family of regular languages that they both accept by REG.

Given any family of languages $\mathcal{L}$ and operators $\odot_1, \ldots, \odot_k$ on languages, we denote the *closure of* L *under* $\odot_1, \ldots, \odot_k$ by

$$\Omega^{\{\odot_1, \ldots, \odot_k\}}(\mathcal{L}), \tag{3}$$

which is the smallest family of languages containing $\mathcal{L}$ that is closed under all of $\odot_1, \ldots, \odot_k$. We use one short form and indicate Ω^{REG} to mean $\Omega^{\{\cup, \cdot, *\}}$ (closed under union, concatenation, and Kleene-$*$).

3 Basic Properties of Quotients, $\lfloor \mathcal{L} \rfloor$, and $\lceil \mathcal{L} \rceil$

Before defining the language families of interest, it will be helpful to establish some properties of quotients.

Proposition 1. *Let* L *be any language and let* $R_1, \ldots, R_n \in \mathsf{REG}$. *There exists* $R \in \mathsf{REG}$ *such that*

$$R_n^{-1}(\cdots(R_2^{-1}(R_1^{-1}L))\cdots) = R^{-1}L. \tag{4}$$

Proposition 2. *For any languages* L, R, *and* S*:*

1. $(R^{-1}L)S^{-1} = R^{-1}(LS^{-1})$.
2. $R^{-1}(L \cup S) = R^{-1}L \cup R^{-1}S$.
3. $(R \cup S)^{-1}L = R^{-1}L \cup S^{-1}L$.
4. $R^{-1}(SL) = (R^{-1}S)L \cup (S^{-1}R)^{-1}L$.

It is known that the right quotient of a regular language with an arbitrary language is regular; [12] contains a non-constructive proof. Because the family of regular languages are closed under reversal, it is evident that the same is true for left quotients as well.

Proposition 3. *For an arbitrary language* L *and a regular language* S, *both* $L^{-1}S$ *and* SL^{-1} *are regular.*

Now we are ready to define the two methods of creating new language families. Consider any family of languages $\mathcal{L}$. Denote by $\lfloor \mathcal{L} \rfloor$ the family of languages

$$\{\, L \in \mathcal{L} \mid R^{-1}L \in \mathcal{L}, \forall R \in \mathsf{REG} \,\}. \tag{5}$$

Proposition 4. *For any family of languages* $\mathcal{L}$, $\lfloor \mathcal{L} \rfloor$ *is closed under left quotient with regular languages, and is the largest subfamily of* $\mathcal{L}$ *where this is true.*

We also consider the closure of $\mathcal{L}$ under left quotient with regular languages, denoted $\lceil \mathcal{L} \rceil$. This is a short-hand for $\Omega^{-1(\mathsf{REG})}(\mathcal{L})$ where $-1(\mathsf{REG})$ is the operation of left quotient with regular languages.

The following characterization is often easier to work with than the above definition.

Proposition 5. *For any family of languages $\mathcal{L}$,*

$$\lceil \mathcal{L} \rceil = \{ R^{-1}L \mid L \in \mathcal{L}, R \in \mathsf{REG} \}. \tag{6}$$

We note the following obvious relationship.

Proposition 6. *Let $\mathcal{L}$ be any family of languages. Then,*

$$\lfloor \mathcal{L} \rfloor \subseteq \mathcal{L} \subseteq \lceil \mathcal{L} \rceil, \tag{7}$$

and equality holds exactly when $\mathcal{L}$ is closed under left quotient with regular languages.

4 Properties of $\lfloor \mathsf{DCFL} \rfloor$

In this section, we are primarily interested in the family of languages $\lfloor \mathsf{DCFL} \rfloor$, the largest subfamily of deterministic context-free languages that is closed under left quotient with regular languages. First, we look at some simple non-regular languages in $\lfloor \mathsf{DCFL} \rfloor$ which also help to better understand the family.

Proposition 7. *The non-regular languages $\{ 0^n 1^n \mid n \geq 0 \}$ and $\{ 0^n 1^{2n} \mid n \geq 0 \}$ are in $\lfloor \mathsf{DCFL} \rfloor$.*

The arguments demonstrating that these two are in $\lfloor \mathsf{DCFL} \rfloor$ are almost identical. This will be useful, as it is known that their union is not in DCFL [8]. Another example of a language in $\lfloor \mathsf{DCFL} \rfloor$ is the following:

Proposition 8. *For every alphabet Σ with $\# \notin \Sigma$, the non-regular language $L = \{ w\#w^{\mathsf{R}} \mid w \in \Sigma^* \}$ is in $\lfloor \mathsf{DCFL} \rfloor$.*

As a final example, let Σ be an alphabet, and define the alphabet $\mathrm{D}(\Sigma) = \{ [_a,]_a \mid a \in \Sigma \}$. The *Dyck language*, $L_{\mathrm{D}}(\Sigma)$, is a language over $\mathrm{D}(\Sigma)$, generated by the context-free grammar

$$G = (\mathrm{D}(\Sigma), \{ S \}, P, S), \tag{8}$$

i.e., S is the only nonterminal symbol, and the productions P are given by

$$S \to S[_a S]_a S \mid \varepsilon, \quad \forall a \in \Sigma. \tag{9}$$

It is known that every CFL L can be written as $\varphi(L_{\mathrm{D}}(\Sigma) \cap R)$ for some alphabet Σ, regular R, and homomorphism φ (stated in the Chomsky-Schützenberger Theorem [12]), and so the Dyck languages are a significant class of non-regular languages.

Proposition 9. *For every alphabet Σ, $L_D(\Sigma) \in \lfloor \mathsf{DCFL} \rfloor$.*

Any of the above propositions imply the following.

Theorem 1. $\mathsf{REG} \subset \lfloor \mathsf{DCFL} \rfloor \subset \mathsf{DCFL}$.

4.1 Closure Properties and Decision Properties

First, closure properties are studied which follow from Proposition 2.

Proposition 10.

1. *$\lfloor \mathsf{DCFL} \rfloor$ is closed under right quotient with regular languages.*
2. *$\lfloor \mathsf{DCFL} \rfloor$ is closed under right concatenation with regular languages.*

Next, by Propositions 8 and 10(2), the language $\{ a^n \# a^n \# \mid n \geq 0 \}$ is in $\lfloor \mathsf{DCFL} \rfloor$. A pumping lemma for DCFL is used below to show that left concatenation by a regular language can produce a non-$\lfloor \mathsf{DCFL} \rfloor$ language.

Lemma 1. *Let $L' = \{ a^n \# a^n \# \mid n \geq 0 \} \in \lfloor \mathsf{DCFL} \rfloor$. Then $L = \{ a, \# \}^* L' \notin \mathsf{DCFL}$.*

Proof. We will use the pumping lemma from [20], which is stated as follows. For every $L \in \mathsf{DCFL}$, there exists a constant C for L such that, for every pair of words $w, w' \in L$, if $w = xy$ and $w' = xz$ with $|x| > C$, and y and z begin with the same letter, then one of the following statements is true:

1. There is a factorization $x = x_1 x_2 x_3 x_4 x_5$ with $x_2 x_4 \neq \varepsilon$ and $|x_2 x_3 x_4| \leq C$ such that for all $i \geq 0$, $x_1 x_2^i x_3 x_4^i x_5 \{ y, z \} \subseteq L$; or,
2. There exist factorizations $x = x_1 x_2 x_3, y = y_1 y_2 y_3$, and $z = z_1 z_2 z_3$ with $x_2 \neq \varepsilon$ and $|x_2 x_3| \leq C$ such that for all $i \geq 0$, $x_1 x_2^i x_3 \{ y_1 y_2^i y_3, z_1 z_2^i z_3 \} \subseteq L$.

Assume by contradiction that $L \in \mathsf{DCFL}$. Let C be this constant from the pumping lemma. Choose $w = a^C \# a^C \#, w' = a^C \# a^{3C+2} \# a^{3C+2} \#$ and

$$x = a^C \# a^{C-1}, \; y = a\#, \; z = a^{2C+3} \# a^{3C+2} \#. \tag{10}$$

Then either statement 1 or 2 must be true.

First suppose 1 is true. Let $x = x_1 x_2 x_3 x_4 x_5$ with $x_2 x_4 \neq \varepsilon$ and $|x_2 x_3 x_4| \leq C$ such that for all $i \geq 0$, $x_1 x_2^i x_3 x_4^i x_5 \{ y, z \} \subseteq L$. Suppose first that the non-empty parts of x_2 and x_4 are both before $\#$ (if one is empty, they can be after). Then setting $i = 0$, $x_1 x_3 x_5 y \notin L$ since it will have only two $\#$ symbols, and fewer a's before the first $\#$ than between the two $\#$'s. Suppose second that the non-empty parts of x_2 and x_4 are both after $\#$. Again setting $i = 0$, $x_1 x_3 x_5 y \notin L$. Next suppose that either x_2 or x_4 contain $\#$. Then again $x_1 x_3 x_5 y \notin L$ as it only has one $\#$ in it. Finally, suppose that x_2 is non-empty and before $\#$ and x_4 is non-empty and after $\#$. Then $x_1 x_3 x_5 z \notin L$ as the number of a's between the first and second $\#$ is fewer than between the second and third (the last). So in every case, 1 cannot be true.

Hence, 2 is true. Let $x = x_1 x_2 x_3, y = y_1 y_2 y_3$, and $z = z_1 z_2 z_3$ with $x_2 \neq \varepsilon$ and $|x_2 x_3| \leq C$ such that for all $i \geq 0$, $x_1 x_2^i x_3 \{ y_1 y_2^i y_3, z_1 z_2^i z_3 \} \subseteq L$. Then since $|x_2 x_3| \leq C$, it follows that x_2 starts, at the earliest, at the $\#$ of x. Assume x_2 contains $\#$. Then $x_1 x_3 y_1 y_3 \notin L$ as it only has one $\#$. Assume finally that x_2 is entirely after the $\#$ in x. Then $x_1 x_3 y_1 y_3 \notin L$ as it has fewer a's between the first and second $\#$ than before the first $\#$.

In every case, there is a contradiction, and so $L \notin \mathsf{DCFL}$. □

Proposition 11. *$\lfloor \mathsf{DCFL} \rfloor$ is not closed under left concatenation with regular languages. Hence, it is not closed under concatenation.*

The same language implies the following:

Corollary 1. *$\lfloor \mathsf{DCFL} \rfloor$ is not closed under reversal.*

In contrast to Proposition 11, if the regular language is over a separate alphabet, then the concatenation remains in $\lfloor \mathsf{DCFL} \rfloor$.

Proposition 12. *Let $L \in \lfloor \mathsf{DCFL} \rfloor$ be a language over Σ_L. Let $R \in \mathsf{REG}$ be a language over Σ_R with $\Sigma_L \cap \Sigma_R = \emptyset$. Then, $RL \in \lfloor \mathsf{DCFL} \rfloor$.*

Because closure under concatenation holds with disjoint alphabets, but not the same alphabet, non-closure under homomorphism immediately follows.

Corollary 2. *$\lfloor \mathsf{DCFL} \rfloor$ is not closed under (non-erasing) homomorphism.*

We aggregate several closure and non-closure results below. The closure results can be shown algebraically.

Proposition 13.

1. *$\lfloor \mathsf{DCFL} \rfloor$ is closed under inverse homomorphism.*
2. *$\lfloor \mathsf{DCFL} \rfloor$ is closed under union with regular languages.*

The following non-closure results can be shown by appealing to the non-closure properties of DCFL under the unmarked operations.

Proposition 14.

1. *$\lfloor \mathsf{DCFL} \rfloor$ is not closed under (marked) union.*
2. *$\lfloor \mathsf{DCFL} \rfloor$ is not closed under (marked) concatenation.*
3. *$\lfloor \mathsf{DCFL} \rfloor$ is not closed under intersection.*

Given point 2 above, the following result is surprising.

Proposition 15. *$\lfloor \mathsf{DCFL} \rfloor$ is closed under marked-$*$.*

There are three major open problems for closure properties; whether $\lfloor \mathsf{DCFL} \rfloor$ is closed under intersection with regular languages, complement, and Kleene-$*$.

Since it is known that the equality problem (given M_1, M_2, is $L(M_1) = L(M_2)$?) is decidable for DCFL [19], it is also decidable for $\lfloor \mathsf{DCFL} \rfloor$. Also the universality problem (given M, is $L(M) = \Sigma^*$?) is decidable for DCFL, as is the regularity problem (given M, is $L(M)$ regular?) [12].

Corollary 3. *The following problems are decidable for $\lfloor \mathsf{DCFL} \rfloor$: the equality problem, the universality problem, and the regularity problem.*

Decidability of the containment problem (given M_1, M_2, is $L(M_1) \subseteq L(M_2)$?) is an open question.

5 Properties of $\lceil \mathsf{DCFL} \rceil$

In this section, we are primarily interested in $\lceil \mathsf{DCFL} \rceil$, the smallest family of languages closed under left quotient with regular languages containing the deterministic context-free languages.

We will find multiple characterizations of $\lceil \mathsf{DCFL} \rceil$. First, we define the concept of a DPDA being preset with a language (we will define this generally over NPDA but use it only for DPDA). Let $M = (Q, \Sigma, \Gamma, \delta, q_0, Z_0, F)$ be a NPDA. Given a language $R \subseteq (\Gamma \setminus Z_0)^*$, the R-preset language accepted by M is

$$L_R(M) = \{ w \mid (q_0, w, Z_0 x) \vdash^* (q_f, \varepsilon, x'), q_f \in F, x \in R \}. \tag{11}$$

So, if M is a DPDA, and it is preset by any word in the preset language R, then M is initially allowed to act nondeterministically (as the preset word is chosen nondeterministically), but then must act deterministically for the remainder of the computation. Let $\mathsf{DCFL}_{\mathsf{REG}}$ be the family of all languages accepted by DPDA preset by regular languages. This notion of presetting a language on a store was previously studied for Turing machines [10] (the notion of preset pushdown automata studied in [17] used a different definition of preset). We will see next that this family is equal to $\lceil \mathsf{DCFL} \rceil$.

Lemma 2. $\lceil \mathsf{DCFL} \rceil \subseteq \mathsf{DCFL}_{\mathsf{REG}}$.

Proof. Let $M = (Q_M, \Sigma, \Gamma, \delta_M, q_{0,M}, Z_0, F_M)$ be a DPDA, and also let $A = (Q_A, \Sigma, \delta_A, q_{0,A}, F_A)$ be a DFA. We will show $L(A)^{-1}L(M) \in \mathsf{DCFL}_{\mathsf{REG}}$. We first build an intermediate NPDA M_1 that simulates both M and A in parallel using states in $Q_M \times Q_A$ on some guessed input $u \in \Sigma^*$ but by only reading ε until some nondeterministically guessed spot, where the state $q = (q_1, q_2)$ at this time is such that $q_2 \in F_A$; at this point it switches to a primed state q_1' (primed version of $q_1 \in Q_M$), then to q_1 (without the paired state q_2) where it will exactly simulate only M until it accepts. These primed states will be used to distinguish the pushdown contents after reading the guessed input u in the store language; let Q_M' be the set of all such states used in this construction.

We know that the store language of this intermediate machine, $S(M_1)$, is regular. Next, we build $M' \in \mathsf{DPDA}$, and we use preset language $R = S(M_1) \cap \Gamma^* Q_M' \in \mathsf{REG}$. Then, M' starts with some word xq' on the pushdown, then it pops q' and switches to state q from where it simulates M. Clearly, M' is deterministic.

We will prove next that $L_R(M') = L(A)^{-1}L(M)$.

Let $w \in L_R(M')$ starting from $xq' \in R$ on the pushdown. Thus, there exists some accepting computation of M_1 that, reading ε, but simulating $u \in \Sigma^*$ leads to x on the stack in state q', then to q, where it leads to a final state of M on $w \in \Sigma^*$. By the construction of M_1, $u \in L(A)$ and $uw \in L(M)$. Hence, $w \in L(A)^{-1}L(M)$.

Let $w \in L(A)^{-1}L(M)$. Then $uw \in L(M)$ and $u \in L(A)$. Consider an accepting computation of $w \in L(M_1)$ where it simulates M and A on u while reading ε, entering pushdown contents x in state (q_1, q_2), switches to q_1', then to q_1, and simulates M on w until it is in a final state. So, $xq' \in R$ and by using the latter part of this accepting computation, w is accepted by M' preset with this word. □

Lemma 3. $\mathsf{DCFL}_{\mathsf{REG}} \subseteq \lceil \mathsf{DCFL} \rceil$.

Proof. Let $M \in \mathsf{DPDA}$ and $A \in \mathsf{DFA}$, and look at $L_{L(A)}(M)$. We build $M' \in \mathsf{DPDA}$, $A' \in \mathsf{DFA}$ where A' accepts $L(A)\$$ ($\$$ a new symbol), and M', on input $u\$w$, simulates A' on $u\$$ while in parallel pushing u to the pushdown verifying that it is accepted by $L(A')$, then it simulates M on w, accepting if M does.

We will show next that $L(A')^{-1}L(M') = L_{L(A)}(M)$.

Let $w \in L(A')^{-1}L(M')$. Then $uw \in L(M'), u \in L(A')$. Hence, $u = u'\$$, where $u' \in L(A)$ and $u'\$w \in L(M')$. Then M' pushes u' to the pushdown and then verifies that $w \in L(M)$. Hence, $w \in L_{L(A)}(M)$.

Let $w \in L_{L(A)}(M)$. We see that M, when started with $u \in L(A)$ on the pushdown, accepts w. Then $u\$ \in L(A')$, and $u\$w \in L(M')$ by the construction of M'. Hence, $w \in L(A')^{-1}L(M')$. □

Putting these together, we see the two families are equal.

Proposition 16. $\mathsf{DCFL}_{\mathsf{REG}} = \lceil \mathsf{DCFL} \rceil$.

Next, we will demonstrate a second characterization using a model of limited-nondeterminism from [14,15]. Intuitively, these machines are only allowed to use determinism except in certain circumstances based on the contents of the pushdown. We use essentially the same notation as [15] (that particular paper equivalently represents the pushdown contents in reverse from our formulation). Given an NPDA $M = (Q, \Sigma, \Gamma, \delta, q_0, Z_0, F)$, let $R \subseteq (\Gamma \setminus Z_0)^*$ be some control language. Then M is called an R-NPDA if:

1. For all $q \in Q, a \in \Sigma \cup \{\varepsilon\}$, and $Z \in \Gamma$, δ can be written as
$$\delta(q, a, Z) = \delta_d(q, a, Z) \cup \delta_{nd}(q, a, Z), \tag{12}$$
where $(Q, \Sigma, \Gamma, \delta_d, q_0, Z_0, F)$ is a DPDA and $(Q, \Sigma, \Gamma, \delta_{nd}, q_0, Z_0, F)$ is an NPDA; and,
2. For all $q, q' \in Q, a \in \Sigma \cup \{\varepsilon\}, w \in \Sigma^*, Z \in \Gamma, x \in \Gamma^*$, and $x' \in (\Gamma \setminus \{Z_0\})^*$, $(q, aw, xZ) \vdash (q', w, xx')$, if $(q', x') \in \delta_{nd}(q, a, Z)$ with $xZ = Z_0x''$ where $x'' \in R$, or if $(q', x') \in \delta_d(q, a, Z)$ with $xZ = Z_0x''$ where $x'' \notin R$.

Hence, nondeterminism is allowed if the pushdown contents (above Z_0) are in R, otherwise it is not. In this paper, we are only concerned with the class $\{\varepsilon\}$-NPDA (called (∞, Z_0)-nPDA in [14]) and the family of languages they accept, $\mathcal{L}(\{\varepsilon\}\text{-}\mathsf{NPDA})$. With this class, machines must operate deterministically unless they are scanning the bottom-of-stack marker Z_0 at which point they can use nondeterminism. The following was shown in [14]:

Proposition 17. *[14]* $\mathcal{L}(\{\varepsilon\}\text{-}\mathsf{NPDA}) = \Omega^{\mathsf{REG}}(\mathsf{DCFL})$ *(the smallest family containing* DCFL *that is closed under union, concatenation, and Kleene-$*$).*

Next, we compare $\mathcal{L}(\{\varepsilon\}\text{-NPDA})$ to $\mathsf{DCFL}_{\mathsf{REG}}$.

Lemma 4. $\mathsf{DCFL}_{\mathsf{REG}} \subseteq \mathcal{L}(\{\varepsilon\}\text{-NPDA})$.

Proof. Let $S \in \mathsf{REG}$ and $M = (Q_M, \Sigma, \Gamma, \delta_M, q_{0,M}, Z_0, F_M) \in \mathsf{DPDA}$, and we are examining $L_S(M)$. Let $A = (Q_A, \Gamma, \delta_A, q_{0,A}, F_A) \in \mathsf{DFA}$ that accepts $S^{\mathsf{R}} \subseteq (\Gamma \setminus \{Z_0\})^*$ and make A such that there is a path from each state to some final state. We build an $\{\varepsilon\}$-NPDA M' as follows. Essentially M' will remember in the finite control a pair (q, b) where q is a state of A and $b \in \Gamma$, and at the same time it will simulate M. At the beginning of the computation (when Z_0 is at the top of the stack and so nondeterminism is allowed), M' switches to remembering some nondeterministically guessed (q, b) where $\delta_A(q_{0,A}, b) = q$. Then when the pushdown of M' has Z_0 at the top, it simulates M as if b were on top of the stack; if it pushes, then M' pushes the same symbol; if it pops, then M' keeps Z_0 but again uses nondeterminism to guess (q', b') such that $\delta_A(q, b') = q'$, and it now instead remembers (q', b'); if it instead keeps the same pushdown contents, then it continues simulating as if b were the topmost symbol; if the pushdown of M' is non-empty (not scanning Z_0), then it will simulate M (which has no branching) deterministically. Optionally, at any nondeterministically guessed time when it has remembered $(q_f, b), q_f \in F_A$ and it has just simulated a pop of b (on Z_0), M' will no longer use A and simply simulate M verbatim including on transitions on Z_0. Then M accepts if the simulated DPDA M is in a final state.

Let $w \in L_S(M)$. There is an accepting computation α of w starting from some $Z_0 x \in Z_0 S$ on the stack. In A, let $x^{\mathsf{R}} = b_1 \cdots b_n, b_i \in \Gamma$ for $1 \leq i \leq n$, and $q_0 = q_{0,A}$ and $\delta_A(q_{i-1}, b_i) = q_i$ for i, $1 \leq i \leq n$. Then M' nondeterministically guesses b_1 and remembers (q_1, b_1). Then M' operates just as α does on all stack symbols that are not the original $b_1 \cdots b_n$ and the top of the stack will be the same in M'. But if the topmost symbol is b_i from the preset contents (which must appear in α in order from $i = 1$ to n) then M' simulates as if b_i were on the top of the stack. If the simulation pops, it will use nondeterminism (on Z_0) to guess the next symbol b_{i+1} which causes (q_{i+1}, b_{i+1}) to be remembered. Continuing in this fashion, when the simulation of M reaches a final state, M' does as well.

Let $w \in L(M')$ on an accepting computation α. When simulating A with Z_0 on top of the stack of M', let $(q_1, b_1), \ldots, (q_n, b_n)$ be the symbols remembered. Either $q_n \in F$ and it stops simulating A, or it does not stop simulating A. In the former case, let $y = \varepsilon$, and in the latter case, because there is a path from every state to a final state, there is a $y \in \Gamma^*$ that leads to a final state from q_n. In either case, $z = b_1 \cdots b_n y \in L(A)$. Then M will accept w with z^{R} preset on the stack. All transitions executed when the stack of M' is Z_0 simulating b_i for i from 1 to n can be executed by M from what remains of the preset pushdown contents, $y^{\mathsf{R}} b_n \cdots b_i$. And, all transitions on Z_0 after A has stopped (if that happens) are simulated verbatim. Similarly, all transitions of $c \neq Z_0$ on the top of the pushdown can occur verbatim in M. □

Lemma 5. $\mathcal{L}(\{\varepsilon\}\text{-NPDA}) \subseteq \mathsf{DCFL}_{\mathsf{REG}}$.

Proof. Let $M \in \{\varepsilon\}$-NPDA. Let n be the number of transitions of M that are defined with Z_0 on the top of the stack where nondeterminism is allowed, and let $b_1, \ldots, b_n$ be new symbols in bijective correspondence with these transitions. Let $\Delta = \{b_1, \ldots, b_n\}$ be a new alphabet. We build a DPDA M_1 preset with Δ^*. Then M_1 operates as follows with a string x preset on the stack: If a symbol $b_i \in \Delta$ is at the top of the stack, then M_1 simulates the transition of M associated with b_i (which is a transition on Z_0) if it can (if transition b_i pushes some $c \notin \Delta$, then M_1 pops b_i then pushes c; there are no transitions that pop Z_0; and if it stays on Z_0, then M_1 pops b_i). This simulation is deterministic since b_i on the stack implies the transition to apply. If a symbol of $\Gamma \setminus \{Z_0\}$ is at the top, then it operates as with M exactly which is also deterministic. Hence M_1 is deterministic.

Let $w \in L(M)$. Given an accepting computation α of M, let $x = b_{i_1}, \ldots, b_{i_n}$ be the sequence of transitions applied in α when Z_0 is at the top of the stack. Then M_1 will start with x^{R} on the stack; M_1 simulates α exactly as described, where α having Z_0 on top for the jth time will have b_{i_j} on top of the stack and executing b_{i_j} will have M_1 to simulate b_{i_j}, and as α pushes symbols above Z_0, they will get pushed above the symbols of Δ^* in M_1. Hence, $w \in L_{\Delta^*}(M_1)$.

Let $w \in L_{\Delta^*}(M_1)$ starting with $x = b_{i_n} \cdots b_{i_1}$ on the pushdown using accepting computation α. M can operate just as α does but instead of popping b_{i_j}, it will have Z_0 on the top of the stack and choose b_{i_j} nondeterministically. It is clear that $w \in L(M)$. □

The paper [14] also showed that the smallest union-closed family containing DPDA is a proper subset of $\mathcal{L}(\{\varepsilon\}$-NPDA), which in turn is a proper subset of CFL. Putting this all together, we obtain:

Proposition 18. *The hierarchy in Fig. 1 is correct.*

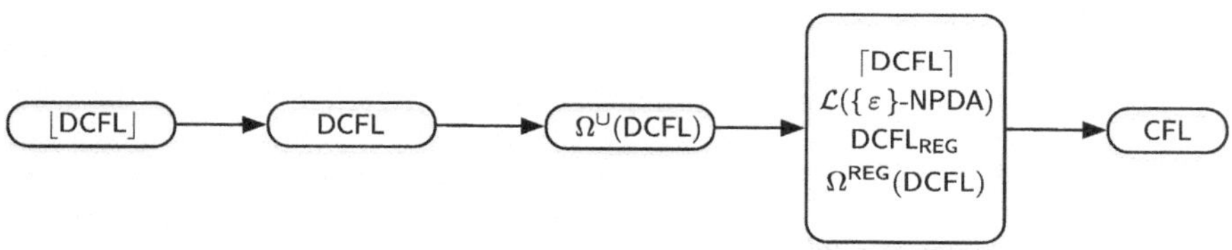

Fig. 1. The image depicts the hierarchy between language families, where those in the same box are equal, and arrows are proper inclusions.

By definition, $\lceil \mathsf{DCFL} \rceil = \Omega^{-1(\mathsf{REG})}(\mathsf{DCFL})$ where $-1(\mathsf{REG})$ means left quotient with regular languages. And so, it is surprising that the smallest language family containing DCFL that is closed under left quotient with regular languages is the same as the smallest family containing DCFL that is closed under all of union, concatenation, and Kleene-$*$. It is also interesting that $\lceil \mathsf{DCFL} \rceil$ is so natural that there are four equivalent methods constructed using quite different formulations involving left quotient, union/concatenation/Kleene-$*$, preset automata, and limited nondeterminism.

In [1], it was shown that all languages in the family $\Omega^{\mathsf{REG}}(\mathsf{DCFL})$ can be recognized in linear time, and it contains all $\mathsf{LR}(k)$ languages, as well as some inherently ambiguous languages. As noted in the introduction, erasing DCFL by homomorphisms gives exactly CFL, which cannot be recognized even in $O(n^2)$ time. However, erasing using left quotient with regular languages still increases the power to contain inherently ambiguous languages, but without increasing recognition time beyond linear time.

Corollary 4. $\lceil \mathsf{DCFL} \rceil$ *languages can be recognized in linear time, and they contain inherently ambiguous languages.*

5.1 Closure Properties and Decision Problems

From the previous characterization, we can use the known closure property results for $\mathcal{L}(\{\varepsilon\}\text{-}\mathsf{NPDA})$ shown in [14,15], summarized as follows:

Proposition 19. $\lceil \mathsf{DCFL} \rceil$ *is closed under union, concatenation, Kleene-$*$, inverse homomorphism, and intersection with regular languages, but is not closed under homomorphism and complement.*

It is shown in [14] that $L = \{ww^{\mathsf{R}} \mid w \in \{a,b\}^+\} \notin \Omega^{\mathsf{REG}}(\mathsf{DCFL})$. So, we have the following stronger result.

Corollary 5. $\lceil \mathsf{DCFL} \rceil$ *is not closed under non-erasing homomorphism.*

We also see the following:

Proposition 20. $\lceil \mathsf{DCFL} \rceil$ *is closed under left and right quotient with regular languages.*

The next result follows from the non-closure properties of CFL.

Proposition 21. $\lceil \mathsf{DCFL} \rceil$ *is not closed under intersection.*

In terms of decision problems, this was studied in [15] where the following were shown (translated here to $\lceil \mathsf{DCFL} \rceil$).

Proposition 22. *The following problems are undecidable for* $\lceil \mathsf{DCFL} \rceil$*: the equality problem, the containment problem, the universality problem, the regularity problem, and whether it accepts a given* $S \in \mathsf{REG}$.

All of these problems except for the containment problem are decidable for DCFL. Many of these problems are undecidable even for the suffix operation (equivalent to a simple left quotient with Σ^*) rather than left quotient by regular languages generally [6].

Table 1. Closure results for ⌊DCFL⌋ and ⌈DCFL⌉. A ✓ means it is closed, ✗ means it is not closed, and ? means it is an open question.

	REG	⌊DCFL⌋	DCFL	⌈DCFL⌉	CFL
complement	✓	?	✓	✗	✗
reversal	✓	✗	✗	?	✓
right concat with REG	✓	✓	✓	✓	✓
left concat with REG	✓	✗	✗	✓	✓
right quotient with REG	✓	✓	✓	✓	✓
left quotient with REG	✓	✓	✗	✓	✓
intersection with REG	✓	?	✓	✓	✓
union with REG	✓	✓	✓	✓	✓
union	✓	✗	✗	✓	✓
concat	✓	✗	✗	✓	✓
Kleene-$*$	✓	?	✗	✓	✓
marked-union	✓	✗	✓	✓	✓
marked-concat	✓	✗	✓	✓	✓
marked-$*$	✓	✓	✓	✓	✓
intersection	✓	✗	✗	✗	✗
inverse homomorphism	✓	✓	✓	✓	✓
non-erasing homomophism	✓	✗	✗	✗	✓

6 Conclusions and Open Problems

In this paper, we performed a more nuanced study of the operation of left quotient by regular languages on more powerful types of automata, with the focus on the deterministic context-free languages (DCFL). We studied the family of all DCFL languages such that the left quotient by regular languages are always in DCFL (denoted by ⌊DCFL⌋). This family is general enough to contain the Dyck languages and to be closed under several operations such as marked-$*$, but not under certain simple operations such as marked union or marked concatenation. We also study the smallest family containing DCFL that is closed under left quotient by regular languages (denoted by ⌈DCFL⌉). Several characterizations of ⌈DCFL⌉ are shown which surprisingly demonstrates that the smallest family containing DCFL that is closed under left quotient with regular languages is equal to the smallest family containing DCFL that is closed under union, concatenation, and Kleene-$*$. They can also all be recognized in linear time despite containing inherently ambiguous languages.

The results regarding closure properties are summarized in Table 1. The open problems are listed in the table as well. One property that seems to be surprisingly difficult is intersection with regular languages for ⌊DCFL⌋. Decidability of the containment problem for $\lfloor \mathsf{DCFL} \rfloor$ also remains open. Examining these language family operations on other deterministic classes is of future interest.

Acknowledgments. Supported, in part, by Natural Sciences and Engineering Research Council of Canada (NSERC) Grant 2022-05092 (I. McQuillan) and scholarship (B. Lockinger).

Disclosure of Interests. The authors have no competing interests to declare that are relevant to the content of this article.

References

1. Bertsch, E., Nederhof, M.J.: Regular closure of deterministic languages. SIAM J. Comput. **29**(1), 81–102 (1999)
2. Brzozowski, J.A.: Derivatives of regular expressions. J. ACM **11**(4), 481–494 (1964)
3. Brzozowski, J.A.: Quotient complexity of regular languages. J. Autom. Lang. Comb. **15**, 71–89 (2010)
4. Chandler, W.J.: Abstract families of deterministic languages. In: Proceedings of the First Annual ACM Symposium on Theory of Computing (STOC 01), pp. 21–30 (1969)
5. Eremondi, J., Ibarra, O.H., McQuillan, I.: Deletion operations on deterministic families of automata. Inf. Comput. **256**, 237–252 (2017)
6. Eremondi, J., Ibarra, O.H., McQuillan, I.: On the density of context-free and counter languages. Int. J. Found. Comput. Sci. **29**(2), 233–250 (2018)
7. Ginsburg, S.: Algebraic and Automata-Theoretic Properties of Formal Languages. North-Holland Publishing Company, Amsterdam (1975)
8. Ginsburg, S., Greibach, S.A.: Deterministic context free languages. Inf. Control **9**(6), 620–648 (1966)
9. Greibach, S.A.: A note on pushdown store automata and regular systems. Proc. Am. Math. Soc. **18**, 263–268 (1967)
10. Greibach, S.A.: One way finite visit automata. Theoret. Comput. Sci. **6**, 175–221 (1978)
11. Hopcroft, J.E., Ullman, J.D.: Deterministic stack automata and the quotient operator. J. Comput. Syst. Sci. **2**, 1–12 (1968)
12. Hopcroft, J.E., Ullman, J.D.: Introduction to automata theory, languages, and computation. Addison-Wesley Series in Computer Science and Information Processing, Addison-Wesley (1979)
13. Ibarra, O.H., McQuillan, I.: On store languages of languages acceptors. Theoret. Comput. Sci. **745**, 114–132 (2018)
14. Kutrib, M., Malcher, A.: Context-dependent nondeterminism for pushdown automata. Theoret. Comput. Sci. **376**(1), 101–111 (2007)
15. Kutrib, M., Malcher, A., Werlein, L.: Regulated nondeterminism in pushdown automata. Theoret. Comput. Sci. **410**(37), 3447–3460 (2009)
16. Kutrib, M., Prigioniero, L.: Kernels of context-free languages. Int. J. Found. Comput. Sci. 1–25 (2024)
17. van Leeuwen, J.: Notes on pre-set pushdown automata, pp. 177–188. Springer Berlin Heidelberg, Berlin, Heidelberg (1974)
18. Nerode, A.: Linear automaton transformations. Proc. Am. Math. Soc. **9**, 541–544 (1958)

19. Sénizergues, G.: $L(A) = L(B)$? Decidability results from complete formal systems. Theoret. Comput. Sci. **251**(1–2), 1–166 (2001)
20. Yu, S.: A pumping lemma for deterministic context-free languages. Inf. Process. Lett. **31**(1), 47–51 (1989)

Checking whether Two Unambiguous Context-Free Grammars Describe the Same Set of Strings of Length n

Vladislav Makarov(✉)

St. Petersburg State University, 7/9 Universitetskaya nab., Saint Petersburg 199034, Russia
vm450@yandex.ru

Abstract. The slice equivalence problem for unambiguous context-free grammars is stated as follows: given two unambiguous context-free grammars G_1 and G_2, tell whether or not they describe the same set of strings of length n. The paper presents an $O(\text{poly}(n, |G_1|, |G_2|))$ randomized algorithm for this problem. If the sets are different, the algorithm constructs the lexicographically smallest string on which they do not agree. The algorithm combines the idea of the Cocke–Kasami–Younger parsing algorithm and the Schwartz–Zippel lemma.

1 Introduction

Proving decidability or undecidability of the equivalence problem for unambiguous context-free grammars is an important open question in formal language theory. The equivalence problem for unambiguous grammars (from here on, I may omit the word "context-free", as long as it does not cause confusion) is stated in the following way:

Problem 1 (The equivalence problem for unambiguous grammars) *You are given two context-free grammars G_1 and G_2. Moreover, both grammars are* guaranteed to be unambiguous *(this is a promise problem). Is there an algorithm to tell whether the languages $L(G_1)$ and $L(G_2)$ are equal?*

The problem appears to be very difficult. So far, only partial progress has been made. Specifically, Semenov [20] proved that the problem is decidable when one of the grammars is regular (and, therefore, describes a regular language). A later result by Chistov and Grigoryev shows that Semenov's approach works in PSPACE [6]. Much later, in a big achievement, Sénizergues [21] proved that the problem is decidable when both languages are described by a deterministic pushdown automaton (DPDA). In grammar terms, this means that both grammars are LR-grammars, which are known to be equivalent to DPDA in power. Later, Stirling [23] and Jančar [16] provided two new algorithms for the same problem, both with a primitive recursive upper bound on the running time.

Supported by Russian Science Foundation, project 23-11-00133.

S.-K. Ko and F. Manea (Eds.): DLT 2025, LNCS 16036, pp. 138–152, 2026.
https://doi.org/10.1007/978-3-032-01475-7_10

Even if we know in advance that the grammars describe different languages, finding the shortest witness of that fact is not easy. It is well-known that the length of the shortest witness grows at least exponentially as a function of the sizes of the original unambiguous grammars. In fact, even the length of the shortest string described by an unambiguous grammar can grow exponentially with the size of the grammar! Indeed, a folklore example of such a grammar is the following unambiguous grammar, generating only the string a^{2^n} (X_1 is the starting nonterminal):

$$
\begin{aligned}
X_1 &\to X_2X_2 \\
X_2 &\to X_3X_3 \\
&\dots \\
X_{n-1} &\to X_nX_n \\
X_n &\to aa
\end{aligned}
$$

Therefore, iterating over all short strings leads to an at least doubly-exponential solution and is completely infeasible. On the other hand, if we could quickly process all strings of some fixed length, the running time would drastically improve.

Hence, it makes sense to consider the *slice equivalence* problem for unambiguous grammars: given two unambiguous grammars G_1 and G_2, tell whether or not they describe the same set of strings of length n.

Definition 1 For a language L, its *n-slice* is the language $\{ w \mid w \in L, |w| = n \}$ of all strings from L of length exactly n. In other words, the n-slice of L is the language $L \cap \Sigma^n$.

Problem 2 (Slice equivalence problem for unambiguous grammars) *Given two unambiguous context-free grammars G_1 and G_2 over an alphabet Σ, tell whether or not they describe the same set of strings of length n. In other words, tell whether or not the n-slices of $L(G_1)$ and $L(G_2)$ are equal.*

The naïve algorithm for this problem needs to iterate over all $|\Sigma|^n$ strings of length n. The main result of this paper is that the slice equivalence problem for unambiguous grammars can be solved in randomized $O(\text{poly}(n, |G_1|, |G_2|))$ time, with one-sided error. More precisely, the algorithm can say that n-slices are equal even if they are different, but with small probability. If the algorithm tells that the slices are different, they are assuredly different. Here, $|G|$ means the total length of all rules of the grammar.

The algorithm works by combining the ideas of the Cocke–Kasami–Younger parsing algorithm and the Schwartz–Zippel lemma. Hence, in a tongue-in-cheek way, I will refer to it as the Cocke–Kasami–Younger–Schwartz–Zippel algorithm (the CKYSZ algorithm for short). Similar ideas have appeared before, for example, in the quasipolynomial [14] and polynomial [18] algorithms for approximately counting strings of given length that are described by a (possibly ambiguous) context-free grammar.

Such an algorithm can be used as a heuristic algorithm for Problem 1. Indeed, let us pick some computable function $\ell(|G_1|, |G_2|)$ — an upper bound on the length of the first string where the languages $L(G_1)$ and $L(G_2)$ differ. Then, for each length n that does not exceed $\ell(|G_1|, |G_2|)$, run the polynomial algorithm for the equivalence of n-slices of $L(G_1)$ and $L(G_2)$. If, for some $n \leqslant \ell(|G_1|, |G_2|)$, the slices are different, we can conclude that the grammars are not equivalent. Otherwise, we cannot conclude anything, but will heuristically assume that any function ℓ that grows quickly enough works.

Hence, we can use the solution of Problem 2 to get a rather fast heuristic algorithm for Problem 1. In general, there was quite a lot of research on the heuristic algorithms for Problem 1.

Another heuristic approach to Problem 1 was suggested by Almeida et al. [1] and is based on the idea of matrix substitution: substituting terminal symbols by real matrices, strings by the product of matrices that correspond to the terminal symbols and languages by the sums of the matrices that corresponds to strings. In the end, a single real matrix is computed for each language. If the resulting matrices are different, then the languages $L(G_1)$ and $L(G_2)$ are assuredly different. If the resulting matrices are equal, then the languages are likely the same. These matrices can be computed either via dynamic programming (for cycle-free grammars) or via iterative methods (when cycles are present) [1].

This heuristic is much better at capturing "global" properties of the languages $L(G_1)$ and $L(G_2)$, compared to the "try all short strings" heuristic and takes care of most cases where trying all short strings fails. However, it still has its limitations. The main obstacle that prevents this heuristic from being a full solution to Problem 1, is the presence of so-called matrix polynomial identities, including the famous Amitsur–Levitsky identity [2]. In their paper, Almeida et al [1] give an example of two languages that cannot be distinguished by substituting 2×2 matrices, regardless of the matrices that are used to substitute each individual terminal symbol. Using Amitsur–Levitsky identity, it is possible to construct similar examples for $d \times d$ matrices, where d is any integer [1,2]. On the other hand, the algorithms suggested in this paper have no issues with polynomial identities. So, in some sense, there is a tradeoff between the "working for *all* string lengths" and the "guaranteed correctness" aspects present here.

Some other heuristic algorithms for Problem 1 have been studied, but they are beyond the scope of this paper. In particular, there is quite a lot of research on the algorithms that can sometimes provide a fully trustworthy proof of the grammar equivalence [12,17] (in contrast to the more expected kind of heuristics, which can only prove that the grammars are not equivalent)!

As mentioned above, Semenov solved [20] the special case of Problem 1. Specifically, the case when one of the grammars is regular. His solution is closely related to the concept of the *generating function* of a language, as introduced by Chomsky and Schützenberger [7]. For a language L, its generating function is the infinite sum $\sum_{n=0}^{+\infty} a_n x^n$, where a_n is the number of strings of length n in the language L and x is a variable (it is known that this sum converges when x is close to zero; hence, it makes sense to treat it as a function of x). The famous Chomsky-Schützenberger enumeration theorem [7] states that, for an unambiguous context-free grammar G, the generating function of the language

$L(G)$ has an important property of being *algebraic* (the exact definition of an algebraic function is not too important for this paper).

The Chomsky-Schützenberger theorem has many interesting consequences. In particular, it can be used to prove that some languages cannot be described by an unambiguous grammar [13]. More importantly for the topic of this paper, the Chomsky-Schützenberger theorem can be combined with Semenov's method to show that the equivalence problem for an unambiguous grammar and a regular language can be reduced to the problem of verifying whether or not two implicitly defined algebraic functions are equal [20].

Hence, Semenov's approach to the aforementioned special case of Problem 1 is a small part of the general study of algebraic function identity testing and solving algebraic systems. By a result of Chistov and Grigoryev [6] algebraic function identity testing is known to be in PSPACE. There are many different approaches to the different special cases of the general problem of solving algebraic systems. However, to the best of my knowledge, all known methods for the algebraic function identity testing require exponential time in the worst case. For a comprehensive report on the different approaches to the algebraic function identity testing, I suggest referring to a survey paper by Ayad [3].

Moreover, it should be noted that Semenov's approach very heavily relies on the fact that one of the languages is regular. Therefore, there is no obvious way to generalize it to solve Problem 1 in its full glory. Of course, one can compute the algebraic functions $f_1(x)$ and $f_2(x)$ that are the generating functions of $L(G_1)$ and $L(G_2)$, respectively. However, in that case, $f_1(x) = f_2(x)$ does not imply that the languages $L(G_1)$ and $L(G_2)$ are equal. Indeed, $f_1(x) = f_2(x)$ implies that, for all nonnegative integers n, the n-slices of $L(G_1)$ and $L(G_2)$ have the *same cardinality*. However, it does not imply that the n-slices of $L(G_1)$ and $L(G_2)$ are equal *as sets*. I am currently unaware of any way to use the general theory of algebraic identity testing to either solve Problem 1 or to get a polynomial-time solution to Problem 2.

As we can see, the problem of computing the generating function of a language is closely related to Problem 1. Similarly, the problem of computing the n-th coefficient of the generating function is closely related to Problem 2. In simpler words, let us consider the problem of *slice size*: find the size of $|L(G) \cap \Sigma^n|$ for a given context-free grammar G. It is important to note that the results of this paper are not directly related to the slice size problem. The reason I am talking about the slice size problem and the algebraic identity testing problem is to give a somewhat comprehensive overview of the related work.

The relationship between the slice equivalence, the slice size and the general equivalence problems is quite complicated. Henceforth, when talking about the computational complexity of said problems, I will assume that the grammar/grammars is/are given as a part of the input. Moreover, for the slice size and the slice equivalence problems, I will also assume that n is given in unary.

For unambiguous grammars, slice size is, in many ways, easier than slice equivalence. Similarly, for unambiguous finite automata (UFAs) slice size is a pretty simple problem, while slice equivalence is harder and general equivalence is harder still, but still solvable in polynomial time by a classic result of Stearns and Hunt [22]. For deterministic finite automata (DFAs), which can be seen as a special case of UFAs, there are even simpler polynomial-time folklore algorithms.

On the other hand, for NFAs slice equivalence is co-NP-complete [11] and slice size is #P-complete [11], making slice size intuitively harder than slice equivalence. Because NFAs can be seen as a special case of context-free grammars, the slice equivalence and the slice size problems remain co-NP-hard and #P-hard, respectively, for context-free grammars as well. On the other hand, if n-slices of $L(G_1)$ and $L(G_2)$ differ, then any string w of length n from the symmetric difference of $L(G_1)$ and $L(G_2)$ is a polynomially verifiable witness of the fact that $L(G_1) \cap \Sigma^n \neq L(G_2) \cap \Sigma^n$. Indeed, to verify w it is enough to run the standard cubic parsing algorithm on w two times: the first time to check, whether w is in $L(G_1)$, and the second time to check, whether w is $L(G_2)$. If the results are different, then w is indeed a string of length n that is in one of the languages, but not in the other one. Hence, slice equivalence for context-free grammars is in co-NP. Similarly, slice size for context-free grammars is in #P, because the definition of #P allows one to run the cubic parsing algorithm on all strings of length n "in parallel". Therefore, the slice equivalence and the slice size problems for context-free grammars are co-NP-complete and #P-complete, respectively.

On the other hand, the general equivalence problem is PSPACE-complete for NFAs [19] and undecidable for context-free grammars [14]. It can be clearly seen that the intuitive expectation of the slice size and the slice equivalence problems becoming much harder in the presence of ambiguity does indeed correspond to the actual reality. It is important to note that it is pretty easy to adapt the algorithm for the slice size of unambiguous grammars to count *the total number of parse trees of all strings of length n* for an arbitrary context-free grammar. However, the presence of ambiguities means that such an algorithm cannot be used to (exactly) compute *the number of strings of length n*.

When grammars are fixed, similar results are known. For example, for any *fixed* (in other words, the grammar is not a part of the input; the whole input is the number n in the unary notation) unambiguous grammar G, the slice size problem is in NC^2 (more precisely, it can be solved by a log-uniform family of Boolean circuits of depth $O(\log n \log\log n)$ and polynomial size [5]). On the other hand, there are known ambiguous context-free grammars, for which the slice size problem is $\#\mathsf{P}_1$-complete [5], where $\#\mathsf{P}_1$ is, informally speaking, the unary equivalent of #P.

2 The Cocke–Kasami–Younger–Schwartz–Zippel Algorithm

Let us solve Problem 2. The solution will be a modification of the standard cubic Cocke–Kasami–Younger parsing algorithm for context-free grammars. Firstly, let us note that both G_1 and G_2 can be assumed to be in the Chomsky normal form. Indeed, transforming a grammar to the Chomsky normal form takes polynomial time and increases the size of the grammar at most polynomially. Here, it is very important that the transformation to the Chomsky's normal form preserves the unambiguity of the grammar [15]. As it happens in many similar situations, it is possible to avoid the reduction overhead by applying Earley's algorithm (a cubic parsing algorithm for arbitrary context-free grammars) [10]. However, for simplicity, let us consider only the case when both grammars are in the Chomsky normal form. Moreover, we may assume without loss of generality that both grammars do not have any useless nonterminals or rules (nonterminals and rules that are not a part of any valid parse tree; removing all useless nonterminals and rules is a standard procedure).

The main idea of the algorithm is the following *indexation trick*. Consider polynomials with integer coefficients in $|\Sigma| \cdot n$ variables $x_{a,i}$, where a ranges over Σ and i ranges over $[1, n]$. Then, we can define the following mapping f from Σ^n to monomials in variables $x_{a,i}$: a single string $w_1 w_2 \dots w_n$ maps to a monomial $f_w := x_{w_1,1} \cdot x_{w_2,2} \cdot \dots \cdot x_{w_n,n}$. Informally speaking, a variable $x_{a,i}$ in the product corresponds to the i-th character of the string being a. Here, it is very important that the variables "record" *both* the character and the index of the position, and not only the character itself. Now, let us lift this mapping f to subsets of Σ^n by linearity. Specifically, let us say that f maps a language $L \subset \Sigma^n$ to the polynomial $f_L := \sum\limits_{w \in L} f_w$ in variables $x_{a,i}$.

It is important to note that these polynomials are commutative. There is no need to substitute matrices or other non-commuting objects in place of variables, because the indexing itself takes care of the noncommutativity of the string concatenation. For example, $f_{ab} = x_{a,1} x_{b,2}$, but $f_{ba} = x_{b,1} x_{a,2} = x_{a,2} x_{b,1} \neq x_{a,1} x_{b,2}$. Hence, different subsets of Σ^n have different images under f: if a string w is present in K, but not in L, then the monomial $f_w = x_{w_1,1} x_{w_2,2} \dots x_{w_n,n}$ is present in f_K, but not in f_L. Moreover, all coefficients of f_K are either 0 or 1: 1 if the corresponding string is in K and 0 otherwise.

Theorem 1 (Cocke–Kasami–Younger–Schwartz–Zippel algorithm) *There is an algorithm that, given unambiguous context-free grammars G_1 and G_2 in the Chomsky normal form, tells whether or not they describe the same set of strings of length n in $O(n^3 \cdot (|G_1| + |G_2|))$ time with bounded one-sided error from randomization. More specifically, the algorithm may only fail by claiming that different sets are equal and does so with probability at most n/p per iteration, where $p \geqslant 2n$ is a prime number, such that the arithmetic operations with integers modulo p are assumed to be performed in $O(1)$ time.*

Proof. Denote the n-slices of $L(G_1)$ and $L(G_2)$ by L_1 and L_2 respectively. Then, $f_{L_1} = f_{L_2}$ if and only if $L_1 = L_2$. Moreover, both f_{L_1} and f_{L_2} are polynomials of degree at most n, because they both are sums of several monomials of degree exactly n. But both f_{L_1} and f_{L_2} can have exponentially many coefficients. Hence, it is inefficient to compute them explicitly. However, as it turns out, we can compute their values at any given point! Let us do that. More precisely, it is necessary to prove the following Lemma 1. For now, let us focus on proving Lemma 1, the proof of Theorem 1 will be resumed after the proof of Lemma 1.

Lemma 1 *For each subset K of Σ^n consider a polynomial $f_K = \sum_{w \in K} (x_{w_1,1} \cdot x_{w_2,2} \cdot \ldots \cdot x_{w_n,n})$ in $|\Sigma| \cdot n$ variables $x_{a,i}$, where a runs over Σ and i runs over the range $[1, n]$. Given an unambiguous grammar G in the Chomsky normal form and a finite field F, it is possible to evaluate the polynomial f_M in a point $y \in F^{|\Sigma| \cdot n}$ in $O(n^3 \cdot |G|)$ field operations, where M is the n-th slice of $L(G)$.*

Proof (Proof of Lemma 1). The proof is an adaptation of the standard cubic Cocke–Kasami–Younger parsing algorithm. As mentioned at the start of this section, we can assume that G contains neither useless nonterminals nor useless rules. Otherwise, we can just remove them, as it will not affect the language $L(G)$. Let us say that the concatenation $K \cdot L$ of two languages K and L is *unambiguous*, if, for every string w from $K \cdot L$, there is only one partition of the form $w = uv$, where $u \in K$ and $v \in L$. If there are no useless rules, then, for every "basic" rule $C \to DE$ of G, the concatenation $L(D)L(E)$ is unambiguous.

For a non-empty subsegment $(\ell, r]$ of $(0, n]$ and a language K, define $f_K^{(\ell,r]} := \sum_{w \in K \cap \Sigma^{r-\ell}} x_{w_1,\ell+1} x_{w_2,\ell+2} \ldots x_{w_{r-\ell},r}$. This is a generalization of f that has been defined before. Indeed, $f_K^{(0,n]} = f_K$ when $K \subset \Sigma^n$.

Then, $f_{K \uplus L}^{(\ell,r]} \equiv f_K^{(\ell,r]} + f_L^{(\ell,r]}$ is always true (from here on, the sign $\uplus$ is used to denote the disjoint union) and $f_{KL}^{(\ell,r]} \equiv \sum_{m=\ell+1}^{r-1} f_K^{(\ell,m]} f_L^{(m,r]}$ is true as long as the concatenation $K \cdot L$ is unambiguous and both K and L do not contain the empty string. The second equation is true because the right-hand side "iterates over" all ways of splitting the string of length $r - \ell$ into two non-empty strings (the unambiguity of the concatenation ensures that each string of length $r - \ell$ contributes to the right-hand side at most once).

Hence, for a nonterminal C of G and a non-empty subsegment $(\ell, r]$ of $(0, n]$, we get the following conditions (here, R is the set of all rules of G):

$$f_{L(C)}^{(\ell,r]} := \begin{cases} \sum\limits_{(C \to a) \in R} x_{a,\ell+1}, & \text{if } r - \ell = 1, \\ \sum\limits_{(C \to DE) \in R} \sum\limits_{m=\ell+1}^{r-1} f_{L(D)}^{(\ell,m]} f_{L(E)}^{(m,r]}, & \text{otherwise.} \end{cases} \tag{1}$$

In the second of the above cases, we implicitly use the unambiguity of the concatenation $L(D) \cdot L(E)$ by iterating over m and using the fact that, due to the unambiguity, each string will contribute to the sum at most once.

Now, let us evaluate $f_{L(C)}^{(\ell,r]}$ in the point y. To do so, apply the above formulas in the order of increasing $r - \ell$. In the end, the value of $f_{L(S)}^{(0,n]}$ in y is exactly what we needed to compute. The total number of field operations is $O(n^3 \cdot |R|) = O(n^3 \cdot |G|)$: for each triple $\ell < m < r$, we iterate over all rules of the form $C \to DE$ in the grammar. □

Proof (Continuation of the proof of Theorem 1). Let us choose a prime p that is at least $2n$. Consider the field F of remainders modulo p. It has exactly p elements. Because all coefficients of f_{L_1} and f_{L_2} are zeroes and ones, f_{L_1} and f_{L_2} are equal as integer polynomials if and only if they are equal when interpeted as polynomials over F. The famous Schwartz–Zippel lemma [8] gives us a way to compare the polynomials without computing all their coefficients explicitly.

Lemma 2 (The Schwartz-Zippel Lemma *[8]) Let $h(t_1, t_2, \dots, t_m)$ be a non-zero polynomial in m variables of total degree d over a field F. If we assign t_1, t_2, …, t_m elements of the field F uniformly and independently at random, then the probability of $h(t_1, t_2, \dots, t_m)$ evaluating to zero is at most $\frac{d}{|F|}$.*

Let us apply the Schwartz-Zippel lemma to the polynomial $f_{L_1} - f_{L_2}$. If the languages L_1 and L_2 are equal, then the polynomial $f_{L_1} - f_{L_2}$ is equal to 0, and, therefore, the polynomials f_{L_1} and f_{L_2} are equal as well. On the other hand, as mentioned in the start of the proof, if the languages L_1 and L_2 are not equal, then the polynomial $f_{L_1} - f_{L_2}$ is a non-zero polynomial of total degree n. Hence, if we evaluate it at a random point from $F^{|\Sigma| \cdot n}$, it will evaluate to zero with probability at most $\frac{n}{|F|}$. Therefore, if L_1 and L_2 are different, then f_{L_1} and f_{L_2} coincide in a random point with probability at most $\frac{n}{|F|}$. Picking F to be the field of residues modulo p completes the proof of Theorem 1. There are no false negatives (equal slices will be always recognized as equal) and the probability of a false positive (different slices being mistakenly recognized as equal) is at most $\frac{n}{p}$. Finally, to make the probability of a mistake at most $\frac{1}{2}$, it is enough to choose $p \geqslant 2n$. □

Example 1. Consider an example with $n = 3$ and the following two grammars:

$$\begin{aligned} S_1 &\to A_1 B_1 \mid B_1 A_1 \\ A_1 &\to A_1 A_1 \mid a \\ B_1 &\to B_1 B_1 \mid b \end{aligned} \qquad \begin{aligned} S_2 &\to A_2 B_2 \mid B_2 A_2 \\ A_2 &\to A_2 A_2 \mid A_2 B_2 \mid a \\ B_2 &\to b. \end{aligned}$$

Neither of these two grammars is actually unambiguous, but, for both of them, all strings of length at most 3 have at most one parse tree. Therefore, they are perfectly fine to use as examples (it is hard to construct an interesting and actually unambiguous example without using too many nonterminals). Specifically, the 3-slice of the first grammar gives us the language $\{aab, abb, baa, bba\}$ and the 3-slice of the second grammar gives us the language $\{aab, abb, baa, bab\}$. These two languages are different (the first one contains the string bba, while the second one contains the string bab). Let us see what the algorithm will do.

By Equation 1,

$f^{(0,1]}_{L(A_1)} = x_{a,1}, f^{(0,1]}_{L(B_1)} = x_{b,1}, f^{(0,1]}_{L(S_1)} = 0, f^{(1,2]}_{L(A_1)} = x_{a,2}, f^{(1,2]}_{L(B_1)} = x_{b,2}, f^{(1,2]}_{L(S_1)} = 0,$

$f^{(2,3]}_{L(A_1)} = x_{a,3}, f^{(2,3]}_{L(B_1)} = x_{b,3}, f^{(2,3]}_{L(S_1)} = 0,$

$f^{(0,2]}_{L(A_1)} = f^{(0,1]}_{L(A_1)} f^{(1,2]}_{L(A_1)} = x_{a,1}x_{a,2}, f^{(0,2]}_{L(B_1)} = f^{(0,1]}_{L(B_1)} f^{(1,2]}_{L(B_1)} = x_{b,1}x_{b,2},$

$f^{(0,2]}_{L(S_1)} = f^{(0,1]}_{L(A_1)} f^{(1,2]}_{L(B_1)} + f^{(0,1]}_{L(B_1)} f^{(1,2]}_{L(A_1)} = x_{a,1}x_{b,2} + x_{b,1}x_{a,2},$

$f^{(1,3]}_{L(A_1)} = f^{(1,2]}_{L(A_1)} f^{(2,3]}_{L(A_1)} = x_{a,2}x_{a,3}, f^{(1,3]}_{L(B_1)} = f^{(1,2]}_{L(B_1)} f^{(2,3]}_{L(B_1)} = x_{b,2}x_{b,3},$

$f^{(1,3]}_{L(S_1)} = f^{(1,2]}_{L(A_1)} f^{(2,3]}_{L(B_1)} + f^{(1,2]}_{L(B_1)} f^{(2,3]}_{L(A_1)} = x_{a,2}x_{b,3} + x_{b,2}x_{a,3},$

$f^{(0,3]}_{L(A_1)} = f^{(0,1]}_{L(A_1)} f^{(1,3]}_{L(A_1)} + f^{(0,2]}_{L(A_1)} f^{(2,3]}_{L(A_1)} = 2x_{a,1}x_{a,2}x_{a,3},$

$f^{(0,3]}_{L(B_1)} = f^{(0,1]}_{L(B_1)} f^{(1,3]}_{L(B_1)} + f^{(0,2]}_{L(B_1)} f^{(2,3]}_{L(B_1)} = x_{3,b} = 2x_{1,b}x_{2,b}x_{3,b},$

$f^{(0,3]}_{L(S_1)} = f^{(0,1]}_{L(A_1)} f^{(1,3]}_{L(B_1)} + f^{(0,2]}_{L(A_1)} f^{(2,3]}_{L(B_1)} + f^{(0,1]}_{L(B_1)} f^{(1,3]}_{L(A_1)} + f^{(0,2]}_{L(B_1)} f^{(2,3]}_{L(A_1)} =$
$x_{a,1}x_{b,2}x_{b,3} + x_{a,1}x_{a,2}x_{b,3} + x_{b,1}x_{a,2}x_{a,3} + x_{b,1}x_{b,2}x_{a,3}.$

It is not hard to see that $L(A_1) = \{a, aa, aaa, \ldots\}$, $L(B_1) = \{b, bb, bbb, \ldots\}$ and $L(S_1) = \{ab, ba, aab, abb, baa, bba, \ldots\}$. Hence, almost all computed values of $f^{(\ell,r]}_{\ldots}$ coincide with what they should be by definition. The only exceptions are $f^{(0,3]}_{L(A_1)}$ and $f^{(0,3]}_{L(B_1)}$, that are both the double of what they should be. Why did it happen? Well, as previously noted, G_1 is an *ambiguous* grammar. In particular, aaa has two parse trees starting from A_1. Similarly, bbb has two parse trees starting from B_1. Hence, aaa and bbb have been counted twice in $f^{(0,3]}_{L(A_1)}$ and $f^{(0,3]}_{L(B_1)}$, respectively.

Now, let us recall that the algorithm *actually* computes only the values modulo some prime p of each $f^{(\ell,r]}_{\ldots}$ in a randomly chosen point y. Let us choose $p = 5$, $y_{a,1} = 1$, $y_{b,1} = 1$, $y_{a,2} = 2$, $y_{b,2} = 4$, $y_{a,3} = 1$, $y_{b,3} = 3$. Then,

$f^{(0,1]}_{L(A_1)}(y) = 1, f^{(0,1]}_{L(B_1)}(y) = 1, f^{(0,1]}_{L(S_1)}(y) = 0, f^{(1,2]}_{L(A_1)}(y) = 2, f^{(1,2]}_{L(B_1)}(y) = 4$

$f^{(1,2]}_{L(S_1)}(y) = 0, f^{(2,3]}_{L(A_1)}(y) = 1, f^{(2,3]}_{L(B_1)}(y) = 3, f^{(2,3]}_{L(S_1)}(y) = 0, f^{(0,2]}_{L(A_1)}(y) = 1 \cdot 2 = 2,$

$f^{(0,2]}_{L(B_1)}(y) = 1 \cdot 4 = 4, f^{(0,2]}_{L(S_1)}(y) = 1 \cdot 4 + 1 \cdot 2 = 1, f^{(1,3]}_{L(A_1)} = 2 \cdot 1 = 2,$

$f^{(1,3]}_{L(B_1)}(y) = 4 \cdot 3 = 2, f^{(1,3]}_{L(S_1)}(y) = 2 \cdot 3 + 4 \cdot 1 = 0,$

$f^{(0,3]}_{L(A_1)}(y) = 1 \cdot 2 + 2 \cdot 1 = 4, f^{(0,3]}_{L(B_1)}(y) = 1 \cdot 2 + 4 \cdot 3 = 4,$

$f^{(0,3]}_{L(S_1)}(y) = 1 \cdot 2 + 2 \cdot 3 + 1 \cdot 2 + 4 \cdot 1 = 4.$

Computing the value of $f^{(0,3]}_{L(S_2)}(y)$ in the same way yields the value of 1, meaning that the languages are indeed different.

Remark 1. It is possible to check whether $L(G_1)$ and $L(G_2)$ differ on a string of length *at most* n (rather than *exactly* n, as in Theorem 1) with exactly the same running time. To do so, run the above procedure, but, in the end, compare $f^{(0,1]}_{L(G_1)}$ with $f^{(0,1]}_{L(G_2)}$, $f^{(0,2]}_{L(G_1)}$ with $f^{(0,2]}_{L(G_2)}$, ..., $f^{(0,n]}_{L(G_1)}$ with $f^{(0,n]}_{L(G_2)}$. If at least one of these pairs consists of two different elements of F, then the languages $L(G_1)$ and $L(G_2)$ differ on a string of length from the range $[1, n]$. The implication works the other way around as well: if the languages differ on a string of length $\ell \in [1, n]$, then it will be detected with a large probability when comparing $f^{(0,\ell]}_{L(G_1)}$ with $f^{(0,\ell]}_{L(G_2)}$. The empty string can be checked separately.

3 Finding the First "Discrepancy"

Let us extend the Cocke–Kasami–Younger–Schwartz–Zippel algorithm in the following reasonable way: when we know that the n-slices are different, return the lexicographically smallest string where the n-slices differ. Another, slightly weaker, problem is to find *any* string where the n-slices differ.

Finding the lexicographically smallest string where the languages differ in *any* polynomial time (that is, without any consideration of the degrees of the polynomials) is a standard problem that can be solved by finding out its characters one-by-one. Indeed, suppose that our alphabet is $\Sigma = \{a_1, a_2, \ldots, a_k\}$, where $a_1 < a_2 < \ldots < a_k$ is the lexicographical ordering of the characters. Let us denote the lexicographically smallest string of length n where the languages $L(G_1)$ and $L(G_2)$ differ by $s = s_1 s_2 \ldots s_n$. It should be noted that we do not actually know s when we start running the algorithm, but discover its characters one-by-one instead. Suppose that we already know some prefix $s_1 s_2 \ldots s_i$ of s. Initially, the known prefix is empty and, therefore, $i = 0$. In the end, $i = n$ and we have finally uncovered the whole string s.

Now, suppose that i is less than n and we need to find out the value of s_{i+1}. Let us use the normal version of CKYSZ algorithm to compare the languages $L(G_1) \cap (s_1 s_2 \ldots s_i a_1 \Sigma^{n-i-1})$ and $L(G_2) \cap (s_1 s_2 \ldots s_i a_1 \Sigma^{n-i-1})$. It can be done, because there is an efficient construction for intersecting a grammar with a regular language (and said construction is known to preserve unambiguity [4]). If the algorithm tells us that they are assuredly different, then we know that $s_{i+1} = a_1$ and can continue the algorithm with a larger known prefix. If they are equal with high probability, then, most likely, s_{i+1} is *not* equal to a_1. Hence, we can use CKYSZ algorithm to compare the languages $L(G_1) \cap (s_1 s_2 \ldots s_i a_2 \Sigma^{n-i-1})$ and $L(G_2) \cap (s_1 s_2 \ldots s_i a_2 \Sigma^{n-i-1})$. Again, we discover with high probability whether or not s_{i+1} and a_2 are equal. If they are equal, we continue the algorithm with a larger known prefix. Otherwise, we check the character a_3, and so on. By repeating the same step for each element of Σ, we will finally discover the actual value of s_{i+1} with high probability.

However, the above approach, while still polynomial-time, can be pretty slow when n is large (because intersecting a grammar with a regular language considerably increases the size of the grammar). Indeed, it uses Problem 2 as a subproblem and needs to solve it up to $|\Sigma| \cdot n$ times (up to $|\Sigma|$ times for each character of the string). Each of said regular languages is recognized by a DFA with $O(n)$ states. Hence, the size of the intersection of G_1 and L grows as $\Theta(n^3 \cdot |G_1|)$ in the worst case (each rule $A \to BC$ in G_1 becomes $\Theta(n^3)$ rules $A_{p\to r} \to B_{p\to q}C_{q\to r}$). Hence, in the worst case, such an algorithm takes $\Omega(n^7 \cdot |\Sigma| \cdot (|G_1| + |G_2|))$ time, which is rather slow.

One way to deal with that is to use the above approach and "bake in" the current considered prefix into the dynamic programming for evaluating the polynomial in the given point (instead of "baking it in" the grammar itself). In this way, we can achieve only a linear overhead for the constant size of the alphabet. Indeed, let us prove a lemma that formalizes this idea.

Lemma 3 *There is a randomized algorithm with bounded one-sided error that, given two unambiguous grammars G_1 and G_2 in Chomsky normal form, a positive integer n and a string t of length at most n, tells whether or not the languages $L(G_1) \cap t\Sigma^{n-|t|}$ and $L(G_2) \cap t\Sigma^{n-|t|}$ are equal in $O(n^3 \cdot (|G_1| + |G_2|))$ time. The algorithm may only fail by claiming that the languages are equal (despite actually being different) and does so with probability at most n/p, where $p \geqslant 2n$ is a prime number, such that the arithmetic operations with integers modulo p are assumed to be perfomed in $O(1)$ time.*

Proof. The proof is very similar to the arguments presented in the proofs of Theorem 1 and Lemma 1. Informally speaking, Lemma 1 "needs to consider" all strings of length n, and, therefore, needs to iterate over all characters of Σ at each of the n positions. On the other hand, in the current situation, the characters at the first $|t|$ positions are already known. Hence, for the first $|t|$ positions, we "have" to fill them with the corresponding character of t, while the remaining position can still be filled by any character of Σ. The important detail here is not the exact form of the language $t\Sigma^{n-|t|}$, but the fact that each of n positions can be "filled independently of others". Let us formalize this idea.

In fact, we will solve a more general problem. Let $T_1, T_2, \ldots, T_n$ be any n subsets of Σ. Then, we will tell, whether or not the languages $L(G_1) \cap (T_1T_2 \ldots T_n)$ and $L(G_2) \cap (T_1T_2 \ldots T_n)$ are equal in $O(n^3 \cdot (|G_1| + |G_2|))$ time. The statement of the lemma corresponds to the case when $T_1 = \{t_1\}$, $T_2 = \{t_2\}$, …, $T_{|t|} = \{t_{|t|}\}$, $T_{|t|+1} = T_{|t|+2} = \ldots = T_{n-1} = T_n = \Sigma$. For a subsegment $(\ell, r]$ of the segment $(0, n]$ and a language $K \subset \Sigma^{r-\ell}$, let us define $g_K^{(\ell,r]} := f_{K\cap(T_{\ell+1}T_{\ell+2}\ldots T_r)}^{(\ell,r]}$. That is, $g_K^{(\ell,r]} = f_{K\cap(T_{\ell+1}T_{\ell+2}\ldots T_r)}^{(\ell,r]} = \sum_{w \in K\cap(T_{\ell+1}T_{\ell+2}\ldots T_r)} (x_{w_{\ell+1},\ell+1} \cdot x_{w_{\ell+2},\ell+2} \cdot \ldots \cdot x_{w_r,r})$. Similarly to Theorem 1, the problem of comparing the languages $L(G_1) \cap (T_1T_2 \ldots T_n)$ and $L(G_2) \cap (T_1T_2 \ldots T_n)$ can be reduced to comparing the polynomials $g_{L(G_1)}^{(0,n]}$ and $g_{L(G_2)}^{(0,n]}$ in a random point.

Again, similarly to Lemma 1, each of said two polynomials can be evaluated independently of the other. Let us focus on evaluating $g_{L(G_1)}^{(0,n]}$. As per usual, we can assume that G_1 has no unnecessary nonterminals and rules. Let R be the set of all rules of G_1. Then, for each nonterminal C of the grammar G_1 and each subsegment $(\ell, r]$ of $(0, n]$ with the length of at least 2 (in other words, a segment with $r - \ell \geqslant 2$), we can see that $L(C) \cap (T_{\ell+1}T_{\ell+2} \ldots T_r) =$

$\biguplus_{m=\ell+1}^{r-1} \biguplus_{(C\to DE)\in R} (L(D) \cap (T_{\ell+1}T_{\ell+2}\dots T_m)) \cdot (L(E) \cap (T_{m+1}T_{m+2}\dots T_r))$. This can be verified either formally, by using the distributivity properties of language operations, or intuitively, by noticing that both the left-hand and the right-hand side are obtained by, essentially, adding independent position-wise restrictions to the left-hand and the right-hand sides of the correct statement $(L(C) \cap \Sigma^{r-\ell}) = \biguplus_{m=\ell+1}^{r-1} \biguplus_{(C\to DE)\in R} (L(D) \cap \Sigma^{m-\ell}) \cdot (L(E) \cap \Sigma^{r-m})$.

Hence, we can use almost the same dynamic programming formulas as in the Lemma 1. The only important difference is how the segments $(\ell, r]$ of length 1 are handled. In Lemma 1, any character of Σ can appear on the ℓ-th position. In the current lemma, only the characters from T_ℓ can appear on the ℓ-th position. Hence, we can compute the following dynamic programming:

$$g_{L(C)}^{(\ell,r]} := \begin{cases} \sum\limits_{(C\to a)\in R \text{ and } a\in T_\ell} x_{a,\ell+1}, & \text{if } r-\ell=1 \\ \sum\limits_{(C\to DE)\in R} \sum\limits_{m=\ell+1}^{r-1} g_{L(D)}^{(\ell,m]} g_{L(E)}^{(m,r]}, & \text{otherwise} \end{cases}$$

After repeating the same dynamic programming for G_2 we can compare the necessary languages. The error probability is at most n/p for the same reasons as in Theorem 1. □

Now, the following theorem is a direct consequence of Lemma 3.

Theorem 2 *There is an algorithm that, given two unambiguous grammars G_1 and G_2, such that the n-slices of $L(G_1)$ and $L(G_2)$ are known to be different, returns the lexicographically smallest string of length n where $L(G_1)$ and $L(G_2)$ differ in $O(n^4 \cdot |\Sigma| \cdot (|G_1| + |G_2|))$ time with probability at least $1/2$ (under the common assumption that the arithmetic operations on integers of absolute value $O(n^2)$ can be performed in $O(1)$ time). Moreover, said algorithm can be modified in a way that yields an $O(n^4 \cdot |\Sigma| \cdot (|G_1| + |G_2|))$ expected time algorithm that returns* some *(not necessary the lexicographically smallest) string of length n where the languages differ, but with no error ($L(G_1)$ and $L(G_2)$ are assuredly different in the returned string, but the running time of the algorithm is a random variable).*

Proof. Let us use the same idea as in the slow algorithm that started this section. As before, let $s = s_1s_2\dots s_n$ be the lexicographically smallest string where the n-slices of $L(G_1)$ and $L(G_2)$ differ. Then, by the same reasoning as before, the problem reduces to at most $|\Sigma| \cdot n$ calls to the procedure from the proof of Lemma 3. Let us call said procedure *Procedure A*. However, per Lemma 3, Procedure A runs in $O(n^3 \cdot (|G_1| + |G_2|))$ time, meaning that the entire algorithm runs in $O(n^4 \cdot |\Sigma| \cdot (|G_1| + |G_2|))$ time. The only remaining detail is the error probability. What could have caused the algorithm to fail? There are n different events that could have done so: we could have "missed" the fact that the languages $L(G_1) \cap s_1\Sigma^{n-1}$ and $L(G_2) \cap s_1\Sigma^{n-1}$ are different, the languages

$L(G_1) \cap s_1 s_2 \Sigma^{n-2}$ and $L(G_2) \cap s_1 s_2 \Sigma^{n-2}$ are different and so on. Hence, by picking $p \geqslant 2n^2$, we can ensure that the probability of each of said events is at most $\frac{n}{p} \leqslant \frac{n}{2n^2} = \frac{1}{2n}$. Hence, the total probability of at least one of said n "bad" events happening is at most $\frac{1}{2n} \cdot n = \frac{1}{2}$.

Now, suppose that we are allowed to find *any* "discrepancy" between the n-slices, not necessarily the lexicographically smallest one. Then, it is possible to achieve a zero-error randomized algorithm. Indeed, let us say that $\Sigma = \{a_1, a_2, \ldots a_k\}$. Suppose that we already know for some string w of length at most n that the languages $L(G_1) \cap w\Sigma^{n-|w|}$ and $L(G_2) \cap w\Sigma^{n-|w|}$ are different. In other words, we know that some discrepancy between the n-slices of $L(G_1)$ and $L(G_2)$ starts with w. Initially, w is empty. If $|w| = n$, then we have "won" and found some string where the n-slices of $L(G_1)$ and $L(G_2)$ differ. Hence, it is enough to find a way to "extend" w by one character.

Now, for each i from 1 to k, run Procedure A to compare the languages $L(G_1) \cap wa_i\Sigma^{n-1-|w|}$ and $L(G_2) \cap wa_i\Sigma^{n-1-|w|}$. If, for some i, Procedure A tells us that the corresponding languages are not equal, we can conclude with complete certainity that some discrepancy between the n-slices of $L(G_1)$ and $L(G_2)$ starts with wa_i. Otherwise, we will run Procedure A for each i from 1 to k again. This process is bound to end eventually. Indeed, we know that w is a prefix of *some* discrepancy between the n-slices of $L(G_1)$ and $L(G_2)$. Let us say that the *next* (after w) character of said discrepancy is a_j. Then, the languages $L(G_1) \cap wa_j\Sigma^{n-1-|w|}$ and $L(G_2) \cap wa_j\Sigma^{n-1-|w|}$ are different. Hence, Procedure A will detect the fact that they are different with high probability. Therefore, on average, we will "extend" w after repeating the process $O(1)$ times, with each iteration of the process calling Procedure A exactly $|\Sigma|$ times.

Therefore, the algorithm makes $O(|\Sigma|)$ calls to Procedure A for each of n times it needs to "extend" w. Hence, the average running time of the algorithm is $O(n^4 \cdot |\Sigma| \cdot (|G_1| + |G_2|))$. By design, the algorithm never makes an error. □

4 Conclusion

It is important to note that the proof of Theorem 1 does not actually use the unambiguity of grammars *that* much. We can use the same procedure as in the proof of Theorem 1 to check whether the grammars (already in Chomsky's normal form; hence, each string has a finite multiplicity) describe the same multiset of strings modulo p (meaning that, for each string w, the multiplicities of w in the two grammars have the same remainder modulo p). Finally, by choosing several different appropriately large values of p, we can, using the Chinese Remainder Theorem, check whether G_1 and G_2 define the same multisets of strings of length n (without any "modulo p"). In general, it is not surprising that the approach extends very well to the multiset settings, as it is based on counting ideas. Similarly, the heuristic approach to Problem 1 by Almeida et al., mentioned in the introduction, also immediately extends to the case of multisets [1].

There are many possible directions for the further research: speeding up all the algorithms used here by using matrix multiplication, trying to avoid a linear slowdown when restoring the actual string where the languages differ, and so on. It is also somewhat interesting to see whether or not we can use both matrix substitution and indexation at the same time to achieve even stronger results.

Moreover, I argue that there is a connection between the slice equivalence and the general equivalence problem. One of the way to look is the following. Famously, for each Turing machine, there is a context-free grammar that describes all strings that *are not* valid computations of the machine (here, some fixed specific encoding of the computations is used) [14]. Therefore, it can be proven that the universality problem is undecidable for context-free grammars. The undecidability of equivalence follows immediately. The same reasoning can be used to prove that slice equivalence is co-NP-complete and slice size is #P-complete [11]. Hence, we can say that "the reason why" universality of context-free grammars is undecidable, slice equivalence is co-NP-complete and slice size is #P-complete, is that context-free grammars can encode computations of non-deterministic Turing machines.

On the other hand, the slice size and the slice equivalence problems for unambiguous grammars can be solved in polynomial time. Hence, it is unlikely that unambiguous grammars can encode computations of Turing machines. This kind of reasoning needs to be used very carefully, as it is only an argument against an existence of an extremely particular way to simulate Turing machines by unambiguous grammars. For example, it could technically happen that unambiguous grammars can encode computation histories of Turing machines, but they encode a computation of a polynomial length by one or several strings of exponential length. Still, I find such a possibility unlikely and I believe that the above observation is a (weak) heuristic evidence in favor of decidability of Problem 1.

References

1. Almeida, J.J., Grande, E., Smirnov, G.: On solving cycle-free context-free grammar equivalence problem using numerical analysis. J. Comput. Lang. **51**, 48–56 (2019)
2. Amitsur, A.S., Levitzki, J.: Minimal identities for algebras. In: Proceedings of American Mathematical Society, vol. 1, pp. 449–463 (1950). https://doi.org/10.1090/S0002-9939-1950-0036751-9
3. Ayad, A.: A survey on the complexity of solving algebraic systems. Int. Math. Forum **5**(7), 333–353 (2010)
4. Bar-Hillel, Y., Perles, M., Shamir, E.: On formal properties of simple phrase-structure grammars. Zeitschrift für Phonetik, Sprachwissenschaft und Kommunikationsforschung **14**, 143–177 (1961)
5. Bertoni, A., Goldwurm, M., Sabadini, N.: The complexity of computing the number of strings of given length in context-free languages. Theor. Comput. Sci. **86**, 325–342 (1991). https://doi.org/10.1016/0304-3975(91)90023-U
6. Chistov, A.L., Grigor'ev, D.Y.: Complexity of quantifier elimination in the theory of algebraically closed fields. In: Chytil, M.P., Koubek, V. (eds.) MFCS 1984. LNCS, vol. 176, pp. 17–31. Springer, Heidelberg (1984). https://doi.org/10.1007/BFb0030287

7. Chomsky, N., Schützenberger, M.P.: Algebraic theory of context-free languages. Stud. Logic Found. Math. **35**, 118–161 (1963). https://doi.org/10.1016/S0049-237X(08)72023-8
8. Demillo, R.A., Lipton, R.J.: A probabilistic remark on algebraic program testing. Inf. Process. Lett. **7**(4), 193–195 (1978). https://doi.org/10.1016/0020-0190(78)90067-4
9. Drensky, V.S.: Free Algebras and PI-Algebras: Graduate Course in Algebra. Springer, Verlag (1996)
10. Earley, J.: An efficient context-free parsing algorithm. Commun. ACM **13**(2), 94–102 (1970)
11. Ellul, K., Shallit, J., Wang, M.: Regular expressions: new results and open problems. J. Autom. Lang. Comb. **10**(4), 407–437 (2005). https://doi.org/10.25596/jalc-2005-407 https://doi.org/10.25596/jalc-2005-407 https://doi.org/10.25596/jalc-2005-407
12. Fischer, B., Lämmel, R., Zaytsev, V.: Comparison of context-free grammars based on parsing generated test data. Softw. Lang. Eng. **6940**, 324–343 (2011). https://doi.org/10.1007/978-3-642-28830-2_18
13. Flajolet, P.: Analytic methods and ambiguity of context-free languages. Theor. Comput. Sci. **49**, 283–309 (1987). https://doi.org/10.1016/0304-3975(87)90011-9
14. Gore, V., Jerrum, M., Kannan, S., Sweedyk, Z., Mahaney, S.R.: A quasi-polynomial-time algorithm for sampling words from a context-free language. Inf. Comput. **134**(1), 59–74 (1997)
15. Hotz, G.: Normal-form transformations of context-free grammars. Acta Cybernet. **4**, 65–84 (1980)
16. Jančar, P.: Decidability of DPDA language equivalence via first-order grammars. In: Proceedings – Symposium on Logic in Computer Science (2012)
17. Madhavan, R., Mayer, M., Gulwani, S., Kuncak, V.: Automating grammar comparison. In: ACM SIGPLAN Notices, vol. 50, issue 10, pp. 183–200 (2015). https://doi.org/10.1145/2858965.2814304
18. Meel, K.S., de Cornet, A.: #CFG and #DNNF admit FPRAS. arXiv preprint arXiv:2406.18224 (2024)
19. Meyer, A.R., Stockmeyer, L.J.: The equivalence problem for regular expressions with squaring requires exponential space. In: 13th Annual Symposium on Switching and Automata Theory (SWAT 1972), pp. 125–129 (1972). https://doi.org/10.1109/SWAT.1972.29
20. Semenov, A.L.: Algorthmic problems for power series and context-free grammars. Sov. Math. **14**, 1319 (1973)
21. Sénizergues, G.: $L(A) = L(B)$? decidability results from complete formal systems. Theor. Comput. Sci. **251**(1–2), 1–166 (2001). https://doi.org/10.1016/S0304-3975(00)00285-1
22. Stearns, R.E., Hunt III, H.B.: On the equivalence and containment problems for unambiguous regular expressions, regular grammars and finite automata. SIAM J. Comput. **14**(3) (1985). https://doi.org/10.1137/0214044
23. Stirling, C.: Decidability of DPDA equivalence. Theor. Comput. Sci. **255**, 1–31 (2001). https://doi.org/10.1016/S0304-3975(00)00389-3

Nondeterminism Makes Unary 1-Limited Automata Concise

Bruno Guillon[1](✉), Luca Prigioniero[2], and Javad Taheri[1]

[1] Université Clermont Auvergne, Clermont Auvergne INP, LIMOS, CNRS, Clermont-Ferrand, France
bruno.guillon@uca.fr

[2] Department of Computer Science, Loughborough University, Loughborough, UK
l.prigioniero@lboro.ac.uk

Abstract. We investigate the descriptional complexity of 1-limited automata (1-LAs) in the unary case. These machines characterize regular languages and are two-way finite automata (2NFAs) extended to allow rewriting each tape cell upon its first visit. We prove exponential lower bounds for the simulations of 1-LAs by deterministic 1-LAs and 2NFAs. These results are derived from a doubly exponential lower bound for the simulation of 1-LAs by one-way deterministic finite automata (1DFAs), and close a question left open in [Pighizzini and Prigioniero. Limited automata and unary languages. Inf. Comput., 266:60–74]. Our results hold even when, besides being unary, the 1-LA is a 2DFA with common-guess (2DFA+**cg**), namely a 1-LA restricted to use nondeterminism only to annotate the tape cells during a write-only initial phase, before performing a read-only deterministic computation.

1 Introduction

Understanding the impact of nondeterminism on the size of the description of computational models is a central issue in descriptional complexity of formal systems. One of the most famous problems in this area is the one posed by Sakoda and Sipser, who conjectured an unavoidable super-polynomial blowup for simulating *two-way* and *one-way nondeterministic finite automata* (2NFAs and 1NFAs) by deterministic two-way finite automata (2DFAs) [12]. A similar question arises in other models characterizing regular languages, and in particular, when considering *1-limited automata* (1-LAs) and their deterministic counterparts (D1-LAs). These computational models extend 2NFAs and 2DFAs by allowing them to rewrite each tape cell *only when it is visited for the first time.* It has been shown that providing two-way finite automata with this limited rewriting capability does not increase their expressive power: 1-LAs characterize regular languages [14]. As in the case of two-way finite automata, the exact cost of the elimination of nondeterminism from 1-LAs remains unknown—see [7] for a recent survey. In [8], the authors showed a tight exponential size cost for the conversion of D1-LAs into one-way deterministic finite automata (1DFAs), as well as a tight doubly-exponential size cost for the conversion of 1-LAs into

S.-K. Ko and F. Manea (Eds.): DLT 2025, LNCS 16036, pp. 153–165, 2026.
https://doi.org/10.1007/978-3-032-01475-7_11

1DFAs. Concerning the cost of determinizing 1-LAs (*i.e.*, simulating 1-LAs with D1-LAs), the just-exposed tight bounds imply a single exponential lower bound, and an doubly-exponential upper bound. Whether a single exponential would be sufficient in all case is still an open problem as of today [7, Problem 2].

It is worth noting that the above-mentioned lower bounds obtained in [8] are witnessed by non-unary languages, *i.e.*, over an alphabet having more than one symbol. In many cases, the cost of simulations between computational models decreases when considering the restricted case of unary languages (see, *e.g.*, [6]). In [9], working on this issue, the authors showed that the smallest 2NFA (and thus 1DFA) simulating a given n-state D1-LA may require a number of states which is exponential in n, even when the input alphabet has a single letter. They however left as an open question whether the double-exponential lower bound for the conversion of 1-LAs into 1DFAs still holds in the unary case [7, Problem 1].

As previously described, 1-LAs are 2NFAs extended with the ability to rewrite each tape cell, during the first visit to that cell. Hence, nondeterminism plays a twofold role in this model: in choosing the symbols that are written onto the tape, and in changing states when reading from it. In [8], the authors attributed the double exponential gap between 1-LAs and 1DFAs to this double role of nondeterminism in 1-LAs. In order to investigate this phenomenon, several restrictions of 1-LAs, which somehow reduce the impact of nondeterminism in writing, have been considered: *forgetting 1-LAs* [10], *once-marking* and *always-marking 1-LAs* [11]. We consider another one, called *two-way automata with common-guess* (2NFA+cg and 2DFA+cg), and we investigate it from a descriptional complexity perspective. The model restricts 1-LAs by syntactically separating an initial write-only phase from a succeeding read-only phase. More precisely, a 2NFA+cg proceeds in two consecutive modes over a given input w: First the machine nondeterministically annotates each letter of w; then it performs a read-only computation over the annotated input. For 2DFAs+cg, this latter phase (but not the preceding one) is required to be deterministic. *Common-guess* is a natural nondeterministic feature in other formalisms, such as *two-way transducers* (*i.e.*, two-way automata that outputs words), and *MSO-transductions* (a formalism describing transformations using monadic second-order logic) [1]. Indeed, nondetermistic MSO-transductions are precisely deterministic MSO-transductions operating over nondeterministically annotated inputs. Concerning 2NFAs+cg and 2DFAs+cg, the two models have already been considered in [2], where simulations of D1-LAs by 2DFAs+cg, and of 1-LAs by 2NFAs+cg, both with polynomial size cost, have been shown. Also, a doubly exponential lower bound for the simulation of 2DFAs+cg (and consequently of 1-LAs) by 1DFAs (and hence an exponential lower bound for the simulation of 2DFAs+cg by D1-LAs or 2NFAs) has been obtained using a two-letter alphabet. Hence, even when restricting the nondeterminism to the write role of 1-LAs, the gap to 1DFAs remains doubly exponential.

In this paper we show that all the above-mentioned lower bounds for the various simulations of 1-LAs still hold when the input alphabet is unary and when furthermore the simulated 1-LA is actually a 2DFA+cg (see Table 1). This

Table 1. Lower bounds for the simulations between variants of unary 1-LAs.

source \ target	1DFA	2NFA	D1-LA
D1-LA	exp	⇐ exp (Thm. 14) ⇓	-
2DFA+cg	expexp (Thm. 21) ⇓	exp ⇓	exp (Thm. 22) ⇓
2NFA+cg	expexp ⇓	exp ⇓	exp ⇓
1-LA	expexp	exp	exp

includes a double-exponential lower bound for the conversion of unary 2DFAs+cg into 1DFAs (Theorem 21 solving [7, Problem 1]), from which an exponential lower bound for the conversion of unary 2DFAs+cg into D1-LAs is derived. We also show an exponential lower bound in size for the transformation of unary D1-LAs into equivalent 2NFAs. Contrary to the analogous bound given on the number of states in [9], our bound relies on D1-LAs that use a constant number of work symbols only (Theorem 14).

2 Preliminaries

In this section, we recall some fundamental definitions and notations used throughout the paper. We assume that the reader is familiar with basic concepts from formal languages and automata theory (see, *e.g.*, [3]). For a set S, $|S|$ denotes its cardinality. Given an alphabet Σ, the set of strings over Σ is denoted by Σ^*. It includes the empty string denoted by ε. The length of a word $w \in \Sigma^*$ is denoted by $|w|$. For a language $\mathsf{L} \subseteq \Sigma^*$, we denote by $\mathrm{Pref}(\mathsf{L})$, $\mathrm{Suff}(\mathsf{L})$, and $\mathrm{Fact}(\mathsf{L})$ the languages of prefixes, suffixes, and factors of (words of) L, respectively.

Definition 1. *A* two-way nondeterministic finite automaton *(*2NFA*) is a tuple $\mathcal{A} = \langle Q, \Sigma, \delta, q_0, F\rangle$, where Q is the finite set of states, Σ is the input alphabet, $q_0 \in Q$ is the initial state, $F \subseteq Q$ is the set of final states, and $\delta : Q \times \Sigma_{\rhd\lhd} \to 2^{Q\times\{-1,+1\}}$ is a nondeterministic transition function with $\Sigma_{\rhd\lhd} = \Sigma \cup \{\rhd, \lhd\}$, where $\rhd, \lhd \notin \Sigma$ are two special symbols called* the left *and* the right endmarker, *respectively.*

In 2NFAs, the input is written on the tape surrounded by the two endmarkers, the left endmarker being at position 0. Hence, on input w, the right endmarker is at position $|w|+1$. In one move, $\mathcal{A}$ reads an input symbol, changes its state, and moves the head one position backward or forward depending on whether δ returns -1 or $+1$, respectively. The head cannot pass the endmarkers. The machine accepts the input if there exists a computation which starts from the initial state q_0 with the head on the cell at position 1 (*i.e.*, scanning the first letter of w if $w \neq \varepsilon$ and scanning $\lhd$ otherwise) and eventually reaches a final state $q_f \in F$ with the head scanning the right endmarker. The language accepted by $\mathcal{A}$ is denoted by $\mathcal{L}(\mathcal{A})$. A 2NFA is said to be *deterministic* (2DFA) whenever $|\delta(q,\sigma)| \leq 1$, for every $q \in Q$ and $\sigma \in \Sigma \cup \{\triangleright, \triangleleft\}$. It is called *one-way* if its head can never move backward, *i.e.*, if no transition returns -1. By *1NFA* and *1DFA* we denote one-way nondeterministic and deterministic finite automata, respectively. We now extend 2NFAs with a limited write ability.

Definition 2. *A* 1-limited automaton *(1-LA, for short) is a tuple* $\mathcal{A} = (Q, \Sigma, \Delta, \delta, q_0, F)$*, where* Q*,* Σ*,* q_0*, and* F *are defined as for 2NFAS, the finite set* Δ *is the* work alphabet *which includes the input alphabet* Σ *but excludes* $\{\triangleright, \triangleleft\}$*, and* δ *is the* nondeterministic transition function*, which maps* $Q \times \Delta_{\triangleright\triangleleft}$ *to subsets of* $Q \times (\Delta_{\triangleright\triangleleft} \setminus \Sigma) \times \{-1, +1\}$*, where* $\Delta_{\triangleright\triangleleft} = \Delta \cup \{\triangleright, \triangleleft\}$*. The transition function* δ *is subject to the following restrictions. For each transition* $(q, \gamma, d) \in \delta(p, \sigma)$*:*

- *if* $\sigma = \triangleright$ *then* $\gamma = \triangleright$ *and* $d = +1$*,*
- *if* $\sigma = \triangleleft$ *then* $\gamma = \triangleleft$ *and* $d = -1$*,*
- *if* $\sigma \in \Sigma$ *then* $\gamma \in \Delta \setminus \Sigma$ *(in particular,* $\sigma \neq \gamma$ *and* $\gamma \notin \{\triangleright, \triangleleft\}$*),*
- *and if* $\sigma \in \Delta \setminus \Sigma$ *then* $\gamma = \sigma$*.*

In one move, according to δ, $\mathcal{A}$ reads a symbol from the tape, changes its state, replaces the symbol just read by a new symbol from the work alphabet, and moves its head one position backward or forward. However, replacing symbols is subject to some restrictions, which, essentially, allow to modify the content of a cell during the first visit only. Technically, the endmarkers are never replaced (first two restrictions) nor used as a replacement of another symbols, while the input symbols from Σ are replaced by work symbols not in Σ (third restrictions), and such symbols can not be replaced in subsequent steps on the cell (fourth restriction). In particular, at any time, both endmarkers $\triangleright$ and $\triangleleft$ occur exactly once on the tape and exactly at the respective left and right boundaries. We say that $\mathcal{A}$ is *deterministic* (D1-LA) if $|\delta(p,\gamma)| \leq 1$ for each $(p,\gamma) \in Q \times \Delta_{\triangleright\triangleleft}$. Acceptance for a 1-LAs is defined exactly as for 2NFAs: the machine accepts an input w if, starting in state q_0 with the head scanning the position 1 of the tape whose contents is $\triangleright w \triangleleft$, it eventually enters a final state $q \in F$ with the head scanning the right endmarker. The language accepted by a given 1-LA $\mathcal{A}$ is denoted by $\mathcal{L}(\mathcal{A})$.

We now extend 2NFAs and 2DFAs with *common-guess*. This feature describes the ability to initially annotate the input $w \in \Sigma^*$ using some annotation symbols from a fixed alphabet Γ, known as the *annotation alphabet*. The annotated word resulting from this initial phase is a word over the product alphabet $\Sigma \times \Gamma$. It is nondeterministically chosen among all the words $v \in (\Sigma \times \Gamma)^*$ such that $\pi_1(v) = w$, where π_1 is the natural projection of $(\Sigma \times \Gamma)^*$ onto Σ^* (in particular $|v| = |w|$).

Definition 3. *A 2NFA (resp. 2DFA) with* common-guess *(2NFA+*cg*, resp. 2DFA+*cg*) is a triplet* $\mathcal{M} = \langle \mathcal{A}, \Sigma, \Gamma \rangle$ *where* Σ *is the input alphabet,* Γ *is the* annotation alphabet*, and* $\mathcal{A}$ *is a 2NFA (resp. 2DFA) over the product alphabet* $\Sigma \times \Gamma$*.*

The language recognized by $\mathcal{M}$ is:

$$\mathcal{L}(\mathcal{M}) = \{w \in \Sigma^* \mid \exists v \in (\Sigma \times \Gamma)^*,\ v \in \mathcal{L}(\mathcal{A}) \text{ and } \pi_1(v) = w\} = \pi_1(\mathcal{L}(\mathcal{A})).$$

It is worth noting that, because the annotation phase is nondeterministic by nature, 2DFAs+cg are nondeterministic devices.

For each model, we evaluate its size as the total number of symbols used to describe it. Hence, under standard representation and denoting by Σ the input alphabet, the *size* of an n-state 2NFA is $\mathcal{O}(n^2|\Sigma|)$, that of an n-state 1-LA with work alphabet Δ is $\mathcal{O}(n^2|\Delta|^2)$, and that of an n-state 2NFA+cg with annotation alphabet Γ is $\mathcal{O}(n^2|\Sigma| \cdot |\Gamma|)$. In our work, we generally consider $|\Sigma|$ as a constant (which is often equal to 1).

By definition, in order to define a 2NFA+cg over Σ, it suffices to describe its underlying 2NFA over $\Sigma \times \Gamma$. Yet, when $\Sigma = \{a\}$ is unary, $\{a\} \times \Gamma$ is isomorphic to Γ. Hence, we will not distinguish the triple $\langle \mathcal{A}, \{a\}, \Gamma \rangle$ in which $\mathcal{A}$ is a 2NFA over Γ from the 2NFA+cg $\langle \mathcal{A}', \{a\}, \Gamma \rangle$ in which $\mathcal{A}'$ is the 2NFA isomorphic to $\mathcal{A}$, obtained by replacing each symbol $\gamma \in \Gamma$ by (a, γ) in the transitions of $\mathcal{A}$. Given a language L, we call *language of lengths of* L the language: $\pi_0(\mathsf{L}) = \{a^\ell \mid \exists u \in \mathsf{L},\ |u| = \ell\}$. The above comment directly gives the following result.

Proposition 4. *If a language $\mathsf{L} \subseteq \Gamma^*$ is recognized by a 2NFA (resp. a 2DFA) $\mathcal{A}$, then the unary language $\pi_0(\mathsf{L})$ is recognized by a 2NFA+cg (resp. a 2DFA+cg) $\mathcal{B}$ with the same state set as $\mathcal{A}$ and annotation alphabet Γ.*

We now briefly discuss how 2NFAs+cg and 2DFAs+cg are related to 1-LAs. In [2], the authors showed that a polynomial increase in size is always sufficient for turning a 1-LA into 2NFA+cg, as well as a D1-LA into 2DFA+cg. Conversely, given an n-state 2NFA+cg $\langle \mathcal{A}, \Sigma, \Gamma \rangle$, a 1-LA with same states and work alphabet $\Sigma \times \Gamma$ can simulate $\mathcal{A}$ by, intuitively, guessing the annotation on the fly as now explained. On input symbol $\sigma \in \Sigma$, in one step, it nondeterministically chooses a symbol $\gamma \in \Gamma$, rewrites the cell with (σ, γ), and simulates one transition of $\mathcal{A}$ over it. On symbol $(\sigma, \gamma) \in \Sigma \times \Gamma$, it simply simulates one step of $\mathcal{A}$ over (σ, γ). Hence 2NFAs+cg and 2DFAs+cg can be seen as particular cases of 1-LAs. Note that, even if $\mathcal{A}$ is a 2DFA, because of the above-described choice of γ, the resulting transition function might be nondeterministic. Eliminating this nondeterminism costs at least exponential in the worst case, as shown in [2]. Yet, in some particular cases, the nondeterminism introduced by the above transformation is useless, and thus a D1-LA of same size can be obtained. For this to be possible, a sufficient condition is that, when entering a cell for the first time, at most one annotation symbol $\gamma \in \Gamma$ for that cell allows the machine to continue its computation. This property is formalized by the concept of Γ-*predictive* 2DFAs.

Definition 5. *Let $\mathcal{A} = \langle Q, \Sigma \times \Gamma, \delta, q_0, F \rangle$ be a 2DFA over $\Sigma \times \Gamma$. We say that $p \in Q$ is Γ-*predictive *if for every $\sigma \in \Sigma$, there exists at most one $\gamma \in \Gamma$ such that $\delta(p, (\sigma, \gamma)) \neq \emptyset$. The 2DFA $\mathcal{A}$ is said Γ-*predictive*, if for every word $w \in \mathcal{L}(\mathcal{A})$, each time the machine enters a cell for the first time during its computation over w, the current state is Γ-predictive. A 2DFA+cg $\langle \mathcal{A}, \Sigma, \Gamma \rangle$ is said *predictive *when $\mathcal{A}$ is Γ-predictive.*

As announced, predictive 2DFAs+cg over Σ can be turned into equivalent D1-LAs with the same number of states and work alphabet $\Sigma \cup (\Sigma \times \Gamma)$.

Proposition 6. *If $\mathcal{A}$ is a Γ-predictive 2DFA over $\Sigma \times \Gamma$, then there exists a D1-LA over Σ with the same state set as $\mathcal{A}$ and work alphabet $\Sigma \cup (\Sigma \times \Gamma)$, which recognizes $\pi_1(\mathcal{L}(\mathcal{A}))$.*

As before, we sometimes do not distinguish a 2DFA over Γ from a 2DFA over $\{a\} \times \Gamma$ since the two alphabets are isomorphic. We may thus say that a 2DFA over Γ is Γ-predictive. This allows us to transform it into a unary D1-LA, according to Proposition 6.

3 Full Binary Sequence

Our main results establish exponential and double-exponential lower bounds for various simulations of 1-LAs, 2DFAs+cg, and D1-LAs by two-way and one-way finite automata. All these results are obtained over a unary alphabet. Yet, as the simulated devices are allowed to rewrite their tape contents, the witness languages actually rely on a particular word b_n over a three-letter alphabet, called *full binary sequence.* The word b_n has length longer than 2^n. Indeed, it consists of consecutive encodings of integers in base 2 over n bits, separated by a special marker $\sharp$. For instance, $b_3 = 000\sharp001\sharp010\sharp011\sharp100\sharp101\sharp110\sharp111\sharp$. Formally, for a fixed positive integer n and for each $i = 0, \ldots, 2^n - 1$, we denote by $b_n(i)$ the binary representation of i over n bits, with the most significant bit on the left. Hence, $\{b_n(i) | 0 \leq i < 2^n\} = \{0,1\}^n$. We define the *full binary sequence* as $b_n = b_n(0)\sharp b_n(1)\sharp \cdots \sharp b_n(2^n - 1)\sharp$. Its length is $2^n(n+1)$. Let us define the language $\mathsf{BinSeq}_n = \{b_n\}$ containing only the word b_n. We shall show that this language can be recognized by a 2DFA with a number of states linear in n. This will be obtained using the following observation.

Fact 1. *For every positive integer n and every $0 \leq i, j \leq 2^n - 1$, it holds that $j = i+1$ if and only if there exist an integer $m \in \mathbb{N}$ and a word $x \in \{0,1\}^*$ such that $b_n(i) = x01^m$ and $b_n(j) = x10^m$.*

As a direct consequence of Fact 1 one can notice that the last symbol of each length-$(n+2)$ factor of b_n is uniquely determined by its $n+1$ preceding symbols. Moreover, it is easy to see that the factor $1^n\sharp$ is the length-$(n+1)$ suffix of b_n. This is formalized in the following. For each symbol $\sigma \in \{0, 1, \sharp\}$, we define the regular language X_σ describing the pattern matching the $n+1$ symbols preceding σ in a factor of b_n, and we define the regular language $X_\$$ describing the pattern matching the length-$(n+1)$ suffix of b_n:

$$X_0 = \left[11^*\sharp(0+1)^+\right] + \left[01^*0(0+1)^*\sharp(0+1)^*\right], \qquad X_\sharp = \sharp(0+1)^+,$$
$$X_1 = \left[01^*\sharp(0+1)^*\right] + \left[11^*0(0+1)^*\sharp(0+1)^*\right], \qquad X_\$ = 1^+\sharp.$$

These four languages are disjoint. The membership of a length-$(n+1)$ factor u of b_n to some X_σ, for $\sigma \in \{0, 1, \sharp, \$\}$, determines the unique symbol σ that can be appended to u to form a factor of $b_n\$$ (where \$ is used to detect the end of the word).

Lemma 7. *Let $\Sigma = \{0, 1, \sharp\}$, and let X_0, X_1, $X_\sharp$, and $X_\$$ be the four disjoint regular languages defined above. For every positive integer n, every length-$(n+1)$ word $u \in \Sigma^*$, and every $\sigma \in \Sigma \cup \{\$\}$, $u\sigma$ is a factor of $b_n\$$ if and only if $u \in X_\sigma$.*

Notably, these languages do not depend on n. Indeed, it can be noticed that their union $X_0 \cup X_1 \cup X_\sharp \cup X_\$$ includes all length-$(n+1)$ factors of b_n for every positive integer n. Conversely, every word w belonging to this union has length at least 2 and is a factor of b_n, where $n = |w| - 1$. Also, one of the consequences of Lemma 7 is that every length-$(n+1)$ factor u of b_n occurs exactly once in b_n, and thus identifies the prefix and the suffix of b_n that ends or starts with u.

Proposition 8. *Let $n > 0$, $\Sigma = \{0, 1, \sharp\}$, $x, y \in \Sigma^*$, and $u, v \in \Sigma^{n+1}$.*

1. *If ux and uy are suffixes of b_n, then $x = y$.*
2. *If xv and yv are prefixes of b_n, then $x = y$.*
3. *If x and y are factors of b_n which both admit u as prefix and v as suffix (possibly with overlapping), then $x = y$.*

As seen before, X_0, X_1, $X_\sharp$ and $X_\$$ are disjoint and contain only words of length larger than 1 over $\Sigma = \{0, 1, \sharp\}$. We define $X_\perp$ to be the regular language of words over Σ that are not prefixes of any word in $X_0 \cup X_1 \cup X_\sharp \cup X_\$$. Each of X_0, X_1, $X_\sharp$, $X_\$$, and $X_\perp$ is a regular language that does not depend on n, and hence is recognized by a 1DFA of constant size. By taking the product of these 1DFAs, we obtain a 1DFA able to recognize each of the above languages (using a specific accepting state q_σ for each $\sigma \in \Sigma \cup \{\$, \perp\}$). Such an automaton, using 11 states only, is given in Fig. 1. We call it F.

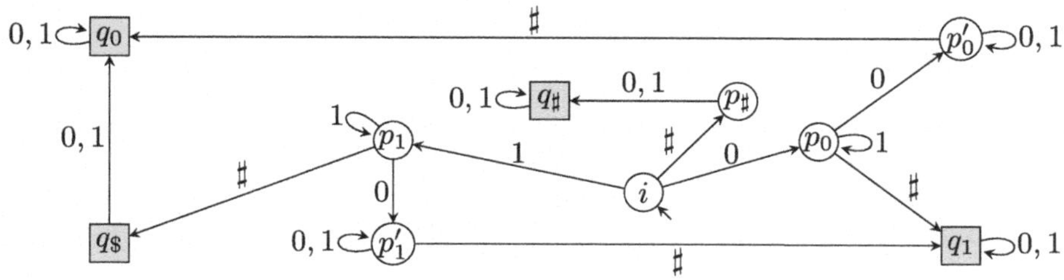

Fig. 1. The 11-state automaton F recognizing X_σ, for every $\sigma \in \{0, 1, \sharp, \$, \perp\}$ using the corresponding state q_σ (represented as a purple-filled rectangle); the trap state $q_\perp$ (entered from every state q_σ or from $p_\sharp$ by reading $\sharp$) is not depicted.

In order to force the length of the recognized factor to be $n+1$, we take the product of F with the canonical $(n+2)$-state 1DFA recognizing Σ^{n+1}. This leads to a $11(n+2)$-state 1DFA F_n that determines, using distinguished final states, whether a given input has length $n+1$ and belongs to X_0, X_1, $X_\sharp$, $X_\$$, or $X_\perp$.

Lemma 9. *Let $n > 0$ and $\Sigma = \{0, 1, \sharp\}$. There exists an $11(n+2)$-state 1DFA F_n that has five distinguished states $r_0, r_1, r_\sharp, r_\$, r_\perp$, such that, for each $u \in \Sigma^*$ and each $\sigma \in \Sigma \cup \{\$, \perp\}$, F_n goes from the initial state to r_σ on u if and only if $|u| = n+1$ and $u \in X_\sigma$.*

Procedure 1: `isFactBinarySeq`$_n$`()`

Check that the input is a factor of the full binary sequence b_n.

while read() $\neq \triangleleft$ **do**
  $\sigma \leftarrow$ `Consecutive`$_n$`()` `// the head is moved` $n+1$ `cells to the right`
  if $\sigma = \bot$ **then reject**
  if $\sigma = \triangleleft$ **then accept**
  if $\sigma = \$$ **and** read() $= \triangleleft$ **then accept**
  if $\sigma \neq$ read() **then reject**
  move the head n cells to the left

As we are interested only in factors with length $n+1$, we may assume that the five distinguished states r_0, r_1, $r_\sharp$, $r_\$$, and $r_\bot$ have no outgoing transition, *i.e.*, they are halting states. We denote by `Consecutive`$_n$`()` the procedure that simulates F_n until either reaching the end of the input, which is marked by $\triangleleft$, or entering one of the states r_σ from $\{r_0, r_1, r_\sharp, r_\$, r_\bot\}$. In the former case, the procedure returns $\triangleleft$, while in the latter case it returns σ. This allows us to design a 2DFA recognizing factors of b_n by implementing Procedure 1.

Lemma 10. *For every $n > 0$, there exists a 2DFA over $\{0, 1, \sharp\}$ with $\mathcal{O}(n)$ states which recognizes* $\text{Fact}(b_n)$, *namely the language of factors of b_n.*

This 2DFA can then be adapted to recognize special factors of b_n. In particular, if we fix the $n+1$ first symbols of the factors to recognize, the 2DFA can be made Γ-predictive, where $\Gamma = \{0, 1, \sharp\}$ is the input alphabet.

Lemma 11. *Let $\Gamma = \{0, 1, \sharp\}$, $n \geq 0$, and $\ell \in \{0, \dots, 2^n(n+1)\}$. Then:*

1. *a Γ-predictive 2DFA with $\mathcal{O}(n)$ states recognizes* $\text{Pref}(b_n) \cap \Gamma^\ell$;
2. *a Γ-predictive 2DFA with $\mathcal{O}(n)$ states recognizes* $\text{Suff}(b_n) \cap \Gamma^\ell$;
3. *a Γ-predictive 2DFA with $\mathcal{O}(n)$ states recognizes* $\{w \in \text{Pref}(b_n) \mid |w| \geq \ell\}$;
4. *a 2DFA with $\mathcal{O}(n)$ states recognizes* $\{w \in \text{Suff}(b_n) \mid |w| \geq \ell\}$.

In particular, $\{b_n\}$ is recognized by a Γ-predictive 2DFA with $\mathcal{O}(n)$ states.

Proof (sketch). For the cases $\ell < n+1$, the results are trivial. Let u be a factor of b_n of length $\ell \geq n+1$. According to Proposition 8, u is uniquely identified by the pair (x, y) formed by its prefix x and its suffix y of length $n+1$. Hence, a Γ-predictive 2DFA recognizing $\{u\}$ can first check that the input starts with x, then check that it is a factor of b_n using the 2DFA from Lemma 10, and finally check that it ends with y. This solves Items 1 and 2, as well as the case $\ell = 2^n(n+1)$ implying $u = b_n$. For Item 3 (resp. Item 4), instead of checking that y is a suffix (resp. x is a prefix) of the input, the machine can check that y (resp. x) occur as a factor in the input. For Item 3 (but not for Item 4), this preserves the Γ-predictiveness. $\square$

4 Exponential Gaps from D1-LAs and 2DFAs+cg to 2NFAs

We show an exponential gap for the simulation of D1-LAs by 2NFAs, which holds even in the unary case as witnessed by $\mathsf{BinSeq}_n = \{b_n\}$. In [9] an exponential gap in state complexity was shown for this conversion. However, the D1-LAs recognizing their witness languages made use of a linear number of annotation symbols, besides the linear number of states, making the gap in size sub-exponential. Our result is stronger in that the witness unary languages are recognized by D1-LAs of linear size.

Lemma 12. *For every $n \in \mathbb{N}$, the unary language $\mathsf{L}_n = \{a^{2^n(n+1)}\}$ is recognized by a 2DFA+cg with $\mathcal{O}(n)$ states and 3 annotation symbols, and a D1-LA with $\mathcal{O}(n)$ states and 3 work symbols.*

Proof. The case $n = 0$ is trivial, so we suppose $n > 0$. According to Lemma 11, there exists a Γ-predictive 2DFA $\mathcal{A}$ over $\Gamma = \{0, 1, \sharp\}$ with $\mathcal{O}(n)$ states which recognizes $\{b_n\}$. Since $\mathsf{L}_n = \pi_0(\{b_n\})$, the statements follow from Propositions 4 and 6, respectively. □

In order to prove that every 2NFA recognizing L_n requires more than 2^n states, we use the following lemma, which is implied by [5, Theorem 3.5].

Lemma 13. *If an n-state 2NFA accepts a word a^m for some $m \geq n$, then there exist $C \geq 1$ and $N \geq 0$ such that it accepts the word a^{m+kC} for every $k \geq N$. In particular, it accepts infinitely many words.*

As a consequence, $2^n(n+1)+1$ states are required (and actually sufficient) for a 2NFA to recognize the *singleton* L_n. Hence, in combination with Lemma 12, we obtain the following result.

Theorem 14. *For each $n \in \mathbb{N}$, there is a unary language L_n such that:*

1. *a 2DFA+cg with $\mathcal{O}(n)$ states and 3 annotation symbols recognizes it;*
2. *a D1-LA with $\mathcal{O}(n)$ states and 3 work symbols recognizes it;*
3. *every 2NFA requires $2^n(n+1)+1$ states to recognize it.*

5 Exponential Gaps from 2DFAs+cg to D1-LAs and 2NFAs

We analyze the size cost for converting 2DFA+cg into 1DFA, 2NFA, and D1-LA. We first present an upper bound for simulating 2DFAs+cg using 1NFAs and 1DFAs, and then expose the corresponding lower bounds.

Proposition 15. *Every n-state 2DFA+cg has equivalent $(n+1)^{n+1}$-state 1NFA, and $2^{(n+1)^{(n+1)}}$-state 1DFA.*

Proof (sketch). First we obtain a 1DFA over $\Sigma \times \Gamma$ using the Shepherdson method [13],[1] then we drop the Γ component in each transition to obtain a 1NFA equivalent to the original 2DFA+cg, that can in turn be converted into a 1DFA using the standard powerset construction. □

[1] that the resulting 1DFA is Γ-predictive if and only if $\mathcal{A}$ is Γ-predictive. Since Γ-predictiveness can be easily tested on one-way devices, this gives an algorithm to decide whether the property holds on 2DFAs.

We now introduce a family of unary languages $(\mathsf{M}_n)_{n\geq 1}$, each of which is succinctly represented by a 2DFA+cg, but requires a double exponential number of states to be recognized by a 1DFA. To define these languages, given an integer $n \geq 1$, first consider the language $\mathsf{IterSuffBinSeq}_n$ over the alphabet $\{0, 1, \sharp, \$\}$ composed of repetitions of some non-empty suffix of b_n followed by a marker $\$$. Formally:

$$\mathsf{IterSuffBinSeq}_n = \{(u\$)^k \mid k \geq 1,\ u \in \mathrm{Suff}(b_n),\ u \neq \varepsilon\}.$$

The length of each word $(u\$)^k$ in this language is clearly divisible by $|u\$| = |u|+1$, which is greater than 1 and less than or equal to $|b_n|+1$. In other words, it has a divisor between 2 and $2^n(n+1)+1$. The unary language M_n is defined as the language of lengths of $\mathsf{IterSuffBinSeq}_n$, *i.e.*,

$$\mathsf{M}_n = \{a^{kd} \mid k \geq 0,\ 1 < d \leq 2^n(n+1)+1\} = \pi_0\,(\mathsf{IterSuffBinSeq}_n).$$

5.1 A Small 2DFA+cg for $\mathsf{IterSuffBinSeq}_n$

By definition of M_n and Proposition 4, in order to define a 2DFA+cg recognizing M_n, it suffices to define a 2DFA recognizing $\mathsf{IterSuffBinSeq}_n$. This will be obtained by first adapting the 2DFA from Lemma 10 so that it recognizes the auxiliary language $(\mathrm{Suff}(b_n)\$)^*$ which is a superset of $\mathsf{IterSuffBinSeq}_n$.

Lemma 16. *For every positive integer n, the language $(\mathrm{Suff}(b_n)\$)^*$ is recognized by a 2DFA with $\mathcal{O}(n)$ states.*

Next, to recognize which word $w \in (\mathrm{Suff}(b_n)\$)^*$ is in $\mathsf{IterSuffBinSeq}_n$, we define a second auxiliary language, named $\mathsf{SamePrefix}_n$, that will also be recognized by a 2DFA of linear size in n. Each word of $\mathsf{SamePrefix}_n$ can be decomposed into factors separated by a marker $\$$, which have the following property: either they are shorter than n and identical, or they are longer than n and they share the same prefix of length n. Formally,

$$\begin{aligned}\mathsf{SamePrefix}_n = {} & \{(u\$)^k \mid k \geq 1, u \in \{0,1,\sharp\}^*, 0 < |u| < n\} \\ & \cup \{ux_1\$ \cdots ux_k\$ \mid k \geq 1, ux_i \in \{0,1,\sharp\}^* \text{ for } i = 1,\ldots,k, |u| = n\}.\end{aligned}$$

The next result states that there is is a small 2DFA recognizing $\mathsf{SamePrefix}_n$.

Lemma 17. *For every positive integer n, the language $\mathsf{SamePrefix}_n$ is recognized by a 2DFA with $\mathcal{O}(n)$ states.*

Proof (sketch). The 2DFA checks for each $i = 1, \ldots, n$ that the i-th symbols of the factors delimited by $\$$ (called *blocks*) are identical. Roughly, the finite control stores a counter i, which is initialized to 1 and incremented after each traversal of the tape until its value is n. Yet, in order to avoid a quadratic number of states, caution is taken to read the i-th symbol of each block: from the left delimiter ($\$$ or $\triangleright$) of a block, the machine moves the head rightward while decrementing the counter i until it equals 0, then reads the symbol (which is the i-th of the block), and finally recovers the counter value by moving the head leftward while increasing i until hitting $\$$ or $\triangleright$. In this way, only a linear number of states in n is sufficient. □

We are now ready to design our small 2DFA recognizing $\mathsf{IterSuffBinSeq}_n$.

Proposition 18. *For every positive integer n, $\mathsf{IterSuffBinSeq}_n$ is recognized by a 2DFA with $\mathcal{O}(n)$ states.*

Proof. The result is a consequence of Lemmas 16 and 17, observing that, by Proposition 8, $\mathsf{IterSuffBinSeq}_n = (\mathrm{Suff}(b_n)\$)^* \cap \mathsf{SamePrefix}_{n+1}$.[2] □

5.2 A Large Minimum 1DFA for $\mathsf{IterSuffBinSeq}_n$

Our goal is now to show that the minimum 1DFA recognizing M_n has a number of states that is doubly exponential in n. In fact, the exact lower bound is expressed using the notion of *primorial.* For a positive integer k, the *primorial of k*, denoted by $k\#$, is the product of all prime numbers less than or equal to k, *i.e.*, $k\# = \prod\{p \in \mathbb{N} \mid p \text{ is prime and } p \leq k\}$. The following result is used to obtain the doubly exponential lower bound.

Lemma 19. *For $n \geq 0$, $(2^n(n+1)+1)\# > 2^{2^n n}$.*

We now recall some notions about the strong structure enjoyed by unary regular languages, see, *e.g.*, [6]. We say that a unary language L over $\{a\}$ is *C-cyclic* for some $C \geq 1$, if for every $p \in \mathbb{N}$, either both or none of a^p and a^{p+C} belong to L. If furthermore, there are no $C' < C$ such that L is C'-cyclic, then it is said *properly C-cyclic.* A language is said *cyclic* if it is C-cyclic for some $C \geq 1$. A unary language L is regular if and only if it is *ultimately cyclic*, namely, for some finite language F and some cyclic language I, $\mathsf{L} = \mathsf{F} \cdot \mathsf{I}$. Also, if L is properly C-cyclic, then the minimum 1DFA recognizing it has C states.

Lemma 20. *Let $n \geq 0$ and $k_n = 2^n(n+1)+1$. The language M_n is properly $(k_n\#)$-cyclic and thus recognized by a minimum $(k_n\#)$-state 1DFA.*

We are now ready to prove the main theorem of the paper:

Theorem 21. *Let $n \in \mathbb{N}$. Then:*

1. *A 2DFA+cg with $\mathcal{O}(n)$ states and 4 annotation symbols recognizes M_n.*
2. *The minimum 1DFA recognizing M_n has $(2^n(n+1)+1)\# > 2^{2^n n}$ states.*

Proof. For $n = 0$ the statement holds trivially. So assume $n \geq 1$. Observe that $\mathsf{M}_n = \pi_0(\mathsf{IterSuffBinSeq}_n)$. By applying Propositions 4 and 18, we conclude that M_n is recognized by a 2DFA+cg with $\mathcal{O}(n)$ states and annotation alphabet $\Gamma = \{0, 1, \sharp, \$\}$. The second statement of the theorem is directly given by Lemmas 19 and 20. □

[2] Given two 2DFAs $\mathcal{A}$ and $\mathcal{B}$ with n and n' states, respectively, an $(n + n' + 1)$-state 2DFA can recognize $\mathcal{L}(\mathcal{A}) \cap \mathcal{L}(\mathcal{B})$ by proceeding as follows. First it simulates $\mathcal{A}$ (using n states), then, from an accepting state of $\mathcal{A}$ on $\lhd$, it moves its head back to the left endmarker (using one fresh state), and finally it simulates $\mathcal{B}$ (using n' states).

As a consequence of Theorem 21, we obtain the following gaps for the other models studied in this paper, as well as for complementing 1-LAs using results from [4]. In particular, we solve [7, Problem 1] about the size cost of the simulation of 1-LA by 1DFA in the unary case. Moreover, by exhibiting an exponential lower bound for the simulation of 1-LA by D1-LA in the unary case, we contribute to the investigation of [7, Problem 2] which asks for the size cost of determinizing 1-LAs. Whether a single exponential is always sufficient for this conversion is still an open problem, in the general and in the unary case (the best-known upper bound being doubly-exponential).

Theorem 22. *For each $n \in \mathbb{N}$, there is a unary language L_n such that:*

1. *a 2DFA+cg with $\mathcal{O}(n)$ states and $\mathcal{O}(1)$ annotation symbols recognizes it;*
2. *a 1-LA with $\mathcal{O}(n)$ states and $\mathcal{O}(1)$ work symbols recognizes it;*
3. *every D1-LA needs at least 2^n states to recognize it;*
4. *every 2NFA needs more than 2^n states to recognize it;*
5. *for sufficiently large n, every 1-LA needs more than 2^n states to recognize the language $\{a\}^* \setminus \mathsf{L}_n$.*

References

1. Bojanczyk, M., Daviaud, L., Guillon, B., Penelle, V.: Which classes of origin graphs are generated by transducers. In: ICALP 2017. LIPIcs, vol. 80, pp. 114:1–114:13. Schloss Dagstuhl - Leibniz-Zentrum für Informatik (2017). https://doi.org/10.4230/LIPIcs.ICALP.2017.114
2. Guillon, B., Prigioniero, L.: Linear-time limited automata. Theoret. Comput. Sci. **798**, 95–108 (2019). https://doi.org/10.1016/j.tcs.2019.03.037
3. Hopcroft, J., Ullman, J.: Introduction to Automata Theory, Languages and Computation. Addison-Wesley (1979)
4. Mera, F., Pighizzini, G.: Complementing unary nondeterministic automata. Theoret. Comput. Sci. **330**(2), 349–360 (2005). https://doi.org/10.1016/j.tcs.2004.04.015
5. Mereghetti, C., Pighizzini, G.: Optimal simulations between unary automata. SIAM J. Comput. **30**(6), 1976–1992 (2001). https://doi.org/10.1137/S009753979935431X
6. Pighizzini, G.: Investigations on automata and languages over a unary alphabet. Int. J. Found. Comput. Sci. **26**(07), 827–850 (2015). https://doi.org/10.1142/S012905411540002X
7. Pighizzini, G.: Limited automata: properties, complexity and variants. In: Hospodár, M., Jirásková, G., Konstantinidis, S. (eds.) DCFS 2019. LNCS, vol. 11612, pp. 57–73. Springer, Cham (2019). https://doi.org/10.1007/978-3-030-23247-4_4
8. Pighizzini, G., Pisoni, A.: Limited automata and regular languages. Int. J. Found. Comput. Sci. **25**(07), 897–916 (2014). https://doi.org/10.1142/S0129054114400140
9. Pighizzini, G., Prigioniero, L.: Limited automata and unary languages. Inf. Comput. **266**, 60–74 (2019). https://doi.org/10.1016/j.ic.2019.01.002
10. Pighizzini, G., Prigioniero, L.: Forgetting 1-limited automata. Electron. Proc. Theor. Comput. Sci. **388**, 95–109 (2023). https://doi.org/10.4204/eptcs.388.10

11. Pighizzini, G., Prigioniero, L.: Once-marking and always-marking 1-limited automata. Int. J. Found. Comput. Sci., 1–24 (2025). https://doi.org/10.1142/S0129054124440015
12. Sakoda, W.J., Sipser, M.: Nondeterminism and the size of two way finite automata. In: STOC 1978, pp. 275–286. ACM (1978). https://doi.org/10.1145/800133.804357
13. Shepherdson, J.C.: The reduction of two-way automata to one-way automata. IBM J. Res. Dev. **3**(2), 198–200 (1959). https://doi.org/10.1147/rd.32.0198
14. Wagner, K., Wechsung, G.: Computational Complexity, Mathematics and its Applications. Springer (1986)

Improved Upper Bounds for Determinizing NIDPDAs with Limited Nondeterminism

Mohammad Zakzok(✉) and Kai Salomaa

School of Computing, Queen's University, Kingston, ON K7L 3N6, Canada
{21mz13,salomaa}@queensu.ca

Abstract. Determinizing a nondeterministic input-driven pushdown automaton (NIDPDA) of size n results in the worst case in a machine of size $2^{\Theta(n^2)}$ (R. Alur, P. Madhusudan, "Adding nesting structure to words", J.ACM 56(3), 2009). A significantly improved determinization upper bound is known for k-path NIDPDAs that can have at most k computations on any input. We improve the known upper bound of determinization of k-path NIDPDAs by, roughly speaking, making the deterministic input-driven machine (DIDPDA) remember the shape of the tree in which the nondeterministic choices were made. Also we provide upper bounds for converting k-path universal input driven pushdown automata to DIDPDAs and establish lower bounds.

Keywords: input driven pushdown automata · measures of nondeterminism · size complexity

1 Introduction

Nondeterminism plays a fundamental role in the theory of computation, and multiple ways to quantify the nondeterminism (of finite automata) have been considered in the literature [8,14,17,23,24]. The *degree of ambiguity* counts the number of accepting computations, the *path size,* a.k.a. *tree width*, counts the number of all computations, and other measures like *guessing* or *branching* [8] are based on the amount of nondeterminism of an accepting computation that uses the least nondeterminism.

In an input-driven pushdown automaton (IDPDA), the current input symbol determines whether the automaton performs a push operation, a pop operation, or does not use the stack. The model was originally introduced by Mehlhorn [19]. Input-driven pushdown automata retain many of the desirable closure and decision properties of finite automata. Most notably, a nondeterministic IDPDA (NIDPDA) always has an equivalent deterministic machine (DIDPDA) with at most $\mathcal{O}(2^{n^2})$ states [4].

In 2004, the model was reintroduced by Alur and Madhusudan [1] under the name of *visibly pushdown automata*. The authors then studied the computational complexity of decision problems for IDPDAs and established a $2^{\Omega(n^2)}$ lower bound for determinizing a NIDPDA of size n, proving that the determinization of NIDPDAs required a blowup of $2^{\Theta(n^2)}$ in the number of states [2]. Jirásková and

S.-K. Ko and F. Manea (Eds.): DLT 2025, LNCS 16036, pp. 166–181, 2026.
https://doi.org/10.1007/978-3-032-01475-7_12

Okhotin [11] would go on to prove that determinization requires at most exactly 2^{n^2} states. These results were further improved by Martynova [18], where the author shows that determinizing a NIDPDA of size n with a fixed size alphabet still requires a 2^{n^2} blowup in the number of states in the worst case and provided a precise upper bound for the number of required stack symbols.

A NIDPDA A is said to be *k-path* if the computation tree of A on any input has at most k leaves, $k \in \mathbb{N}$. The NIDPDA A has the finite path property if A is k-path for some $k \in \mathbb{N}$. An analogous notion of limited nondeterminism for finite automata has been considered under the names "leaf size" [10] or "tree width" [23]. Okhotin and Salomaa [22] show that it is decidable whether a given NIDPDA has the k-path property, and for a fixed k this can be decided in polynomial time. More recently, Han et al. [9] have shown that deciding the finite path property of a NIDPDA can be done in polynomial time.

As the k-path property limits NIDPDAs, it is expected that the determinization of k-path NIDPDAs would be more succinct than the determinization of general NIDPDAs. Indeed, it has been shown in [22] that a k-path NIDPDA A with n states and m stack symbols can be converted to a DIDPDA with

$$\sum_{i=1}^{k} i^i \cdot (n+1)^i \text{ states and } \sum_{i=1}^{k} m^i \text{ stack symbols.} \tag{1}$$

This paper improves the upper bound (1) for the size of a DIDPDA equivalent to A to use $\sum_{i=1}^{k}(i-1)! \cdot (n+1)^i$ states (without changing the number of stack symbols). Our construction is based on having the DIDPDA remember the shape of the full binary tree that represents the nondeterministic choices in a computation tree of A, including the *history* in which the tree was constructed. In addition, we give an alternative upper bound that further reduces the number of states of the equivalent DIDPDA, but increases the number of stack symbols.

Also, we give upper bounds for the determinization of a *universal* IDPDA. Universal pushdown automata are a special case of alternating machines, where the machine allows a multivalued transition function and accepts an input only if all computations accept. Alternating pushdown automata are considerably more powerful than nondeterministic pushdown automata [16] whereas an alternating IDPDA always has an equivalent DIDPDA [3].

Lastly, we prove a new lower bound for the determinization k-path NIDPDAs that deals with the previously unexplored case where the number of paths is larger than the number of states. We also establish symmetric lower bounds for determinizing k-path UIDPDAs.

2 Preliminaries

In the following, Σ is a finite alphabet, Σ^* is the set of strings over Σ, and ε is the empty string. The cardinality of a finite set S is $|S|$ and the power set of S is 2^S. The set of natural numbers is $\mathbb{N}$.

The input alphabet of an input-driven pushdown automaton is a tuple $(\Sigma_{+1}, \Sigma_{-1}, \Sigma_0)$ where the components of Σ_{+1}, Σ_{-1}, and Σ_0 are disjoint finite sets. Elements of Σ_{+1}, Σ_{-1}, and Σ_0 are called, respectively, *left brackets, right brackets,* and *neutral symbols.* A string is well-matched if every left bracket has a matching right bracket, and vice versa.

Definition 1. [2,21] *A deterministic input-driven pushdown automaton (DIDPDA) is a 7-tuple* $A = (\Sigma, Q, \Gamma, q_0, \bot, [\delta_a]_{a\in\Sigma}, F)$*, where*

- $\Sigma = (\Sigma_{+1}, \Sigma_{-1}, \Sigma_0)$ *is an input alphabet split into three disjoint parts,*
- Q *is a finite set of internal states of the automaton,* q_0 *is the initial state, and* $F \subseteq Q$ *is the set of accepting states,*
- Γ *is a finite stack alphabet, and a special symbol* $\bot \notin \Gamma$ *denotes an empty stack,*
- *for a neutral symbol* $c \in \Sigma_0$*, the state change is described by a partial function* $\delta_c\colon Q \to Q$*,*
- *for each left bracket symbol* $< \in \Sigma_{+1}$*, the behaviour of the automaton is described by a partial function* $\delta_<\colon Q \to Q \times \Gamma$ *which, for a given current state, provides the next state and the symbol to be pushed onto the stack,*
- *for every right bracket symbol* $> \in \Sigma_{-1}$*, the behaviour is described by a partial function* $\delta_>\colon Q \times (\Gamma \cup \{\bot\}) \to Q$ *specifying the next state, assuming that the given stack symbol is popped from the stack (or that the stack is empty).*

A configuration of A is a triple (q, w, x)*, where* $q \in Q$ *is the state,* $w \in \Sigma^*$ *is the remaining input and* $x \in \Gamma^*$ *is the stack contents. For each configuration, the next configuration is defined as follows.*

$$
\begin{array}{rll}
(q, cw, x) \vdash_A (\delta_c(q), w, x), & & \textit{for } c \in \Sigma_0,\ q \in Q \\
(q, {<}w, x) \vdash_A (q', w, \gamma x), & & \textit{for } {<} \in \Sigma_{+1},\ q \in Q,\ \delta_<(q) = (q', \gamma) \\
(q, {>}w, \gamma x) \vdash_A (\delta_>(q, \gamma), w, x), & & \textit{for } {>} \in \Sigma_{-1},\ q \in Q,\ \gamma \in \Gamma \\
(q, {>}w, \epsilon) \vdash_A (\delta_>(q, \bot), w, \epsilon), & & \textit{for } {>} \in \Sigma_{-1}
\end{array}
$$

Intuitively, the automaton ignores the stack on neutral symbols, pushes onto the stack on left bracket symbols, pops a symbol from the stack on right bracket symbols if it is non-empty, and pops an empty stack indicator $\bot$ *on a right bracket symbol when the stack is empty. In all cases,* $w \in \Sigma^*$ *is the remaining input and* $x \in \Gamma^*$ *is a string of symbols in the stack.*

The language recognized by A is the set of all strings $w \in \Sigma^*$*, on which the automaton, having begun its computation in the configuration* (q_0, w, ϵ)*, eventually reaches a configuration of the form* (q, ϵ, x)*, with* $q \in F$ *and with any stack contents* $x \in \Gamma^*$*. Note that since the transition functions are partial, A does not need to reach the end of w before halting and rejecting.*

We extend the definition to include DIDPDAs with more than one initial state. A k-entry DIDPDA has k initial states. The DIDPDA accepts the input if the computation from some initial state accepts the input.

A *nondeterministic IDPDA* (NIDPDA) [2,21] $A = (\Sigma, Q, \Gamma, Q_0, \bot, [\delta_a]_{a\in\Sigma}, F)$ is an extension of a deterministic machine where the initial state is replaced by a set of initial states $Q_0 \subseteq Q$ and each transition function is allowed to be multiple-valued: namely, $\delta_c\colon Q \to 2^Q$ for a neutral symbol $c \in \Sigma_0$, $\delta_<\colon Q \to 2^{Q\times\Gamma}$ for a left bracket $< \in \Sigma_{+1}$ and $\delta_>\colon Q \times (\Gamma \cup \{\bot\}) \to 2^Q$ for a right bracket $> \in \Sigma_{-1}$. a NIDPDA may have multiple computations on an input string, and a string is accepted if one computation originating from one of the initial states of Q_0 accepts.

We also consider universal IDPDAs. The syntactic definition of a *universal IDPDA* (UIDPDA) A is the same as the definition of a NIDPDA. An UIDPDA A accepts a string w if all computations of A on w accept. A *k-entry universal DIDPDA* is a DIDPDA with k initial states that accepts a string w if the computation originating from each initial state accepts w. Note that universal machines are a special case of alternating machines [12,16].

NIDPDAs employing limited nondeterminism were first considered in [22]. The various computations of NIDPDAs and UIDPDAs can be modeled as a rooted computation tree. Computation trees of finite automata have been used, e.g. in [6,7,10,23], in research pertaining to the simulation of automata, as it is often the case that efficient exploration of the computation tree of a nondeterministic automaton is key to simulating it using a deterministic automaton.

Let $A = (\Sigma, Q, \Gamma, Q_0, \bot, [\delta_a]_{a\in\Sigma}, F)$ be a NIDPDA (or a UIDPDA), and let w be an input string. The root of the *computation tree of* A *on* w, $t_{A,w}$, is labeled by a special symbol τ_0. The children of the root are labeled by the $|Q_0|$ configurations corresponding to the different initial states, (q', w, ϵ), $q' \in Q_0$. Generally, in $t_{A,w}$, the children of a configuration (q, u, x) are labeled by the possible next configurations (q', u', x') such that $(q, u, x) \vdash_A (q', u', x')$. The leaves of $t_{A,w}$ are labeled by configurations for which the transition relation of A is undefined. The NIDPDA (or UIDPDA) A is said to be a *k-path* NIDPDA/UIDPDA if any computation tree of A has at most k leaves [22], and A is *finite path* if it is k-path for some $k \in \mathbb{N}$.

To prove lower bounds for the size of a DIDPDA we use the notion of 1-separator sets used in [22]. More general methods for establishing lower bounds are considered in [20,22].

2.1 Full Binary Trees

The proof of the main result in Sect. 3 requires a certain level of understanding of *full binary trees* and some of their properties [5]. To refresh the memory of the reader, we provide the definition of full binary trees and showcase some important facts about them. For a discussion of binary search trees and their variations, see Knuth [15].

Definition 2. [5] *A full binary tree is a rooted binary tree where every node has exactly two or zero children.*

Fact 1. *For any rooted tree* t*, we can create a full binary tree* t' *such that:*

(i) Each node u in t that has exactly one child is not in t'.
(ii) If a node $u \in t'$ is an ancestor of $v \in t'$, then u is also an ancestor of v in t.
(iii) t' has the same number of leaves as t.

Note that rooted trees are typically used to represent the computation of an automaton, including the computations of NIDPDAs discussed earlier. Because of this, the process of producing a binary tree based on a rooted tree is important for our proofs.

Given any rooted tree, we can create its "representation" as a full binary tree using a fairly simple construction. First, we remove all nodes with exactly one child, granting their only child to their parent node. Then, we reduce the number of children each node has by introducing intermediary nodes between it and its children. For each node u with k children $c_1, c_2 \ldots c_k$, we create a full "almost complete" binary tree rooted in u, with $c_1, c_2 \ldots c_k$ as the leaves, by introducing a number of nodes in-between them, filling the tree from top to bottom. We then plug this tree back into its original position in the tree. This new full binary tree has the same number of leaf nodes, but typically has a higher number of intermediary nodes.

Now, consider the following method of counting the number of ways to produce a full binary tree with k leaves. A full binary tree with 1 or 2 leaves has only one possible production method. To create a full binary tree with 3 leaves, there are 2 possible nodes that can be the parent of the two newly added nodes, and so we have a total of 2 possible production methods, and so on. More generally, we can describe the total number of ways to generate a k-leaf full binary tree using the following function:

$$f(1) = 1, \quad f(2) = 1, \quad f(k) = (k-1) \cdot f(k-1) \tag{2}$$

Clearly this function is just $(k-1)!$ for $k \geq 1$. A weakness of generating all possible trees using this method is that we may produce many duplicates of the same shape. However, this process also enables us to track the unique *history* of every generated tree. If we label each internal node with an index between 1 and $k-1$ indicating the order of the creation of the nodes, we are able to deduce which tree shapes of $(k-1), (k-2), \ldots 1$ leaves were used to produce it. We call this an *indexed* full binary tree. Examples of indexed full binary trees are shown in Figs. 1 and 2.

We define the sequence of tree shapes used to generate a k-leaf full binary tree t_k as the *history* of t_k, denoted as $H(t_k) = (t_{k-1}, t_{k-2}, \ldots t_1)$. Note that $H(t_k)$ is defined for both indexed and non-indexed full binary trees, but $H(t_k)$ can be deduced only from t_k itself if it is indexed.

Now, consider the total number of unique non-indexed full binary trees of size k. This is nothing but the total number of unique k leafed full binary tree shapes, which has a known relation to Catalan numbers. As shown in [26], there are C_k shapes of full binary trees with $k+1$ leaf nodes, where $C_k = \frac{1}{k+1}\binom{2k}{k}$ is the kth Catalan number. Moreover, it has been shown that the Catalan number C_k asymptotically grows as $\frac{4^k}{k^{3/2}\sqrt{\pi}}$ [5, 26]. We use this description of the total number of possible full binary trees in an alternative upper bound construction.

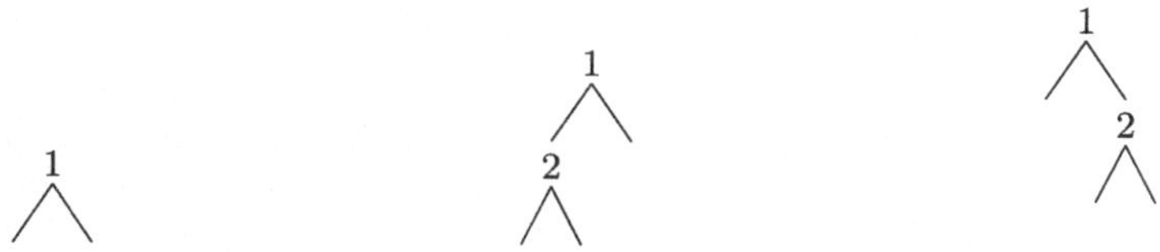

(a) The only possible tree with 2 leaves.

(b) The first possible tree with 3 leaf nodes.

(c) The second option of a tree of with 3 leaf nodes.

Fig. 1. Examples of indexed full binary trees.

(a) One possible indexing of this shape of a 4 leafed tree.

(b) Alternative indexing of the same shape of a 4 leafed tree.

Fig. 2. Examples of indexed full binary trees with the same shape.

Fact 2. *A full binary tree with* $k+1$ *leaves has at most* $C_k = \frac{1}{k+1}\binom{2k}{k}$ *tree shapes, which asymptotically grows as* $\frac{4^k}{k^{3/2}\sqrt{\pi}}$.

3 Determinizing *k*-Path NIDPDAs and UIDPDAs

With the construction used above to showcase Fact 1, we are able to produce an indexed full binary tree representation of the computation tree of a NIDPDA A on a string w, $t_{A,w}$. This tree is useful as it succinctly represents nondeterministic choices of A. We call this tree the *history tracking binary normal form (htb normal form)* of $t_{A,w}$. Note that the internal nodes in the htb normal form of $t_{A,w}$ are only labelled by the index, and no longer contain any information about the exact configuration that the node represents.

Lemma 5 of [22] gives an upper bound for the size of a DIDPDA equivalent to a given k-path NIDPDA A. The state of the DIDPDA, roughly speaking, is an ℓ-tuple of states of A, $\ell \leq k$, where each state is indexed by the "number" of the computation it represents. Here, instead, we use a construction that "remembers" the parent-child structure of the computation branches of the NIDPDA being simulated. That is, the state of the simulating DIDPDA stores the htb normal form of the computation tree of the NIDPDA and the sequence of current states.

For the proof of Lemma 3 we introduce the following notation. Let A be a k-path NIDPDA with state set Q of cardinality n and let T_i be the set of all indexed full binary trees with i leaves. We define S_i^A to be the set of all possible lists of size i formed by elements in $Q \cup \{-\}$ (where "$-$" is the failure marker). The cardinality of S_i^A is $(n+1)^i$. Lastly, T_k^A is the set of all possible tuples (t_i, s_i), for $1 \le i \le k$, such that t_i is a tree in T_i and s_i is a list in S_i^A. The cardinality of T_k^A is $\sum_{i=1}^{k}(i-1)! \cdot (n+1)^i$. Examples of elements of T_k^A are shown in Figs. 3, 4, and 5.

Lemma 3. *Let* $A = (\Sigma, Q, Q_0, \Gamma, [\delta_a]_{a \in \Sigma}, F)$ *be a* k*-path NIDPDA. We can construct a DIDPDA with set of states* T_k^A *and stack alphabet* $\Gamma^{\le k}$*, recognizing* $L(A)$*.*

Proof. Define a DIDPDA $B = (\Sigma, T_k^A, q_0', \Gamma^{\le k}, [\delta_a']_{a \in \Sigma}, F')$ where the initial state q_0' of B is $(t_s, [q_1, q_2, \ldots q_s])$ where t_s is an indexed full binary tree with exactly s leaves and $q_1, \ldots, q_s$ are the initial states of A. Note that t_s can be any indexed full binary tree with s leaves, but we choose it to be the almost complete full binary tree produced by always adding leaves from left to right. Lastly, a state $(t_m, [q_1, q_2, \ldots q_m])$ is in F' if there exists $1 \le i \le m$ such that $q_i \in F$.

Unlike general nondeterministic pushdown automata, NIDPDAs are determinizable. This is because their branches act in sync when pushing, popping, or ignoring the stack, as the input symbol is solely responsible for deciding the appropriate stack action. As a result, the crux of constructions determinizing NIDPDAs is defining the transition function on neutral, left bracket, and right bracket symbols. As such, we define the transition function of B in all three cases and show how B is able to keep track of all the branches explored by A.

Consider a prefix w of the input seen so far and let t_m be an indexed htb normal form computation tree of A on w with m leaves, $1 \le m \le k$. For $c \in \Sigma_0$ and $q_1, \ldots, q_m \in Q \cup \{-\}$, the transition δ_c' is defined as

$$\delta_c'(t_m, [q_1, q_2, \ldots q_m]) = (t_\ell, [p_1, p_2, \ldots p_\ell]), \tag{3}$$

where the indexed htb normal form tree t_ℓ and states $p_1, \ldots, p_\ell$ are obtained as follows. Let r_1 be the leaf labeled tree obtained from t_m by assigning elements $q_1, \ldots, q_m$ to label the leaves of t_m, where the leafs are labeled from left to right by elements $q_1, \ldots, q_m$. Note that this method of labeling leaves of a tree with elements of a list is used repeatedly from this point on, but we simply refer to it as "matching" or "labeling" elements of a list to a tree.

The tree r_1 is expanded to a tree r_2 by assigning to each leaf node u labeled by q_i, $1 \le i \le m$, $|\delta_c(q_i)|$ leaves labeled by the sequence of elements in $\delta_c(q_i)$. If q_i is "$-$", we do not attach successors to u. Note that since u is no longer a leaf node, it is not labeled in r_2. The tree r_2 is converted to an indexed full binary t_ℓ, as outlined in Fact 1. The sequence $p_1, \ldots, p_\ell$ consists of the sequence of elements of $\delta_c(q_1), \ldots, \delta_c(q_m)$, that is,

$$\ell = \sum_{i=1}^{m} \max\{|\delta_c(q_i)|, 1\}.$$

Note that if $\delta_c(q_i) = \emptyset$, the leaf labeled by q_i will be replaced by "$-$". A specific scenario illustrating the operation of δ'_c is shown in Example 31.

The correctness of the construction is based on the following. If t_m with leaves labeled by $q_1, \ldots, q_m$ is the htb normal form of a computation tree of A on string w (according to the procedure of Fact 1), then t_ℓ with leaves labeled by $p_1, \ldots, p_\ell$ is the htb normal form of a computation tree of A obtained from the computation tree of A on the string wc.

We define the transition function on left bracket symbols, which behaves very similarly, but also pushes a list of symbols on the stack, one for each followed branch so far. Consider a state $(t_m, [q_1, q_2, \ldots q_m])$ where the transition of A in a state q_i on a left bracket symbol $< \in \Sigma_{+1}$ is defined as

$$\delta_<(q_i) = \{(p_1^i, \gamma_1^i), \ldots, (p_{\ell_i}^i, \gamma_{\ell_i}^i)\}, \text{ for } i = 1, \ldots, m.$$

Then, the transition function $\delta'_<$ of B is defined as

$$\begin{gathered}\delta'_<(t_m, [q_1, \ldots, q_m]) = \\ ((t_\ell, [p_1^1, \ldots, p_{\ell_1}^1, \ldots, p_1^m, \ldots, p_{\ell_m}^m]), [\gamma_1^1, \ldots, \gamma_{\ell_1}^1, \ldots, \gamma_1^m, \ldots, \gamma_{\ell_m}^m]).\end{gathered} \tag{4}$$

Note that the states q_i, $i = 1, \ldots, m$, label leaves of t_m by virtue of their ordering. Here, t_ℓ is the indexed htb normal form tree obtained from t_m by expanding the leaves that correspond to states q_i, where $\delta_<(q_i)$ has more than one element, and then converting the resulting tree to a full binary tree, as in the transitions for the neutral symbols described earlier. The states $p_{\ell_i}^i$ label the leaves of t_ℓ by virtue of their ordering. Note that in (4), the action of A on the failure symbol "$-$" is interpreted as $\delta_<(-) = \{(-, \gamma_0)\}$, where $\gamma_0 \in \Gamma$ can be any stack symbol. A specific scenario illustrating the operation of $\delta'_<$ is shown in Example 32.

The correctness of the construction is justified similarly as for the transitions on neutral symbols. If t_m is a htb normal form of a computation tree of A on string w (that has reached states $q_1, \ldots, q_m$ after processing w), the tree t_ℓ is a htb normal form of a computation tree of A on string $w <$. Also, $\gamma_1^1, \ldots \gamma_{\ell_m}^m$ are the stack symbols pushed by the branches of the computation of A, whose tree shape is represented by t_ℓ, while reading the last $<$ symbol.

The operation of B on the right bracket symbols is the most crucial part of the construction. The states of B store an indexed htb normal form of a computation tree of A, and based on that tree, the transition of B "finds out" what the computation tree of A looked like at the most recent point that the current top stack symbol was pushed. The transition pops a list of symbols and ensures that every state-stack symbol pair is accurately matched.

Consider a right bracket symbol $> \in \Sigma_{-1}$, a tree $t_m \in T_m$, a list of elements $q_1, \ldots, q_m \in Q \cup \{-\}$, and a list of stack symbols $\gamma_1, \ldots, \gamma_u \in \Gamma$, $1 \leq u \leq m$. We define

$$\delta'_>((t_m, [q_1, q_2, \ldots q_m]), [\gamma_1, \gamma_2, \ldots \gamma_u]) = (t_\ell, [p_1, p_2, \ldots p_\ell]). \tag{5}$$

The tree t_ℓ and the states p_i, $1 \leq i \leq \ell$ are obtained as follows. The indices of t_m yield the history of the tree t_m, $H(t_m) = (t_{m-1}, t_{m-2}, \ldots, t_1)$. From the history we select the tree t_u that describes the htb normal form of the computation tree of A at the most point which the symbol $[\gamma_1, \ldots, \gamma_u]$ was pushed. By overlaying t_u on t_m, we know, for each state q_i, $i = 1, \ldots, m$, which of the stack symbols γ_{x_i}, $1 \leq x_i \leq u$, q_i's ancestor pushed onto the stack. Note that if the stack is empty, then B just operates as if it had popped a list with a singular element, $[\perp]$.

Let r_1 be the leaf labeled tree obtained from t_m by assigning the symbols $q_1, \ldots, q_m$ as labels of the leaves. The tree r_1 is expanded to a tree r_2 by assigning to each leaf u labeled by q_i, $|\delta_>(q_i, \gamma_{x_i})|$ leaves labeled by the elements of $\delta_>(q_i, \gamma_{x_i})$. If q_i is "$-$", the computation of that branch has failed and we do not attach successors to u. The tree r_2 is converted to a full binary t_ℓ, as outlined in Fact 1. The sequence $p_1, \ldots, p_\ell$ consists of the sequence of elements of $\delta_>(q_1, \gamma_{x_1}), \ldots, \delta_>(q_m, \gamma_{x_m})$, that is,

$$\ell = \sum_{i=1}^{m} \max\{|\delta_>(q_i, \gamma_{x_i})|, 1.$$

A specific scenario illustrating the operation of $\delta'_>$ is shown in Example 33. The correctness of the construction is based on the following. If t_m with leaves labeled by $q_1, \ldots, q_m$ is the htb normal form of a computation tree of A on a prefix w of the input, then, because γ_{x_i} was the symbol pushed by the predecessor of q_i, the tree t_ℓ with leaves labeled by $p_1, \ldots, p_\ell$ is htb normal form of a computation tree of A on a prefix $w >$ after reading the right bracket symbol $>$.

Finally, as B is able to continuously keep track of which states of A are reachable by the end of any string w, it can "know" whether w is in $L(A)$ based on whether the list of reached states contains a state in F or not. Thus, the choice of the final states of B guarantees that B accepts exactly the strings in $L(A)$. □

As an aid to the reader, the Examples 31, 32, and 33 respectively provide illustrations of how B operates when reading neutral, left bracket, and right bracket symbols. Note that the transition function of B operates fairly similarly on neutral and left bracket symbols, but operates in a more complex manner on right bracket symbols.

Example 31. *Consider the specific scenario, where A's computation on string w_1 reaches the set of states $\{q_1, q_2, q_3\}$, and then encounters a neutral symbol c_1. Suppose that A's relevant transitions on c_1 are:*

$$\delta_{c_1}(q_1) = \{p_1, p_2, p_3\} \qquad \delta_{c_1}(q_2) = \textit{undefined} \qquad \delta_{c_1}(q_3) = \{p_4\}$$

The resulting transition $\delta'_{c_1}(t_{w_1}, [q_1, q_2, q_3])$ of B would be $(t_{w_1c_1}, [p_1, p_2, p_3, -, p_4])$, where t_{w_1} and $t_{w_1c_1}$ are respectively the indexed htb normal form computation trees of A on w_1 and w_1c_1. More specifically, consider that the tree t_{w_1} was the indexed tree shown in Fig. 3a, where each of the states q_1, q_2, and q_3 label a leaf by virtue of their ordering. The resulting tree $t_{w_1c_1}$ would be a tree shown in Fig. 3b, where each of the elements in $[p_1, p_2, p_3, -, p_4]$ also label a leaf

(a) Visualization of state $(t_{w_1}, [q_1, q_2, q_3])$.

(b) Visualization of state $(t_{w_1 c_1}, [p_1, p_2, p_3, -, p_4])$.

Fig. 3. Visualizations of states of the simulating automaton.

by virtue of their ordering. (Note that since $\delta_{c_1}(q_1) = \{p_1, p_2, p_3\}$ *is unordered, the order of the first three leaves could be different.)*

Example 32. *Consider the specific scenario, where A's computation on string* w_2 *reaches the set of states* $\{q_1, q_2, q_3, q_4\}$*, and then encounters a left bracket symbol* $<_1$*. Suppose that A's relevant transitions on* $<_1$ *are:*

$$\delta_{<_1}(q_1) = \{(p_1, \gamma_1), (p_2, \gamma_2)\} \qquad \delta_{<_1}(q_3) = \{(p_4, \gamma_4)\}$$
$$\delta_{<_1}(q_2) = \{(p_3, \gamma_3)\} \qquad \delta_{<_1}(q_4) = \{(p_5, \gamma_5), (p_6, \gamma_6)\}$$

The resulting transition of B would be

$$\begin{aligned} &\delta'_{<_1}(t_{w_2}, [q_1, q_2, q_3, q_4]) = \\ &((t_{w_2<_1}, [p_1, p_2, p_3, p_4, p_5, p_6]), [\gamma_1, \gamma_2, \gamma_3, \gamma_4, \gamma_5, \gamma_6]), \end{aligned} \tag{6}$$

where t_{w_2} *and* $t_{w_2<_1}$ *are respectively the indexed htb normal form computation trees of A on* w_2 *and* $w_2 <_1$.

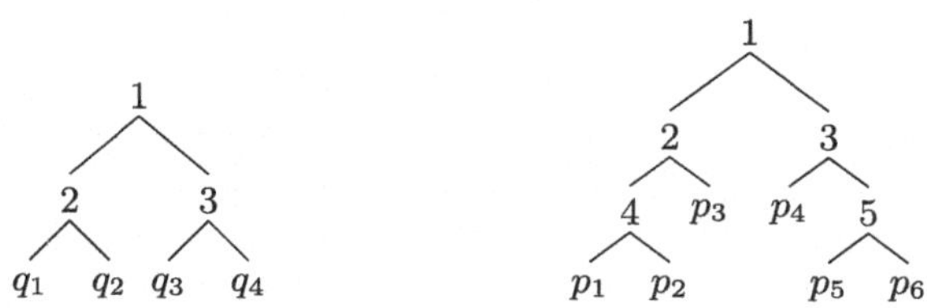

(a) Visualization of state $(t_{w_2}, [q_1, q_2, q_3, q_4])$.

(b) Visualization of state $(t_{w_2<_1}, [p_1, p_2, p_3, p_4, p_5, p_6])$.

Fig. 4. Visualizations of states of the simulating automaton.

More specifically, consider that the tree t_{w_2} *was the indexed tree shown in Fig. 4a, where each of the states* $q_1, \ldots, q_4$ *label a leaf by virtue of their ordering.*

The resulting tree $t_{w_2<_1}$ *would be a tree shown in Fig. 4b, where each of the elements in* $[p_1, \ldots, p_6]$ *also label a leaf by virtue of their ordering. Finally, notice that by using* $t_{w_2<_1}$ *and the length of* $[\gamma_1, \ldots, \gamma_6]$*, we are easily able to match each branch to the symbol in* $[\gamma_1, \ldots, \gamma_6]$ *that it pushed.*

Example 33. *Consider the specific scenario where A's computation on a string* w_3 *reaches 11 branches in states* $q_1, q_2, \ldots q_{11}$*, and then encounters a right bracket symbol* $>_1$*. Suppose that each of the branches, represented by their states below, pop the following symbols from the stack:*

branch in state	q_1, q_2, q_3	q_4	q_5	q_6	q_7, q_8	q_9	q_{10}, q_{11}
stack symbol	γ_1	γ_2	γ_3	γ_4	γ_5	γ_6	γ_7

Notice that some of the branches have previously pushed the same symbol. In general, this could either be because A's transition function pushed the same symbol while in different branches or because the branches share an ancestor configuration that had pushed that symbol. Our example deals exclusively with the latter scenario, to showcase how B matches symbols to branches in its simulation. Suppose that A's relevant transitions on $>_1$ *are:*

$$\begin{array}{lll} \delta_{>_1}(q_1, \gamma_1) = \{p_1\} & \delta_{>_1}(q_5, \gamma_3) = \{p_5\} & \delta_{>_1}(q_9, \gamma_6) = \{p_9\} \\ \delta_{>_1}(q_2, \gamma_1) = \{p_2\} & \delta_{>_1}(q_6, \gamma_4) = \{p_6\} & \delta_{>_1}(q_{10}, \gamma_7) = \{p_{10}, p_{11}\} \\ \delta_{>_1}(q_3, \gamma_1) = \{p_3\} & \delta_{>_1}(q_7, \gamma_5) = \{p_7\} & \delta_{>_1}(q_{11}, \gamma_7) = \{p_{12}, p_{13}\} \\ \delta_{>_1}(q_4, \gamma_2) = \{p_4\} & \delta_{>_1}(q_8, \gamma_5) = \{p_8\} & \end{array}$$

The resulting transition of B would be

$$\delta'_{>_1}((t_{w_3}, [q_1, q_2, q_3, q_4, q_5, q_6, q_7, q_8, q_9, q_{10}, q_{11}]), [\gamma_1, \gamma_2, \gamma_3, \gamma_4, \gamma_5, \gamma_6, \gamma_7]) = \\ (t_{w_3>_1}, [p_1, p_2, p_3, p_4, p_5, p_6, p_7, p_8, p_9, p_{10}, p_{11}, p_{12}, p_{13}])$$

where t_{w_3} *and* $t_{w_3>_1}$ *are respectively the indexed htb normal form computation trees of A on* w_3 *and* $w_3 >_1$*. More specifically, consider that the tree* t_{w_3} *was the indexed tree shown in Fig. 5a, where each of the states* $q_1, \ldots, q_{11}$ *label a leaf by virtue of their ordering. The resulting tree* $t_{w_3>_1}$ *would be the tree shown in Fig. 5b, where each of the state* $p_1, \ldots, p_{13}$ *also label leaf by virtue of their ordering.*

Notice that the length of list of symbols popped from the stack indicates the number of leafs of the tree when the symbol was pushed. Since we were able to find out the shape of the tree when it had exactly 7 leaves, we are able to deduce which of the current branches shared an ancestor configuration, and consequently match each branch to its correct symbol in the list. Lastly, note that if t_{w_3} *was not indexed, it would be much more difficult, or perhaps impossible, to discern which 7-leafed subtree of* t_{w_3} *was used in building it without knowing more information.*

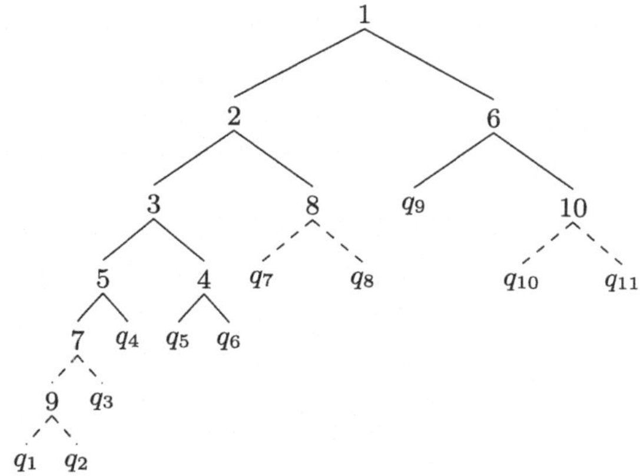

(a) Visualization of state $(t_{w_3}, [q_1, q_2, q_3, q_4, q_5, q_6, q_7, q_8, q_9, q_{10}, q_{11}])$. The shape of the tree that existed when it had 7 leaves is in solid lines.

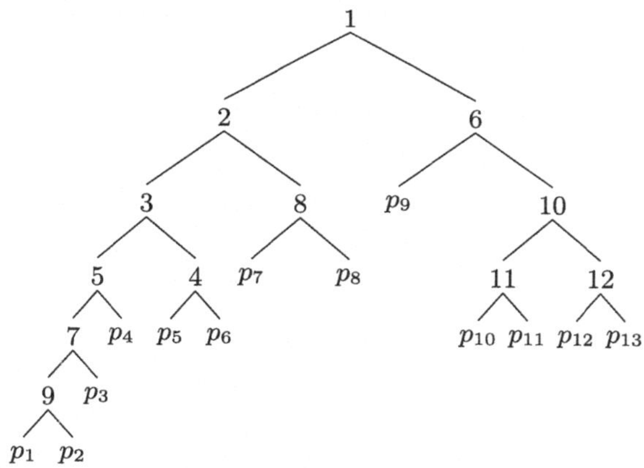

(b) Visualization of state $\left(t_{w_3>1}, [p_1, p_2, p_3, p_4, p_5, p_6, p_7, p_8, p_9, p_{10}, p_{11}, p_{12}, p_{13}]\right)$.

Fig. 5. Visualizations of states of the simulating automaton.

Theorem 4. *For any k-path NIDPDA A with n states and m stack symbols, $L(A)$ is recognized by a DIDPDA B with $\sum_{i=1}^{k}(i-1)!\cdot(n+1)^i$ states and $\sum_{i=1}^{k} m^i$ stack symbols.*

Theorem 4 follows from Lemma 3. It improves the upper bound for the number of states from [22], roughly speaking, by replacing the factor i^i in the sum by $(i-1)!$. The data structure of the history tracking binary normal form of a computation tree allows the DIDPDA to store all necessary information about the computation tree of the NIDPDA with a smaller number of states.

3.1 Alternative Upper Bound and Constructions for UIDPDAs

We consider an alternative construction that further reduces the number of states of the simulating DIDPDA at the cost of increasing the stack alphabet size. The construction works in a fairly similar manner, remembering in the state a *non-indexed* full binary tree representing the decisions of the NIDPDA. To make up for the missing information, the DIDPDA pushes the current tree onto the stack.

According to Fact 2, there are $C_{i-1} = \frac{1}{i}\binom{2i-2}{i-1}$ (non-indexed) full binary trees with i leaves. This yields the following bound.

Theorem 5. *For any k-path NIDPDA A with n states and m stack symbols, L(A) is recognized by a DIDPDA with* $\sum_{i=1}^{k} C_{i-1} \cdot (n+1)^i$ *states and* $\sum_{i=1}^{k} C_{i-1} \cdot m^i$ *stack symbols, where* C_i *is the i-th Catalan number.*

To conclude this section we consider determinization of UIDPDAs. We are able to use nearly identical constructions to simulate UIDPDAs with DIDPDAs. The bounds for UIDPDAs are slightly better than the bounds in Theorem 4 and Theorem 5 for NIDPDAs. This is due to the fact that a string is rejected by a UIDPDA if only one branch rejects it, which enables us to stop the simulation when any branch halts and rejects.

Our first construction uses the same ideas from Theorem 4, but functions slightly differently in the way that it accepts or rejects strings.

Theorem 6. *For any k-path UIDPDA A with n states and m stack symbols, L(A) is recognized by a DIDPDA B with* $\sum_{i=1}^{k} (i-1)! \cdot n^i$ *states and* $\sum_{i=1}^{k} m^i$ *stack symbols.*

Similarly, an alternative construction uses the same ideas as in Theorem 5, using non-indexed full binary trees in the states, and pushing the non-indexed tree onto the stack.

Theorem 7. *For any k-path UIDPDA A with n states and m stack symbols, L(A) is recognized by a DIDPDA B with* $\sum_{i=1}^{k} C_{i-1} \cdot n^i$ *states and* $\sum_{i=1}^{k} C_{i-1} \cdot m^i$ *stack symbols, where* C_i *is the i-th Catalan number.*

4 Lower Bounds

It is known that there exists a language recognized by a k-entry DIDPDA with n states and k stack symbols for which any DIDPDA recognizing the same language needs at least $(n+1)^k - 1$ states [22]. Since a k-entry DIDPDA is a k-path NIDPDA, the result implies a lower bound for the determinization of k-path NIDPDAs. A limitation is that the construction requires that the number of paths k of the NIDPDA to be at most the number of states n.

To address this gap, we develop a lower bound construction that applies to situations where the number of paths is larger than the number of states. As a by-product, we also give an analogous lower bound for determinization of UIDPDAs. The lower bounds are established using 1-separator sets.

Lemma 8. *For every $k \geq 1$ and $n \geq 1$, there exists an alphabet $\Sigma^{k,n}$ and a language $\hat{L}_{k,n}$ over $\Sigma^{k,n}$ recognized by a kn-path NIDPDA A with $k+n$ states, n initial states, and n stack symbols such that any DIDPDA recognizing $\hat{L}_{k,n}$ has at least 2^{kn} states.*

We also prove a similar lower bound for determinizing k-path UIDPDAs using a nearly identical argument.

Lemma 9. *For every $k \geq 1$ and $n \geq 1$, there exists an alphabet $\Sigma^{k,n}$ and a language $\hat{M}_{k,n}$ over $\Sigma^{k,n}$ recognized by a kn-path UIDPDA A with $k+n$ states, n initial states, and n stack symbols such that any DIDPDA recognizing $\hat{M}_{k,n}$ has at least 2^{kn} states.*

To conclude we give a lower bound for determinizing k-entry universal DIDPDAs, which is analogous to the lower bound shown for determinizing k-entry DIDPDAs in [22].

Lemma 10. *For every $k \geq 1$ and $n \geq k$, there exists an alphabet $\Sigma^{k,n}$ and a language $\hat{P}_{k,n}$ over $\Sigma^{k,n}$ recognized by a k-entry universal DIDPDA with n states and k stack symbols, such that any DIDPDA for $\hat{P}_{k,n}$ needs $n^k - n$ states.*

5 Conclusion

We have improved the determinization upper bound for NIDPDAs with limited nondeterminism. The improved construction of a DIDPDA relies on the use of a different data structure that more efficiently stores the history of nondeterministic choices of the simulated nondeterministic machine in a full binary tree. However, the upper bound still does not match the known lower bounds either from [22] or from this paper.

A possible improved lower bound construction should use k-path NIDPDAs where all the nondeterministic choices cannot be made at the start of the computation, for general values of k and n, as the lower bound produced in [22] was shown to be tight for k-entry DIDPDAs. Producing a better lower bound would likely require an innovative improvement of the 1-separator set technique, as well as the conceptualization of a language that accurately represents the limitations of various versions of IDPDAs. Here, ideas explored e.g. in [13,25] could be useful.

References

1. Alur, R., Madhusudan, P.: Visibly pushdown languages. In: ACM Symposium on Theory of Computation, pp. 202–211 (2004)
2. Alur, R., Madhusudan, P.: Adding nesting structure to words. J. ACM **56**(3) (2009)
3. Bozzelli, L.: Alternating automata and temporal fixpoint calculus for visibly pushdown automata. In: Concurrency Theory, CONCUR'07. Lecture Notes in Computer Science, vol. 4703, pp. 476–491. Springer (2007)

4. von Braunmühl, B., Verbeek, R.: Input driven languages are recognized in log n space. Ann. Discrete Math. **24**, 1–20 (1985)
5. Cormen, T., Leiserson, C., Rivest, R.: Introduction to Algorithms. The MIT Press (1999)
6. Geffert, V., Kapoutsis, C.A., Zakzok, M.: Complement for two-way alternating automata. Acta Inform. **58**(5), 463–495 (2021)
7. Geffert, V., Kapoutsis, C.A., Zakzok, M.: Improved complement for two-way alternating automata. Acta Inform (2022)
8. Goldstine, J., Kintala, C.M.R., Wotschke, D.: On measuring nondeterminism in regular languages. Inf. Comput. **86**(2), 179–194 (1990)
9. Han, Y.S., Ko, S.K., Salomaa, K.: Deciding path size of nondeterministic (and input-driven) pushdown automata. Theoret. Comput. Sci. **939**, 170–181 (2023)
10. Hromkovič, J., Seibert, S., Karhumäki, J., Klauck, H., Schnitger, G.: Communication complexity method for measuring nondeterminism in finite automata. Inf. Comput. **172**, 202–217 (2002)
11. Jirásková, G., Okhotin, A.: Towards exact state complexity bounds for input-driven pushdown automata. In: Hoshi, M., Seki, S. (eds.) Developments in Language Theory, pp. 441–452. Springer International Publishing, Cham (2018)
12. Kapoutsis, C., Zakzok, M.: Alternation in two-way finite automata. Theoret. Comput. Sci. **870**, 75–102 (2021)
13. Kapoutsis, C.A.: Size complexity of two-way finite automata. In: Diekert, V., Nowotka, D. (eds.) Developments in Language Theory, pp. 47–66. Springer, Berlin Heidelberg, Berlin, Heidelberg (2009)
14. Keeler, C., Salomaa, K.: Structural properties of NFAs and growth rates of nondeterminism measures. Inf. Comput. **284**, 104690 (2022)
15. Knuth, D.E.: Seminumerical Algorithms, volume 3 of The Art of Computer Programming. Addison-Wesley (1973)
16. Ladner, R., Lipton, R., Stockmeyer, L.: Alternating pushdown and stack automata. SIAM J. Comput. **13**(1), 135–155 (1984)
17. Leung, H.: On finite automata with limited nondeterminism. Acta Informatica **35**, 595–624 (1998)
18. Martynova, O.: Exact descriptional complexity of determinization of input-driven pushdown automata. In: Implementation and Application of Automata: 28th International Conference, CIAA 2024, pp. 249–260. Springer-Verlag, Berlin, Heidelberg (2024)
19. Mehlhorn, K.: Pebbling mountain ranges and its application to DCFL recognition. In: Automata, Languages and Programming, 1980, pp. 422–435. Lecture Notes in Computer Science, Springer (1980)
20. Okhotin, A., Piao, X., Salomaa, K.: Descriptional complexity of input-driven pushdown automata. In: Languages Alive. Lecture Notes Computer Science, vol. 7300, pp. 186–206. Springer (2012)
21. Okhotin, A., Salomaa, K.: Complexity of input-driven pushdown automata. SIGACT News **45**(2), 47–67 (2014)
22. Okhotin, A., Salomaa, K.: Input-driven pushdown automata with limited nondeterminism. In: Shur, A.M., Volkov, M.V. (eds.) Developments in Language Theory, pp. 84–102. Springer International Publishing, Cham (2014)
23. Palioudakis, A., Salomaa, K., Akl, S.G.: State complexity of finite tree width NFAs. J. Autom. Lang. Comb. **17**(2–4), 245–264 (2012)
24. Ravikumar, B., Ibarra, O.H.: Relating the type of ambiguity of finite automata to the succinctness of their representation. SIAM J. Comput. **18**(6), 1263–1282 (1989)

25. Sakoda, W.J., Sipser, M.: Nondeterminism and the size of two way finite automata. In: Proceedings of the Tenth Annual ACM Symposium on Theory of Computing, pp. 275–286. STOC '78, Association for Computing Machinery, New York, NY, USA (1978). https://doi.org/10.1145/800133.804357
26. Stanley, R.P., Fomin, S.: Trees and the Composition of Generating Functions, pp. 1–158. Cambridge Studies in Advanced Mathematics, Cambridge University Press (1999)

Pattern Mining Under Simon's Congruence

Sungmin Kim and Yo-Sub Han(✉)

School of Computing, Yonsei University, Seoul 03722, Republic of Korea
{rena_rio,emmous}@yonsei.ac.kr

Abstract. Given two strings u and v and an integer k, we say u and v are Simon's congruent with respect to k if they have the same set of subsequences of length at most k. We study the complete pattern mining problem for Simon's congruence, where the problem is to find the substrings of a given text $\mathtt{T}$ that maximizes the number of congruent substrings of the text, for each possible value of k. We design new data structures that capture the equivalence classes with respect to $\sim_k$ for substrings of the text. We then propose an $O(|\mathtt{T}|^2 \log^2 |\mathtt{T}|)$-time algorithm for fixed-sized alphabets using the new data structures.

1 Introduction

Given a long string called the text $\mathtt{T}$ and a string equivalence relation $\equiv$, the frequent pattern mining problem for $\equiv$ is to construct a pattern that satisfies $\equiv$ with the most number of substrings of the text [1,2,8]. Pattern mining algorithms have been found useful in the area of bioinformatics [1,10,21], market analysis [3,14], mobility analysis [22] and sequential data handling [19,20]. Especially, pattern mining problems related to subsequences have drawn attention since they are closely related with sequential data [4,15,25]. Ji et al. [15] proposed pattern mining methods for mining contrasting subsequences between two texts, which can be used to compare the given texts in bioinformatical contexts. Albarrak et al. [4] used constrained subsequences to increase the efficiency of pattern mining methods.

We investigate another pattern mining problem related to subsequences. For an integer k, two strings u and v are k-congruent, denoted by $u \sim_k v$, if u and v have the same set of subsequences with length at most k. This congruence is called Simon's congruence [24], and has been recently discussed in the context of pattern matching [16]. We say that a substring u of the text $\mathtt{T}$ is a $\sim_k$-*match* of $\mathtt{P}$ if $u \sim_k \mathtt{P}$. For an integer k, the pattern mining problem for $\sim_k$ is to compute a pattern with the greatest number of substrings in a given text that matches the pattern under $\sim_k$.

Problem 1 (Simon's congruence pattern mining (SCPM)). Given a text $\mathtt{T}$ and an integer k, find a substring t of $\mathtt{T}$ with the most number of $\sim_k$-matches in $\mathtt{T}$.

S.-K. Ko and F. Manea (Eds.): DLT 2025, LNCS 16036, pp. 182–196, 2026.
https://doi.org/10.1007/978-3-032-01475-7_13

Since $\sim_k$ is a symmetric relation, we can interpret any matching substring of $\mathtt{T}$ for a given pattern as the pattern itself. For example, let $\mathtt{T} = baabba$ and $k = 2$. Then, we have $\mathtt{T}[1:6] \sim_2 \mathtt{T}[2:6] \sim_2 \mathtt{T}[3:6] \sim_2 \mathtt{T}[1:5] \sim_2 \mathtt{T}[1:4]$, which is the most number of $\sim_2$-matches. Letting $\mathtt{P} = \mathtt{T}[1:6]$ gives us a pattern with the most number of $\sim_2$-matches. However, pattern mining problems excel at finding patterns where we are not fully aware of the significance of the text [12]. Therefore, it is natural that we do not know which value of k should be chosen before constructing a representative pattern. This leads us to define the *complete pattern mining problem* with respect to Simon's congruence, where we consider all possible values of k.

Problem 2 (Complete Simon's congruence pattern mining (CSCPM)). Given a text $\mathtt{T}$, find the substrings of $\mathtt{T}$ with the greatest number of $\sim_k$-matches in $\mathtt{T}$ for each $k \leq |\mathtt{T}|$.

Note that k is no greater than $|\mathtt{T}|$ because the most frequent patterns for $k \geq |\mathtt{T}|$ will all be equivalent by the definition of Simon's congruence. Returning to our running example, a solution of Problem 2 would be $\mathtt{T}[2:4]$ with 13 matches for $k = 1$, $\mathtt{T}[1:6]$ with 5 matches for $k = 2$, and $\mathtt{T}[4:4]$ with 3 matches for all $k \geq 3$. One naive solution of Problem 1 would be to enumerate all substrings of $\mathtt{T}$ and count the number of occurrences of each congruence class. Computing a unique representative string for the congruence class of a given string takes time linear in the length of the string [6,11]. If we use a trie to store the representative strings of all substrings of the text, this method takes $O(|\mathtt{T}|^3)$ time. Finally, repeating this method for all $k \leq |\mathtt{T}|$ gives us an overall $O(|\mathtt{T}|^4)$ solution for Problem 2. We contribute a $O(|\Sigma||\mathtt{T}|^2 \log^2 |\mathtt{T}|)$ solution for Problem 2, where $|\Sigma|$ is the size of the alphabet. Furthermore, in pattern mining, it is usual to fix the size of the alphabet to some constant [5,7,9,18]. For example, DNA analysis in bioinformatics is one of the most studied applications of pattern mining, where the context uses the set of nucleobases with size four, which is a fixed-sized alphabet. Then, if the alphabet is a fixed-sized alphabet, our algorithm's running time becomes $O(|\mathtt{T}|^2 \log^2 |\mathtt{T}|)$.

2 Preliminaries

2.1 Strings, Sets and Maps

For two integers i and j with $i \leq j$, we denote the set of integers $\{i, i+1, \ldots, j\}$ as $[i, j]$. Let Σ be a finite ordered alphabet. The *length* $|w|$ of a string w is the number of characters in w. For an integer $i \in [1, |w|]$, we denote the ith character of w as $w[i]$, and call i an *index* of w. For a string w, an index i of w and a character σ, we define $\mathtt{next}_\sigma(w, i)$ as $\min(\{j \in [i, |w|] \mid w[j] = \sigma\} \cup \{\infty\})$. Also, let $\mathtt{sfx}_\sigma(w, i)$, $\delta(w)$ and w_{-j} denote the suffixes of w starting at $\mathtt{next}_\sigma(w, i) + 1$, starting at index 2 and the suffix of w with length j, respectively. The alphabet *induced by* w is the set of characters that are used at least once in w; we denote the alphabet induced by w with $\mathtt{alph}(w)$. A string u is a *subsequence* of a string w if u can be obtained by deleting some characters in w, which we denote as $u \preceq_s w$. If u appears contiguously in w, then u is a *substring* of w. We denote $w[i]w[i+1]\cdots w[j] = w[i:j]$, and denote the set of all substrings of w as $\mathtt{Sub}_w$. For a string w and an integer k, the set $\mathbb{S}_k(w) = \{u \in \Sigma^{\leq k} \mid u \preceq_s w\}$ is the *subsequence set* of length at most k. Then, two strings u and v satisfy *Simon's congruence* for an integer k if $\mathbb{S}_k(u) = \mathbb{S}_k(v)$, which we denote as $u \sim_k v$. Finally, the supremum of the k's allowing $u \sim_k v$ is denoted as $\mathtt{sup}(u, v)$.

Given a set of intervals and a point, the *interval stabbing problem* is to fetch all intervals that contain the point [23]. For a subset X of a set S and a strict total order $<$ on S, let us denote $\min(<, X)$ as the minimum of all elements in X with respect to $<$. For an equivalence relation $\equiv$ on a set S, let us denote the *equivalence class* of $x \in S$ as $[x]_\equiv$, and the *quotient* of S by $\equiv$ as $S/\equiv$. A *map* is a finite set of tuples in the form of (key, value), where the set cannot contain different tuples $(\varphi, v_1), (\varphi, v_2)$ with $v_1 \neq v_2$. We assume that all values are strings and use an array to implement maps. For a map S, we denote the value corresponding to the key φ as $S[\varphi]$. If there are no tuples in S with the key φ, then we define $S[\varphi] = \mathtt{NULL}$. The set of keys φ of S with $S[\varphi] \neq \mathtt{NULL}$ is denoted as $S.\mathtt{keys}$. Note that a subset of a map is also a map. A string v is a *value string* of S if v is a value for some key of S. For a string w, let $\mathtt{M}_w$ be the map with keys $1, 2, \ldots, |w|$ and $\mathtt{M}_w[\varphi] = w[1:\varphi]$ for all $\varphi \leq |w|$. Intuitively, $\mathtt{M}_w$ is the map of all prefixes of w.

2.2 Trees, Simon-Trees and P-Connections

Let T be a rooted tree. The *size* $|T|$ of T is the number of nodes in T. For a node n of T, the *arity* $\mathtt{ar}(n)$ of n is the number of children of n. The node n is a *leaf* if it has arity 0, otherwise n is an *internal node*. If the arity of all internal nodes of T is at most 2, then T is a *binary tree*. A binary tree is *full* if all nodes have arity 0 or 2. The *depth* of n is the length of the simple path from n to the root, which we denote as $\mathtt{d}(n)$. For two nodes n_1 and n_2, n_2 is a *sibling* of n_1 if they have the same parent. If T is ordered, the *sibling order* of a child c of n is the number of siblings of c that precedes c plus one. Note that the children of a node will have a unique sibling order between 1 and the number of children. We denote the sibling order of c as $\mathtt{ord}(c)$. The *coordinate* $\mathtt{crd}(n)$ of n is a $\mathtt{d}(n)$-tuple $(\mathtt{ord}(a_1), \mathtt{ord}(a_2), \ldots, \mathtt{ord}(n))$, where $r, a_1, a_2, \ldots, n$ is the simple path from the root r to n. A node a is an *ancestor* of a node n if a appears on the simple path from the root to n. Conversely, n is a *descendant* of a if a is an ancestor of n. Note that a node can be an ancestor of itself, since n always appears on any path from the root to n. For two nodes n_1 and n_2, the *lowest common ancestor* (LCA) of n_1 and n_2 is the node with the maximum depth among the nodes that are both ancestors of n_1 and n_2. Let $<_C$ be the dictionary ordering on the set of coordinates; $(x_1, x_2, \ldots, x_k) <_C (y_1, y_2, \ldots, y_l)$ if there exists an $i \leq \min(k, l)$ such that $x_i < y_i$ and $x_j = y_j$ for all $j < i$, or $k < l$ and $x_j = y_j$ for all $j \leq k$. The relation $<_T$ is a strict total order on the set of nodes of T, where we have $n_1 <_T n_2$ if $\mathtt{crd}(n_1) <_C \mathtt{crd}(n_2)$ for nodes n_1 and n_2. For a binary tree T, let $\sqsubset_A$ be a relation on the set of nodes of T where $n_1 \sqsubset_A n_2$

if n_2 appears as a descendant of the child of n_1 with sibling order 2, or n_1 appears as a descendant of the child of n_2 with sibling order 1. The relation $\sqsubset_T$ is defined as the transitive closure of $\sqsubset_A$. Then, $\sqsubset_T$ is a strict total order on the set of nodes of T. For example, let n_1 and n_2 be nodes of a binary tree T with coordinates (2) and $(2, 1, 2)$, respectively. Then, we have $n_1 <_T n_2$ because $(2) <_C (2, 1, 2)$, while we have $n_2 \sqsubset_T n_1$ because n_2 is a descendant of the child of n_1 with coordinates $(2, 1)$. Let $\leq_T$, $\leq_C$, $\sqsubseteq_T$ and $\sqsubseteq_A$ be the reflexive versions of $<_T$, $<_C$, $\sqsubset_T$ and $\sqsubset_A$, respectively. The *pre-order traversal* of T traverses the nodes of T in increasing order of $<_T$. The *pre-order successor* and the *pre-order predecessor* of a node n of T is the node that is visited right after visiting n and just before visiting n in the pre-order traversal of T, respectively. For a node n of T and an integer $k \leq \mathtt{d}(n)$, the k-successor of n is the least node n' in T with respect to $<_T$ satisfying $n <_T n'$ and $\mathtt{d}(n') \leq k$. If such a node does not exist, the k-successor of n is a virtual node of T with coordinates (∞). The *in-order traversal* of T traverses the nodes of T in increasing order of $\sqsubset_T$, and the *rank* $\mathtt{rank}(n)$ of a node n of a binary tree T is the number of nodes n' of T that satisfy $n' \sqsubset_T n$. A binary tree is *order-statistic* if we can compute $\mathtt{rank}(n)$ from n and n from $\mathtt{rank}(n)$ in logarithmic time [17].

For a string w, the *Simon-tree* $\mathtt{ST}_w$ of w captures the equivalence classes of suffixes of w with respect to $\sim_k$ for all $k \leq |w|$, and is defined as the following [13]. First, there are exactly $|w|$ leaf nodes, where the string w_{-i} for an integer $i \leq |w|$ corresponds to the ith smallest leaf with respect to $<_{\mathtt{ST}_w}$, which we denote as $\mathtt{lf}_w(i)$. Second, there are no nodes with arity 1. Finally, for indices i and j of w, the depth of $\mathtt{LCA}(\mathtt{lf}_w(i), \mathtt{lf}_w(j))$ is $\mathtt{sup}(w_{-i}, w_{-j})$. See Fig. 1 for an example of a Simon-tree. For a given string w and an integer k, we define the *end-of-block index* $\mathtt{EOB}_w(k)$ at depth k as $\max\{j \in [1 : |w|] \mid w \sim_k w[j : |w|]\}$. Furthermore, for a string w and some large integer N greater than $|w|$, we define the *padded coordinates* of w with size N as the tuple $\mathtt{pc}_N(w)$ defined as $\mathtt{pc}_N(w)[i] = \mathtt{crd}(\mathtt{lf}_w(|w|))[i]$ for all $i \leq \mathtt{d}(\mathtt{lf}_w(|w|))$ and $\mathtt{pc}_N(w)[i] = 1$ for all $\mathtt{d}(\mathtt{lf}_w(|w|)) + 1 \leq i \leq N$. For example, the padded coordinates of $w = baab$ for $N = 3$ is $(2, 3, 1)$, since $\mathtt{crd}(\mathtt{lf}_w(|w|)) = (2, 3)$. Two strings w and u with length at most N are *P-connected at level* k if $\mathtt{pc}_N(w)[i] = \mathtt{pc}_N(u)[i]$ for all $i \leq k$. We denote $w\mathcal{P}_k u$ if w and u are P-connected at level k. Otherwise, we denote $w\overline{\mathcal{P}_k}u$. If $w\mathcal{P}_N u$, we say that w and u are *boundlessly* P-connected and we simply write $w\mathcal{P}u$. For example, consider the strings $w_1 = baab$, $w_2 = abbaa$ and $w_3 = baabba$. Then, $\mathtt{crd}(\mathtt{lf}_{w_1}(|w_1|)) = \mathtt{crd}(\mathtt{lf}_{w_2}(|w_2|)) = (2, 3)$ and $\mathtt{crd}(\mathtt{lf}_{w_3}(|w_3|)) = (2, 3, 3)$. Therefore, w_1 and w_2 are boundlessly P-connected, while w_1 and w_3 are P-connected at level 2 but not P-connected at level 3 or higher. For two strings w and u, Gawrychowski et al. [13] computed $\mathtt{sup}(w', u')$ for all pairs of suffixes w' and u' of w and u in linear time based on the fact that a string pair satisfying $\sim_k$ is also P-connected at level k. Specifically, they first identify P-connected suffixes of w and u using the Simon-trees of w and u, and then select exactly the suffix pairs that satisfy $\sim_k$ based on the following lemma. Note that $\mathtt{alph}(w) = \mathtt{alph}(u)$ if and only if $w \sim_1 u$.

Lemma 1 (Gawrychowski et al. [13]). *For two strings w and u and an integer $k \geq 2$, $w \sim_k u$ if and only if $w\mathcal{P}_k u$, $w \sim_{k-1} u$ and all characters $\sigma \in \Sigma$ satisfy* $\mathtt{sfx}_\sigma(w, \mathtt{EOB}_k(w)) \sim_{k-1} \mathtt{sfx}_\sigma(u, \mathtt{EOB}_k(u))$.

3 Identifying P-Connections for Multiple Strings

3.1 Simon-Tree Variants for Multiple Strings

Our solution uses a similar tactic as Gawrychowski et al. [13]. We first construct the Simon-trees, then identify the P-connections between all suffixes at all levels and finally refine the P-connections into $\sim_k$. However, there are several non-trivial difficulties in directly applying their pairwise algorithm. The main difficulty is that there are now $|\mathtt{T}|^2$ pairs of strings, whereas the original algorithm works for a single pair of strings. On the one hand, ordering the prefixes of $\mathtt{T}$ so that we only require a subquadratic number of runs of the pairwise algorithm is non-trivial. On the other hand, the pairwise algorithm only works when there is only one other string to match against. Therefore, rather than directly applying their algorithm, our goal is to adapt core ideas from the pairwise algorithm to CSCPM. The following Lemma 2 describes a hierarchical structure of $\sim_k$ with respect to k, which allow us to use Lemma 1 in the context of CSCPM.

Lemma 2. *For three strings w, u and v, let $\mathtt{sup}(u, v)$ be the maximum value among $\mathtt{sup}(w, u)$, $\mathtt{sup}(w, v)$ and $\mathtt{sup}(u, v)$. We have $\mathtt{sup}(w, u) = \mathtt{sup}(w, v)$.*

Therefore, the maximum k values allowing Simon's congruence between multiple strings can be stored as a tree, where each string corresponds to a leaf. The LCA of a pair of leaves holds the maximum k value allowing Simon's congruence. A full binary tree is an *S-Simon-tree*, or $\mathtt{SST}_S$, of a map S if it satisfies the followings:

- there are exactly $\sum_{\varphi \in S.\mathtt{keys}} |S[\varphi]|$ leaf nodes, where $S[\varphi]_{-i}$ corresponds to a unique leaf $\mathtt{slf}_S(\varphi, i)$ for $\varphi \in S.\mathtt{keys}$ and $i \leq |S[\varphi]|$,
- for two keys φ_1, φ_2 in S and integers $i_1 \leq |S[\varphi_1]|$ and $i_2 \leq |S[\varphi_2]|$, the label of $\mathtt{LCA}(\mathtt{slf}_S(\varphi_1, i_1), \mathtt{slf}_S(\varphi_2, i_2))$ is $\mathtt{sup}(S[\varphi_1]_{-i_1}, S[\varphi_2]_{-i_2})$,
- the label of a node is no less than the label of its parent.

The initial "S" stands for S-connections, which is another name for $\sim_k$ relations between suffixes [13]. For any pair w, u of suffixes of value strings in S, we can retrieve $\mathtt{sup}(w, u)$ from the LCA of the corresponding leaves. Thus, if the set of value strings of S is the set of prefixes of $\mathtt{T}$, we can efficiently solve CSCPM. We construct such an S-Simon-tree by the following idea. Let N be the length of the longest value string of S. Now, for a subset X of S, let $\hat{X}$ be a map with $\hat{X}.\mathtt{keys} = X.\mathtt{keys}$, where $X[\varphi] = \delta(\hat{X}[\varphi])$ for all $\varphi \in \hat{X}.\mathtt{keys}$, and the value strings of $\hat{X}$ are boundlessly P-connected. We further assume that all padded coordinates of value strings in S strictly precede the padded coordinates of value strings in $\hat{X}$ with respect to $<_C$. Our strategy is to incrementally construct an S-Simon-tree for $\hat{S} = \hat{X} \cup (S \setminus X)$ from $\mathtt{SST}_S$. Since the set of all suffixes of the values of $\hat{S}$ is equal to the union of $\hat{X}$ and the set of all suffixes of S, we add to $\mathtt{SST}_S$ the $|\hat{X}|$ leaf nodes that correspond to the value strings of $\hat{X}$ and the $|\hat{X}|$ internal nodes allowing the tree to be full and binary. Here, we compute another full binary tree with $|\hat{X}|$ leaves which are exactly the leaves we need to insert into $\mathtt{SST}_S$. This tree, called the *D-Simon-tree* $\Delta\mathtt{ST}_{\hat{X}}$ for $\hat{X}$, satisfies the properties of an S-Simon-tree except it captures congruence classes of only the value strings of $\hat{X}$, not the suffixes of value strings of $\hat{X}$. Specifically, each leaf corresponds to a key in $\hat{X}$, the LCA of two leaves corresponding to $\varphi_1, \varphi_2 \in \hat{X}.\mathtt{keys}$ is labeled $\mathtt{sup}(\hat{X}[\varphi_1], \hat{X}[\varphi_2])$, and the labels of internal nodes are no greater than the labels of its children internal nodes. The initial "D" comes from the word "difference", with emphasis on the fact that $\Delta\mathtt{ST}_{\hat{X}}$ represents the strings in $\hat{S} \setminus S$.

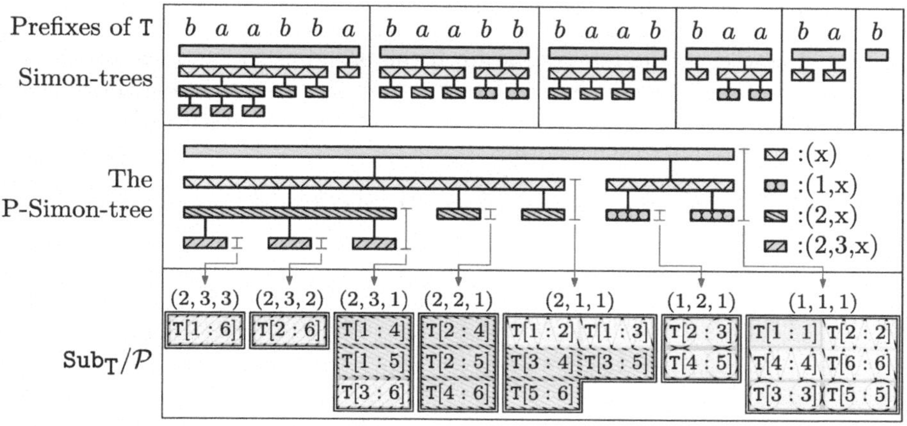

Fig. 1. Consider the text $\mathtt{T} = baabba$ and the map $\mathtt{M_T}$. We illustrate the Simon-trees for each prefix of $\mathtt{T}$ as well as the P-Simon-tree of $\mathtt{M_T}$. The nodes are drawn right-to-left with respect to the sibling order to align with the strings. We can compute the equivalence classes in $\mathtt{Sub_T}/\mathcal{P}$ from the P-Simon-tree. The padded coordinates of size $N = 3$ are written above each equivalence class.

Assume that we can partition $\mathtt{Sub_T}$ of $\mathtt{T}$ into its quotient $\mathtt{Sub_T}/\mathcal{P}$ and order these sets according to $<_C$ by the padded coordinates of the strings. We first make a D-Simon-tree for each $X \in \mathtt{Sub_T}/\mathcal{P}$, merge the D-Simon-tree with the S-Simon-tree made from the equivalence classes that precede the current equivalence class, and repeat until we have added all substrings of $\mathtt{T}$ to the S-Simon-tree. Then it remains to compute $\mathtt{Sub_T}/\mathcal{P}$. We identify the equivalence classes and their relative order under $<_C$ using yet another variation of the Simon-tree, which is called the *P-Simon-tree* $\mathtt{PST}_S$ of S. The P-Simon-tree of S is an overlay of the Simon-trees of strings in S. Specifically, there exists a node in the P-Simon-tree if and only if there is a node in the Simon-tree of some string in S with the same coordinate. Therefore, the number of nodes in $\mathtt{PST}_S$ is no greater than the sum of $|\mathtt{ST}_w|$ for each value string w of S. Additionally, each node of $\mathtt{PST}_S$

carries a set of leaves of the Simon-trees with the same coordinate. For example, in Fig. 1, $\mathtt{lf}_{\mathtt{T}[1:6]}(4)$ is stored in the P-Simon-tree node with coordinates $(2, 3, 1)$, while $\mathtt{lf}_{\mathtt{T}[1:5]}(5)$ and $\mathtt{lf}_{\mathtt{T}[1:4]}(4)$ are stored in the P-Simon-tree node with coordinates $(2, 3)$. Then, an enumeration of the union of the Simon-tree leaf sets for nodes visited in a descending path of a pre-order traversal on $\mathtt{PST}_S$ gives us the equivalence classes of $\mathtt{Sub}_{\mathtt{T}}/\mathcal{P}$ in that order. Returning to our example, the path in the P-Simon-tree from the node with coordinates $(2, 3)$ to the node with the coordinates $(2, 3, 1)$ is a descending path in its pre-order traversal. Note that $\mathtt{T}[3 : 6]$, $\mathtt{T}[1 : 5]$ and $\mathtt{T}[1 : 4]$, which are the substrings corresponding to the Simon-tree leaves stored in the two P-Simon-tree nodes, are all boundlessly P-connected.

3.2 Constructing the P-Simon-Tree

We first characterize the relation between the ordering $<_C$ and the depth of a node. Observation 1 gives us the ordering between i-successors of a single node, while Lemma 3 allows us to order the nodes in multiple trees with respect to $<_C$ by simply checking their depth.

Observation 1 *For any node n in an ordered tree T, we have $s_{\mathtt{d}(n)} \leq_T s_{\mathtt{d}(n)-1} \leq_T \cdots \leq_T s_1$, where s_i is the i-successor of n.*

Lemma 3. *Given two ordered trees T_1 and T_2 and some coordinate c, let n_1 and n_2 be nodes in T_1 and T_2 satisfying*

$$\mathtt{crd}(n_1), \mathtt{crd}(n_2) \leq_C c \leq_C \mathtt{crd}(s_1) <_C \mathtt{crd}(s_2),$$

where s_1 and s_2 are the pre-order successors of n_1 and n_2, respectively. Then, for all nodes b_1 with $n_1 <_{T_1} b_1 <_{T_1} f_1$ such that f_1 is the $\mathtt{d}(s_2)$-successor of n_1, we have $c \leq_C \mathtt{crd}(b_1) <_C \mathtt{crd}(s_2)$ and $\mathtt{d}(b_1) > \mathtt{d}(s_2)$.

We now describe the algorithm that computes the P-Simon-tree of S. Let Q be an array of N sets of Simon-tree nodes where $Q[i]$ is the set that corresponds to depth i. The idea is to construct nodes of $\mathtt{PST}_S$ in increasing order of $<_{\mathtt{PST}_S}$, and maintain Q at each node n of $\mathtt{PST}_S$ so that the Simon-tree nodes in $Q[i]$ have depth i and have the same coordinates as the i-successor of n using Lemma 3. At the beginning of the construction, we start by computing the Simon-trees $\mathtt{ST}_{S[\varphi]}$ for all $\varphi \in S.\mathtt{keys}$. Then, we initialize $Q[i] \leftarrow \emptyset$ for all $i \leq N$ and put in $Q[1]$ the pre-order successors of the roots of the Simon-trees for all value strings of S. We construct a tree T that only contains the root, and let the current node $\mathtt{cur}$ be the root of T. This completes the preparation for computing the P-Simon-tree. We then repeat the following process until every $Q[i]$ is empty. Let d be the depth of $\mathtt{cur}$. First, we scan the sets $Q[d+1]$ through $Q[1]$ in order until we find the first non-empty set; let d' be the index of that set in Q. We then move to the ancestor of $\mathtt{cur}$ with depth $d' - 1$, and add a child to that node with the greatest sibling order. We move $\mathtt{cur}$ to the newly created node, and move all Simon-tree nodes in $Q[d']$ to a set Q'. For each Simon-tree node n in Q', we first check if n is a leaf; if so, we add the node to the list of Simon-tree leaves of $\texttt{cur}$. Then, we access the pre-order successor s of n and put s into $Q[\texttt{d}(s)]$ if it exists. This ends the procedure for each iteration of the loop. When $Q[i]$ is empty for all $i \leq N$, we return the tree T.

Lemma 4. *Given a set S of m strings of length at most N, we can construct the P-Simon-tree of S in $O(mN)$ time.*

Finally, given a P-Simon-tree of S, we can traverse $\texttt{PST}_S$ in pre-order and return the set of Simon-tree leaves encountered in each descending path. This procedure enumerates the boundlessly P-connected equivalence class of all the suffixes of the strings in S in increasing order of their coordinates. The overall time to enumerate the boundlessly P-connected equivalence classes is also $O(mN)$ from traversing the P-Simon-tree and regrouping the Simon-tree leaves.

Corollary 1. *Given a text $\texttt{T}$, we can enumerate the sets in $\texttt{Sub}_\texttt{T}/\mathcal{P}$ in increasing order with respect to $<_C$ in overall $O(|\texttt{T}|^2)$ time.*

4 Constructing and Merging D-Simon-Trees

For the map S mapping integers to strings, let $\hat{X}$ be a map of integers mapping to strings, where there exists a subset X of S satisfying $X.\texttt{keys} = \hat{X}.\texttt{keys}$, $X[\varphi] = \delta(\hat{X}[\varphi])$ for all $\varphi \in \hat{X}.\texttt{keys}$, and the value strings of $\hat{X}$ are boundlessly P-connected. We first give an algorithm that constructs a D-Simon-tree $\Delta\texttt{ST}_{\hat{X}}$ from an S-Simon-tree $\texttt{SST}_S$ in Sect. 4.1, and then give a construction for $\texttt{SST}_{\hat{X}\cup\{S\setminus X\}}$ from a D-Simon-tree $\Delta\texttt{ST}_{\hat{X}}$ and an S-Simon-tree $\texttt{SST}_S$ in Sect. 4.2.

4.1 Constructing a D-Simon-Tree

We characterize $\Delta\texttt{ST}_{\hat{X}}$ with respect to the nodes in $\texttt{SST}_S$. For a value string w of S, let k be the largest integer such that the $k+1^{\text{th}}$ element of $\texttt{crd}(w)$ is not 1. Then, w and $\delta(w)$ are P-connected at level k but not at level $k+1$. Note that k only depends on the coordinates of w. Therefore, for two value strings w and u of $\hat{X}$, we have $k = \texttt{sup}(w, \delta(w)) = \texttt{sup}(u, \delta(u))$. We now have three cases, which we describe in Lemma 5.

Lemma 5. *For two boundlessly P-connected strings w and u, let $k = \texttt{sup}(w, \delta(w)) = \texttt{sup}(u, \delta(u))$. If $\texttt{alph}(w) \neq \texttt{alph}(u)$, then $\texttt{sup}(w, u) = 0$. If $[\texttt{alph}(w) = \texttt{alph}(u)] \wedge [\texttt{sup}(\delta(w), \delta(u)) < k]$, then $\texttt{sup}(w, u) = \texttt{sup}(\delta(w), \delta(u))$. Otherwise, $\texttt{sup}(w, u) = \min_{\sigma\in\Sigma} \texttt{sup}(\texttt{sfx}_\sigma(w, 1), \texttt{sfx}_\sigma(u, 1)) + 1$.*

The idea is to split $\hat{X}$ using a list of *splitting actions* for each case of Lemma 5. From now, let us denote the leaves of $\texttt{SST}_S$ that correspond to $\texttt{sfx}_\sigma(S[\varphi], 1)$ as $L_\sigma[\varphi]$, and the leaves of $\texttt{SST}_S$ that correspond to $\delta(S[\varphi])$ as $L_\delta[\varphi]$ for all $\varphi \in S.\texttt{keys}$. Then, for two value strings w and u of S, the $\texttt{sup}(\texttt{sfx}_\sigma(w, 1), \texttt{sfx}_\sigma(u, 1)) + 1$ values can be fetched from LCAs of nodes in L_σ and the $\texttt{sup}(\delta(w),$

$\delta(u)$) can be fetched from the LCAs of nodes in L_δ. Therefore, the splitting actions corresponding to the second and third cases of Lemma 5 can be represented as nodes in the S-Simon-tree for S. Let us call the list of all the LCAs of two different nodes in L_σ for all $\sigma \in \Sigma$ the *dependency list* $\texttt{dep}_\Sigma$ *for* Σ. On the other hand, let us call the list of all the LCAs of nodes in L_δ as the *dependency list* $\texttt{dep}_\delta$ *for* δ. Now, let us interpret $\texttt{alph}(\hat{X}[\varphi])$ as a binary string of length $|\Sigma|$, where the character corresponding to σ is 1 if $\sigma \in \texttt{alph}(\hat{X}[\varphi])$, and 0 otherwise. If we can sort $\hat{X}$ into a list with respect to the dictionary ordering on the binary strings for each induced alphabet, then the value strings inducing the same alphabet will appear contiguous in this array. In other words, the splitting actions for the first case of Lemma 5 can be represented as unique integer values. Let us call this list of integer values $\texttt{dep}_\alpha$, the *dependency list for* α.

For now, we assume that we already have $\texttt{dep}_\alpha$, $\texttt{dep}_\Sigma$ and $\texttt{dep}_\delta$. First, we partition $\hat{X}$ into $\mathcal{Z} = \{Z^1, Z^2, \ldots, Z^t\}$ using $\texttt{dep}_\alpha$ so that each Z^i induces a unique alphabet. Then, we compute $k = \texttt{sup}(w, \delta(w))$ by accessing the Simon-tree of some value string w of $\hat{X}$. Now, we split sets in $\mathcal{Z}$ using splitting actions of $\texttt{dep}_\delta$ with labels strictly less than k in increasing order of their labels. Finally, we further split sets in $\mathcal{Z}$ using the splitting actions in $\texttt{dep}_\Sigma$ in increasing order of their labels until every set in $\mathcal{Z}$ is a singleton. During the splitting procedure, we maintain a tree T where the leaves correspond to sets in $\mathcal{Z}$. Then, if some set $Z \in \mathcal{Z}$ is split into two sets Z_1 and Z_2 using a splitting action a in $\texttt{dep}_\alpha$, $\texttt{dep}_\delta$ or $\texttt{dep}_\Sigma$, we label the leaf corresponding to Z as 0, the label for a, or the label for a plus one, respectively. Finally, we add two children to that node, each corresponding to Z_1 and Z_2. The resulting T is the D-Simon-tree of $\hat{X}$, where the correctness follows from Lemma 5.

Computing the Dependency Lists. We first recall that $\texttt{dep}_\alpha$ can be obtained by sorting $\hat{X}$ and then listing all indices where the adjacent elements do not induce the same alphabet. Lemma 6 proves that we can compute $\texttt{dep}_\alpha$ in $O(|\Sigma||\hat{X}|)$ time. Now, recall that a splitting action for $x \in \Sigma \cup \{\delta\}$ is the LCA of two different leaves in L_x. Although there are $O(|L_x|^2)$ leaf pairs, Lemma 7 implies that there are only $|L_x| - 1$ splitting actions for each x.

Lemma 6. *Assume that we are given* $\texttt{alph}(\hat{X}[\varphi])$*, for all* $\varphi \in \hat{X}.\texttt{keys}$*. We can compute* $\texttt{dep}_\alpha$ *and partition* $\hat{X}$ *into subsets where the value strings of each subset induce the same alphabet in* $O(|\Sigma||\hat{X}|)$ *time.*

Lemma 7. *For an ordered binary tree* T*, let* L *be a subset of the leaves of* T *and let* n *be the LCA of some two nodes in* L*. Then, for the leaf* $l_1 \in L$ *with the maximum rank less than* $\texttt{rank}(n)$ *and the leaf* $l_2 \in L$ *with the minimum rank greater than* $\texttt{rank}(n)$*, we have* $\texttt{LCA}(l_1, l_2) = n$*.*

Thus, we can obtain all splitting actions for x if we first sort the leaves in L_x by their ranks and then retrieve the LCAs of pairs of consecutive leaves. We first compute the ranks of the leaves in L_x for all $x \in \Sigma \cup \{\delta\}$. We assign each key $\varphi \in \hat{X}$.keys a $|\Sigma| + 1$-vector

$$\overrightarrow{\varphi} = (\texttt{rank}(L_{\sigma_1}[\varphi]), \texttt{rank}(L_{\sigma_2}[\varphi]), \ldots, \texttt{rank}(L_{\sigma_{|\Sigma|}}[\varphi]), \texttt{rank}(L_\delta[\varphi])),$$

where we access each element with the notation $\overrightarrow{\varphi}[x] = \texttt{rank}(L_x[\varphi])$ for all $x \in \Sigma \cup \{\delta\}$. We note that $\overrightarrow{\varphi}$ depends on $\texttt{SST}_S$ and therefore must be recomputed whenever we modify $\texttt{SST}_S$. Now that we have $\overrightarrow{\varphi}$ for each $\varphi \in \hat{X}$.keys, we initialize $\texttt{dep}_\delta$ and $\texttt{dep}_\Sigma$ as empty lists and we repeat the following for all $x \in \Sigma \cup \{\delta\}$. First, we sort the set of $\overrightarrow{\varphi}$'s with respect to $\overrightarrow{\varphi}[x]$ in increasing order. Then, for each φ_1 and φ_2 pair which are consecutive in the sorted list, we retrieve the splitting action $\texttt{LCA}(L_x[\varphi_1], L_x[\varphi_2])$ and put it in the corresponding dependency list. Finally, when the iteration for x is over, we sort $\texttt{dep}_\delta$ and $\texttt{dep}_\Sigma$ in increasing order with respect to the labels.

Lemma 8. *Given an order-statistic S-Simon-tree* $\texttt{SST}_S$ *for* S *and the lists* L_x *for all* $x \in \Sigma \cup \{\delta\}$, *we can construct the dependency lists* $\texttt{dep}_\Sigma$, $\texttt{dep}_\alpha$ *and* $\texttt{dep}_\delta$ *in* $O(|\Sigma||\hat{X}| \log |\texttt{SST}_S|)$ *time.*

Splitting the set $\hat{X}$ for each splitting action Given a splitting action a_x for $x \in \Sigma \cup \{\delta, \alpha\}$ and a partition $\mathcal{Z}$ of $\hat{X}$, we explain how to locate the sets in $\mathcal{Z}$ that need to be split by a_x, and how to split those sets efficiently. For $x \in \Sigma \cup \{\delta\}$, we recall that any pair l_1, l_2 of leaves of $\texttt{SST}_S$ in L_x with $\texttt{rank}(l_1) < \texttt{rank}(l_2)$ and $a_x = \texttt{LCA}(l_1, l_2)$ satisfy $\texttt{rank}(l_1) < \texttt{rank}(a_x) < \texttt{rank}(l_2)$. Therefore, we split each $Z \in \mathcal{Z}$ into two sets Z_1 and Z_2 so that $\texttt{rank}(L_x[\varphi_1]) < \texttt{rank}(a_x)$ for $\varphi_1 \in Z_1$.keys, and $\texttt{rank}(L_x[\varphi_2]) > \texttt{rank}(a_x)$ for $\varphi_2 \in Z_2$.keys. We can also easily locate which sets to be split by finding the Z's satisfying $\min_{\varphi \in Z.\texttt{keys}} \overrightarrow{\varphi}[x] < \texttt{rank}(a) < \max_{\varphi \in Z.\texttt{keys}} \overrightarrow{\varphi}[x]$.

We now explain the splitting procedure. We first handle all the splitting actions in $\texttt{dep}_\alpha$ using Lemma 6. Let the resulting partition of $\hat{X}$ be $\mathcal{Z}$. We then construct a binary tree, where each leaf corresponds to a set in $\mathcal{Z}$ and all the internal nodes have label 0 and have two children. Now, before we start splitting the sets for any splitting action for $x \in \Sigma \cup \{\delta\}$, we first initialize $|\Sigma| + 1$ instances $\texttt{Int}_x$ of the interval stabbing problem for all $x \in \Sigma \cup \{\delta\}$, where each $\texttt{Int}_x$ contains intervals $[\min\{\overrightarrow{\varphi}[x] \mid \varphi \in Z.\texttt{keys}\}, \max\{\overrightarrow{\varphi}[x] \mid \varphi \in Z.\texttt{keys}\}]$ for each $Z \in \mathcal{Z}$. Then, for each $Z \in \mathcal{Z}$, we construct $|\Sigma| + 1$ binary search trees $\texttt{BST}(Z, x)$ for all $x \in \Sigma \cup \{\delta\}$, where each binary search tree $\texttt{BST}(Z, x)$ contains the keys φ in Z.keys sorted using $\overrightarrow{\varphi}[x]$. Now, whenever a splitting action a for $x \in \Sigma \cup \{\delta\}$ arrives, we issue an interval stabbing query to $\texttt{Int}_x$ with the value $\texttt{rank}(a)$. Then, for each stabbed interval, we access the subset Z corresponding to that interval, and remove the intervals corresponding to Z in all the interval stabbing instances. We then split Z into two sets Z_1 and Z_2 so that $\texttt{rank}(L_x[\varphi_1]) < \texttt{rank}(a_x)$ for $\varphi_1 \in Z_1$.keys, and $\texttt{rank}(L_x[\varphi_2]) > \texttt{rank}(a_x)$ for $\varphi_2 \in Z_2$.keys. This is done by choosing the smaller set Z' between Z_1 and Z_2, moving all elements of the smaller set from each binary search tree $\texttt{BST}(Z, x')$ into a new binary search tree $\texttt{BST}(Z', x')$, and renaming the set Z to the larger set between Z_1 and Z_2. Finally, we add the intervals $[\min\{\overrightarrow{\varphi}[x'] \mid \varphi \in Z_1.\texttt{keys}\}, \max\{\overrightarrow{\varphi}[x'] \mid \varphi \in Z_1.\texttt{keys}\}]$ and

$[\min\{\overrightarrow{\varphi}[x'] \mid \varphi \in Z_2.\texttt{keys}\}, \max\{\overrightarrow{\varphi}[x'] \mid \varphi \in Z_2.\texttt{keys}\}]$ to all the interval stabbing instances. This ends the splitting procedure.

Lemma 9. *Given two maps S and $\hat{X}$ where $S[\varphi] = \delta(\hat{X}[\varphi])$ for all $\varphi \in \hat{X}.\texttt{keys}$ and the value strings in $\hat{X}$ are all boundlessly P-connected, we can construct the D-Simon-tree $\Delta\texttt{ST}_{\hat{X}}$ in $O(|\Sigma||\hat{X}|(\log^2|\hat{X}| + \log|\texttt{SST}_S|))$ time if we have an S-Simon-tree for S and the dependency lists $\texttt{dep}_\Sigma$, $\texttt{dep}_\alpha$ and $\texttt{dep}_\delta$.*

4.2 Merging a D-Simon-Tree with an S-Simon-Tree

In Sect. 4.1, we have constructed a D-Simon-tree for $\hat{X}$ and the value $k = \texttt{sup}(w, \delta(w))$ for some value string w of $\hat{X}$. For a suffix u of a value string of S and a value string w of $\hat{X}$, observe that the length of the longest common prefix of $\texttt{crd}(u)$ and $\texttt{crd}(w)$ is at most k. Therefore, we have $\texttt{sup}(w, u) \leq \texttt{sup}(w, \delta(w))$, which allows the leaf corresponding to w to be inserted into an S-Simon-tree as a child of a node labeled k on the path from the root to the leaf corresponding to $\delta(w)$. We recall that the LCA of leaves in $\Delta\texttt{ST}_{\hat{X}}$ corresponding to two keys φ_1 and φ_2 in $\hat{X}.\texttt{keys}$ is labeled $\texttt{sup}(\hat{X}[\varphi_1], \hat{X}[\varphi_2])$. Thus, we simply move the subtrees of $\Delta\texttt{ST}_{\hat{X}}$ with root labeled a number no less than k as a subtree of nodes with label k in $\texttt{SST}_S$ to obtain $\texttt{SST}_{\hat{X}\cup(S\setminus X)}$.

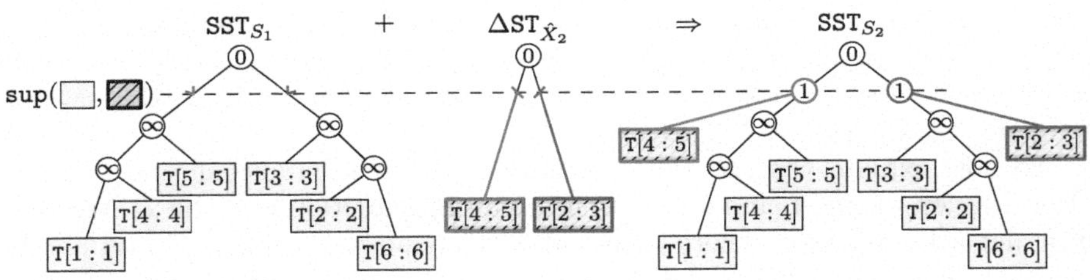

Fig. 2. For the running example $\texttt{T} = baabba$, we illustrate a merging step. We merge the S-Simon-tree for S_1 with value strings $\texttt{T}[1,1], \texttt{T}[2,2], \texttt{T}[3,3], \texttt{T}[4,4], \texttt{T}[5,5], \texttt{T}[6,6]$ with the D-Simon-tree for $\hat{X}_2$ with value strings $\texttt{T}[2:3]$ and $\texttt{T}[4:5]$. Note that the subtree of $\Delta\texttt{ST}_{\hat{X}_2}$ containing the leaf corresponding to $\texttt{T}[4:5]$ is transplanted to the path from the root to the leaf $\texttt{slf}_{S_1}(5,1)$, which corresponds to the string $\texttt{T}[5:5]$.

Thus, we can merge $\texttt{SST}_S$ and $\Delta\texttt{ST}_{\hat{X}}$ to obtain $\texttt{SST}_{\hat{X}\cup(S\setminus X)}$ through the following procedure. First, we perform breadth-first traversal on $\Delta\texttt{ST}_{\hat{X}}$ to obtain the set of nodes R where, for each $r \in R$, the label of the parent of r is less than k and the label of r is at least k. Then, for each node in R, we retrieve a key $\varphi_r \in \hat{X}.\texttt{keys}$, where r is an ancestor of the leaf of $\Delta\texttt{ST}_{\hat{X}}$ corresponding to φ. Now, for each retrieved key φ_r, we find the lowest ancestor a_r of $\texttt{slf}_S(\varphi_r, |S[\varphi_r]|)$ with label less than k, insert a node b_r with label k between a_r and the first child of a_r, and then transplant the subtree rooted at r in $\Delta\texttt{ST}_{\hat{X}}$ as a subtree rooted at b_r. This ends the construction of $\texttt{SST}_{\hat{X}\cup(S\setminus X)}$ from $\texttt{SST}_S$ and $\Delta\texttt{ST}_{\hat{X}}$.

Lemma 10. *Given an order-statistic S-Simon-tree* $\texttt{SST}_S$ *for* S *and a D-Simon-tree for* $\hat{X}$*, we can construct an order-statistic S-Simon-tree for* $\hat{X} \cup (S \setminus X)$ *in* $O(|\hat{X}| \log |\texttt{SST}_S| + \eta_{\hat{X}})$ *time, where* $\eta_{\hat{X}}$ *is the time it takes to find the lowest ancestor of* $\texttt{slf}_S(\varphi, |S[\varphi]|)$ *with label less than* k *for all* $\varphi \in \hat{X}.\texttt{keys}$.

5 The Pattern Mining Algorithm for Simon's Congruence

We are now ready to solve the complete Simon's congruence pattern mining problem. We first explain how to construct the S-Simon-tree representing all substrings of the text. Given a text $\texttt{T}$, we first construct the map $\texttt{M}_\texttt{T}$. Then, we construct the P-Simon-tree $\texttt{PST}_{\texttt{M}_\texttt{T}}$ using Lemma 4 and enumerate sets in $\texttt{Sub}_\texttt{T}/\mathcal{P}$ in increasing order with respect to $<_C$ using Corollary 1. Let $p = |\texttt{Sub}_\texttt{T}/\mathcal{P}|$, and let P_j be the set of leaf nodes of Simon-trees for the equivalence class with rank j with respect to $<_C$. For each P_j, we define the map $\hat{X}_j$ where $\hat{X}_j.\texttt{keys} = \{\varphi \le |\texttt{T}| \mid \text{there is a leaf of } \texttt{ST}_{\texttt{T}[1:\varphi]} \text{ in } P_j\}$ and $\hat{X}_j[\varphi] = \texttt{T}[1:\varphi]_{-i}$ for every $\texttt{lf}_{\texttt{T}[1:\varphi]}(i) \in P_j$. If $j = 1$, then we construct the S-Simon-tree $\texttt{SST}_{S_1}$ between characters in Σ. Otherwise, we construct the D-Simon-tree $\Delta\texttt{ST}_{\hat{X}_j}$ from $\texttt{SST}_{S_{j-1}}$ using Lemmas 6, 8 and 9. We construct the S-Simon-tree $\texttt{SST}_{S_j}$ from $\texttt{SST}_{S_{j-1}}$ and $\Delta\texttt{ST}_{\hat{X}_j}$ using Lemma 10. We repeat this process for all P_j, and the resulting S-Simon-tree is equal to $\texttt{SST}_{\texttt{M}_\texttt{T}}$. Here, we can use shortcut pointers in each $\texttt{SST}_{S_j}$ so that $\sum_j \eta_{\hat{X}_j}$ is at most $O(|\texttt{T}|^2)$.

Lemma 11. *Given a text* $\texttt{T}$*, we can construct an order-statistic S-Simon-tree* $\texttt{SST}_{\texttt{M}_\texttt{T}}$ *in* $O(|\Sigma||\texttt{T}|^2 \log^2 |\texttt{T}|)$ *time.*

Finally, we describe the last step of the pattern mining algorithm. We prepare a $|\texttt{T}|$-by-$|\texttt{T}|$ array A, where each cell holds a node of $\texttt{SST}_S$. All the cells are initialized with NULL. From an S-Simon-tree $\texttt{SST}_S$ for S, we first compute for each node the number of leaves of the subtree rooted at that node and a pointer to one of the leaves by a simple bottom-up propagation. Then, we identify the set D of nodes that are labeled with a different number compared to their parent by a simple breadth-first search on $\texttt{SST}_S$. For each node n in D, let the label of n be k_n and the label of the parent of n be k_p. We assign n to the cell $A[k_p + 1][k_n]$ whenever $A[k_p + 1][k_n]$ is NULL or the subtree rooted at the node in $A[k_p + 1][k_n]$ has less leaves than the subtree rooted at n. In other words, we update $A[k_p + 1][k_n]$ so that it holds the node with more leaves in its subtree. Finally, for each $k \le |\texttt{T}|$, we locate the cell that has the maximum integer value among all cells $A[k_1][k_2]$ with $k_1 \le k \le k_2$ and find the node n_k stored in that cell. We report the substring of $\texttt{T}$ that corresponds to a leaf of the subtree rooted at n_k as the pattern with the most number of $\sim_k$-matches.

Lemma 12. *Given an S-Simon-tree* $\texttt{SST}_{\texttt{M}_\texttt{T}}$*, we can report for all* $k \le |\texttt{T}|$ *the substrings of* $\texttt{T}$ *with the most number of* $\sim_k$*-matches in overall* $O(|\texttt{T}|^2)$ *time.*

By Lemmas 11 and 12, we have Theorem 1. We also observe that our results do not rely on the fact that the matches occur in a single string. Therefore,

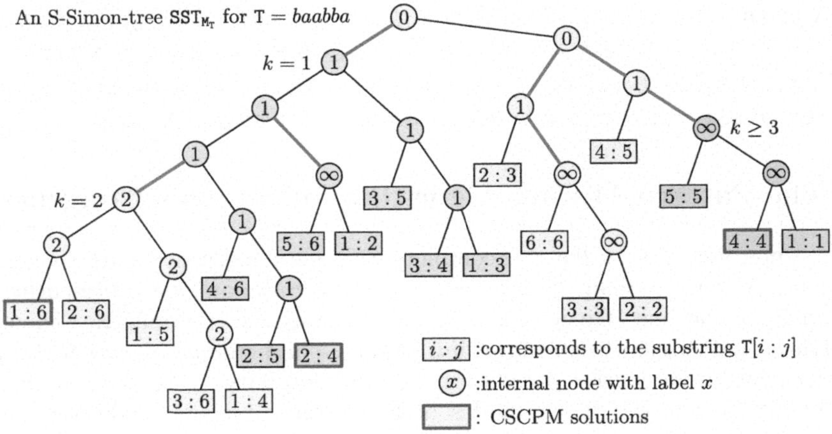

Fig. 3. The S-Simon-tree $\mathtt{SST}_{\mathtt{M_T}}$ for the running example $\mathtt{T} = baabba$. We first identify the edges between nodes with different labels. These edges are marked in bold. Then, for each subtree rooted at the children of those edges, we count the number of leaves. For each k, the roots of the subtrees with the most number of leaves are marked with the value of k. Recall that $\mathtt{T}[2:4]$, $\mathtt{T}[1:6]$ and $\mathtt{T}[4:4]$ were the substrings of $\mathtt{T}$ with the greatest number of $\sim_k$-matches for $k = 1, 2$ and 3, respectively, where they each match 13, 5 and 3 substrings of T.

we can extend our results to the pattern mining setting where we are given multiple texts in Theorem 2 by letting S as the map mapping a unique integer to each prefix in the given set of texts. Note that this pattern mining variant is not as simple as concatenating all strings in M using some delimiter and then running an algorithm for Problem 2.

Theorem 1. *Given a text* T*, we can find the substrings of* T *with the greatest number of* $\sim_k$ *matches in* T *for each* $k \leq |\mathtt{T}|$ *in* $O(|\Sigma||\mathtt{T}|^2 \log^2 |\mathtt{T}|)$ *time.*

Theorem 2. *Given a set* $\mathcal{T}$ *of texts of total length* M*, we can find for each* $k \leq \max_{\mathtt{T}\in\mathcal{T}} |\mathtt{T}|$ *the substrings of the strings in* $\mathcal{T}$ *with the greatest number of* $\sim_k$*-matches in all texts of* $\mathcal{T}$ *in* $O(|\Sigma|M^2 \log^2 M)$ *time.*

6 Conclusions

We have solved the complete Simon's congruence pattern mining problem for a given text T in $O(|\mathtt{T}|^2 \log^2 |\mathtt{T}|)$ time for fixed-sized alphabets. Although this is a significant improvement compared to existing methods for mining patterns, we think there is room for improvement in logarithmic order. For future work, we are interested in matching and mining patterns for the k-binomial equivalence or weighted versions of Simon's congruence. Finally, for two strings w and u and an integer k, we propose the problem of computing the longest subsequence of w that is $\sim_k$-congruent to another subsequence of u. Note that this problem is an extension of the longest congruent subsequence problem proposed by Kim et al. [16], which is to find for a given integer k the longest subsequence of a given text that is $\sim_k$-congruent to a given pattern.

References

1. Abouelhoda, M., Ghanem, M.: String mining in bioinformatics. In: Scientific Data Mining and Knowledge Discovery - Principles and Foundations, pp. 207–247. Springer (2010)
2. Aggarwal, C.C.: An introduction to frequent pattern mining. In: Frequent Pattern Mining, pp. 1–17. Springer (2014)
3. Agrawal, R., Imielinski, T., Swami, A.N.: Mining association rules between sets of items in large databases. In: Proceedings of the ACM SIGMOD International Conference on Management of Data, pp. 207–216 (1993)
4. Albarrak, A.M., Al-Maskari, S., Ibrahim, I.A., Almars, A.M.: Efficiently mining constrained subsequence patterns. In: Proceedings of the 14th International Conference on Advanced Data Mining and Applications, pp. 3–16 (2018)
5. Amir, A., Nor, I.: Real-time indexing over fixed finite alphabets. In: Proceedings of the Nineteenth Annual ACM-SIAM Symposium on Discrete Algorithms, pp. 1086–1095 (2008)
6. Barker, L., Fleischmann, P., Harwardt, K., Manea, F., Nowotka, D.: Scattered factor-universality of words. In: Proceedings of the 24th International Conference on Developments in Language Theory, pp. 14–28 (2020)
7. Brodnik, A., Jekovec, M.: Sliding suffix tree. Algorithms **11**(8), 118 (2018)
8. Calders, T.: Introduction to pattern mining. In: Tutorial Lectures of the Third European Summer School on Business Intelligence, vol. 172 of Lecture Notes in Business Information Processing, pp. 1–32. Springer (2013)
9. Chen, Y., Chen, B., Wu, Y.: On the multiple pattern string matching in DNA databases. SN Comput. Sci. **4**(3), 260 (2023)
10. Cho, D., Han, Y., Kim, H.: Frequent pattern mining with non-overlapping inversions. In: Proceedings of the 9th International Conference on Language and Automata Theory and Applications, pp. 121–132 (2015)
11. Fleischer, L., Kufleitner, M.: Testing Simon's congruence. In: Proceedings of the 43rd International Symposium on Mathematical Foundations of Computer Science, pp. 62:1–62:13 (2018)
12. Fournier-Viger, P., et al.: Pattern mining: current challenges and opportunities. In: Proceedings for the International Workshop on Database Systems for Advanced Applications, pp. 34–49 (2022)
13. Gawrychowski, P., Kosche, M., Koß, T., Manea, F., Siemer, S.: Efficiently testing Simon's congruence. In: Proceedings of the 38th International Symposium on Theoretical Aspects of Computer Science, pp. 34:1–34:18 (2021)
14. George, A., Binu, D.: An approach to products placement in supermarkets using prefixspan algorithm. J. King Saud Univ. Comput. Inf. Sci. **25**(1), 77–87 (2013)
15. Ji, X., Bailey, J., Dong, G.: Mining minimal distinguishing subsequence patterns with gap constraints. Knowl. Inf. Syst. **11**(3), 259–286 (2007)
16. Kim, S., Ko, S., Han, Y.: Simon's congruence pattern matching. Theoret. Comput. Sci. **994**, 114478 (2024)
17. Kreher, D.L., Stinson, D.R.: Combinatorial algorithms: generation, enumeration, and search. SIGACT News **30**(1), 33–35 (1999)

18. Kucherov, G., Nekrich, Y.: Full-fledged real-time indexing for constant size alphabets. Algorithmica **79**(2), 387–400 (2017)
19. Kum, H., Pei, J., Wang, W., Duncan, D.: Approxmap: approximate mining of consensus sequential patterns. In: Proceedings of the Third SIAM International Conference on Data Mining, pp. 311–315 (2003)
20. Lai, F., Chen, G., Gan, W., Sun, M.: Mining frequent temporal duration-based patterns on time interval sequential database. Inf. Sci. **665**, 120421 (2024)
21. Liao, V.C., Chen, M.: Efficient mining gapped sequential patterns for motifs in biological sequences. BMC Syst. Biol. **7**(S-4), S7 (2013)
22. Sakouhi, T., Akaichi, J.: Clustering-based multidimensional sequential pattern mining of semantic trajectories. Int. J. Data Min. Model. Manag. **16**(2), 148–175 (2024)
23. Schmidt, J.M.: Interval stabbing problems in small integer ranges. In: Proceedings of the 20th International Conference on Algorithms and Computation, vol. 5878 of Lecture Notes in Computer Science, pp. 163–172. Springer (2009)
24. Simon, I.: Piecewise testable events. In: Proceedings of the 2nd GI Conference on Automata Theory and Formal Languages, pp. 214–222 (1975)
25. Wu, Y., Wang, Y., Liu, J., Yu, M., Liu, J., Li, Y.: Mining distinguishing subsequence patterns with nonoverlapping condition. Clust. Comput. **22**, 5905–5917 (2019)

A General Information Extraction Framework Based on Formal Languages

Markus L. Schmid(✉)

Humboldt-Universität zu Berlin, Berlin, Germany
MLSchmid@MLSchmid.de

Abstract. For a terminal alphabet Σ and an attribute alphabet Γ, a (Σ, Γ)-extractor is a function that maps every string over Σ to a table with a column per attribute and with sets of positions of w as cell entries. This rather general information extraction framework extends the well-known document spanner framework, which has intensively been investigated in the database theory community over the last decade. Moreover, our framework is based on formal language theory in a particularly clean and simple way. In addition to this conceptual contribution, we investigate closure properties, different representation formalisms and the complexity of natural decision problems for extractors.

Keywords: Information Extraction · Regular Languages · Context-Free Languages

1 Introduction

Over roughly the last decade, the data query paradigm of so-called *information extraction* has received a lot of attention in database theory. In a nutshell, information extraction is the task to extract from a text document (string, word, sequence, etc.) a relational table of structured data. The most famous instance of information extraction are so-called *document spanners* (introduced in [8]), which are based on *spans*, e. g., $(3, 6)$ is a span of $w = \mathtt{abaabcba}$ referring to the factor $w[3..6] = \mathtt{aabc}$. A document spanner has a fixed set of variables and from any string w it extracts a table with a column per variable and with spans of w as cell entries (every row of the table is considered to be a result tuple of the spanner). We refer to the papers [2,25,28] for general information about document spanners, and [3,4,6,7,10–16,19–24,27,29,30] for recent results. Document spanners are attractive from a formal languages point of view, since representations and algorithmic techniques are based on classical concepts from formal languages and automata theory.

We propose a more general information extraction framework, which properly extends that of document spanners. Our motivation is as follows:

1. Several deep algorithmic techniques developed for document spanners also work without additional effort in a more general setting (e. g., [3,4,16,20]), so it makes sense to formally define and investigate this setting.

S.-K. Ko and F. Manea (Eds.): DLT 2025, LNCS 16036, pp. 197–212, 2026.
https://doi.org/10.1007/978-3-032-01475-7_14

2. Our general framework embeds into classical formal language theory in an even cleaner way, i. e., the equivalence of classes of extractors and classes of formal languages is more explicit. Hence, our framework may serve as an interface especially tailored to formal language and automata theorists.
3. Our information extraction framework seems to occupy an interesting area between language descriptors and transducers; this is worth investigating in pure terms of formal language theory (in contrast to the work in database theory, which focuses on solving data management tasks).

Intuitive Explanation. We consider strings over a finite *terminal alphabet* Σ as data objects that we want to query. Our queries – called *extractors* – will extract a table whose cells contain sets of positions of the terminal string:

$w = \texttt{abacadcdd}$ $\Longrightarrow$

x	y	z
$\{4,7,9\}$	$\{8,9\}$	$\{1,3\}$
$\{4,7\}$	$\{6\}$	$\{1,5\}$
$\emptyset$	$\{2,5\}$	$\{5\}$
$\{7\}$	$\emptyset$	$\{7\}$

Here, $\Gamma = \{x, y, z\}$ is the set of attributes that label the columns. The table can contain the empty set as entries (as illustrated above) and it can also be completely empty (i. e., it has no rows).

This extends the setting of document spanners by replacing spans (i. e., exactly two positions of the string) by arbitrary sets of positions (thus, the setting obviously still covers document spanners).

Contributions of this Work. Our main conceptual contribution is the introduction of the general information extraction framework (Sect. 2). We define several operators on extractors in Sect. 3. In Sect. 4, we show that our extractors have a convenient formulation as formal languages, and their operators translate into natural language operations. In the rest of the paper, we focus on classes of extractors that can be described by finite automata and context-free grammars. We investigate closure properties, different representation formalisms (Sect. 5), and the complexity of several natural decision problems (Sect. 6).

Due to space restrictions, all formal proofs are omitted (note that the full version [26] contains all technical details).

1.1 Basic Definitions

Let REG and CFL denote the classes of regular and context-free languages, respectively. We use nondeterministic finite automata (NFA), deterministic finite automata (DFA) and context-free grammars (CFG) as commonly defined (see, e. g., [17]). By $\mathcal{P}(A)$ we denote the power set of a set A. For a string w, we use $w[i]$ for $i \in \{1, 2, \ldots, |w|\}$ to denote the ith letter of w, and $w[i..j]$ for $i, j \in \{1, 2, \ldots, |w|\}$ with $i \leq j$ to denote the factor $w[i]w[i+1] \ldots w[j]$. We will generally use the symbol $\perp$ for signifying "undefined".

2 Formal Definition of the Framework

Let Σ be a finite *terminal alphabet* and let Γ be a finite *attribute alphabet*, and every $x \in \Gamma$ is called an *attribute symbol.* For complexity considerations, we let Σ be constant. The set Γ will play the role of attributes of the extracted tables as explained above, i. e., the columns will be labelled by the attribute symbols from Γ. Formally, we represent the row of a table extracted from $w \in \Sigma^*$ as a *Γ-tuple (for w)*, which is a function $t : \Gamma \to \mathcal{P}(\{1, 2, \ldots, |w|\})$. For convenience, we also call $t(x)$ the *x-entry* of t, and for the sake of presentation, we sometimes assume a fixed order $\preceq$ on Γ and then represent t in tuple notation, i. e., as $(t(x), t(y), t(z))$, where $x \preceq y \preceq z$. Moreover, we denote by $t^{\emptyset}_{\Gamma}$ the *empty Γ-tuple*, i. e., $t^{\emptyset}_{\Gamma}(x) = \emptyset$ for every $x \in \Gamma$; note that $t^{\emptyset}_{\Gamma}$ is the only Γ-tuple for ε. A *Γ-table (for w)* is any (possibly empty) set of Γ-tuples for w, and a *(Σ, Γ)-extractor* is a total function E that maps every terminal string $w \in \Sigma^*$ to a Γ-table for w. The *support* of E is $\{w \in \Sigma^* \mid E(w) \neq \emptyset\}$. The *empty Γ-extractor* $E^{\emptyset}_{\Gamma}$ is defined by $E^{\emptyset}_{\Gamma}(w) = \emptyset$ for every $w \in \Sigma^+$ and $E^{\emptyset}_{\Gamma}(w)(\varepsilon) = \{t^{\emptyset}_{\Gamma}\}$; note that $E^{\emptyset}_{\Gamma}$ is different from the Γ-extractor that maps every string (including ε) to $\emptyset$.

$E_1(w)$		
x	y	z
$\{2,8\}$	$\{2,3\}$	$\{7,9,10\}$
$\{2,8\}$	$\{2,3\}$	$\{9,10\}$
$\{2,8\}$	$\{2,3\}$	$\{7,10\}$
$\{2,8\}$	$\{2,3\}$	$\{7,9\}$
$\{2,8\}$	$\{2,3\}$	$\{10\}$
$\{2,8\}$	$\{2,3\}$	$\{9\}$
$\{2,8\}$	$\{2,3\}$	$\{7\}$
$\{2,8\}$	$\{2,3\}$	$\emptyset$

$E_2(w)$	
A	B
$\{1,7\}$	$\{7,10\}$
$\{1,10\}$	$\{7,10\}$
$\{5,7\}$	$\{7,10\}$
$\{5,10\}$	$\{7,10\}$
$\{1,7\}$	$\{7\}$
$\{1,10\}$	$\{7\}$
$\{5,7\}$	$\{7\}$
$\{5,10\}$	$\{7\}$
$\{1,7\}$	$\{10\}$
$\{1,10\}$	$\{10\}$
$\{5,7\}$	$\{10\}$
$\{5,10\}$	$\{10\}$
$\{1,7\}$	$\emptyset$
$\{1,10\}$	$\emptyset$
$\{5,7\}$	$\emptyset$
$\{5,10\}$	$\emptyset$

Fig. 1. The tables extracted from $w = \mathtt{baaabacadcb}$ by E_1 and E_2 from Example 1.

Note that every Γ-tuple for w and every Γ-table for w depend on Γ and $|w|$, but not on the actual content of w, i. e., not on the terminal alphabet Σ. On the other hand, (Σ, Γ)-extractors depend on Γ *and* Σ, because Σ^* is their domain. Nevertheless, whenever Σ or Γ is clear from the context or negligible, then we may drop them from our notations, i. e., we just talk about tuples and tables, or Γ-extractors or just extractors. For convenience, we represent a Γ-table T by listing its Γ-tuples (in any order) in tuple notation.

Example 1. Let $\Sigma = \{\mathtt{a}, \mathtt{b}, \mathtt{c}, \mathtt{d}\}$, let $\Gamma_1 = \{x, y, z\}$ and let $\Gamma_2 = \{A, B\}$. Let E_1 be the Γ_1-extractor that maps every $w \in \Sigma^*$ to the set of Γ_1-tuples with an

x-entry $\{i, j\}$, where i is the first and j the last occurrence of $\mathtt{a}$ in w (or $\emptyset$ if w does not contain any occurrences of $\mathtt{a}$), a y-entry that contains all starting positions of factors $\mathtt{a}^\ell$ with $\ell \geq 2$, and a z-entry that contains some occurrences of $\mathtt{c}$ and $\mathtt{d}$. Let E_2 be the Γ_2-extractor that maps every $w \in \Sigma^*$ to the set of Γ_2-tuples with an A-entry $\{i, j\}$, where $w[i] = b$ and $w[j] = c$, and a B-entry that contains occurrences of c (i. e., it is a subset of the set of all occurrences of $\mathtt{c}$). For $w = \mathtt{baaabacadcb}$, the Γ_1-table $E_1(w)$ and the Γ_2-table $E_2(w)$ (where we assume that $x \preceq y \preceq z$) is shown in Fig. 1.

3 Operations on Extractors

In this section, we define several operators on extractors based on typical operators for manipulating relational tables (i. e., operators from relational algebra), but we also consider the concatenation and Kleene star, which is motivated by our formulation of the framework purely in terms of formal languages (Sect. 4).

Set Operations. If E_1 and E_2 are Γ-extractors and $\odot \in \{\cup, \cap, \setminus\}$, then $E_1 \odot E_2$ is the Γ-extractor defined by $(E_1 \odot E_2)(w) = E_1(w) \odot E_2(w)$ for every $w \in \Sigma^*$, and $\neg E_1$ is a Γ-extractor defined by $\neg E_1(w) = \overline{E_1(w)}$ for every $w \in \Sigma^*$. Note that for the complement $\overline{T_1}$ and the set difference $T_1 \setminus T_2$ for Γ-tables T_1, T_2 for w, we consider the set of all Γ-tuples for w as the universe.

We can also lift these set operations to a Γ_1-extractor E_1 and a Γ_2-extractor E_2 with $\Gamma_1 \neq \Gamma_2$ by simply padding every Γ_1-tuple with $\emptyset$-entries for all $x \in \Gamma_2 \setminus \Gamma_1$ and every Γ_2-tuple with $\emptyset$-entries for all $x \in \Gamma_1 \setminus \Gamma_2$, e. g.,

x	y
$\{1, 4\}$	$\{2, 3\}$
$\{2, 5\}$	$\{4, 6, 7\}$

$\cup$

y	z
$\{5\}$	$\{2, 4, 7\}$
$\{6\}$	$\{3\}$

$=$

x	y	z
$\{1, 4\}$	$\{2, 3\}$	$\emptyset$
$\{2, 5\}$	$\{4, 6, 7\}$	$\emptyset$
$\emptyset$	$\{5\}$	$\{2, 4, 7\}$
$\emptyset$	$\{6\}$	$\{3\}$

Join Variants. In the following, let Γ_1 and Γ_2 be two attribute alphabets, let E_1 and E_2 be (Σ, Γ_1)- and (Σ, Γ_2)-extractors, respectively, let T_1 and T_2 be Γ_1- and Γ_2-tables for w_1 and w_2, respectively, and let t_1 and t_2 be Γ_1- and Γ_2-tuples for w_1 and w_2, respectively.

The product $\times$ and the join operation $\bowtie$ are well-known operators on tables, so they can be lifted to operators on extractors, i. e., for every $\circ \in \{\times, \bowtie\}$, we have $(E_1 \circ E_2)(w) = E_1(w) \circ E_2(w)$ for every $w \in \Sigma$ (note that $E_1 \times E_2$ is only defined if $\Gamma_1 \cap \Gamma_2 = \emptyset$).[1]

Recall that the join $t_1 \bowtie t_2$ is only defined if the tuples agree on their common attributes from $\Gamma_1 \cap \Gamma_2$, i. e., $t_1(x) = t_2(x)$ for every $x \in \Gamma_1 \cap \Gamma_2$. However, in our setting of information extraction, the entries of tuples are all subsets of $\mathbb{N}$.

[1] If $\Gamma_1 \cap \Gamma_2 = \emptyset$, then product and join are the same, so we shall not consider the product in the rest of the paper.

Thus, we could as well define $(t_1 \bowtie t_2)(x) = t_1(x) \odot t_2(x)$ for some $\odot \in \{\cup, \cap, \setminus\}$, which motivates the following generalised join variants.

For $\odot \in \{\cup, \cap, \setminus\}$, the *$\odot$-join* $t_1 \bowtie_\odot t_2$ is a $(\Gamma_1 \cup \Gamma_2)$-tuple defined by

$$(t_1 \bowtie_\odot t_2)(x) = \begin{cases} t_1(x) & \text{if } x \in \Gamma_1 \setminus \Gamma_2, \\ t_2(x) & \text{if } x \in \Gamma_2 \setminus \Gamma_1, \\ t_1(x) \odot t_2(x) & \text{if } x \in \Gamma_1 \cap \Gamma_2. \end{cases}$$

Observe that the normal join is therefore given by $t_1 \bowtie t_2 = \bot$ if $t_1(x) \neq t_2(x)$ for some $x \in \Gamma_1 \cap \Gamma_2$, and $t_1 \bowtie t_2 = t_1 \bowtie_\cup t_2$ otherwise.

Each $\circ \in \{\bowtie_\cup, \bowtie_\cap, \bowtie_\setminus, \bowtie)$ extends to tables and extractors in the obvious way, i. e., the $(\Gamma_1 \cup \Gamma_2)$-table $T_1 \circ T_2$ is defined by $T_1 \circ T_2 = \{t_1 \circ t_2 \mid t_1 \in T_1, t_2 \in T_2\}$, and the $(\Gamma_1 \cup \Gamma_2)$-extractor $E_1 \circ E_2$ is defined by $(E_1 \circ E_2)(w) = E_1(w) \circ E_2(w)$ for every $w \in \Sigma$.

Let us discuss some examples of these join operations on tables. The following is an example for the normal join $\bowtie$:

u	v	w
$\{1,4\}$	$\{2,3\}$	$\{7\}$
$\emptyset$	$\{4,6,7\}$	$\{1\}$
$\{5\}$	$\{2\}$	$\{4,5\}$

$\bowtie$

u	v	x
$\{5\}$	$\{2\}$	$\emptyset$
$\emptyset$	$\{4,6,7\}$	$\{4,5\}$
$\{5\}$	$\{2\}$	$\{1,2,7\}$
$\{1,4\}$	$\{1,2,3\}$	$\{7\}$

$=$

u	v	w	x
$\emptyset$	$\{4,6,7\}$	$\{1\}$	$\{4,5\}$
$\{5\}$	$\{2\}$	$\{4,5\}$	$\emptyset$
$\{5\}$	$\{2\}$	$\{4,5\}$	$\{1,2,7\}$

As an example for the generalised join operations, let T_1 be an $\{x,y\}$-table with $T_1 = ((\{1,2\},\{4\}),(\emptyset,\{1\}))$ and let T_2 be an $\{x,z\}$-table with $T_2 = ((\{2,3\},\{3,7\}),(\{4\},\{1,3\}))$ (we assume that $x \preceq y \preceq z$). Then we have:

$T_1 \bowtie_\cup T_2$		
x	y	z
$\{1,2,3\}$	$\{4\}$	$\{3,7\}$
$\{1,2,4\}$	$\{4\}$	$\{1,3\}$
$\{2,3\}$	$\{1\}$	$\{3,7\}$
$\{4\}$	$\{1\}$	$\{1,3\}$

$T_1 \bowtie_\cap T_2$		
x	y	z
$\{2\}$	$\{4\}$	$\{3,7\}$
$\emptyset$	$\{4\}$	$\{1,3\}$
$\emptyset$	$\{1\}$	$\{3,7\}$
$\emptyset$	$\{1\}$	$\{1,3\}$

$T_1 \bowtie_\setminus T_2$		
x	y	z
$\{1\}$	$\{4\}$	$\{3,7\}$
$\{1,2\}$	$\{4\}$	$\{1,3\}$
$\emptyset$	$\{1\}$	$\{3,7\}$
$\emptyset$	$\{1\}$	$\{1,3\}$

Concatenation and Kleene-Star. The $(\Gamma_1 \cup \Gamma_2)$-tuple $t_1 \cdot t_2$ for $w_1 \cdot w_2$ is defined by

$$(t_1 \cdot t_2)(x) = \begin{cases} t_1(x) & \text{if } x \in \Gamma_1 \setminus \Gamma_2, \\ \{i + |w_1| \mid i \in t_2(x)\} & \text{if } x \in \Gamma_2 \setminus \Gamma_1, \\ t_1(x) \cup \{i + |w_1| \mid i \in t_2(x)\} & \text{if } x \in \Gamma_1 \cap \Gamma_2. \end{cases}$$

As an example, let $\Gamma_1 = \{u,v,w\}$ and $\Gamma_2 = \{u,v,x\}$ (with assumed order $u \preceq v \preceq w \preceq x$ for the tuple notation), let $w_1 = \mathtt{aba}$ and $w_2 = \mathtt{bacd}$, and let $t_1 = (\{1\},\{1,2\},\emptyset)$ be a Γ_1- and $t_2 = (\{1,3\},\{2\},\{4\})$ a Γ_2-tuple for w_1 and w_2, respectively. Then, intuitively speaking, the concatenation operation first shifts all values of t_2 by $|w_1| = 3$, i. e., we get an intermediate Γ_2-tuple $t'_2 = (\{4,6\},\{5\},\{7\})$, which is then combined with t_1 in the same way as the $\cup$-join operation works, i. e., $t_1 \cdot t_2 = t_1 \bowtie_\cup t'_2 = (\{1,4,6\},\{1,2,5\},\emptyset,\{7\})$. Also note that if $t_1 = t^\emptyset_{\Gamma_1}$ and $w_1 = \varepsilon$, then $t_1 \cdot t_2 = t_2$.

The $(\Gamma_1 \cup \Gamma_2)$-table $T_1 \cdot T_2$ for $w_1 \cdot w_2$ is defined by $T_1 \cdot T_2 = \{t_1 \cdot t_2 \mid t_1 \in T_1, t_2 \in T_2\}$, and the $(\Gamma_1 \cup \Gamma_2)$-extractor $E_1 \cdot E_2$ is defined by $(E_1 \cdot E_2)(w) = \bigcup_{w=w_1 \cdot w_2} E_1(w_1) \cdot E_2(w_2)$ for every $w \in \Sigma^*$. Note that the empty Γ-extractor $E^{\emptyset}_{\Gamma}$ satisfies $E^{\emptyset}_{\Gamma} \cdot E = E$ for any Γ-extractor E.

Proposition 1. *For $i \in \{1, 2, 3\}$, let t_i be a Γ_i-tuple for w_i, let T_i be a Γ_i-table for w_i, and let E_i be a Γ_i-extractor. Then we have $(t_1 \cdot t_2) \cdot t_3 = t_1 \cdot (t_2 \cdot t_3)$, $(T_1 \cdot T_2) \cdot T_3 = T_1 \cdot (T_2 \cdot T_3)$ and $(E_1 \cdot E_2) \cdot E_3 = E_1 \cdot (E_2 \cdot E_3)$.*

Due to this associativity, we can lift the concatenation to a Kleene-star operator in the usual way. For a Γ-extractor E, we define $(E)^0 = E^{\emptyset}_{\Gamma}$ and $(E)^k = (E)^{k-1} \cdot E$ for every $k \geq 2$. Finally, we set $E^* = \bigcup_{k \geq 1} (E)^k$. Note that we always have $(E^*)(\varepsilon) = \{t^{\emptyset}_{\Gamma}\}$ for every Γ-extractor E.

Other Unary Operators. There are several other natural unary operators that could be defined, e. g., the projection, which removes columns and is a crucial operator in relational algebra. See the full version [26] for several reasonable unary operators on extractors.

4 Extractors as Formal Languages

We now define the framework of information extraction introduced in Sect. 2 purely in terms of formal languages.

Let Σ and Γ be fixed. For every set $X \subseteq \Gamma$ of attributes and $b \in \Sigma$, we call X_b a *Σ-signed Γ-marker*, where $\mathsf{sign}(X_b) := b$ is the *sign* of X_b and X is the *marker set* of X_b. We let $\Delta_{\Sigma,\Gamma}$ be the finite alphabet of all Σ-signed Γ-markers, and strings over $\Delta_{\Sigma,\Gamma}$ are called *Σ-signed Γ-marker strings*. For a Σ-signed Γ-marker string W of length n, the *sign* of W is defined by $\mathsf{sign}(W) = \mathsf{sign}(W[1])\mathsf{sign}(W[2]) \ldots \mathsf{sign}(W[n])$. By a slight abuse of notation, we also use the typical set notations directly on Σ-signed Γ-markers. In particular, we write $x \in X_b$ or $x \notin X_b$ to express $x \in X$ or $x \notin X$, respectively, or, for any $\odot \in \{\cup, \cap, \setminus\}$, we write $X_b \odot Y_b$ to denote $(X \odot Y)_b$.

For the sake of a simpler notation, we shall only mention Σ and Γ if they are not clear from the context. In particular, we shall use the terms *markers* and *marker strings* and keep in mind that they are always Σ-signed.

A central observation is that every pair (w, t) of a string $w \in \Sigma^*$ and tuple t for w, is uniquely represented by the marker string $W_{w,t}$ of length $|w|$, where, for every $i \in \{1, 2, \ldots, |w|\}$, $W_{w,t}[i] = X_{w[i]}$ with $X = \{x \in \Gamma \mid i \in t(x)\}$.

As an example, consider $w = \mathtt{baaabacadcb}$ from Example 1, and let s_i be the $\{x, y, z\}$-tuple for w given by the ith row of the table $E_1(w)$ in Fig. 1, and let t_i be the $\{A, B\}$-tuple for w given by the ith row of the table $E_2(w)$ in Fig. 1.

Then we have:

w	$=$	b	a	a	a	b	a	c	a	d	c	b
W_{w,s_1}	$=$	$\emptyset_\mathsf{b}$	$\{x,y\}_\mathsf{a}$	$\{y\}_\mathsf{a}$	$\emptyset_\mathsf{a}$	$\emptyset_\mathsf{b}$	$\emptyset_\mathsf{a}$	$\{z\}_\mathsf{c}$	$\{x\}_\mathsf{a}$	$\{z\}_\mathsf{d}$	$\{z\}_\mathsf{c}$	$\emptyset_\mathsf{b}$
W_{w,s_7}	$=$	$\emptyset_\mathsf{b}$	$\{x,y\}_\mathsf{a}$	$\{y\}_\mathsf{a}$	$\emptyset_\mathsf{a}$	$\emptyset_\mathsf{b}$	$\emptyset_\mathsf{a}$	$\{z\}_\mathsf{c}$	$\{x\}_\mathsf{a}$	$\emptyset_\mathsf{d}$	$\emptyset_\mathsf{c}$	$\emptyset_\mathsf{b}$
W_{w,t_1}	$=$	$\{A\}_\mathsf{b}$	$\emptyset_\mathsf{a}$	$\emptyset_\mathsf{a}$	$\emptyset_\mathsf{a}$	$\emptyset_\mathsf{b}$	$\emptyset_\mathsf{a}$	$\{A,B\}_\mathsf{c}$	$\emptyset_\mathsf{a}$	$\emptyset_\mathsf{d}$	$\{B\}_\mathsf{c}$	$\emptyset_\mathsf{b}$
W_{w,t_8}	$=$	$\emptyset_\mathsf{b}$	$\emptyset_\mathsf{a}$	$\emptyset_\mathsf{a}$	$\emptyset_\mathsf{a}$	$\{A\}_\mathsf{b}$	$\emptyset_\mathsf{a}$	$\{B\}_\mathsf{c}$	$\emptyset_\mathsf{a}$	$\emptyset_\mathsf{d}$	$\{A\}_\mathsf{c}$	$\emptyset_\mathsf{b}$

Conversely, every marker string W uniquely represents $(\mathsf{sign}(W), [\![W]\!])$, where $\mathsf{sign}(W)$ is W's sign as defined above (so a string over Σ) and $[\![W]\!]$ is a tuple for $\mathsf{sign}(W)$ defined by $[\![W]\!](x) = \{i \mid x \in W[i]\}$ for every $x \in \Gamma$. Observe, for example, that $[\![W_{w,s_1}]\!]$ is exactly the first row of $E_1(w)$ from Fig. 1. As a special case, the Γ-marker string ε satisfies $\mathsf{sign}(\varepsilon) = \varepsilon$ and $[\![\varepsilon]\!] = t^{\emptyset}_{\Gamma}$.

Hence, there is a one-to-one correspondence between marker strings and pairs (w,t), where $w \in \Sigma^*$ and t is a tuple for w.

We next consider marker languages, i. e., sets of marker strings. For such a marker languages L, it is convenient to define, for every $w \in \Sigma^*$, the set $\mathsf{sl}_w(L) = \{W \in L \mid \mathsf{sign}(W) = w\}$, which we call the *$w$-signed slice of L.* Obviously, $L = \bigcup_{w\in\Sigma^*} \mathsf{sl}_w(L)$ and this is a disjoint union.

Every marker language L uniquely represents an extractor $[\![L]\!]$ defined by $[\![L]\!](w) = \{[\![W]\!] \mid W \in \mathsf{sl}_w(L)\}$ for every $w \in \Sigma^*$, and, conversely, for every extractor E the unique marker language $L_E = \{W_{w,t} \mid w \in \Sigma^*, t \in E(w)\}$ describes E in the sense that $[\![L_E]\!] = E$ (observe that for every $w \in \Sigma^*$ the w-signed slice of L_E is $\mathsf{sl}_w(L_E) = \{W_{w,t} \mid t \in E(w)\}$). In particular, the tables $[\![L]\!](w)$ extracted by $[\![L]\!]$ uniquely correspond to the signed slices of L. A special case is $[\![\{\varepsilon\}]\!] = E^{\emptyset}_{\Gamma}$ or, equivalently, $L_{E^{\emptyset}_{\Gamma}} = \{\varepsilon\}$.

We can now conveniently define classes of extractors via the corresponding classes of marker language.

Definition 1. *Let $\mathcal{L}$ be a language class, let Σ be a terminal alphabet and let Γ be an attribute alphabet. Then $\mathcal{E}^{\Sigma,\Gamma}_{\mathcal{L}} = \{[\![L]\!] \mid L \in \mathcal{L} \wedge L \subseteq \Delta^*_{\Sigma,\Gamma}\}$ is the set of (Σ,Γ)-extractors represented by Σ-signed Γ-marker languages from $\mathcal{L}$.*

For example, $\mathcal{E}^{\Sigma,\Gamma}_{\mathcal{L}}$ with $\mathcal{L} \in \{\mathsf{REG}, \mathsf{CFL}, \mathsf{CSL}, \mathsf{RE}\}$ are all well-defined classes of extractors, or we could consider the class of extractors represented by marker languages from the complexity classes NL, P, NP, etc. Extractors represented by undecidable marker languages are also covered by our definition.

4.1 Extractor Operations as Language Operations

As a consequence of our observations above, we can interpret all the extractor operations from Sect. 3 directly as operations on the corresponding marker languages. This can be done in an implicit way, e. g., for marker languages L_1 and L_2 we simply say that $L_1 \bowtie L_2$ is defined as the unique marker language with

$[\![L_1 \bowtie L_2]\!] = [\![L_1]\!] \bowtie [\![L_2]\!]$, but it will be helpful to also define them explicitly as operators on marker strings that are then lifted to marker languages.

The following can be concluded from the fact that Boolean operations are well-defined for signed marker languages.

Proposition 2. *Let L_1 and L_2 be Γ_1- and Γ_2-marker languages, respectively. Then $L_1 \odot L_2$ is a $(\Gamma_1 \cup \Gamma_2)$-marker language with $[\![L_1 \odot L_2]\!] = [\![L_1]\!] \odot [\![L_2]\!]$ for every $\odot \in \{\cup, \cap, \setminus\}$. Moreover, $\overline{L_1}$ is a Γ_1-marker language with $[\![\overline{L_1}]\!] = \neg[\![L_1]\!]$.*

We will now define operations on marker strings, which can then be lifted to marker languages in a straightforward way.

Let us first note that for a Γ_1-marker string W_1 and a Γ_2-marker string W_2, their concatenation $W_1 \cdot W_2$ is a $(\Gamma_1 \cup \Gamma_2)$-marker string with $\mathsf{sign}(W_1 \cdot W_2) = \mathsf{sign}(W_1) \cdot \mathsf{sign}(W_2)$. In order to define the different join variants for maker-strings, we first do this on the level of markers. Let $X_b \in \Delta_{\Gamma_1}$ and $Y_b \in \Delta_{\Gamma_2}$ (note that $\mathsf{sign}(X_b) = \mathsf{sign}(Y_b) = b$). Then, for every $\odot \in \{\cup, \cap, \setminus\}$, $X_b \bowtie_\odot Y_b$ is the $(\Gamma_1 \cup \Gamma_2)$-marker with sign b and marker set Z, where,

- for every $x \in \Gamma_1 \setminus \Gamma_2$, $x \in Z \Leftrightarrow x \in X$,
- for every $x \in \Gamma_2 \setminus \Gamma_1$, $x \in Z \Leftrightarrow x \in Y$,
- for every $x \in \Gamma_1 \cap \Gamma_2$, $x \in Z \Leftrightarrow x \in X \odot Y$.

Next, let W_1 be a Γ_1-marker string, let W_2 be a Γ_2-marker string, and let $\mathsf{sign}(W_1) = \mathsf{sign}(W_2) = w \in \Sigma^*$. For every $\odot \in \{\cup, \cap, \setminus\}$, $W_1 \bowtie_\odot W_2$ is the $(\Gamma_1 \cup \Gamma_2)$-marker string with sign w, where, for every $i \in \{1, 2, \ldots, |w|\}$, $(W_1 \bowtie_\odot W_2)[i] = W_1[i] \bowtie_\odot W_2[i]$. Moreover, we define $W_1 \bowtie W_2$ as $W_1 \bowtie W_2 = \bot$ if there is some $i \in \{1, 2, \ldots, |w|\}$ such that $W_1[i] \cap \Gamma_1 \cap \Gamma_2 \neq W_2[i] \cap \Gamma_1 \cap \Gamma_2$(observe that this is equivalent to $[\![W_1]\!](x) \neq [\![W_2]\!](x)$ for some $x \in \Gamma_1 \cap \Gamma_2$), and $W_1 \bowtie W_2 = W_1 \bowtie_\cup W_2$ otherwise. Observe that $W_1 \circ W_2$ with $\circ \in \{\bowtie, \bowtie_\cup, \bowtie_\cap, \bowtie_\setminus\}$ is only defined if $\mathsf{sign}(W_1) = \mathsf{sign}(W_2)$, whereas $W_1 \cdot W_2$ is also defined in the case that $\mathsf{sign}(W_1) \neq \mathsf{sign}(W_2)$.

The next proposition, which is not difficult to prove, states that all these operations on marker strings correspond to the respective operations on tuples.

Proposition 3. *Let W_1 be a marker string and let W_2 be a marker string. Then $[\![W_1 \cdot W_2]\!] = [\![W_1]\!] \cdot [\![W_2]\!]$. If further $\mathsf{sign}(W_1) = \mathsf{sign}(W_2)$, then $[\![W_1 \circ W_2]\!] = [\![W_1]\!] \circ [\![W_2]\!]$ for every $\circ \in \{\bowtie, \bowtie_\cup, \bowtie_\cap, \bowtie_\setminus\}$.*

We can now lift these operations to marker languages. For any Γ_1-marker language L_1, any Γ_2-marker language L_2 and every $\circ \in \{\bowtie, \bowtie_\cup, \bowtie_\cap, \bowtie_\setminus\}$, let $L_1 \circ L_2 = \bigcup_{w \in \Sigma^*} \{W_1 \circ W_2 \mid W_1 \in \mathsf{sl}_w(L_1), W_2 \in \mathsf{sl}_w(L_2)\}$, and let $L_1 \cdot L_2 = \{W_1 \cdot W_2 \mid W_1 \in L_1, W_2 \in L_2\}$. Moreover, let the Kleene star be defined as the usual language operation, i.e., $L_1^* = \bigcup_{k \geq 0} L_1^k$.

Finally, with the help of Proposition 3, we can conclude that the operations on marker languages correspond to the respective operations on extractors.

Proposition 4. *Let L_1 and L_2 be Γ_1- and Γ_2-marker language, respectively, and let $\circ \in \{\bowtie, \bowtie_\cup, \bowtie_\cap, \bowtie_\setminus, \cdot\}$. Then $L_1 \circ L_2$ is a $(\Gamma_1 \cup \Gamma_2)$-marker language with $[\![L_1 \circ L_2]\!] = [\![L_1]\!] \circ [\![L_2]\!]$, and L_1^* is a Γ_1-marker language with $[\![L_1^*]\!] = ([\![L_1]\!])^*$.*

5 Regular and Context-Free Extractors

In this section (as well as in Sect. 6), we consider the classes $\mathcal{E}^{\Sigma,\Gamma}_{\mathsf{REG}}$ and $\mathcal{E}^{\Sigma,\Gamma}_{\mathsf{CFL}}$ of *regular* and *context-free* (Σ, Γ)*-extractors*, i. e., extractors represented by regular and context-free marker languages, which can therefore be represented by NFAs and CFGs (and other equivalent description mechanisms). Those classes inherit several nice properties from the classes of regular and context-free languages.

5.1 Closure Properties of Regular Extractors

Regular extractors are closed under the extractor operations mentioned in Sect. 3, which can be shown by using the interpretation of extractor operations as operations on marker languages (see Sect. 4.1) and then showing that regular marker languages are closed under these operations.

The closure properties of regular marker languages under Boolean operations, concatenation and Kleene star directly follow from the known closure properties of regular languages. Proving this for the different join variants is also not difficult, if we argue on the level of NFAs.

Proposition 5. *Let* $L_1 \in \mathsf{REG}_{\Delta_{\Sigma,\Gamma_1}}$, $L_2 \in \mathsf{REG}_{\Delta_{\Sigma,\Gamma_2}}$ *and* $\circ \in \{\bowtie, \bowtie_\cup, \bowtie_\cap, \bowtie_\setminus\}$. *Then* $L_1 \circ L_2 \in \mathsf{REG}_{\Delta_{\Sigma,\Gamma_1 \cup \Gamma_2}}$.

The closure properties of $\mathcal{E}^{\Sigma,\Gamma}_{\mathsf{REG}}$ can now be easily concluded, e. g., let $E_1, E_2 \in \mathcal{E}^{\Sigma,\Gamma}_{\mathsf{REG}}$ be represented by marker languages L_1 and L_2, then $L_1 \bowtie L_2$ is regular (Proposition 5) and $E_1 \bowtie E_2 = [\![L_1]\!] \bowtie [\![L_2]\!] = [\![L_1 \bowtie L_2]\!]$ (Proposition 4).

Proposition 6. *Let* $E_1 \in \mathcal{E}^{\Sigma,\Gamma_1}_{\mathsf{REG}}$ *and* $E_2 \in \mathcal{E}^{\Sigma,\Gamma_2}_{\mathsf{REG}}$. *Then* $E_1 \odot E_2 \in \mathcal{E}^{\Sigma,\Gamma_1 \cup \Gamma_2}_{\mathsf{REG}}$, *for every* $\odot \in \{\cup, \cap, \setminus, \cdot, \bowtie, \bowtie_\cup, \bowtie_\cap, \bowtie_\setminus\}$. *Moreover,* $\neg E_1, (E_1)^* \in \mathcal{E}^{\Sigma,\Gamma_1}_{\mathsf{REG}}$.

5.2 Representations of Regular Extractors

Regular extractors can be represented by any kind of regular language description mechanisms (since we only have to represent a regular marker language). However, there are two other ways of representing regular extractors – as algebraic expressions over atomic extractors, and as MSO-formulas over Σ-strings.

Let us start with the algebraic representation. For an arbitrary set of (Σ, Γ)-extractors $\mathcal{E}$ and a subset Λ of the extractor operations defined in Sect. 3, a Λ*-expression over atoms from* $\mathcal{E}$ is a valid algebraic expression that uses operations from Λ and atoms from $\mathcal{E}$ (here "valid" means that every operation in the expression is well-defined for its arguments). By $\mathcal{E}^{\Lambda}$ we denote the set of all (Σ, Γ)-extractors that can be described by a Λ-expression over atoms from $\mathcal{E}$. Let Λ_{full} be the set of all the extractor-operators discussed in Sect. 3, i. e., the set operations $\cup, \cap, \setminus, \neg$, the join variants $\bowtie, \bowtie_\cup, \bowtie_\cap, \bowtie_\setminus$, concatenation and Kleene star, and projection.

Due to the closure properties discussed in Sect. 5.1, we know that:

Proposition 7. $(\mathcal{E}^{\Sigma,\Gamma}_{\mathsf{REG}})^{A_{\mathsf{full}}} = \mathcal{E}^{\Sigma,\Gamma}_{\mathsf{REG}}$.

For every marker $X_b \in \Delta_{\Sigma,\Gamma}$, let $E_{X,b}$ be the (Σ,Γ)-extractor that is undefined for every $w \in \Sigma^* \setminus \{b\}$, and $E_{X,b}(b) = t$ with $t(x) = \{1\}$ for every $x \in X$ and $t(y) = \emptyset$ for every $y \in \Gamma \setminus X$. Equivalently, $E_{X,b} = [\![X_b]\!]$. We denote such extractors as *atomic* (Σ,Γ)*-extractors*, and we let $\mathbb{A}_{\Sigma,\Gamma} = \{E_{X,b} \mid X_b \in \Delta_{\Sigma,\Gamma}\}$ be the set of all atomic (Σ,Γ)-extractors.

Obviously, every $\{\cup, \cdot, *\}$-expression over atoms from $\mathbb{A}_{\Sigma,\Gamma}$ describes a regular extractor (due to the closure properties). However, it can also be shown that every regular extractor can be represented as a $\{\cup, \cdot, *\}$-expression over atoms from $\mathbb{A}_{\Sigma,\Gamma}$. This follows from the fact that its corresponding marker language (since it is regular) can be represented as a $\{\cup, \cdot, *\}$-expression over single markers. Thus, by an inductive application of Proposition 4, we obtain the desired expression from $(\mathbb{A}_{\Sigma,\Gamma})^{\{\cup,\cdot,*\}}$.

Lemma 1. $(\mathbb{A}_{\Sigma,\Gamma})^{\{\cup,\cdot,*\}} = \mathcal{E}^{\Sigma,\Gamma}_{\mathsf{REG}}$.

Let us now come to the representation by formulas of monadic second order logic (MSO for short). To this end, we interpret strings over some alphabet A as relational structures in the usual way, i. e., as relational structures $w = (\{1, 2, \ldots, n\}, <, (P_a)_{a \in A})$, where $<$ is the linear order on $\{1, 2, \ldots, n\}$ and the P_a are unary relations that describe a partition of $\{1, 2, \ldots, n\}$, i. e., $P_a \cap P_{a'} = \emptyset$ for every $a, a' \in A$ with $a \neq a'$, and $\bigcup_{a \in A} P_a = \{1, 2, \ldots, n\}$.

For any alphabet A, formulas of *monadic second order logic for* A*-strings* (MSO_A) are just MSO-formulas over the signature $(<, (P_a)_{a \in A})$. For an MSO_A-sentence ϕ and a string $w \in A^*$, we write $w \models \phi$ to denote that ϕ holds in the string w, and $L(\phi) = \{w \mid w \models \phi\}$ is the language over A described by ϕ. For an MSO_A-formula $\phi(X_1, X_2, \ldots, X_k)$, a string w and sets $S_1, S_2, \ldots, S_k \subseteq \{1, 2, \ldots, |w|\}$, we write $w \models \phi(S_1, \ldots, S_k)$ to denote that the formula ϕ holds in the string w if the variable X_i is set to S_i for every $i \in \{1, 2, \ldots, k\}$. Hence, we can interpret each MSO_A-formula $\phi(X_1, \ldots, X_k)$ as defining a *result set* $[\![\phi]\!](w) = \{(S_1, S_2, \ldots, S_k) \mid w \models \phi(S_1, \ldots, S_k)\}$. In the following, we consider monadic second order logic for Σ-strings and $\Delta_{\Sigma,\Gamma}$-strings (i. e., marker strings).

Let $\Gamma = \{\gamma_1, \gamma_2, \ldots, \gamma_m\}$ be an attribute alphabet. The Γ-MSO_Σ *formulas* are the MSO_Σ formulas of the form $\phi(X_{\gamma_1}, X_{\gamma_2}, \ldots, X_{\gamma_m})$, i. e., MSO_Σ formulas with exactly one free set-variable for each attribute. Since every $(S_1, S_2, \ldots, S_k) \in [\![\phi]\!](w)$ for a Γ-MSO_Σ formula $\phi(X_{\gamma_1}, X_{\gamma_2}, \ldots, X_{\gamma_m})$ can be interpreted as a Γ-tuple, the result set $[\![\phi]\!](w)$ can be interpreted as a Γ-table. Consequently, any Γ-MSO_Σ formula $\phi(X_{\gamma_1}, X_{\gamma_2}, \ldots, X_{\gamma_m})$ represents a (Σ,Γ)-extractor $[\![\phi]\!]$. Finally, $\mathcal{E}^{\Sigma,\Gamma}_{\mathsf{MSO}}$ denotes the set of (Σ,Γ)-extractors described by Γ-MSO_Σ formulas.

Now in order to see that Γ-MSO_Σ formulas describe exactly the set of regular Γ-extractors, we first have to show that any such formula ϕ can be transformed into a sentence that describes the marker language of the extractor $[\![\phi]\!]$, and vice versa. Then, the well-known Büchi–Elgot–Trakhtenbrot theorem [18, Chapter 7] implies that the marker language is necessarily regular.

Lemma 2. $\mathcal{E}^{\Sigma,\Gamma}_{\mathsf{REG}} = \mathcal{E}^{\Sigma,\Gamma}_{\mathsf{MSO}}$.

Hence, we have shown that regular extractors are also characterised by MSO-formulas and algebraic expressions using union, concatenation and Kleene star.

Theorem 1. $\mathcal{E}^{\Sigma,\Gamma}_{\mathsf{REG}} = \mathcal{E}^{\Sigma,\Gamma}_{\mathsf{MSO}} = (\mathbb{A}_{\Sigma,\Gamma})^{\{\cup,\cdot,*\}}$.

5.3 Closure Properties of Context-Free Extractors

From the known closure properties of context-free languages, we can conclude the following closure properties of context-free extractors.

Proposition 8. *Let* $E_1 \in \mathcal{E}^{\Sigma,\Gamma_1}_{\mathsf{CFL}}$ *and* $E_2 \in \mathcal{E}^{\Sigma,\Gamma_2}_{\mathsf{CFL}}$. *Then* $E_1 \cup E_2, E_1 \cdot E_2 \in \mathcal{E}^{\Sigma,\Gamma_1 \cup \Gamma_2}_{\mathsf{CFL}}$ *and* $(E_1)^* \in \mathcal{E}^{\Sigma,\Gamma_1}_{\mathsf{CFL}}$.

The non-closure of context-free languages under intersection and complement implies the following non-closure properties for context-free extractors.

Proposition 9. *Let* $E_1 \in \mathcal{E}^{\Sigma,\Gamma_1}_{\mathsf{CFL}}$ *and* $E_2 \in \mathcal{E}^{\Sigma,\Gamma_2}_{\mathsf{CFL}}$. *Then* $E_1 \cap E_2$, $E_1 \setminus E_2$, $\neg E_1$ *and* $E_1 \circ E_2$ *for* $\circ \in \{\bowtie, \bowtie_\cup, \bowtie_\cap, \bowtie_\setminus\}$ *are not necessarily context-free.*

6 Computational Problems

We will now investigate certain natural computational problems for extractors, and we will mainly concentrate on regular and context-free extractors, which, if not stated otherwise, are represented as NFAs and CFGs, respectively. For convenience, we simply write $|E|$ to denote the size of the NFA or CFG that represents the extractor E. We first discuss several problems that trivially reduce to well-known formal language problems on the marker languages.

The *tuple membership problem* (for a class $\mathcal{E}$ of extractors) is to decide for a given extractor $E \in \mathcal{E}$, a string w and a tuple t for w whether $t \in E(w)$.

Obviously, checking $t \in E(w)$ boils down to checking $W_{w,t} \in L_E$, which means that the tuple membership problem for regular or context-free extractors inherits the upper bounds of the membership problem for regular or context-free languages (and similarly for any other language class). Moreover, we can interpret any $L \subseteq \Sigma^*$ as a Σ-signed $\emptyset$-marker language and then check $w \in L$ by checking $t^{\emptyset}_{\Gamma} \in [\![L]\!](w)$. Hence, the conditional lower bounds for the membership problem for regular and context-free languages (see [1,5]) carry over to the tuple membership problem for regular and context-free extractors.

In principle, we could also compute the full set $E(w)$ by testing $W_{w,t} \in L_E$ for every Γ-tuple t for w. This is of course rather inefficient, since there are $2^{|w||\Gamma|}$ different Γ-tuples for w that we have to consider. However, for regular and context-free extractors, we can do much better, which has been thoroughly investigated in terms of an enumeration problem (i. e., we wish to enumerate all elements from $E(w)$ without repetition and with a guaranteed upper bound on the delay between two elements). See, e. g., [3,16,20] for regular extractors and [4] for context-free extractors (in addition, there are several papers investigating the enumeration problem for regular and context-free document spanners (e. g. [9, 21, 22]), which are a subset of our regular and context-free extractors).

Let us move on to more complex decision problems. For (Σ, Γ)-extractors E_1, E_2, we write $E_1 \subseteq E_2$ if and only if $E_1(w) \subseteq E_2(w)$ for every $w \in \Sigma^*$ (note that $E_1 = E_2 \iff E_1 \subseteq E_2 \wedge E_2 \subseteq E_1$). The *containment* and *equivalence problem* is to decide for given (Σ, Γ)-extractors E_1, E_2 whether $E_1 \subseteq E_2$ or $E_1 = E_2$, respectively, and the *emptiness problem* is to decide for a given (Σ, Γ)-extractor E whether $E(w) \neq \emptyset$ for some $w \in \Sigma^*$. These decision problems are obviously identical to the corresponding problems on the marker languages.

Observation 1. For (Σ, Γ)-marker languages L_1 and L_2, we have that $[\![L_1]\!] \subseteq [\![L_2]\!]$ if and only if $L_1 \subseteq L_2$, $[\![L_1]\!] = [\![L_2]\!]$ if and only if $L_1 = L_2$, and $[\![L_1]\!](w) \neq \emptyset$ for some w if and only if $L_1 \neq \emptyset$.

This implies, for example, that the containment problem for regular extractors is PSPACE-complete, the emptiness problem for regular extractors is in P, the equivalence problem of context-free extractors is undecidable, etc.

6.1 Table Problems

Let us next define the so-called *table problems*. The *table containment*, *table equivalence* and *table disjointness problem* is to decide for given (Σ, Γ)-extractors E_1, E_2 and $w \in \Sigma^*$ whether $E_1(w) \subseteq E_2(w)$, $E_1(w) = E_2(w)$ or $E_1(w) \cap E_2(w) \neq \emptyset$, respectively, and the *table emptiness problem* is to decide for a given (Σ, Γ)-extractor E and $w \in \Sigma^*$ whether $E(w) = \emptyset$.

Unlike the problems from above, the table problems are not already covered by known language problems on the marker languages. Instead, they can be seen as problems on the w-slice of a marker languages, since $E_1(w) \subseteq E_2(w)$, $E_1(w) = E_2(w)$, $E_1(w) \cap E_2(w) \neq \emptyset$ and $E(w) = \emptyset$ if and only if $\mathsf{sl}_w(L_{E_1}) \subseteq \mathsf{sl}_w(L_{E_2})$, $\mathsf{sl}_w(L_{E_1}) = \mathsf{sl}_w(L_{E_2})$, $\mathsf{sl}_w(L_{E_1}) \cap \mathsf{sl}_w(L_{E_2}) \neq \emptyset$ and $\mathsf{sl}_w(L_E) = \emptyset$, respectively.

The slices of any marker language are always finite sets of strings, but this does not necessarily mean that the table problems are easy, since slices have in general exponential size. For the mere decidability of the table problems, the decidability of the tuple membership problem (which we can assume for most reasonable classes of extractors) is a sufficient condition.

Theorem 2. *If the tuple membership problem is decidable for a class $\mathcal{E}$ of extractors, then the table problems for $\mathcal{E}$ are decidable.*

We now investigate the table problems for regular and context-free extractors.

Regular Extractors. We first observe that the table disjointness and emptiness problems can be solved efficiently by exploiting the NFA representation.

Theorem 3. *For regular extractors, the table disjointness and table emptiness problem can be solved in time $O(|E_1||E_2||w|)$ and $O(|E||w|)$, respectively.*

We can complement these upper bounds with conditional lower bounds, i. e., if $E(w) \neq \emptyset$ can be checked in time $O((|E||w|)^{1-\epsilon})$ or $E_1(w) \cap E_2(w) \neq \emptyset$ can be checked in time $O((|E_1||E_2||w|)^{1-\epsilon})$, then the strong exponential time hypothesis fails (see the full version [26] for more details).

In contrast to table disjointness and emptiness, the table containment and equivalence problems are intractable. However, the complexity changes if extractors are represented by DFAs instead of NFAs.

Theorem 4. *The table containment and table equivalence problem for regular extractors is coNP-complete, even if* $|\Sigma| = 1$, $|\Gamma| = 2$, *both* E_1 *and* E_2 *have finite support, and* E_1 *is given by a DFA. The table containment problem for regular extractors can be solved in time* $O(|E_1||E_2||w|)$, *provided that* E_2 *is given by a DFA, and the table equivalence problem for regular extractors can be solved in time* $O(|E_1||E_2||w|)$, *provided that both* E_1 *and* E_2 *are given by DFAs.*

Context-Free Extractors. The table emptiness problem for context-free extractors can be solved as efficiently as the membership problem for CFGs.[2]

Theorem 5. *The table emptiness problem for context-free extractors can be solved in time* $O(|E||w|^3)$.

We next observe that the table disjointness problem, which is tractable for regular extractors, becomes intractable for context-free extractors.

Theorem 6. *The table disjointness problem for context-free extractors is NP-complete, even for extractors with finite support and if* $|\Sigma| = 1$, *but it can be solved in polynomial time, provided that one of the two extractors is regular.*

Finally, we consider the table containment and table equivalence problem for context-free extractors.

Theorem 7. *The table containment and equivalence problem for context-free extractors is coNP-complete, but the table containment problem for context-free extractors can be solved in polynomial time, provided that* E_2 *is given by a DFA.*

7 Conclusions

This paper attempts to answer the following question: Given the fact that the work on information extraction in database theory is based on concepts and techniques of classical formal language theory, is there a unifying framework rooted purely in formal language theory (independent on specific data management tasks)? In particular, such a framework should be robust in the following sense: While existing information extraction techniques in database theory are somehow "based on regular languages", but ultimately designed in an ad-hoc

[2] The upper bound mentioned in Theorem 5 can be improved by using Valiants parsing algorithm; we mention the CYK-based bound for simplicity.

way to solve a specific data management task at hand, our framework should allow to simply replace "regular languages" by just any language class.

There are several aspects of this approach that might be beneficial for the future. Most importantly, we might identify new theoretical questions and worthwhile research tasks in formal language theory, which will allow us to progress this traditional field of theoretical research. For example, decision problems like membership, inclusion, equivalence, universality etc. are well-investigated for many language classes, but the table problems (see Sect. 6.1) are a new set of decision problems that do not arise from classical considerations in formal languages (intuitively speaking, table problems are problems concerned with "finite slices" of languages). Moreover, while undecidability is a very common obstacle for problems on formal languages, the table problems are all trivially decidable as long as the membership problem of the underlying class of marker languages is decidable (a property shared by virtually all useful language classes). This can be a new play area for complexity theoretical and algorithmic research within formal language theory.

Another observation is that regular extractors can also be seen as a restricted form of transducers (called *annotation transducers* or *annotation automata* in [16,20]), i. e., a transducer that can either erase an input symbol, or replace it by a marker set. However, in order to get the complete information of the corresponding Γ-tuple, we would also need along with the marker set the position of the input symbol in the overall input string (which, technically, requires an unbounded output alphabet). Hence, we are dealing with a setting that is close to, but not really covered by existing models in the theory of transducers.

In summary, while techniques from formal languages can be beneficially exploited for data management tasks, one might also go in the other direction and ask whether these tools based on formal languages imply some interesting new theoretical questions to be investigated in the broader, more general setting of formal language theory.

Finally, there might also be some practical implications. Arguably, extractor classes based on languages that are strictly more powerful than context-free languages likely suffer from intractability. However, there are many well-investigated subregular language classes as well as language classes sandwiched between regular and context-free languages. All of those are possible candidates for practically relevant extractor classes.

Acknowledgments. The author is supported by the German Research Foundation (Deutsche Forschungsgemeinschaft, DFG) – project number 522576760 (gefördert durch die Deutsche Forschungsgemeinschaft (DFG) – Projektnummer 522576760).

References

1. Abboud, A., Backurs, A., Vassilevska Williams, V.: If the current clique algorithms are optimal, so is Valiant's parser. SIAM J. Comput. **47**(6), 2527–2555 (2018). https://doi.org/10.1137/16M1061771

2. Amarilli, A., Bourhis, P., Mengel, S., Niewerth, M.: Constant-delay enumeration for nondeterministic document spanners. SIGMOD Rec. **49**(1), 25–32 (2020). https://doi.org/10.1145/3422648.3422655
3. Amarilli, A., Bourhis, P., Mengel, S., Niewerth, M.: Constant-delay enumeration for nondeterministic document spanners. ACM Trans. Database Syst. **46**(1), 2:1–2:30 (2021). https://doi.org/10.1145/3436487
4. Amarilli, A., Jachiet, L., Muñoz, M., Riveros, C.: Efficient enumeration for annotated grammars. In: PODS '22: International Conference on Management of Data, Philadelphia, PA, USA, June 12 - 17, 2022, pp. 291–300 (2022). https://doi.org/10.1145/3517804.3526232
5. Backurs, A., Indyk, P.: Which regular expression patterns are hard to match? In: IEEE 57th Annual Symposium on Foundations of Computer Science, FOCS 2016, 9-11 October 2016, Hyatt Regency, New Brunswick, New Jersey, USA, pp. 457–466 (2016). https://doi.org/10.1109/FOCS.2016.56
6. Doleschal, J., Kimelfeld, B., Martens, W.: The complexity of aggregates over extractions by regular expressions. Log. Methods Comput. Sci. **19**(3) (2023). https://doi.org/10.46298/LMCS-19(3:12)2023
7. Doleschal, J., Kimelfeld, B., Martens, W., Peterfreund, L.: Weight annotation in information extraction. In: 23rd International Conference on Database Theory, ICDT 2020, March 30-April 2, 2020, Copenhagen, Denmark, pp. 8:1–8:18 (2020). https://doi.org/10.4230/LIPIcs.ICDT.2020.8
8. Fagin, R., Kimelfeld, B., Reiss, F., Vansummeren, S.: Document spanners: a formal approach to information extraction. J. ACM **62**(2), 12:1–12:51 (2015)
9. Florenzano, F., Riveros, C., Ugarte, M., Vansummeren, S., Vrgoc, D.: Constant delay algorithms for regular document spanners. In: Proceedings of PODS'18 (2018)
10. Florenzano, F., Riveros, C., Ugarte, M., Vansummeren, S., Vrgoc, D.: Efficient enumeration algorithms for regular document spanners. ACM Trans. Database Syst. **45**(1), 3:1–3:42 (2020). https://doi.org/10.1145/3351451
11. Freydenberger, D.: A logic for document spanners. Theory Comput. Syst. **63**(7), 1679–1754 (2019). https://doi.org/10.1007/s00224-018-9874-1
12. Freydenberger, D., Holldack, M.: Document spanners: from expressive power to decision problems. Theory Comput. Syst. **62**(4), 854–898 (2018)
13. Freydenberger, D.D., Kimelfeld, B., Peterfreund, L.: Joining extractions of regular expressions. In: Proceedings of the 37th ACM SIGMOD-SIGACT-SIGAI Symposium on Principles of Database Systems, Houston, TX, USA, June 10-15, 2018, pp. 137–149 (2018). https://doi.org/10.1145/3196959.3196967
14. Freydenberger, D.D., Thompson, S.M.: Dynamic complexity of document spanners. In: 23rd International Conference on Database Theory, ICDT 2020, March 30-April 2, 2020, Copenhagen, Denmark, pp. 11:1–11:21 (2020). https://doi.org/10.4230/LIPIcs.ICDT.2020.11
15. Freydenberger, D.D., Thompson, S.M.: Splitting spanner atoms: a tool for acyclic core spanners. In: 25th International Conference on Database Theory, ICDT 2022, March 29 to April 1, 2022, Edinburgh, UK (Virtual Conference), pp. 10:1–10:18 (2022). https://doi.org/10.4230/LIPIcs.ICDT.2022.10
16. Gawrychowski, P., Manea, F., Schmid, M.L.: Revisiting weighted information extraction: a simpler and faster algorithm for ranked enumeration. Proc. ACM Manag. Data **2**(5), 222:1–222:19 (2024). https://doi.org/10.1145/3695840
17. Hopcroft, J.E., Motwani, R., Ullman, J.D.: Introduction to automata theory, languages, and computation, 3rd edn. Addison-Wesley, Pearson international edition (2007)

18. Libkin, L.: Elements of Finite Model Theory. Texts in Theoretical Computer Science. An EATCS Series, Springer (2004). https://doi.org/10.1007/978-3-662-07003-1, http://www.cs.toronto.edu/%7Elibkin/fmt
19. Maturana, F., Riveros, C., Vrgoc, D.: Document spanners for extracting incomplete information: expressiveness and complexity. In: Proceedings of the 37th ACM SIGMOD-SIGACT-SIGAI Symposium on Principles of Database Systems, Houston, TX, USA, June 10-15, 2018, pp. 125–136 (2018). https://doi.org/10.1145/3196959.3196968
20. Muñoz, M., Riveros, C.: Constant-delay enumeration for SLP-compressed documents. Log. Methods Comput. Sci. **21**(1) (2025). https://doi.org/10.46298/LMCS-21(1:17)2025
21. Peterfreund, L.: The Complexity of Relational Queries over Extractions from Text. Ph.D. thesis (2019)
22. Peterfreund, L.: Enumerating grammar-based extractions. Discret. Appl. Math. **341**, 372–392 (2023). https://doi.org/10.1016/J.DAM.2023.08.014
23. Peterfreund, L., ten Cate, B., Fagin, R., Kimelfeld, B.: Recursive programs for document spanners. In: 22nd International Conference on Database Theory, ICDT 2019, March 26-28, 2019, Lisbon, Portugal, pp. 13:1–13:18 (2019)
24. Peterfreund, L., Freydenberger, D.D., Kimelfeld, B., Kröll, M.: Complexity bounds for relational algebra over document spanners. In: Proceedings of the 38th ACM SIGMOD-SIGACT-SIGAI Symposium on Principles of Database Systems, PODS 2019, Amsterdam, The Netherlands, June 30 - July 5, 2019, pp. 320–334 (2019)
25. Schmid, M.L.: The information extraction framework of document spanners - a very informal survey. In: SOFSEM 2024: Theory and Practice of Computer Science - 49th International Conference on Current Trends in Theory and Practice of Computer Science, SOFSEM 2024, Cochem, Germany, February 19-23, 2024, Proceedings, pp. 3–22 (2024). https://doi.org/10.1007/978-3-031-52113-3_1
26. Schmid, M.L.: A general information extraction framework based on formal languages. CoRR abs/2505.15605 (2025). https://doi.org/10.48550/arXiv.2505.15605, https://doi.org/10.48550/arXiv.2505.15605
27. Schmid, M.L., Schweikardt, N.: Spanner evaluation over SLP-compressed documents. In: PODS'21: Proceedings of the 40th ACM SIGMOD-SIGACT-SIGAI Symposium on Principles of Database Systems, Virtual Event, China, June 20-25, 2021, pp. 153–165 (2021). https://doi.org/10.1145/3452021.3458325
28. Schmid, M.L., Schweikardt, N.: Document spanners - A brief overview of concepts, results, and recent developments. In: PODS '22: International Conference on Management of Data, Philadelphia, PA, USA, June 12 - 17, 2022, pp. 139–150 (2022). https://doi.org/10.1145/3517804.3526069
29. Schmid, M.L., Schweikardt, N.: Query evaluation over SLP-represented document databases with complex document editing. In: PODS '22: International Conference on Management of Data, Philadelphia, PA, USA, June 12 - 17, 2022, pp. 79–89 (2022). https://doi.org/10.1145/3517804.3524158
30. Schmid, M.L., Schweikardt, N.: Refl-spanners: a purely regular approach to non-regular core spanners. Log. Methods Comput. Sci. **20**(4) (2024). https://doi.org/10.46298/LMCS-20(4:16)2024

Deciding Sparseness of Regular Languages of Finite Trees and Infinite Words

Kord Eickmeyer(✉) and Georg Schindling

Technische Universität Darmstadt, Darmstadt, Germany
{eickmeyer,schindling}@mathematik.tu-darmstadt.de

Abstract. We study the notion of sparseness for regular languages over finite trees and infinite words. A language of trees is called sparse if the relative number of n-node trees in the language tends to zero, and a language of infinite words is called sparse if it has measure zero in the Bernoulli probability space. We show that sparseness is decidable for regular tree languages and for regular languages of infinite words. For trees, we provide characterisations in terms of forbidden subtrees and tree automata, leading to a linear time decision procedure. For infinite words, we present a characterisation via infix completeness and give a novel proof of decidability. Moreover, in the non-sparse case, our algorithm computes a measurable subset of accepted words that can serve as counterexamples in almost-sure model checking. Our findings have applications to automata based model checking in formal verifications and XML schemas, among others.

Keywords: regular languages · trees · infinite words · monadic second-order logic · automata · sparseness

1 Introduction

We call a language of finite objects *sparse* if the relative number of n-element objects in the language tends to zero: For a language $L \subseteq \Sigma^*$ of words we set $\mathbb{P}_n(L) := |L \cap \Sigma^n|/|\Sigma^n|$ and call L sparse if $\lim_{n\to\infty} \mathbb{P}_n(L) = 0$. Likewise, if L_T is a language of Σ-labelled trees and $\mathcal{B}_n^\Sigma$ is the set of all such trees with n nodes, we set $\mathbb{P}_n(L_T) := |L_T \cap \mathcal{B}_n^\Sigma|/|\mathcal{B}_n^\Sigma|$ and call L_T sparse if $\lim_{n\to\infty} \mathbb{P}_n(L_T) = 0$. For a language $L_\omega \subseteq \Sigma^\omega$ of infinite words we turn Σ^ω into a Bernoulli probability space and call L_ω sparse if it has measure 0.

Regular languages play an important rôle both in applied and in theoretical computer science. Apart from appearing in practical applications such as pattern matching, one reason for their importance from a theoretical perspective is that they allow for several seemingly unrelated characterisations, in particular by finite monoids, finite automata, and as being definable in monadic second-order logic (MSO). Many problems that are undecidable or at least computationally hard for general classes of languages are computable for regular languages. In particular the emptiness problem and, more generally, the subset relation between regular languages is efficiently decidable.

S.-K. Ko and F. Manea (Eds.): DLT 2025, LNCS 16036, pp. 213–228, 2026.
https://doi.org/10.1007/978-3-032-01475-7_15

Naturally, the techniques used in dealing with regular languages have been adapted to other scenarios. Two of the most important and successful generalisations have been to languages of finite trees and languages of infinite words, also called ω-languages. In both these settings, the correspondence between regular languages, automata, and definability in MSO still holds, yielding similar algorithmic properties as for regular languages of finite words.

We provide a characterisation of sparse regular tree languages via *forbidden subtrees* and *tree automata* (Theorem 5). As a consequence, we deduce that sparseness of regular tree languages is decidable in linear time (Corollary 2). We also observe that our characterisation by forbidden subtrees does not hold for non-uniform models of random binary trees, specifically the random binary search tree model.

A major motivation for regular ω-languages, on the other hand, is in model checking of reactive systems [3]. In this setting, we are given a formal specification of a system (e.g. some piece of hardware or software) and a specification of properties that the system should have or not have. If both the system and the property are specified in linear temporal logic (LTL), model checking essentially reduces to checking whether a certain regular ω-language is empty or not. Theorem 6 extends this to *almost sure model checking*, i.e. the question whether a certain system *almost surely* satisfies a given property. If the system fails to do so, it is even possible to compute a counterexample, in this case a set of ω-words of strictly positive measure that almost surely violates the condition to be checked.

Note that because the complement of a regular language is again regular, and L is sparse if and only if its complement has (asymptotic) probability 1, the same algorithms can be used to check whether a language of trees or ω-words has (asymptotic) probability 1.

Techniques and Related Work

Sparseness of regular languages was investigated by Sin'ya [20]. He showed that a regular language $L \subseteq \Sigma^*$ over a finite alphabet Σ is sparse if, and only if, L has an excluded factor, i.e. $\Sigma^* x \Sigma^* \cap L = \emptyset$ for some $x \in \Sigma^*$. The proof of this characterisation was simplified by Koga in [15]. We build on their insights and generalize them significantly to handle the richer settings of regular tree languages and ω-languages.

The trees we are most interested in are *rooted*, *ordered*, *unranked* trees, i.e. trees with a distinguished root node in which the number of children of a node is unbounded and for every node there is a linear order on the set of its children. The terms *siblinged* or *planar* are sometimes used for what we call ordered, e.g. in [4,11]. We state our results on regular tree languages in terms of binary trees with distinguished left and right children, cf. Sect. 2.2. This allows for a simpler definition of tree automata and comes at no loss of generality because ordered unranked trees are bijectively MSO-interpretable in binary trees, a folklore technique that is spelled out, for example, in [12, Sec. 10.2].

Because regular languages exactly capture the expressive power of monadic second-order logic (MSO), the (asymptotic) probability of regular languages can be rephrased as the (asymptotic) probability that a random structure satisfies a certain MSO sentence. This has been well studied: For every first-order logic (FO) sentence with relational signature, the probability that a random labelled structure satisfies the sentence converges to either 0 or 1 (this was proved independently in [13] and [10], cf. [9, Ch. 4]), a result that has been extended by Spencer [21] to Erdős-Rényi graphs with more general edge probabilities. The fact that every sentence is satisfied asymptotically with probability 0 or 1 is usually called 0-1 *law*. For monadic second-order logic (MSO), McColm [17] showed a 0-1 law for MSO on labelled trees. Note that McColm (crucially) considers trees without a distinguished root, while our trees are rooted. More recently, Malyshkin and Zhukovskii [16] showed a 0-1 law for MSO on finite trees with probability measures other than uniform probabilities, namely uniform and preferential attachment. Further results on the limit behaviour of MSO sentences on various graph classes can be found in [7] and [2]. These results are not directly applicable to our setting. In fact, there can be no 0-1 law for regular languages: Both in trees and in ω-words there is some designated element (such as the first position in a word or the root of a tree), and the set of all structures in which this element gets a specific label is a regular set with (asymptotic) density strictly between 0 and 1.

Recently, Niwiński et al. [18] proved that the exact probability of acceptance for MSO formulae in infinite random trees is computable. However, this can not be used to decide sparseness for languages of finite trees or infinite words.

A necessary and sufficient condition for an ω-language to have strictly positive measure has already been proved by Staiger [23]. However, Staiger does not address the question of decidability of this property, and his proof uses deep results from topology and Kolmogorov complexity, namely a connection between the *subword complexity* of an ω-word ξ and the *Hausdorff dimension* of the regular ω-languages containing ξ. Courcoubetis et al. [6] treated the sparseness problem for ω-languages under the name *probabilistic emptyness* and showed it to be decidable with methods similar to ours. By relating non-sparseness to infix completeness, our (shorter) proof has the additional feature of giving an exact characterisation of non-sparse ω-languages. Furthermore, the algorithm we present in Theorem 6 is the first to provide a concise representation of a set M of ω-words that is almost a subset of L, in the sense that $\mathbb{P}(M \setminus L) = 0$. This is extremely valuable in the context of LTL model checking, as it amounts to a non-negligible set of counterexamples.

2 Preliminaries

2.1 Languages and Automata

By Σ we denote a finite alphabet assuming $|\Sigma| > 1$ to avoid trivialities. For $k \geq 0$ we denote by Σ^*, Σ^k, and $\Sigma^{\leq k}$ the sets of all strings, those of length k, and of length at most k, respectively. The empty string is denoted by ε. A *language* is any subset $L \subseteq \Sigma^*$. Concatenation of strings $u, v \in \Sigma^*$ is denoted by $u \cdot v$ or just uv, and for a language L we set $L^* := \{u_1 \cdots u_k \mid k \geq 0, u_1, \ldots, u_k \in L\}$.

An *ω-word* over the alphabet Σ is an infinite sequence $a_0 a_1 a_2 \cdots$ of letters $a_i \in \Sigma$. The set of all ω-words is denoted by Σ^ω, and an ω-language is any subset $L \subseteq \Sigma^\omega$. A (finite) word $u \in \Sigma^*$ and an ω-word $v \in \Sigma^\omega$ may be concatenated to $uv \in \Sigma^\omega$, and for a language $U \subseteq \Sigma^*$ and an ω-language $V \subseteq \Sigma^\omega$ we set

$$U^\omega := \{u_1 u_2 u_3 \cdots \mid u_1, u_2, \ldots \in U\} \qquad \text{and} \qquad UV := \{uv \mid u \in U \text{ and } v \in V\}.$$

A word $u \in \Sigma^*$ is a *prefix* of $v \in \Sigma^* \cup \Sigma^\omega$, written $u \preceq v$, if $v = uw$ for some $w \in \Sigma^* \cup \Sigma^\omega$.

A *deterministic finite automaton (DFA)* over the alphabet Σ is a tuple $\mathcal{A} = (\Sigma, Q, q_0, \delta, A)$ consisting of a finite set Q of *states*, a designated *initial state* $q_0 \in Q$, a *transition function* $\delta\colon Q \times \Sigma \to Q$ and a set $A \subseteq Q$ of *accepting states*. We extend δ to a function $\hat{\delta}\colon Q \times \Sigma^* \to Q$ by setting $\hat{\delta}(q, \varepsilon) := q$ and $\hat{\delta}(q, aw) := \hat{\delta}(\delta(q, a), w)$ for $a \in \Sigma$ and $w \in \Sigma^*$. The automaton $\mathcal{A}$ *accepts* a word $w \in \Sigma^*$ if $\hat{\delta}(q_0, w) \in A$. The *language accepted by* $\mathcal{A}$ is the set of words accepted by it. A language is called *regular* if it is accepted by some DFA.

For states $q, q' \in Q$ we say that q' is *reachable* from q, written $q \rightsquigarrow q'$, if $\hat{\delta}(q, w) = q'$ for some word $w \in \Sigma^*$. If q' is reachable from the initial state q_0 we just call q' reachable. The relation $\rightsquigarrow$ is obviously reflexive and transitive, so the relation $q \approx q' \Leftrightarrow (q \rightsquigarrow q'$ and $q' \rightsquigarrow$ q) is an equivalence relation. Its equivalence classes are called the *reachability classes* of $\mathcal{A}$. A reachability class C is called *closed* if $\delta(q, a) \in C$ for all $q \in C$ and $a \in \Sigma$

There are various generalisations of finite automata to ω-words, resulting in a robust concept of regular ω-languages, cf. [24]. We use the following characterisation:

Theorem 1 (cf. [22], Thm. 3.2). *An ω-language $L \subseteq \Sigma^\omega$ is regular if and only if there is a $k \geq 1$ and regular languages $U_i, V_i \subseteq \Sigma^*$ for $i = 1, \ldots, k$ such that $L = \bigcup_{i=1}^k U_i V_i^\omega$.*

2.2 Tree Languages

We will mostly be concerned with *binary trees* in the sense of Knuth [14, 2.3]. These trees have a distinguished root node, and every node may have a left and/or a right child. We formalise this as a prefix-closed language $T \subseteq D^*$ over the alphabet $D = \{l, r\}$. The empty word ε denotes the root of the tree T, a node $u \in T$ may have children $ul, ur \in T$ and is the *parent* of these.

For a finite alphabet Σ and binary tree T we call a function $\lambda\colon T \to \Sigma$ a *Σ-labelling* of T and the pair (T, λ) a Σ-labelled binary tree. We denote the set of all finite Σ-labelled binary trees by $\mathcal{B}^\Sigma$. We often just write $T \in \mathcal{B}^\Sigma$ and refer to the labelling as λ_T when necessary. A set $L \subseteq \mathcal{B}^\Sigma$ is called a *tree language*. Note that $\mathcal{B}^\Sigma$ contains the empty tree $\emptyset$.

For $S, T \in \mathcal{B}^\Sigma$ we say that S is a *subtree* of T and write $S \preceq T$ if there exists $u \in D^*$ such that $uD^* \cap T = uS$ and $\lambda_S(v) = \lambda_T(uv)$ for all nodes $v \in S$. Each node $u \in T$ induces a subtree $T(u)$ of T consisting of u and all its descendants. We write T_l for the *left subtree* $T(l)$ and T_r for the *right subtree* $T(r)$ of the root of T, respectively.

For two trees $S, T \in \mathcal{B}^\Sigma$ and $a \in \Sigma$ we define the Σ-labelled binary tree $\mathrm{conc}_a(S,T)$ which consists of the root labelled with a and has S and T as left and right subtrees respectively.

For a tree language $L \subseteq \mathcal{B}^\Sigma$ and $T \in \mathcal{B}^\Sigma$ we define the tree languages

$$LT := \bigcup_{a\in\Sigma} \{\mathrm{conc}_a(S,T) \mid S \in L\}$$

$$LT^{-1} := \{S \in \mathcal{B}^\Sigma \mid \text{There exists } a \in \Sigma \text{ with } \mathrm{conc}_a(S,T) \in L\}$$

and $TL, T^{-1}L$ are defined analogously. For trees $T_1, \ldots, T_k \in \mathcal{B}^\Sigma$ we inductively set

$$L[T_1]^{-1} := LT_1^{-1} \cup T_1^{-1}L$$

$$L[T_k, \ldots, T_1]^{-1} := (L[T_{k-1}, \ldots, T_1]^{-1})T_k^{-1} \cup T_k^{-1}(L[T_{k-1}, \ldots, T_1]^{-1}),$$

so $L[T_k, \ldots, T_1]^{-1}$ is the language of all trees that can be concatenated successively with $T_k, \ldots, T_1$ to obtain a tree from L.

A *tree automaton* is a tuple $\mathcal{A} = (\Sigma, Q, \Delta, A)$ consisting of a finite set Q of *states*, a finite alphabet Σ, a *transition relation* $\Delta \subseteq (Q \cup \{\bot\}) \times (Q \cup \{\bot\}) \times \Sigma \times Q$, and a set $A \subseteq Q$ of *accepting* states. Given a Σ-labelled binary tree T, a *run* of $\mathcal{A}$ on T is a function $d \colon D^* \to (Q \cup \{\bot\})$ such that

- if $u \notin T$ then $d(u) = \bot$, and
- if $u \in T$ then $(d(ul), d(ur), \lambda_T(u), d(u)) \in \Delta$.

A run d is called *accepting* if $d(\varepsilon) \in A$ and we say that an automaton $\mathcal{A}$ *accepts* a tree T if there is an accepting run of $\mathcal{A}$ on T.

For states $q \in Q$ and $q' \in Q \cup \{\bot\}$, we say that q' is 1-step reachable from q (written $q \rightsquigarrow_1 q'$) if $(q, \tilde{q}, a, q') \in \Delta$ or $(\tilde{q}, q, a, q') \in \Delta$ for some $a \in \Sigma$ and $\tilde{q} \in Q \cup \{\bot\}$. The reflexive transitive closure $\rightsquigarrow := \rightsquigarrow_1^*$ is called reachability, and a state $q \in Q$ is called *reachable* if $\bot \rightsquigarrow q$. Equivalently, q is reachable if and only if there exists a tree $T \in \mathcal{B}^\Sigma$ and a run d of $\mathcal{A}$ on T with $d(\varepsilon) = q$. We call $\mathcal{A}$ *reduced* if every state is reachable.

2.3 Random Trees and ω-Words

For $n \in \mathbb{N}$ we denote by $\mathcal{B}_n^\Sigma$ and $\mathcal{B}_{<n}^\Sigma$ the set of Σ-labelled binary trees of size n and size strictly less than n respectively. We consider $\mathcal{B}_n^\Sigma$ as a finite discrete probability space equipped with the uniform distribution $\mathbb{P}_n$ by setting $\mathbb{P}_n(T) = \frac{1}{|\mathcal{B}_n^\Sigma|}$ for $T \in \mathcal{B}_n^\Sigma$. Note that $|\mathcal{B}_n^\Sigma| = C_n \cdot |\Sigma|^n$, where $C_n = \frac{1}{n+1}\binom{2n}{n} \approx \frac{4^n}{n\sqrt{\pi n}}$ is the n-th *Catalan number*, cf. [14, 2.3.4.4].

The *asymptotic density* (or *asymptotic probability*) of a tree language $L \subseteq \mathcal{B}^\Sigma$ is defined as $\mathbb{P}_{\lim}(L) := \lim_{n\to\infty} \mathbb{P}_n(L \cap \mathcal{B}_n^\Sigma) = \lim_{n\to\infty} \left(|L \cap \mathcal{B}_n^\Sigma| / (C_n \cdot |\Sigma|^n)\right)$, given the limit exists. We set $\overline{\mathbb{P}}_{\lim}(L) := \limsup_{n\to\infty} \mathbb{P}_n(L \cap \mathcal{B}_n^\Sigma)$, and one easily checks that for all $L_1, L_2 \subseteq \mathcal{B}^\Sigma$, the following hold:

$$\begin{aligned} \overline{\mathbb{P}}_{\lim}(L) = 0 \Leftrightarrow \mathbb{P}_{\lim}(L) = 0 \qquad & \overline{\mathbb{P}}_{\lim}(L_1 \cup L_2) \leq \overline{\mathbb{P}}_{\lim}(L_1) + \overline{\mathbb{P}}_{\lim}(L_2) \\ L_1 \subseteq L_2 \Rightarrow \overline{\mathbb{P}}_{\lim}(L_1) \leq \overline{\mathbb{P}}_{\lim}(L_2) \qquad & \overline{\mathbb{P}}_{\lim}(\mathcal{B}^\Sigma) = 1. \end{aligned} \tag{1}$$

We use standard terminology from probability theory, cf. [25]. We turn the set Σ^ω of ω-words over the finite alphabet Σ into a probability space by making each of the projections $\pi_i \colon \Sigma^\omega \to \Sigma, w_1 w_2 \ldots \mapsto w_i$ measurable. By $\mathbb{P}$ we denote the probability measure for which the projections are iid random variables with $\mathbb{P}(w_i = a) = |\Sigma|^{-1}$ for every $i \geq 1$ and $a \in \Sigma$. Note that for any $U, V \subseteq \Sigma^*$, $UV^\omega = \bigcap_{k\geq 1} \bigcup_{\ell \geq k} \{w_1 w_2 \ldots | w_1 \ldots w_\ell \in UV^*\}$ is measurable, so by Theorem 1, every ω-regular $L \subseteq \Sigma^\omega$ is measurable and the probability (or measure) $\mathbb{P}(L)$ is well-defined.

We review some basic facts about discrete-time Markov chains with a finite state space, cf. [19]: Fix a finite set I of *states* and for every $i, j \in I$ a *transition probability* $p_{ij} \geq 0$ such that $\sum_{j\in I} p_{ij} = 1$ for every $i \in I$. A *Markov chain* with state space I and transition probabilities $P = (p_{ij})_{i,j\in I}$ is a sequence $(X_t)_{t\in\mathbb{N}}$ of random variables taking values in I such that $\mathbb{P}(X_{t+1} = j \mid X_t = i) = p_{ij}$ for every $t \geq 0$ and $i, j \in I$. The probability distribution of X_0 is called *initial distribution* of the chain. The initial distribution and the transition probabilities P together determine the joint distributions of the X_t by

$$\mathbb{P}(X_0 = i_0, \ldots, X_t = i_t) = \mathbb{P}(X_0 = i_0) \cdot p_{i_0 i_1} \cdots p_{i_{t-1} i_t}.$$

If $\mathbb{P}(X_0 = j) = \delta_{ij}$ we say that the chain is started in state i and denote the resulting probability distribution by $\mathbb{P}_i$. With this definition, $\mathbb{P}(X_{s+t} = j \mid X_s = i) = \mathbb{P}_i(X_t = j)$. A state $i \in I$ is called *recurrent* if $\mathbb{P}_i(X_t = i \text{ for infinitely many t}) = 1$. We say that a state $i \in I$ *leads to* a state $j \in I$, written $i \to j$, if $\mathbb{P}_i(X_t = j \text{ for some } t \in \mathbb{N}) > 0$, and that i and j *communicate* (written $i \leftrightarrow j$) if both $i \to j$ and $j \to i$. Then $\leftrightarrow$ is an equivalence relation on I and its equivalence classes are just called *classes* of states. A class $C \subseteq I$ is called *closed* if $i \in C$ and $i \to j$ imply $j \in C$. We need the following theorem:

Theorem 2 (cf. [19, Thm. 1.5.6]). *If $C \subseteq I$ is a closed class, every $i \in C$ is recurrent.*

Corollary 1. *If $C \subseteq I$ is a closed class and $i \in C$, then $\mathbb{P}(X_t = i \text{ infinitely often} \mid X_t \in C \text{ for some } t) = 1$ for every $s \in \mathbb{N}$.*

3 Characterising Sparseness of Regular Tree Languages

In this section we exactly characterise regular tree languages with asymptotic density 0 by excluded factors, namely *forbidden subtrees*. This generalises Sin'ya's result for regular languages. The well-known *infinite monkey theorem* states that a language of finite words $L \subseteq \Sigma^*$ has asymptotic density 1 if $\Sigma^* x \Sigma^* \subseteq L$

for some $x \in \Sigma^*$. This has been generalised to tree languages by Asada et al. [1, Thm. 2.13], who prove that contexts of up to logarithmic size appear asymptotically almost surely in certain regular tree languages. We only need a weaker version stated in Theorem 3, and give a short proof of it. In Theorem 5 we then show that, for a *regular* tree language, the existence of a forbidden subtree is a necessary condition for sparseness.

Theorem 3. $\mathbb{P}_{\lim}(\{T \in \mathcal{B}^\Sigma \mid S \preceq T\}) = 1$ *for every nonempty* $S \in \mathcal{B}^\Sigma$.

Proof. We fix a nonempty tree $S \in \mathcal{B}^\Sigma$ and examine the asymptotic behaviour of the sequence

$$a_n := |\{T \in \mathcal{B}_n^\Sigma \mid S \not\preceq T\}|$$

following the approach of [11, Example III.41]. By $f_{n,k}$ we denote the number of Σ-labelled binary trees of size n that contain S as a subtree at k different positions, and let

$$f(u, z) := \sum_{n,k \geq 0} f_{n,k} z^n u^k$$

be its bivariate generating function. In particular $a_n = f_{n,0}$ and $f(0, z) = \sum_n a_n z^n$ is the generating function of the sequence $(a_n)_{n \geq 0}$.

By $\omega(T)$ we denote the number of distinct occurrences of S in a tree $T \in \mathcal{B}^\Sigma$. Then $\omega(\emptyset) = 0$ and since S can occur in the left or right subtree, or be the whole tree T, we get $\omega(T) = \omega(T_l) + \omega(T_r) + [T = S]$, where $[T = S]$ is 1 if $T = S$ and 0 otherwise. for $T \neq \emptyset$. For the complex function $u \mapsto u^{\omega(T)}$ this can be rewritten as

$$u^{\omega(T)} = u^{\omega(T_l)} u^{\omega(T_r)} u^{[T=S]} = u^{\omega(T_l)} u^{\omega(T_r)} + [T = S](u - 1), \tag{2}$$

and justifying algebraic manipulations as in [11, A.5.], we get:

$$\begin{aligned}
f(u, z) &= \sum_{n=0}^{\infty} z^n \sum_{k=0}^{\infty} u^k f_{n,k} = \sum_{n=0}^{\infty} z^n \sum_{T \in \mathcal{B}_n^\Sigma} u^{\omega(T)} \\
&\overset{(2)}{=} \sum_{n=0}^{\infty} z^n \sum_{T \in \mathcal{B}_n^\Sigma} \left([T = S](u - 1) + u^{\omega(T_l)} u^{\omega(T_r)} \right) \\
&= \sum_{n=0}^{\infty} z^n \sum_{T \in \mathcal{B}_n^\Sigma} [T = S](u - 1) + \sum_{n=0}^{\infty} z^n \sum_{T \in \mathcal{B}_n^\Sigma} u^{\omega(T_l)} u^{\omega(T_r)} \\
&= z^m (u - 1) + 1 + \sum_{n=1}^{\infty} z^n \sum_{T \in \mathcal{B}_n^\Sigma} u^{\omega(T_l)} u^{\omega(T_r)}
\end{aligned}$$

for $m := |S|$. We set $f_n(u) := \sum_{T \in \mathcal{B}_n^\Sigma} u^{\omega(T)} = \sum_{k=0}^{\infty} u^k f_{n,k}$ and get

$$\begin{aligned} f(u,z) - z^m(u-1) - 1 &= \sum_{n=1}^{\infty} z^n \sum_{T \in \mathcal{B}_n^\Sigma} u^{\omega(T_l)} u^{\omega(T_r)} \\ &= \sum_{n=1}^{\infty} z^n \sum_{j=0}^{n-1} |\Sigma| \Big(\sum_{T_l \in \mathcal{B}_j^\Sigma} u^{\omega(T_l)} \Big) \Big(\sum_{T_r \in \mathcal{B}_{n-1-j}^\Sigma} u^{\omega(T_r)} \Big) \\ &= z|\Sigma| \sum_{n=1}^{\infty} z^{n-1} \sum_{j=0}^{n-1} \Big(\sum_{T_l \in \mathcal{B}_j^\Sigma} u^{\omega(T_l)} \Big) \Big(\sum_{T_r \in \mathcal{B}_{n-1-j}^\Sigma} u^{\omega(T_r)} \Big) \\ &= z|\Sigma| \sum_{n=0}^{\infty} \sum_{j=0}^{n} z^j f_j(u) \cdot z^{n-j} f_{n-j}(u) = z|\Sigma| f(u,z)^2. \end{aligned}$$

Solving this quadratic equation for $f(u,z)$ gives two candidate solutions

$$f(u,z) = \frac{1 \pm \sqrt{1 - 4z|\Sigma| - 4|\Sigma| z^{m+1}(u-1)}}{2z|\Sigma|},$$

and since $f(1, \frac{z}{|\Sigma|})$ is the generating function of the Catalan numbers, subtracting the square root gives the right solution. The generating function $f(0,z)$ of the sequence $(a_n)_{n \geq 0}$ is now given by

$$f(0,z) = \frac{1 - \sqrt{1 - 4z|\Sigma| + 4|\Sigma| z^{m+1}}}{2z|\Sigma|}.$$

The function $f(0,z)$ is analytic at 0 by extending it to $f(0,0) = 1$. The radius around 0, where $f(0,z)$ is analytic is exactly the radius R, for which the polynomial $p(z) := 1 - 4|\Sigma|z + 4|\Sigma|z^{m+1}$ is non-zero (cf. analyticity of $\sqrt{1-z}$). Considering the reciprocal polynomial of p yields $R > \frac{1}{4|\Sigma|}$.

Finally, by [11, Theorem IV.7 (Exponential Growth Formula)] there exist a subexponential factor $(\eta_n)_n$ such that $a_n = R^{-n}\eta_n$. Also, by Stirling's formula there exists a subexponential factor $(\theta_n)_n$ such that $C_n = 4^{-n}\theta_n$. In total, this yields

$$\frac{|\{T \in \mathcal{B}_n^\Sigma \mid S \not\preceq T\}|}{|\mathcal{B}_n^\Sigma|} = \frac{a_n}{C_n |\Sigma|^n} = \frac{R^{-n}\eta_n}{4^n \theta_n |\Sigma|^n} = \Big(\frac{1}{4R|\Sigma|}\Big)^n \frac{\eta_n}{\theta_n}.$$

Since the factors η_n and θ_n are subexponential the sequence converges to 0 as n tends to infinity. □

In order to show the converse for *regular* tree languages, we lift the proof from [15] to the case of tree languages. First, we derive bounds for the asymptotic density of specific tree languages.

Lemma 1. *For $L \subseteq \mathcal{B}^\Sigma$ and $T, T_1, \ldots, T_k \in \mathcal{B}^\Sigma$ the following hold:*

1. $\overline{\mathbb{P}}_{\lim}(LT) = \frac{1}{|\Sigma|^{|T|}4^{|T|+1}}\overline{\mathbb{P}}_{\lim}(L)$
2. $\overline{\mathbb{P}}_{\lim}(LT^{-1}), \overline{\mathbb{P}}_{\lim}(T^{-1}L) \leq |\Sigma|^{|T|}4^{|T|+1}\overline{\mathbb{P}}_{\lim}(L)$
3. $\overline{\mathbb{P}}_{\lim}(L[T_k, \ldots, T_1]^{-1}) \leq 2^k|\Sigma|^{\sum_i |T_i|}4^{\sum_i |T_i|+k}\overline{\mathbb{P}}_{\lim}(L)$

Proof. By

$$\begin{aligned}\mathbb{P}_n(LT \cap \mathcal{B}_n^\Sigma) &= \frac{|(L \cap \mathcal{B}_{n-|T|-1}^\Sigma)T|}{|\mathcal{B}_n^\Sigma|} = \frac{|L \cap \mathcal{B}_{n-|T|-1}^\Sigma|}{|\mathcal{B}_{n-|T|-1}^\Sigma|}\frac{|\Sigma| \cdot |\mathcal{B}_{n-|T|-1}^\Sigma|}{|\mathcal{B}_n^\Sigma|} \\ &= \mathbb{P}_{n-|T|-1}(L \cap \mathcal{B}_{n-|T|-1}^\Sigma)\frac{C_{n-|T|-1}|\Sigma|^{n-|T|}}{C_n|\Sigma|^n} \\ &= \mathbb{P}_{n-|T|-1}(L \cap \mathcal{B}_{n-|T|-1}^\Sigma)\frac{1}{|\Sigma|^{|T|}}\frac{C_{n-|T|-1}}{C_n},\end{aligned}$$

taking the limes superior on both sides together with the identity $\lim_{n\to\infty} \frac{C_{n-k}}{C_n} = \frac{1}{4^k}$ proves the first part. For the second part note that $(LT^{-1})T \subseteq L$ and hence by (1) together with the first part we obtain

$$\overline{\mathbb{P}}_{\lim}(L) \geq \overline{\mathbb{P}}_{\lim}\big((LT^{-1})T\big) = \frac{1}{|\Sigma|^{|T|}4^{|T|+1}}\overline{\mathbb{P}}_{\lim}(LT^{-1}),$$

and likewise for $\overline{\mathbb{P}}_{\lim}(T^{-1}L)$. The third part follows by induction.

Lemma 2. *Let $\mathcal{A} = (Q, \Sigma, \Delta, A)$ be a tree automaton. For every reachable state $q \in Q$ there exists a tree $T \in \mathcal{B}^\Sigma$ with $|T| \leq 2^{|Q|} - 1$ such that a run of $\mathcal{A}$ on T ends in q.*

Proof (sketch). Note that any binary tree with at least $2^{|Q|}$ nodes must have a path of length (counting the number of edges along the path) at least $|Q|$. By the pigeonhole principle, some state $q \in Q$ must appear more than once on this path, and the tree can be pruned.

If a binary tree S occurs as a subtree in a regular language L, then there is a tree $T \in L$ with $S \preceq T$ that is "not much larger" than S:

Lemma 3. *For every regular tree language $L \subseteq \mathcal{B}^\Sigma$ there exists $n \in \mathbb{N}$ such that for all $S \in \mathcal{B}^\Sigma$ and $T \in L$ with $S \preceq T$ there exist $T_1, \ldots, T_k \in \mathcal{B}_{<2^n}^\Sigma$ with $k \leq n$ such that $S \in L[T_k, \ldots, T_1]^{-1}$.*

Proof. Let $\mathcal{A} = (\Sigma, Q, \Delta, A)$ be a tree automaton recognising the language L. For $T \in L$ and $S \in \mathcal{B}^\Sigma$ with $S \preceq T$ there exists $w = w_1 \cdots w_\ell \in T$ with $S = T(w)$. Let $d\colon D^* \to Q \cup \{\bot\}$ be a run of $\mathcal{A}$ on T and $d_i = d(w_1 \cdots w_i) \in Q$ be state of $\mathcal{A}$ at node $w_1 \cdots w_i$ in this run, for $i = 0, \ldots, \ell$. Then $d_0 \in A$ (because $\mathcal{A}$ accepts T) and $d_i \rightsquigarrow_1 d_{i-1}$ for $i = 1, \ldots, \ell$. If $d_i = d_j$ for some $i < j$ we remove that subsequence $d_i, \ldots, d_{j-1}$ and repeat this process until we get a sequence $d'_0, \ldots, d'_m$ with $m < |Q|$.

Note that $d_S : D^* \to Q \cup \{\bot\}$ with $d_S(v) := d(wv)$ is a run of $\mathcal{A}$ on S with $d_S(\epsilon) = d(w) = d_\ell = d'_n$, and our reduced sequence d' still satisfies $d'_i \rightsquigarrow_1 d'_{i-1}$ for $i = 1, \ldots, m$. The result, with $n = |Q|$, now follows from Lemma 2 and the definition of $\rightsquigarrow_1$.

We use the previous lemma to show that if a regular tree language L is sparse, then it must already admit a *forbidden subtree*. That is, a fixed tree S which does not occur as subtree of any tree in L.

Theorem 4. *Let $L \subseteq \mathcal{B}^{\Sigma}$ be a regular tree language with $\mathbb{P}_{\lim}(L) = 0$. Then there exists $S \in \mathcal{B}^{\Sigma}$ such that $\{T \in \mathcal{B}^{\Sigma} \mid S \preceq T\} \cap L = \emptyset$.*

Proof. We argue by contraposition and assume that for all $S \in \mathcal{B}^{\Sigma}$ we have $\{T \in \mathcal{B}^{\Sigma} \mid S \preceq T\} \cap L \neq \emptyset$. That is, for every $S \in \mathcal{B}^{\Sigma}$ there exists $T \in L$ with $S \preceq T$. By Lemma 3, we infer that for every $S \in \mathcal{B}^{\Sigma}$ there exist $k \leq n$ and $T_1, \ldots, T_k \in \mathcal{B}^{\Sigma}_{<2^n}$ such that $S \in L[T_k, \ldots, T_1]^{-1}$. This in turn is equivalent to

$$\mathcal{B}^{\Sigma} \subseteq \bigcup_{k=1}^{n} \bigcup_{T_1,\ldots,T_k \in \mathcal{B}^{\Sigma}_{<2^n}} L[T_k, \ldots, T_1]^{-1}.$$

To show that $\mathbb{P}_{\lim}(L) > 0$ we use Lemma 1 to get

$$\begin{aligned} 1 = \overline{\mathbb{P}}_{\lim}(\mathcal{B}^{\Sigma}) &\leq \overline{\mathbb{P}}_{\lim}\Big(\bigcup_{k=1}^{n} \bigcup_{T_1,\ldots,T_k \in \mathcal{B}^{\Sigma}_{<2^n}} L[T_k, \ldots, T_1]^{-1}\Big) \\ &\leq \sum_{k=1}^{n} \sum_{T_1,\ldots,T_k \in \mathcal{B}^{\Sigma}_{<2^n}} \overline{\mathbb{P}}_{\lim}(L[T_k, \ldots, T_1]^{-1}) \\ &\leq \sum_{k=1}^{n} \sum_{T_1,\ldots,T_k \in \mathcal{B}^{\Sigma}_{<2^n}} 2^k \prod_{j=1}^{k} |\Sigma|^{|T_j|} 4^{|T_j|+1} \overline{\mathbb{P}}_{\lim}(L) \end{aligned}$$

□

This converse of the infinite monkey theorem for tree languages provides a characterisation of sparse tree languages in terms of *forbidden subtrees.* In order to also obtain such a characterisation in terms of tree automata akin to [20], we give the following definition.

Definition 1. *Let $\mathcal{A} = (\Sigma, Q, \Delta, A)$ be a tree automaton. A set of states $V \subseteq Q$ is called a* sink *if for all $q \in V$ and all $q_l, q_r \in Q \cup \{\bot\}, a \in \Sigma$ it holds that $\delta(q_l, q, a), \delta(q, q_r, a) \subseteq V$. That is, V is a sink exactly if every run of $\mathcal{A}$ remains in V once it entered a state in V.*

Putting everything together we get:

Theorem 5. *Let L be a regular tree language and $\mathcal{A} = (\Sigma, Q, \Delta, A)$ be a reduced tree automaton recognising L. Then the following assertions are equivalent:*

1. $\mathbb{P}_{\lim}(L) = 0$
2. *There exists a tree $S \in \mathcal{B}^{\Sigma}$ such that $\{T \in \mathcal{B}^{\Sigma} \mid S \preceq T\} \subseteq \mathcal{B}^{\Sigma} \setminus L$*
3. *$\mathcal{A}$ has a sink $V \subseteq Q$ with $V \cap A = \emptyset$*

Proof. Theorems 3 and 4 together imply 1 $\Leftrightarrow$ 2.

2 $\Rightarrow$ 3: Let $S \in \mathcal{B}^{\Sigma}$ such that $\{T \in \mathcal{B}^{\Sigma} \mid S \preceq T\} \subseteq \mathcal{B}^{\Sigma} \setminus L$. Let $q_0 \in Q$ be a state in which $\mathcal{A}$ ends after reading S. Then $q_0 \in Q \setminus A$ since $\mathcal{A}$ has to reject S because all trees which contain S as subtree have to be rejected. Therefore, all $q_l, q_r \in Q \cup \{\bot\}$ and $a \in \Sigma$ satisfy $\delta(q_l, q_0, a) \subseteq Q \setminus A$ and $\delta(q_0, q_r, a) \subseteq Q \setminus A$ because otherwise one could construct a tree (since every state is reachable) which contains S as subtree and is accepted by $\mathcal{A}$. Hence, there exists a sink $S \subseteq Q \setminus A$ which contains q_0.

3 $\Rightarrow$ 2: Let $V \subseteq Q \setminus A$ be a sink of $\mathcal{A}$. There exists a tree S on which $\mathcal{A}$ ends in a state of V. If $\mathcal{A}$ runs on any tree containing S as a subtree, $\mathcal{A}$ cannot leave the sink and thus cannot leave $Q \setminus A$. This yields $\{T \in \mathcal{B}^{\Sigma} \mid S \preceq T\} \subseteq L$. □

Remark 1. By duality, Theorem 5 also provides a characterisation of regular tree languages L with asymptotic density 1: $\mathbb{P}_{\text{lim}}(L) = 1$ if and only if there exists a tree S such that $S \preceq T \Rightarrow T \in L$, and this is the case if and only if an automaton recognising L has a sink consisting of accepting states.

Unranked Trees. The concept of regular tree language can easily be extended to trees of unbounded degree, also called unranked trees. We get a similar characterisation of sparse tree languages, namely that a regular tree language L is sparse if and only if some tree S does not appear as a subtree of any tree in L

Proving Non-regularity. As a consequence of this characterisation, we obtain a simple linear time algorithm for deciding sparseness (or denseness) of regular tree languages akin to the algorithm in [20]. Note that in contrast to the characterisation given there, here we do not need to require that a sink is a strongly connected component. If there is a sink $V \subseteq Q$ with $V \subseteq Q \setminus A$, there cannot be a sink $V' \subseteq A$.

Corollary 2. *Let $L \subseteq \mathcal{B}^{\Sigma}$ be a regular tree language. There is an algorithm deciding whether L has asymptotic density 0 or 1 in time $O(n)$, where n is the number of states of a given deterministic tree automaton $\mathcal{A}$ recognising L.*

In another direction, Theorem 5 shows a sufficient condition for the non-regularity of tree languages:

Corollary 3. *Let $L \subseteq \mathcal{B}^{\Sigma}$ be a tree language with $\mathbb{P}_{\text{lim}}(L) = 0$. If L does not have a forbidden subtree, then L is not regular.*

Example 1. We call a binary tree $T \in \mathcal{B}^{\Sigma}$ *symmetric* if the left and right subtree obtained from the root are isomorphic. Let L_{sym} be the set of all symmetric binary trees, then for every $a \in \Sigma$ and $S \in \mathcal{B}^{\Sigma}$ we have $\text{conc}_a(S, S) \in L_{\text{sym}}$ and thus L_{sym} does not have a forbidden subtree. However, for the asymptotic density of L_{sym} we get $|L_{\text{sym}} \cap \mathcal{B}^{\Sigma}_{2n}| = 0$ and

$$\lim_{n\to\infty} \frac{|L_{\text{sym}} \cap \mathcal{B}^{\Sigma}_{2n+1}|}{|\mathcal{B}^{\Sigma}_{2n+1}|} = \lim_{n\to\infty} \frac{C_n|\Sigma|^{n+1}}{C_{2n+1}|\Sigma|^{2n+1}} = \lim_{n\to\infty} \frac{C_n}{C_{2n+1}|\Sigma|^{n}} = 0,$$

which yield $\mathbb{P}_{\text{lim}}(L_{\text{sym}}) = 0$. By Corollary 3 we infer that L_{sym} is not a regular tree language.

Binary Search Trees. Another prominent model for random binary trees are *random binary search trees.* In a random binary search tree of size n, the size $|T_l|$ of the left-hand subtree of the root is distributed uniformly on $\{0, \dots, n-1\}$, and conditioned on $|T_l|$ (and therefore $|T_r| = (n-1)-|T_l|$), the subtrees T_l and T_r are independent random binary search trees. These trees appear in the analysis of algorithms such as Quicksort and Find, cf. [8]. Fixing, e.g., the left subtree of the root to be an empty tree one easily obtains a regular tree language that is sparse in this model but does not exclude any subtree.

4 Infinite Words

We first review Sin'ya's result and Koga's simplified proof [15] of it and then see how it can be extended to infinite words.

Definition 2. *For a language $L \subseteq \Sigma^*$, the* infix language $\text{infix}(L)$ *is defined as*

$$\text{infix}(L) := \{w \in \Sigma^* \mid xwy \in L \textit{ for some } x, y \in \Sigma^*\}. \tag{3}$$

The language is said to be infix complete *if* $\text{infix}(L) = \Sigma^*$.

In [20], Sin'ya proved that a regular language L has asymptotic density strictly larger than 0 if and only if it is infix complete. Koga's simplified proof in [15] hinges on the fact that for regular languages, the length of the prefix x and the suffix y in (3) may be bounded uniformly in L, independent of w. We need a slight strengthening of this in that the prefix x may actually be assumed to depend only on L, not on v. We prove this by giving an equivalent condition on DFAs accepting the language L:

Lemma 4. *Let $\mathcal{A} = (\Sigma, Q, q_0, \delta, A)$ be a deterministic finite automaton in which every state is reachable. Then $L(\mathcal{A})$ is infix complete if, and only if, some closed reachability class contains an accepting state. In this case there is a word $x \in \Sigma^*$ and a $k \geq 0$ such that for every $v \in \Sigma^*$ there is a $y \in \Sigma^{\leq k}$ with $xvy \in L$.*

Proof. Let us first assume that some closed reachability class C contains an accepting state $q \in C \cap A$. Since all states are assumed to be reachable, there is a string $x \in \Sigma^*$ with $\hat{\delta}(q_0, x) = q$. Now for every $v \in \Sigma^*$, $q_v := \hat{\delta}(q, v) \in C$, and because q is reachable from every state in C there is a $y_v \in \Sigma^*$ with $\hat{\delta}(q_v, y_v) = q$ and therefore

$$\hat{\delta}(q_0, xvy_v) = \hat{\delta}(q, vy_v) = \hat{\delta}(q_v, y_v) = q \in A,$$

so $xvy \in L(\mathcal{A})$. Since C is finite and the string y depends only on $q_v \in C$ we may assume $|y_v| \leq k$ for some k depending only on $\mathcal{A}$.

For the converse direction, we adversarially construct a string v such that for every $q \in Q$, $\hat{\delta}(q, v)$ is an element of some closed reachability class. Then for all $x, y \in \Sigma^*$, $\hat{\delta}(q, xvy)$ is an element of a closed reachability class, and if no such class contains an accepting state we get $xvy \notin L(\mathcal{A})$, so $L(\mathcal{A})$ is not infix complete.

To construct v we first pick, for every state $q \in Q$, a string $v_q \in \Sigma^*$ such that $\hat{\delta}(q, v_q)$ is an element of some closed reachability class (such a v_q must exist because Q is finite). We enumerate the states as $Q = \{q_0, \ldots, q_r\}$ and set

$$v_0 := \varepsilon \quad \text{and} \quad v_{i+1} := v_i v_{\hat{\delta}(q_i, v_i)} \text{ for } i = 0, \ldots, r.$$

Then v_{r+1} has the desired property: If the automaton starts in some state q_i, then v_i takes it to some state $q = \hat{\delta}(q_i, v_i)$, and from there $v_{\hat{\delta}(q_i,v_i)} = v_q$ takes it to some closed reachability class C. Since this can not be left, also $\hat{\delta}(q_i, v_{r+1}) \in C$. □

Corollary 4. *Given a DFA $\mathcal{A}$, it is decidable whether $L(\mathcal{A})$ is infix complete.*

In fact, for a regular language L the language $\text{infix}(L)$ is again regular, and one can easily compute a DFA for $\text{infix}(L)$ given a DFA for L.

An ω-word $\xi \in \Sigma^\omega$ is called *rich* if for every $v \in \Sigma^*$ there are $x \in \Sigma^*$ and $y \in \Sigma^\omega$ such that $\xi = xvy$ [23]. Our main result on the density of regular ω-languages now reads:

Theorem 6. *Let $L \subseteq \Sigma^\omega$ be a regular ω-language. Then $\mathbb{P}(L) > 0$ if and only if L contains a rich ω-word. Moreover, given a suitable description of L it is decidable whether $\mathbb{P}(L) > 0$ or not.*

Proof. Let $L = \bigcup_i U_i V_i^\omega$ with regular $U_i, V_i \subseteq \Sigma^*$. Obviously L contains a rich ω-word if, and only if, one of the languages $U_i V_i^\omega$ does, which is the case if, and only if, V_i^ω contains such a word. On the other hand, $\mathbb{P}(L) \leq \sum_{i=1}^k \mathbb{P}(U_i V_i^\omega)$, so $\mathbb{P}(L) > 0$ if and only if $\mathbb{P}(U_i V_i^\omega) > 0$ for some i. Therefore without loss of generality we may assume that $L = UV^\omega$ for regular $U, V \subseteq \Sigma^*$.

V^ω contains a rich ω-word if and only if V^* is infix complete: Since Σ^* is countable, if V^* is infix complete we may take a word $w_v = xvy \in V$ for each $v \in \Sigma^*$ and concatenate these words to get a rich ω-word in V^ω. On the other hand, if $\xi \in V^\omega$ is rich, then every $w \in \Sigma^*$ is contained in some finite prefix of ξ, which in turn is a prefix of some word in V^*.

We now show the following: If $U, V \subseteq \Sigma^*$ are regular languages and V^* is infix complete there is a word $x \in \Sigma^*$ such that

$$\mathbb{P}\big(\zeta \in L \mid x \preceq \zeta\big) = 1 \qquad \text{or, equivalently,} \qquad \mathbb{P}\big(\{\zeta \in \Sigma^\omega \mid x \preceq \zeta\} \setminus L\big) = 0$$

Since $(V^*)^\omega = V^\omega$ we may actually assume that V itself is infix complete. Pick a DFA $\mathcal{A} = (Q, q_0, \delta, A)$ with $L(\mathcal{A}) = V$. By Lemma 4 there is a closed reachability class $C \subseteq Q$ that contains an accepting state $\tilde{q} \in C \cap A$. Let $w = a_1 \cdots a_\ell \in \Sigma^*$ be any word such that $\hat{\delta}(q_0, w) \in C$. We define a new automaton $\mathcal{A}' = (\Sigma, Q', q_0, \delta', A)$ by $Q' := Q \cup \{q_1', \ldots, q_\ell'\}$ and

$$\delta'(q,a) = \delta(q,a) \text{ if } q \notin \{\tilde{q}, q'_1, \dots, q'_\ell\},$$

$$\delta'(\tilde{q},a) = \begin{cases} q_1 & \text{if } a = a_1 \\ \delta(\tilde{q},a) & \text{if } a \neq a_1 \end{cases}$$

$$\delta'(q'_i,a) = \begin{cases} q'_{i+1} & \text{if } a = a_{i+1} \text{ and } i < \ell \\ \hat{\delta}(\tilde{q}, a_1 \cdots a_i a) & \text{if } a \neq a_{i+1} \end{cases}$$

$$\delta'(q'_\ell, a) = \delta(\hat{\delta}(q_0, w), a)$$

Then $C \cup \{q'_1, \dots, q'_\ell\}$ is a closed reachability class in $\mathcal{A}'$ and if $\zeta = z_0 z_1 \cdots \in \Sigma^\omega$ is an ω-word such that $w \preceq \zeta$ and $\hat{\delta}'(q_0, z_0 \cdots z_k) = q'_\ell$ for infinitely many $k \geq 0$ (i.e. when reading the ω-word ζ, the automaton $\mathcal{A}'$ passes through the state q'_ℓ infinitely often), then $\zeta \in V^\omega$. In fact, let $k_1 < k_2 < \cdots$ be chosen such that $\hat{\delta}'(q_0, z_0 \cdots z_{k_i}) = q'_\ell$ for every $i \geq 1$. Then each of the subwords

$$z_0 \cdots z_{k_1 - \ell}, \quad z_{k_1-\ell+1} \cdots z_{k_2-\ell}, \quad z_{k_2-\ell+1} \cdots z_{k_3-\ell}, \quad \cdots$$

is in V, because it starts with the prefix w and then takes the automaton $\mathcal{A}$ to the (accepting) state $\tilde{q}$.

Now, if $\zeta = \zeta_1 \zeta_2 \zeta_3 \cdots$ is a random ω-word, the random variables $(X_t)_{t \geq 0}$ with $X_t = \hat{\delta}'(q_0, \zeta_1 \cdots \zeta_t)$ are a Markov chain with state space Q', started in q_0 and with transition probabilities $p_{q,q'} = |\Sigma|^{-1}$ if $\delta(q,a) = q'$ for some $a \in \Sigma$, and $p_{q,q'} = 0$ otherwise. In particular, the reachability relation $\to$ on the state space of this Markov chain is the same as the reachability relation $\rightsquigarrow$ of the automaton $\mathcal{A}'$, and the closed classes of the Markov chain are exactly the closed reachability classes of $\mathcal{A}'$. Thus q'_ℓ is recurrent by Theorem 2, and with Cor. 1 we get

$$\mathbb{P}(X_t = q'_\ell \text{ infinitely often} \mid w \preceq \zeta) = \mathbb{P}(X_t \in C \text{ for some } t \geq 0 \mid w \preceq \zeta) = 1,$$

because if $w \preceq \zeta$ then $X_\ell = \hat{\delta}(q_0, w) = \tilde{q} \in C$. Finally, if $u \in U$ is any word in U then $\mathbb{P}(\zeta \in L \mid uw \preceq \zeta) = 1$. □

5 Conclusion

We gave decidable characterisations for sparseness of regular tree languages and of regular ω-languages, both in terms of excluded subtrees and in terms of automata accepting these languages.

By [18], sparseness is decidable also for regular languages of *infinite* trees, allowing for probabilistic model checking not just for linear temporal logic, but also for computation tree logic (CTL), opening the route to many further applications in model checking (cf. [3, Ch. 6]). However, Niwinski et al. solve the more general problem of exactly computing the measure of a regular infinite tree language, at prohibitively high computational cost. Focussing on sparseness is likely to allow for more efficient algorithms, as exemplified by the linear time algorithm from Corollary 2.

Several known graph properties, such as being series-parallel or, more generally, having bounded tree-width, imply that graphs with these properties can be encoded in labelled trees in such a way that the original graph can be MSO interpreted in the tree (cf. [12, Ch. 11.4]). Our results on sparseness of regular tree languages can therefore be translated into sparseness conditions for MSO-definable properties of graphs in these graph classes. However, since a given graph may in general be interpretable in many different trees, this yields sparseness with respect to some non-uniform notion of density, prompting for a closer investigation.

Another interesting direction for future work would be more general probabilistic models. While our methods (as well as those in [6] and [18]) easily generalise from independent letters to Markov chains, it is not immediately clear if this can be done also for strings or trees generated by hidden Markov models (HMMs, cf. [5]) because in this case, reachability classes of the automaton are no longer the same as communicating classes of the resulting Markov chain.

References

1. Asada, K., Kobayashi, N., Sin'ya, R., Tsukada, T.: Almost every simply typed lambda-term has a long beta-reduction sequence. Log. Methods Comput. Sci. **15**(1) (2019). https://doi.org/10.23638/LMCS-15(1:16)2019
2. Atserias, A., Kreutzer, S., Noy Serrano, M.: On zero-one and convergence laws for graphs embeddable on a fixed surface. In: 45th International Colloquium on Automata, Languages, and Programming (ICALP 2018): July 9-13, 2018, Prague, Czech Republic, pp. 1–14. Schloss Dagstuhl-Leibniz-Zentrum für Informatik (2018)
3. Baier, C., Katoen, J., Larsen, K.: Principles of Model Checking. MIT Press (2008)
4. Benedikt, M., Segoufin, L.: Towards a characterization of order-invariant queries over tame graphs. J. Symb. Log. **74**(1), 168–186 (2009)
5. Byung-Jun, Y.: Hidden markov models and their applications in biological sequence analysis. Curr. Genom. **10**(6), 402–415 (2009)
6. Courcoubetis, C., Yannakakis, M.: The complexity of probabilistic verification. J. ACM (JACM) **42**(4), 857–907 (1995)
7. Dawar, A., Kopczynski, E.: Logical properties of random graphs from small addable classes. CoRR abs/1707.02081 (2017). http://arxiv.org/abs/1707.02081
8. Devroye, L.: A note on the height of binary search trees. J. ACM (JACM) **33**(3), 489–498 (1986)
9. Ebbinghaus, H.D., Flum, J.: Finite Model Theory, 2nd edn. Perspectives in Mathematical Logic, Springer (1999)
10. Fagin, R.: Probabilities on finite models. J. Symbolic Logic **41**(1), 50–58 (1976)
11. Flajolet, P., Sedgewick, R.: Analytic Combinatorics. Cambridge University Press (2009). https://doi.org/10.1017/CBO9780511801655
12. Flum, J., Grohe, M.: Parameterized Complexity Theory. Texts in Theoretical Computer Science. An EATCS Series, Springer (2006)
13. Glebskij, Y., Kogan, D., Liogon'kij, M., Talanov, V.: Range and degree of realizability of formulas in the restricted predicate calculus. Cybernetics **5**, 142–154 (1969)
14. Knuth, D.E.: The Art of Computer Programming: Fundamental Algorithms, vol. I. Addison-Wesley, 3rd edn. (1997)

15. Koga, T.: On the density of regular languages. Fundam. Informaticae **168**(1), 45–49 (2019). https://doi.org/10.3233/FI-2019-1823
16. Malyshkin, Y., Zhukovskii, M.: MSO 0–1 law for recursive random trees. Stat. Probab. Lett. **173**, 109061 (2021). https://doi.org/10.1016/j.spl.2021.109061, https://www.sciencedirect.com/science/article/pii/S0167715221000237
17. McColm, G.L.: MSO zero-one laws on random labelled acyclic graphs. Discret. Math. **254**(1), 331–347 (2002). https://doi.org/10.1016/S0012-365X(01)00375-2, https://www.sciencedirect.com/science/article/pii/S0012365X01003752
18. Niwiński, D., Parys, P., Skrzypczak, M.: The probabilistic Rabin tree theorem*. In: 2023 38th Annual ACM/IEEE Symposium on Logic in Computer Science (LICS), pp. 1–13 (2023)
19. Norris, J.R.: Markov Chains. Cambridge University Press (1997)
20. Sin'ya, R.: An automata theoretic approach to the zero-one law for regular languages: Algorithmic and logical aspects. In: Esparza, J., Tronci, E. (eds.) Proceedings of the Sixth International Symposium on Games, Automata, Logics and Formal Verification, GandALF 2015, Genoa, Italy, 21-22nd September 2015. EPTCS, vol. 193, pp. 172–185 (2015). https://doi.org/10.4204/EPTCS.193.13
21. Spencer, J.: The strange logic of random graphs, vol. 22. Springer Science & Business Media (2013)
22. Staiger, L.: ω-languages. In: Handbook of Formal Languages: Volume 3 Beyond Words, pp. 339–388. Springer Berlin Heidelberg, Berlin, Heidelberg (1997)
23. Staiger, L.: Rich ω-words and monadic second-order arithmetic. In: Nielsen, M., Thomas, W. (eds.) Computer Science Logic, pp. 478–490. Springer, Berlin Heidelberg, Berlin, Heidelberg (1998)
24. Thomas, W.: Languages, automata, and logic. In: Handbook of Formal Languages: Volume 3 Beyond Words, pp. 389–455. Springer Berlin Heidelberg, Berlin, Heidelberg (1997)
25. Williams, D.: Probability with Martingales. Cambridge University Press (1991)

Subregular Expressions with Two Operations

Martin Kutrib[1(✉)], Priscilla Raucci[1,2], and Matthias Wendlandt[1]

[1] Institut für Informatik, Universität Giessen, Arndtstr. 2, 35392 Giessen, Germany
{kutrib,matthias.wendlandt}@informatik.uni-giessen.de,
priscilla.raucci@uni-giessen.de

[2] Dipartimento di Informatica "G. Degli Antoni", Università degli Studi di Milano, via Celoria 18, 20133 Milano, Italy

Abstract. Subregular expressions are closely related to regular expressions, but instead of relying solely on the classical operations of union, concatenation, and Kleene star, they explore subsets of operations under which regular languages remain closed. Previous work has examined various combinations of these operations, including complement and intersection, alongside the standard trio. In this work, we investigate several unexplored combinations of operations, focusing specifically on three language families, namely, subregular expressions with complement and Kleene star, subregular expressions with concatenation and complement, and subregular expressions with intersection and Kleene star. We examine basic properties of these families of languages and analyze how these families relate to each other. Further, we distinguish between the unary and general alphabet cases, and explore the possibility of a hierarchical structure based on the number of operations used.

1 Introduction

It is well known that regular languages can be described using regular expressions, typically defined through the fundamental operations of union, concatenation, and the Kleene star. In addition to these, regular languages are closed under several other operations, such as intersection and complement, meaning that these operations can be applied without extending the expressive power of the expressions [10].

While union, concatenation, and Kleene star are commonly taken as the basic operations for defining regular languages, one may naturally ask what happens if we choose a different set of basic operations? Do these alternative sets characterize subsets of the regular languages? How do these families relate to each other?

One significant example of such a variation is the class of star-free languages [6]. This language family is defined using union, concatenation, and complement, excluding the Kleene star. It is relevant that at least one operation (either Kleene star or complement) must be present to allow the construction of infinite languages. Star-free languages, while still infinite in scope,

S.-K. Ko and F. Manea (Eds.): DLT 2025, LNCS 16036, pp. 229–243, 2026.
https://doi.org/10.1007/978-3-032-01475-7_16

form a proper subset of the regular languages. For instance, languages like $(aa)^*$ cannot be expressed within this framework. The star-free languages have elegant characterizations, for instance, through aperiodic syntactic monoids [21], permutation-free deterministic finite automata [15], and loop-free alternating finite automata [20]. Building on this idea, regular expressions where the classical operations are removed or substituted by intersection and complement were also explored, analyzing their expressive power and structural properties.

In general, the study of relations between several subregular language families originates in [8]. The subfamilies are well motivated by their representations as finite automata or regular expressions. Just to mention a few of them: finite languages (are accepted by acyclic finite automata), definite languages [2,17] (can be realized by a register and a combinational circuit), star-free languages or regular non-counting languages [6,15] (which can be described by regular(-like) expressions using only union, concatenation, and complement and which have nice characterizations in terms of aperiodic monoids and permutation-free DFA [15]), combinational languages (are accepted by automata modeling combinational circuits), locally testable languages (where the set of factors of a given length obtained from a word uniquely determines whether or not the word belongs to the language), star languages [3], and (two-sided) comet languages [4].

In analogy to star-free expressions, the broader study of subregular expressions involves trading one of the regular operations for another, leading to a diverse landscape of language families. For example, concatenation-free languages equipped with the operations of union, Kleene star, and complementation have been studied extensively [13]. In particular, a characterization of unary concatenation-free languages by the Boolean closure of certain sets of languages has been derived. The characterization is then used to show that concatenation-free languages form a proper subset of the regular languages. Furthermore, the position of the family of concatenation-free languages in the subregular hierarchy is considered It turns out that there are concatenation-free languages that do not belong to any other of the subregular families.

Additionally, subregular expressions are studied in which either one of the regular operations is omitted or replaced by complementation or intersection [14]. In particular, a finite strict subregular hierarchy is shown that reveals that the operation complementation is generally stronger than intersection.

In [16] union-free expressions have been considered, where the union operation is omitted without replacement, but where the literals are only letters from the alphabet. As a consequence, all languages described by expressions as studied in [16] are either infinite or contain a single word only.

Here, as in previous works, to ensure that non-trivial languages can be represented, we assume that any finite set of words can be used as literals in the base set of expressions.

By exploring various sets of allowable operations, many significant results regarding the complexity of decision problems have been uncovered, for example, in [7,11,18,19,23,24]. A survey on the complexity of regular(-like) expressions can be found in [9].

In the following, we investigate several unexplored combinations of operations, where only two operations are used: subregular expressions with complement and Kleene star, subregular expressions with concatenation and complement, and subregular expressions with intersection and Kleene star. We examine how these families relate to each other, highlighting both their expressive power and limitations. Additionally, we distinguish between the unary and general alphabet cases, and explore whether a hierarchy emerges based on the number and type of operations employed. The structure of the paper is as follows: Sect. 3 focuses on regular expressions with complement and Kleene star, Sect. 4 examines those with concatenation and complement, Sect. 5 addresses the case of intersection and Kleene star, and the final section provides a comparative analysis of all the language classes discussed.

2 Definitions and Preliminaries

We write Σ^* for the set of all words over the finite alphabet Σ, we denote the *empty word* as λ and let $\Sigma^+ = \Sigma^* \backslash \{\lambda\}$. Given a word $w \in \Sigma^*$, we say that $|w|$ is its *length*. We use $\subseteq$ for *inclusions* and $\subset$ for *strict inclusions*. The *complement* of a language L over alphabet Σ is again a language over alphabet Σ which is denoted by $\overline{L}$. The *family of finite languages* is denoted by FIN.

The *regular expressions* over an alphabet Σ and the languages they describe are defined inductively in the usual way: $\emptyset$ and every word (of length one) $v \in \Sigma$ are regular expressions, and when s and t are regular expressions, then $(s \cup t)$, $(s \cdot t)$, and (s^*) are also regular expressions. The language $L(r)$ defined by a regular expression r is defined as follows: $L(\emptyset) = \emptyset$, $L(v) = \{v\}$, $L(s \cup t) = L(s) \cup L(t)$, $L(s \cdot t) = L(s) \cdot L(t)$, and $L(s^*) = L(s)^*$.

Since the regular languages are closed under many more operations, the approach to add operations like intersection ($\cap$) and complementation ($\overline{}$) does not increase their expressive power. However, replacing operations by others may decrease the expressive power. So, in general, $\mathrm{RE}(\Sigma, \Lambda, \Phi)$, where $\Lambda \subset \Sigma^*$ is a finite set of initial words, and Φ is a set of (regularity preserving) operations, denotes all regular(-like) expressions over Λ using only operations from Φ. The family of languages described by the regular expressions $\mathrm{RE}(\Sigma, \Lambda, \Phi)$, for some Σ and finite $\Phi \subset \Sigma^*$, is denoted $\mathscr{L}_\Phi$. Hence $\mathrm{RE}(\Sigma, \Sigma, \{\cup, \cdot, *\})$ refers to the set of all ordinary regular expressions, and $\mathrm{RE}(\Sigma, \Sigma, \{\cup, \cdot, \overline{}\})$ refers to the star-free languages.

In [12–14] different variants of subregular expressions have been investigated. Here we study the expressive power of some further variants where only two operations from $\{\cap, \overline{}, \cdot, *\}$ are available. Also we briefly discuss expressions with Boolean operations. Clearly, $\mathrm{RE}(\Sigma, \Lambda, \{\cup, \cap, \overline{}\})$, $\mathrm{RE}(\Sigma, \Lambda, \{\cup, \overline{}\})$, and $\mathrm{RE}(\Sigma, \Lambda, \{\cap, \overline{}\})$ describe the same family of languages $\mathscr{L}_{\cup,\cap,-}$.

The family of languages $\mathscr{L}_{*,-}$ described by expressions $\mathrm{RE}(\Sigma, \Lambda, \{*, \overline{}\})$ seems to be special since both operations are unary. The superset of expressions with concatenation, union and complement form the star-free languages, but the expressions $\mathrm{RE}(\Sigma, \Lambda, \{\cdot, \overline{}\})$ with solely concatenation and complement forming the family of languages $\mathscr{L}_{\cdot,-}$ have not been investigated as expressions $\text{RE}(\Sigma, \Lambda, \{\cap, *\})$ with solely intersection and star forming the family of languages $\mathscr{L}_{\cap,*}$.

Due to the absence of concatenation in some classes, here, as in previous works [12–14] we provide initially a *finite set* of words. In the absence of union, applying the Kleene star to a set nevertheless means to apply it to the set and not only to its elements.

For convenience, parentheses in regular expressions are sometimes omitted, where it is understood that the unary operations complementation and star have a higher priority than binary operations.

Before we start with the investigation of the classes discussed, we shortly consider subregular expressions with the operations of union, intersection and complement. Given the following properties:

- The complement of a cofinite set is finite.
- The union of two finite sets is finite.
- The union of a finite set and a cofinite set is cofinite.
- The union of two cofinite sets is cofinite.
- The intersection of two finite sets is finite.
- The intersection of a finite set and a cofinite set is finite.
- The intersection of two cofinite sets is cofinite.

It follows that every language that can be described using only the operations of union, intersection, and complement, starting from finite sets, is either finite or cofinite.

Theorem 1. *Every language $L \in \mathscr{L}_{\cup,\cap,-}$ is either a finite or a cofinite language.*

3 Regular Expressions with Complement and Kleene Star

In this section we focus on the expressivity of regular expression formulated only using complement and Kleene star, that is $\text{RE}(\Sigma, \Lambda, \{\overline{}, *\})$, which express the class of languages $\mathscr{L}_{*,-}$. Since the allowed operation are both unary, there is no way of describing a binary tree of operations. The only meaningful method of composing languages in this context is to alternately apply the complement and Kleene star operations to the initial finite language in a continuous sequence. The sole variation arises from the choice of which operation is applied first. In the case of subregular expressions, starting from a set of words and using the operations of star and complement, it only makes sense to apply these two operations alternately. Thus, a basic question is whether alternately applying these operations leads to an infinite hierarchy of language classes or whether the hierarchy eventually collapses.

If a finite set of words $F \subseteq \Sigma^*$ that is first complemented, yielding the cofinite language $\overline{F} = \Sigma^* \setminus F$ and then the Kleene star is applied, the resulting language $\overline{F}^*$ is again cofinite. Taking the complement of this language yields the finite language $\overline{\overline{F}^*} \subseteq \Sigma^*$. Thus, when considering multiple alternations of applying the Kleene star and the complement, it suffices to consider only the case starting with the star operation. Starting with the star captures the essential behavior of these alternations without loss of generality.

Let L_0 be a finite set of words. Then we write $L_1 = L_0^*$, $L_2 = \overline{L_0^*}$, $L_3 = \overline{L_0^*}^*$, and so on.

We start with results about unary $\mathscr{L}_{*,-}$.

Example 1. Let us take the finite language $L_0 = \{a^2\}$. Then we build the language L by repeatedly applying Kleene star and complement. We first obtain as result of the operation $L_2 = \overline{L_0^*}$ which is the set of unary strings with odd length, containing also the word a. Therefore when we star L_2 we obtain as result Σ^*. After this first phase, by the use of complement and star, we will always obtain the repetition:

$$L_3 = \Sigma^*, \quad L_4 = \overline{\Sigma^*} = \emptyset, \quad L_5 = \emptyset^* = \{\lambda\}, \quad L_6 = \overline{\lambda} = \Sigma^+ \quad L_7 = \Sigma^*.$$

■

We obtain the same result for any finite starting set $L_0 \subset \{a\}^*$, since either the language itself or its complement contains the word consisting of a single a. In particular, when $a \in L_0$, the collapse of the hierarchy begins with the first application of the Kleene star. Starting at the latest from L_3, the alternation between the languages Σ^*, $\emptyset$, $\{\lambda\}$, and Σ^+ begins.

Proposition 1. *Given a set of words $L_0 \subseteq \{a\}^*$, then, for all $i \geq 0$, if $a \in L_0$ language $L_{4\cdot i+1} = \Sigma^*$, $L_{4\cdot i+2} = \emptyset$, $L_{4\cdot i+3} = \{\lambda\}$, and $L_{4\cdot i+4} = \Sigma^+, i \geq 0$. If $a \notin L_0$, then, for all $i \geq 0$, language $L_{4\cdot i+3} = \Sigma^*$, $L_{4\cdot i+4} = \emptyset$, $L_{4\cdot i+5} = \{\lambda\}$, and $L_{4\cdot i+6} = \Sigma^+$.*

As argued above, when starting the generation with complementation, we obtain the same result. However, the number of preliminary steps may be smaller. Use as example the starting language $L_0 = \{aa\}$, from which we obtain $L_2 = \overline{L_0}^* = \Sigma^*$, and the second starting language $L_0 = \{a\}$ from which we get $L_4 = \Sigma^*$.

Theorem 2. *In the case of unary languages, all the languages in $\mathscr{L}_{*,-}$ can be expressed using at most three complements and three stars, starting with a finite language L_0.*

We turn to the general case.

The dot-depth hierarchy, introduced by Brzozowski and Cohen [6], classifies star-free regular languages based on the complexity of combining Boolean operations (union, intersection, complement) and concatenation (represented by a dot). Each level in the hierarchy corresponds to languages that require a specific number of alternations between Boolean operations and concatenations to be described. It has been proven that the hierarchy is infinite and strict [5], thus, each higher level contains languages that cannot be expressed at any lower level.

For subregular expressions, when starting with a set of words and using the star and complement operations, it is only logical to apply these two operations in alternation. A fundamental question according to the dot depth hierarchy is whether alternating these operations can continuously change the language at each step, or if the process eventually stabilizes. In the sequel, we show that the process collapses in such a way that, from a certain point onward, the languages repeat periodically.

Theorem 3. *Given a finite set of words L_0, then language L_i is equal to L_{i+4}, for $i \geq 3$.*

Proof. In the following, we consider a subregular expression from the class $\mathrm{RE}(\Sigma, \Lambda, \{^{\overline{}}, *\})$. Thus we have language $L_0 = \Lambda$.

We first show some basic relationships for all $i \geq 0$:

$L_{2 \cdot i} \subseteq L_{2 \cdot i+1}$: Since $L_{2 \cdot i+1}$ is the starred language $L_{2 \cdot i}$, the inclusion follows.

$L_5 \subseteq L_1$: L_2 is the complement of L_1 and L_2 is a subset of L_3. Moreover, L_4 is the complement of L_3, thus, we have $L_4 \subseteq L_1$. Since L_1 is Kleene star closed, we conclude that for each subset of L_1 the star of the subset is a subset of L_1. So, the star of L_4 is a subset of L_1. This implies $L_5 \subseteq L_1$.

$L_2 \subseteq L_6$: Since the complement of L_2 which is L_1, is a superset of L_5 and L_5 is the complement of L_6, we have $L_2 \subseteq L_6$.

$L_6 \subseteq L_3$: If a word w does not belong to L_3 then it must belong to L_4. So, it belongs to L_5 as well. Hence w cannot belong to L_6. So, we have $w \notin L_3$ implies $w \notin L_6$ which is equivalent to $w \in L_6$ implies $w \in L_3$.

Now, let $w \in \Sigma^*$ be an arbitrary word. The last relationship from above shows that $w \in L_6$ implies $w \in L_3$. Furthermore, since L_3 is Kleene star closed, we conclude that $w \in L_7$ implies $w \in L_3$. This shows $L_7 \subseteq L_3$.

Conversely, a relationship from above shows $L_2 \subseteq L_6$. So, we have $L_2^* \subseteq L_6^*$ and hence, $L_3 \subseteq L_7$.

This shows $L_3 = L_7$ and We conclude L_i is equal to L_{i+4}, for $i \geq 3$. ■

For languages over an arbitrary alphabet we have the same result as for unary alphabets, that is, three complements and three stars are enough.

Corollary 1. *All the languages in $\mathscr{L}_{*,-}$ can be expressed using at most three complements and three stars, starting with a finite language L_0.*

Example 2. Let us take the finite language $L_0 = \{aa, bb, c\}$. We first build the star $L_1 = L_0^*$ which is the set of words over the alphabet $\Sigma = \{a, b, c\}$ in which the numbers of consecutive a's and the numbers of consecutive b's are even.

Therefore, the complement $L_2 = \overline{L_0^*}$ contains all words over Σ in which there is at least one a-block or b-block whose length is odd. Note that no word in L_2 consists solely of c's, since such words (including the empty word) are part of L_1.

Now consider the Kleene star of the complement $L_3 = \overline{L_0^*}^*$. This set consists of all words over Σ, except for the words c^+. This is because any word in Σ^* that contains at least one block with an even number of a's or b's can be expressed as a concatenation of segments with an odd number of a's or b'.

Thus, we conclude

$$\begin{aligned} L_4 &= \{c^n \mid n \geq 1\}, \\ L_5 &= \{c^n \mid n \geq 0\}, \\ L_6 &= \Sigma^* \setminus \{c^n \mid n \geq 0\}, \\ L_7 &= L_3. \end{aligned}$$

■

4 Regular Expressions with Concatenation and Complement

This section analyzes the class of languages defined by regular expressions using only the operations of concatenation and complement $\mathscr{L}_{.,-}$. This family of languages is undoubtedly closely related to the star-free languages, with the key difference being the exclusion of the union operation.

We begin by examining this class in the unary case and obtain the following characterization.

Proposition 2. *Every unary language in $\mathscr{L}_{.,-}$ is either finite or cofinite.*

Proof. We use an inductive argument.

As basis, it is easy to see that we have only finite sets (which are given by Λ) and cofinite sets (obtained by the complement of finite sets) as possible generated languages.

Now, we assume that the languages generated are either finite or cofinite.

For the inductive step, we observe that any further application of the two possible operations, that is complementation and concatenation, generate again a finite or cofinite language.

This is trivial for the complementation operation, which generate finite language from cofinite and cofinite from finite ones.

For what regards the concatenation we observe that:

- the concatenation of two finite languages is a finite language,
- the concatenation of two unary cofinite languages is a cofinite language,
- the concatenation of a finite language and a cofinite language in the case of unary language is a cofinite language,
- the concatenation of a cofinite language and a finite language in the case of unary language is again a cofinite language.

Since every expression starts with a finite set of words, it can be concluded that every unary language in $\mathscr{L}_{.,-}$ is either finite or cofinite. ■

We now turn to languages over general alphabets and consider the following language:

$$L = \{a\, \Sigma^* a\} \cup \{b\, \Sigma^* b\}$$

over the alphabet $\Sigma = \{a, b\}$. We can observe that L is a star-free language since it can be generated by the expression $\{a \cdot \overline{\emptyset} \cdot a\} \cup \{b \cdot \overline{\emptyset} \cdot b\}$. However it is not in the family of languages $\mathscr{L}_{\cdot,-}$.

Lemma 1. *The language L does not belong to $\mathscr{L}_{\cdot,-}$.*

Proof. In contrast to the assertion, we assume that L can be defined by some expression in $\mathrm{RE}(\Sigma, \Lambda, \{\cdot, \overline{}\})$ and let us consider its expression tree. Clearly, applying the complementation operator twice needs not to be considered.

First, we consider the case where the root node is labeled with concatenation, which corresponds to attaching two languages L_1 and L_2. Consider some prefix u with $uv \in \{a\, \Sigma^* a\}$, where $u \in \Sigma^+$ and $v \in \Sigma^*$, such that $u \in L_1$. Furthermore, consider a suffix v' with $u'v' \in \{b\, \Sigma^* b\}$, where $u' \in \Sigma^*$ and $v' \in \Sigma^+$, such that $v' \in L_2$. Then, the word uv' belongs to the concatenation $L_1 L_2$ but begins with an a and ends with a b, a contradiction. Similarly, a contradiction is obtained if the concatenation is $L_2 L_1$. Thus, all words of L must lie entirely in either L_1 or L_2, and the concatenation does not simplify the expression.

So, we next consider the case where the root node is labeled with complementation. Then the complementation is applied to some language L_3 which contains all words starting with an a and ending with a b, all words starting with a b and ending with an a, as well as the empty word λ. It does not contain any word starting and ending with an a or starting and ending with a b. Moreover, L_3 is infinite, which means that L_3 is obtained by concatenating two languages L_4 and L_5, since nested complementation cannot occur at all. So, we can apply a similar argument as above. Consider a prefix u with $uv \in \{a\, \Sigma^* b\}$, where $u \in \Sigma^+$ and $v \in \Sigma^*$, such that $u \in L_4$, and a suffix v' with $u'v' \in \{b\, \Sigma^* a\}$, where $u' \in \Sigma^*$ and $v' \in \Sigma^+$, such that $v' \in L_5$. Then, uv' belongs to the concatenation $L_4 L_5 = L_3$ but begins with an a and ends with an a, a contradiction. Similarly, a contradiction is obtained if the concatenation is $L_5 L_4$.

Since there is no further option, it can be concluded that the language L cannot be generated in $\mathscr{L}_{\cdot,-}$. ■

From Lemma 1 and the observation that $\mathscr{L}_{\cdot,-}$ is a subset of the set of star-free language, we obtain that the inclusion of the two classes is proper:

Theorem 4. *The family of star-free languages is a proper superset of the family of languages $\mathscr{L}_{\cdot,-}$.*

From this result we can easily obtain that infinitely many languages $L_w = \{\, w^* \mid |w| > 1 \,\}$ are not in $\mathrm{RE}(\Sigma, \Lambda, \{\cdot, \overline{}\})$, since they are not a star-free languages.

5 Regular Expressions with Intersection and Kleene Star

Here we turn to the combination of the two operations intersection and star. In order to explore the structure of the languages that are described by expressions in $\mathrm{RE}(\Sigma, \Lambda, \{\cap, *\})$, we consider their expression trees. The leaves of such trees are labeled with finite subsets of words from Λ, thus by languages from FIN. Assume now that the predecessor node of a leave is labeled by the intersection. Then the result of applying the intersection is again a finite language. So, we can replace the (sub-)tree rooted in the node labeled by the intersection by a leave attached with this finite language. Repeating this process prunes the tree either to a single node to which a finite language is attached or to a tree where each predecessor of a leave is labeled by star (cf. Fig. 1).

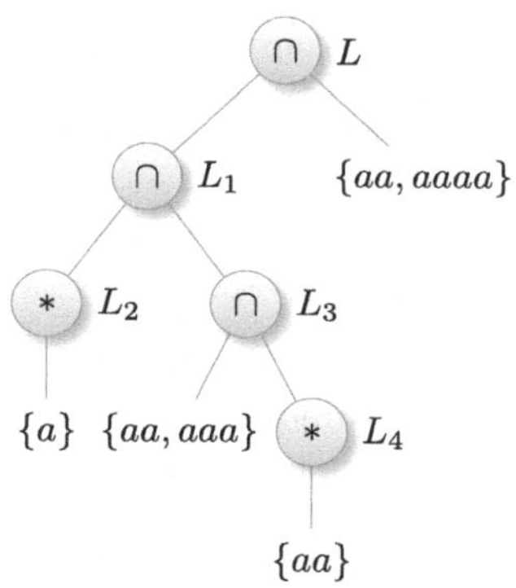

Fig. 1. Expression tree for the expression $r = (\{a\}^* \cap (\{aa, aaa\} \cap \{aa\}^*)) \cap \{aa, aaaa\}$. Since there is a path from the leaf $\{aa, aaa\}$ to the root on which no star occurs, the represented language is finite, it is $L = \{aa\}$.

So, for expressions describing infinite languages, we assume that any predecessor of a leave is labeled by star. For the further exploration of the structure of the languages described by expressions in $\mathrm{RE}(\Sigma, \Lambda, \{\cap, *\})$, we show the next proposition.

Proposition 3. *Let L_1 and L_2 be finite languages over a common alphabet Σ. Then, $(L_1^* \cap L_2^*)^* = (L_1^* \cap L_2^*)$.*

Proof. Let $w \in (L_1^* \cap L_2^*)^*$. Then the word w can be written as concatenation $w = v_1 v_2 \cdots v_k$, where $v_i \in L_1^* \cap L_2^*$ for $1 \leq i \leq k$. Moreover, any v_i can be written as concatenation $v_i = x_1 x_2 \cdots x_\ell$, where $x_i \in L_1$ for $1 \leq i \leq \ell$, and as concatenation $v_i = y_1 y_2 \cdots y_m$, where $y_i \in L_2$ for $1 \leq i \leq m$, as well. This implies $w \in L_1^*$ and $w \in L_2^*$. Hence, we have $w \in L_1^* \cap L_2^*$.

Finally, the trivial implication $w \in L_1^* \cap L_2^*$ implies $w \in (L_1^* \cap L_2^*)^*$ shows the proposition. ∎

By Proposition 3 and the commutativity and associativity of intersection, we now conclude that any language in $\mathscr{L}_{\cap,*}$ is either finite or $L(r)$ for an expression $r \in \mathrm{RE}(\Sigma, \Lambda, \{\cap, *\})$ that is of the form $\mathrm{FIN}^* \cap \mathrm{FIN}^* \cap \cdots \cap \mathrm{FIN}^*$. As immediate corollary we obtain some languages not belonging to $\mathscr{L}_{\cap,*}$. A language is letter-bounded if it is a subset of $a_1^* a_2^* \cdots a_m^*$, where $a_1, a_2, \ldots, a_m$ are pairwise distinct symbols.

Corollary 2. *1. Any language in $\mathscr{L}_{\cap,*}$ not containing the empty word is finite.*
2. No infinite non-unary letter-bounded language belongs to $\mathscr{L}_{\cap,*}$.
3. The language $a^* \cup b^*$ does not belong to $\mathscr{L}_{\cap,*}$.

More general, we have the following lemma.

Lemma 2. *Let L be an infinite language in $\mathscr{L}_{\cap,*}$. Then $w \in L$ implies $w^+ \subseteq L$.*

Proof. By Proposition 3, we may assume that L has a representation as $L_1^* \cap L_2^* \cap \cdots \cap L_k^*$. So, if w belongs to L, it belongs to all L_i^*, for $1 \leq i \leq k$. Therefore, for all $j \geq 1$, also w^j belongs to all L_i^*. This implies $w^j \subseteq L$. ■

On the other hand, there are interesting and non-trivial languages in the family $\mathscr{L}_{\cap,*}$.

Example 3. Let L_b be the language $\#_1(ab)^+\#_2$. This means the words in L_b are repetitions of the factor ab with a unique left endmarker $\#_1$ and a unique right endmarker $\#_2$. By Lemma 2, L_b cannot belong to $\mathscr{L}_{\cap,*}$. However, L_b^* does belong to $\mathscr{L}_{\cap,*}$.

More precisely, L_b^* has the representation $L_1^* \cap L_2^*$, where $L_1 = \{\#_1 a, ba, b\#_2\}$ and $L_2 = \{\#_1, ab, \#_2\}$.

The only possibility for a prefix of a non-empty word $w \in L_1^* \cap L_2^*$ is $\#_1 a$. By the words in L_2, now a b must follow. Then there may be further a's followed by b's. Eventually, the symbol $\#_2$ must appear. So far, $w \in L_1^* \cap L_2^*$ if and only if $w \in L_b$. But now w can be extended to the right by another word from L_b and so on. ■

Having the representation of Proposition 3, as before the natural question arises whether there is a hierarchy of language classes depending on the number of intersections. For the general case we have at least two levels. Whether the hierarchy collapses at level two is currently open. But for unary languages the hierarchy collapses immediately.

Let us first show that one intersection is needed for certain languages in $\mathscr{L}_{\cap,*}$. Here, we use $L_d = L_1^* \cap L_2^*$ with $L_1 = \{ab, a\}$ and $L_2 = \{ab, aa, ba, bb\}$ as witness.

Proposition 4. *The language L_d does not have a representation of the form L^*, where L is finite.*

Proof. Essentially, the non-empty words in L_1^* are the words over $\{a, b\}$ beginning with an a that do not have the factor bb. The non-empty words in L_2^* are the words over $\{a, b\}$ that have an even length.

Assume in contrast to the assertion that there is a finite language L such that $L_d = L^*$. Clearly, L may contain only words of even lengths. Moreover, all words in L have to begin with the letter a. Since both aa and ab belong to L_d they belong to L as well. Next, we consider the word $w_1 = aaba \in L_d$. The word ba does not belong to L. Therefore, the word w_1 must belong to L. We continue inductively and consider the words $w_i = aa(ba)^i$, for all $i \geq 1$. Let $w_i \in L$. Again, the word ba does not belong to L. So, w_{i+1} has to belong to L. This implies that L cannot be finite, a contradiction. ∎

Now we turn to the unary languages in the family $\mathscr{L}_{\cap,*}$. As mentioned before, in this case the hierarchy on the number of intersections collapses immediately. To show this result we recall a well-known useful fact which is related to number theory and Frobenius numbers. Let $t, d \geq 0$ be two integers. An *arithmetic progression with offset t and period d* is the set $\{\, t + x \cdot d \mid x \geq 0 \,\}$. See [22] for a survey on Frobenius numbers. The next lemma follows immediately from a result in [1].

Lemma 3. *Let $0 < c_1 < c_2 < \cdots < c_r \leq n$ be positive integers. Then the set of integers $z > n^2$, that can be written as a non-negative integer linear combination of the c_i, $1 \leq i \leq r$, is the arithmetic progression with offset t and period d $\{\, t + x \cdot d \mid x \geq 0 \,\}$, where t is the least integer greater than n^2 that is a multiple of $d = \gcd(c_1, c_2, \ldots, c_r)$.*

Example 4. 1. The language $\{aa, aaa\}^* \in \mathscr{L}_{\cap,*}$ consists of the empty word and all unary words with length $n \geq 2$. We have $c_1 = 2$, $c_2 = 3$, and $d = 1$. The arithmetic progression starts already before $n^2 = 9$.
2. The language $\{aaa, aaaa, aaaaa\}^* \in \mathscr{L}_{\cap,*}$ consists of the empty word and all unary words with length $n \geq 3$. We have $c_1 = 3$, $c_2 = 4$, $c_3 = 5$, and $d = 1$. Again, the arithmetic progression starts already before $n^2 = 25$. ∎

Theorem 5. *Let L_1 and L_2 be finite unary languages. Then either $L_1^* \cap L_2^*$ is finite or there is a finite unary language L such that $L^* = L_1^* \cap L_2^*$.*

Proof. Since all finite languages belong to $\mathscr{L}_{\cap,*}$, for the rest of the proof we assume that $L_1^* \cap L_2^*$ is infinite. Therefore, let $L_1 = \{u_1, u_2, \ldots, u_k\}$, for some $k \geq 1$, with $|u_1| < |u_2| < \cdots < |u_k|$ and, similarly, $L_2 = \{v_1, v_2, \ldots, v_\ell\}$, for some $\ell \geq 1$, with $|v_1| < |v_2| < \cdots < |v_\ell|$.

Without loss of generality, we may assume $|u_k| \geq |v_\ell|$. We apply Lemma 3 to L_1 and set $c_1 = |u_1|, c_2 = |u_2|, \ldots, c_k = |u_k|$, $d_1 = \gcd(c_1, c_2, \ldots, c_k)$, and $n = |u_k|$. In this way, we obtain the lengths of all words in L_1^* beyond n^2 as $P_1 = \{\, t_1 + x \cdot d_1 \mid x \geq 0 \,\}$, where t_1 is a multiple of d_1.

Doing the same with L_2, we set $c_1 = |v_1|, c_2 = |v_2|, \ldots, c_k = |v_\ell|$, and $d_2 = \gcd(c_1, c_2, \ldots, c_\ell)$. So, the lengths of all words in L_2^* beyond n^2 are given by $P_2 = \{\, t_2 + x \cdot d_1 \mid x \geq 0 \,\}$, where t_2 is a multiple of d_2.

If the intersection of P_1 and P_2 is empty, the language $L_1^* \cap L_2^*$ is finite and we are done. Otherwise, we set $d = \operatorname{lcm}\{d_1, d_2\}$ and define t to be the least integer in the intersection $P_1 \cap P_2$. So, t is a multiple of d, say $t = y \cdot d$.

Now we start the construction of L by setting $L = \{ a^{t+i \cdot d} \mid 0 \leq i < y \}$. In this way, L^* contains exactly all words that are belonging to the intersection of $P_1 \cap P_2$. However, words with lengths below t remain to be considered. To this end, we take all the finitely many words of the intersection $L_1^* \cap L_2^*$ that are shorter than t and add it to L. In this way, now all words from $L_1^* \cap L_2^*$ are contained in L^*. On the other hand, no extra word can belong to L^*, since all words that can be obtained by concatenations of words in L do belong to $L_1^* \cap L_2^*$ as well. ∎

Theorem 5 says that any unary language in $\mathscr{L}_{\cap,*}$ is either finite or can be represented as L^*, where L is a finite language.

6 Comparisons of the Language Families

In this section, we classify the family of languages definable by subregular expressions with complement and Kleene star, the family of languages definable by subregular expressions with concatenation and complementation, as well as those definable by subregular expressions with intersection and Kleene star. We conduct a detailed analysis of the expressive power of each family, comparing them pairwise.

We first investigate the relationship between $\mathscr{L}_{\cap,*}$ and $\mathscr{L}_{*,-}$.

Theorem 6. *In the unary case the family of languages $\mathscr{L}_{\cap,*}$ is a proper subset of the family of languages $\mathscr{L}_{*,-}$.*

Proof. Since every infinite unary language in $\mathscr{L}_{\cap,*}$ can be described by a starred set (cf. Theorem 5), it follows that every unary language L in $\mathscr{L}_{\cap,*}$ can also be described by an expression in $\mathscr{L}_{*,-}$, since the Kleene star is available in $\mathscr{L}_{*,-}$.

Assume, for contradiction, that the family of unary languages in $\mathscr{L}_{\cap,*}$ is equal to the family of unary languages in $\mathscr{L}_{*,-}$. Consider the simple unary cofinite language $L = \overline{\{aa\}}$, which belongs to $\mathscr{L}_{*,-}$. According to Theorem 5, every infinite unary language in $\mathscr{L}_{\cap,*}$ must be describable by a starred set. Therefore, the symbol a must be included in the generating set, as it is a member of L and cannot be further decomposed. However, if a is in the language and the expression is starred, then aa must also be included in the language which contradicts the assumption that $aa \notin L$. Hence, $\mathscr{L}_{\cap,*} \neq \mathscr{L}_{*,-}$ in the unary case. ∎

Theorem 7. *The family of languages $\mathscr{L}_{\cap,*}$ is incomparable to the family of languages $\mathscr{L}_{*,-}$.*

Proof. It has already been shown that there exists a unary language in $\mathscr{L}_{*,-}$ that is not in $\mathscr{L}_{\cap,*}$. Consider the language $L_d = L_1^* \cap L_2^*$, with $L_1 = \{ab, a\}$ and $L_2 = \{ab, aa, ba, bb\}$, from the previous section, which is described by an expression in $\mathscr{L}_{\cap,*}$. The language L_d can be informally described as the set of all words of even length that start with an a and do not contain consecutive b's.

Assume, for contradiction, that L_d can be described by an expression in $\mathscr{L}_{*,}-$. As shown in Proposition 4, L_d cannot be described by the star of a finite set. We proceed by reasoning along the structure of the expression tree. Note that in the class $\mathscr{L}_{*,}-$, the operations of star and complement must alternate.

Suppose the topmost operation is complement, that is, $\overline{L} = L_d$. Then L must contain all words of odd length, and it must be constructed using a star operation. Consider a word $w = uv, w \in L_d$ where u and v are of odd length. Then $u, v \in L$, and since L is described by a star, their concatenation uv must also be in L, implying $w \in L$, contradicting the assumption that $L_d = \overline{L}$.

Therefore, the topmost operation must be the star: $L^* = L_d$. Since L_d cannot be represented as the star of a finite set, L must be an infinite set of words of even length. Then $\overline{L'} = L$ must contain all words of odd length. Again, since L' is formed by a star, and $w \in L$, there must exist factors u, v of odd length in L' such that $uv = w$. But this implies $w \in L'$, contradicting that $L' = \overline{L}$.

Hence, L_d cannot be described by any expression in $\mathscr{L}_{*,}-$, demonstrating the incomparability between the classes $\mathscr{L}_{*,}-$ and $\mathscr{L}_{\cap,*}$. ■

We turn to consider the comparison of $\mathscr{L}_{\cap,*}$ and $\mathscr{L}_{\cdot,}-$.

Theorem 8. *The family of languages $\mathscr{L}_{\cap,*}$ is incomparable to the family of languages $\mathscr{L}_{\cdot,}-$. This is also true in the unary case.*

Proof. Clearly, the family of languages $\mathscr{L}_{\cdot,}-$ is a subset of the star-free languages. Therefore, the language $(aa)^*$ cannot be described by an expression using only concatenation and complement, and thus does not belong to $\mathscr{L}_{\cdot,}-$.

Conversely, the simple unary cofinite language $L = \overline{\{aa\}}$, which belongs to $\mathscr{L}_{\cdot,}-$, cannot be described in $\mathscr{L}_{\cap,*}$, as shown in Theorem 6. ■

The last pairwise comparison is between $\mathscr{L}_{\cdot,}-$ and $\mathscr{L}_{*,}-$.

Theorem 9. *In the unary case the family of languages $\mathscr{L}_{\cdot,}-$ is a proper subset of the family of languages $\mathscr{L}_{*,}-$.*

Proof. As shown in Proposition 2, every unary language described by $\mathscr{L}_{\cdot,}-$ is either finite or cofinite. Clearly, every finite (or cofinite) language can be described in $\mathscr{L}_{*,}-$, since the complement operation is allowed. Thus $\mathscr{L}_{\cdot,}-$ is a subset of $\mathscr{L}_{*,}-$. It remains to be shown that the inclusion is proper. To this end, we consider the language $L = (aa)^*$. It can be described by an expression using only Kleene star and complement, and thus belongs to $\mathscr{L}_{*,}-$. Since L is not star-free, it leads to the conclusion that, in the unary case, the family of languages $\mathscr{L}_{\cdot,}-$ is a proper subset of the family of languages $\mathscr{L}_{*,}-$. ■

Theorem 10. *The family of languages $\mathscr{L}_{\cdot,}-$ is incomparable to the family of languages $\mathscr{L}_{*,}-$.*

Proof. We have already a language that belongs to $\mathscr{L}_{*,}-$ but not to $\mathscr{L}_{\cdot,}-$.

The converse remains to be shown. Consider the language $L = ab^+$. This language can be described by the expression $(a\,b \cdot \overline{(\overline{\emptyset} \cdot a \cdot \overline{\emptyset})})$, and thus belongs to $\mathscr{L}_{\cdot,}-$.

Assume for contradiction that L can be generated by an expression using only complementation and star. Then the outermost operation must be either the star or the complement.

First, assume that the outermost operation is the star. Since $ab \in L$, this would imply that $abab \in L$ as well, which contradicts the definition of L. Therefore, the topmost operation must be the complement, i.e., $\overline{L'} = L$.

In this case, L' must contain all words not in L, including all words a^* and b^*, in particular the words a and b. Since L' is infinite, it must be constructed using a star, that is, $L' = L''^*$. However, if $a, b \in L'$, and L' is of the form L''^*, then it must also contain the word ab. But $ab \in L'$ implies $ab \notin L = \overline{L'}$, which contradicts the definition of L.

Hence, L cannot be described by an expression using only star and complement, and therefore does not belong to $\mathscr{L}_{*,-}$. ■

Future Research

We investigated several previously unexplored subregular expressions, in particular: subregular expressions with complement and Kleene star, subregular expressions with concatenation and complement, and subregular expressions with intersection and Kleene star. We examined basic properties of the resulting language families and analyzed how these families relate to each other.

For the class of subregular expressions with complement and Kleene star, we identified a finite hierarchical structure based on the number of operations used in the expressions that collapses at level six at the latest. As future work, we aim to explore whether a similar hierarchy exists for the other two combinations: subregular expressions with concatenation and complement, and subregular expressions with intersection and Kleene star.

It also remains to systematically compare them to existing language classes in the literature, to better understand their relative positions in the landscape of subregular language families.

References

1. Brauer, A.: On a problem of partitions. Amer. J. Math. **64**, 299–312 (1942)
2. Brzozowski, J.A.: Canonical regular expressions and minimal state graphs for definite events. In: Mathematical Theory of Automata, pp. 529–561. Polytechnic Institute of Brooklyn (1962)
3. Brzozowski, J.A.: Roots of star events. J. ACM **14**, 466–477 (1967)
4. Brzozowski, J.A., Cohen, R.S.: On decompositions of regular events. J. ACM **16**, 132–144 (1969)
5. Brzozowski, J.A., Knast, R.: The dot-depth hierarchy of star-free languages is infinite. J. Comput. Syst. Sci. **16**, 37–55 (1978)
6. Cohen, R.S., Brzozowski, J.A.: Dot-depth of star-free events. J. Comput. Syst. Sci. **5**, 1–16 (1971)

7. Fürer, M.: The complexity of the inequivalence problem for regular expressions with intersection. In: International Colloquium on Automata, Languages and Programming (ICALP 1980). LNCS, vol. 85, pp. 234–245. Springer (1980)
8. Havel, I.M.: The theory of regular events II. Kybernetica **6**, 520–544 (1969)
9. Holzer, M., Kutrib, M.: The complexity of regular(-like) expressions. Int. J. Found. Comput. Sci. **22**, 1533–1548 (2011)
10. Hopcroft, J.E., Ullman, J.D.: Introduction to Automata Theory, Languages, and Computation. Addison-Wesley, Reading, Massachusetts (1979)
11. Hunt III, H.B.: The equivalence problem for regular expressions with intersections is not polynomial in tape. Tech. Rep. TR 73-161, Department of Computer Science, Cornell University, Ithaca, New York (1973)
12. Kutrib, M., Wendlandt, M.: Expressive capacity of concatenation freeness. In: Drewes, F. (ed.) Implementation and Application of Automata (CIAA 2015). LNCS, vol. 9223, pp. 199–210. Springer (2015)
13. Kutrib, M., Wendlandt, M.: Concatenation-free languages. Theor. Comput. Sci. p. to appear (2017)
14. Kutrib, M., Wendlandt, M.: Expressive capacity of subregular expressions. RAIRO Inform. Théor. **52**, 201–218 (2018). https://doi.org/10.1051/ita/2018014
15. McNaughton, R., Papert, S.: Counter-Free Automata. No. 65 in Research Monographs, MIT Press (1971)
16. Nagy, B.: Union-free regular languages and 1-cycle-free-path automata. Publicationes Mathematicae Debrecen **68**, 183–197 (2006)
17. Perles, M., Rabin, M.O., Shamir, E.: The theory of definite automata. IEEE Trans. Elect. Comput. **EC-12**, 233–243 (1963)
18. Petersen, H.: Decision problems for generalized regular expressions. In: Descriptional Complexity of Automata, Grammars and Related Structures (DCAGRS 2000), pp. 22–29. London, Ontario (2000)
19. Petersen, H.: The membership problem for regular expressions with intersection is complete in LOGCFL. In: Theoretical Aspects of Computer Science (STACS 2002). LNCS, vol. 2285, pp. 513–522. Springer (2002)
20. Salomaa, K., Yu, S.: Alternating finite automata and star-free languages. Theor. Comput. Sci. **234**, 167–176 (2000)
21. Schützenberger, M.P.: On finite monoids having only trivial subgroups. Inform. Control **8**, 190–194 (1965)
22. Shallit, J.: The Frobenius problem and its generalizations. In: Ito, M., Toyama, M. (eds.) Developments in Language Theory (DLT 2008). LNCS, vol. 5257, pp. 72–83. Springer (2008)
23. Stockmeyer, L.J.: The Complexity of Decision Problems in Automata Theory and Logic. Ph.D. thesis, Massachusetts Institute of Technology, Cambridge, Massachusetts (1974)
24. Stockmeyer, L.J., Meyer, A.R.: Word problems requiring exponential time. In: ACM (ed.) Proceedings of the Fifth Annual ACM Symposium on Theory of Computing (STOC 1973), pp. 1–9. ACM Press, New York, NY, USA (1973)

Mapped Exponent and Asymptotic Critical Exponent of Words

Eva Foster[1], Aleksi Saarela[2], and Aleksi Vanhatalo[2](✉)

[1] School of Computing and Mathematical Sciences, Birkbeck College, University of London, London, UK

[2] Department of Mathematics and Statistics, University of Turku, Turku, Finland

aevanh@utu.fi

Abstract. We study how much injective morphisms can increase the repetitiveness of a given word. This question has a few possible variations depending on the meaning of "repetitiveness". We concentrate on fractional exponents of finite words and asymptotic critical exponents of infinite words. We characterize finite words that, when mapped by injective morphisms, can have arbitrarily high fractional exponent. For infinite words, alongside other results, we show that the asymptotic critical exponent grows at most by a constant factor (depending on the size of the alphabet) when mapped by an injective morphism. For both finite and infinite words, the binary case is better understood than the general case. This is a shortened version of the full paper.

1 Introduction

The study of different variations of repetitions and the study of morphisms are two of the most common and old areas in combinatorics on words, dating back to Thue's original papers in 1906 [17]. Since then, properties of repetitions and morphisms have been studied extensively, both as separate areas and in relation to each other. A common viewpoint is to follow Thue's footsteps and use morphisms to generate words with desired properties, one of which is often a variation of avoiding repetitions, see [3,10] for some examples.

A variation which is relevant for this article is to consider infinite words where small (in length) exceptions to repetition-freeness are allowed. For example, there exist infinite binary words that have a small number of short square factors but are otherwise square-free [6]. Many articles study similar questions, e.g., [9]. Taking this idea further leads to the concept of asymptotic critical exponent (ACE). It is known that there are binary infinite words with ACE equal to one [1,2]. The asymptotic critical exponent has also been considered (sometimes under different names) in other papers, such as [4,18].

E. Foster—Work completed whilst affiliated with School of Computer Science and Mathematics, Keele University, United Kingdom

A. Saarela and A. Vanhatalo—Supported by the Research Council of Finland under grants 339311, 346566 and 359921.

S.-K. Ko and F. Manea (Eds.): DLT 2025, LNCS 16036, pp. 244–260, 2026.
https://doi.org/10.1007/978-3-032-01475-7_17

In this paper, instead of using morphisms to generate words with few repetitions, we study questions of the following type: How good are morphisms at introducing some form of repetition into a given word? How good are words at resisting being transformed in this way by morphisms? If the morphisms in question are not restricted in any way, the answer is trivial, as we can just map all words to a unary alphabet. However, maybe a bit surprisingly at the first glance, restricting our attention to injective morphisms leads to many interesting results.

We consider two variations of this topic: one about finite and the other about infinite words. The research on finite words was started by the second author in [15]. Explicitly he asked, given a word, what natural numbers can appear as exponents of that word after it has been mapped by arbitrary, nonperiodic or injective morphisms. In this article, we study a fairly direct extension of the problem, where instead of integer exponents, we consider all rational exponents. The main question we then try to answer is the following: For which words w is the fractional exponent of $h(w)$ bounded by a finite number when h runs through all injective morphisms? We also introduce some extremal example words for which the optimal value of such a bound is either large or arbitrarily close to one.

The second variation is an attempt to formalize a similar problem for infinite words. Morphisms, even injective ones, can trivially introduce large repetitions as factors, so studying ordinary repetition-freeness is not interesting here. The concept of asymptotic critical exponent, however, is well-suited for study. Among other things, we are able to prove an upper bound for how large the ratio of the ACE of $h(w)$ and the ACE of w can be when h is an injective morphism. We also give examples of words where this ratio is large, and examples of words such that the ACE of $h(w)$ is exactly one for all injective morphisms h. In the case of a binary alphabet, we obtain particularly strong results.

This is a shortened version of the full paper, and thus the rather long proofs of Theorems 11 and 12, as well as some minor theorems, are omitted. They can be found in the full version of this paper [8] (preprint).

1.1 Preliminaries

Preliminaries are divided between two subsections. In the first we give formal definition of words as general sequences. This is needed in order to carefully prove two lemmas, but there is no harm in skipping to the second subsection, in which we give the problem statement and fix the common notation used. While most of the notation used is defined below, we expect the reader to be familiar with some of the basic notation and terminology of combinatorics on words (e.g., alphabets, finite and infinite words, concatenation). For a general reference, see [12].

Compactness Properties of Sequences and Formal Definition of Words: We are mostly interested in finite and right-infinite words. However, at one point, we use a previous result on bi-infinite words. To use it, we need to formalize a way to take a limit of finite words that results in bi-infinite words. To do this, we use the notion of general sequences. This chapter presents the formal definition of these sequences and words, but we mostly use more common notations. Naturally, these definitions are in use only when needed in the section about infinite words, specifically in Lemma 8 and Lemma 9.

We need four different types of sequences: finite, shifted finite, right-infinite and bi-infinite. Formally, sequences over a set X are defined as functions $f : A \to X$ in the following way:

- If $A = \{0, \dots, n\}$ for some $n \in \mathbb{N}$, then f is a *finite sequence* and we write
$$f = (f(0), \dots, f(n)).$$
- If $A = \{-m, \dots, n\}$ for some $m, n \in \mathbb{N}$, then f is a *shifted finite sequence* and we write
$$f = (f(-m), \dots, f(-1)|f(0), \dots, f(n)).$$
- If $A = \mathbb{N}$, then f is a *right-infinite sequence* (or just *infinite sequence*) and we write
$$f = (f(0), f(1), f(2), \dots).$$
- If $A = \mathbb{Z}$, then f is a *bi-infinite sequence* and we write
$$f = (\dots, f(-2), f(-1)|f(0), f(1), f(2), \dots).$$

Instead of $f(i)$ we often write f_i. We can also write $f = (f(i))_{i \in A}$ or $f = (f(i))$ if A is clear from the context. The *length* of f is $|f| = |A|$, where it naturally might be that $|f| = \infty$. By $|f|_x$ where $x \in X$ we mean the size of the set $\{i \in A \mid f(i) = x\}$, that is, the number of elements x in the sequence f. The set of all functions $A \to X$ is denoted by X^A. The definition of shifted finite sequences is nonstandard and is only used for the purpose of constructing bi-infinite sequences as limits.

If f is a right-infinite sequence, t a bi-infinite sequence and $g : \mathbb{Z} \to \mathbb{Z}$ is strictly increasing, then

$$f \circ g = (f(g(0)), f(g(1)), f(g(2)), \dots)$$

is a *subsequence* of f, and

$$t \circ g = (\dots, t(g(-2)), t(g(-1))|t(g(0)), t(g(1)), t(g(2)), \dots)$$

is a subsequence of t. More specifically, these can be called the *g-subsequences* of f and t respectively. Similarly if w is a finite of shifted finite sequence and $g : B \to A$ is strictly increasing and such that $w \circ g$ is also a finite or a shifted finite sequence, then $w \circ g$ is called a (g-)subsequence of w.

Instead of using the sequence notation that was defined above, words (sequences over finite X) can of course be written in a more usual way as follows:

$$\begin{aligned}
(a_0, \dots, a_n) &= a_0 \cdots a_n, \\
(a_{-m}, \dots, a_{-1}|a_0, \dots, a_n) &= a_{-m} \cdots a_{-1}|a_0 \cdots a_n, \\
(a_0, a_1, a_2, \dots) &= a_0 a_1 a_2 \cdots, \\
(\dots, a_{-2}, a_{-1}|a_0, a_1, a_2, \dots) &= \cdots a_{-2} a_{-1}|a_0 a_1 a_2 \cdots.
\end{aligned}$$

The set of all words that are finite sequences over X is denoted by X^*. This includes empty word, that is the unique function $\varepsilon : \emptyset \to X$. The basic concepts on words, such as factor, prefix, suffix and morphism, naturally work on sequences that are words. We define these with the standard notation in the second subsection.

A sequence $(u_i)_{i\in\mathbb{N}}$ of finite or infinite sequences over X *converges* to a right-infinite sequence w (or, in other words, has a *limit* w) if for all $k \in \mathbb{N}$ there exists a bound L_k such that $w(k) = u_i(k)$ for all $i > L_k$. It is possible to turn the set of all finite and right-infinite sequences into a metric space so that the above definition of a limit is compatible with the definition of a limit in the metric space, see, e.g., [13]. The following well-known lemma is a consequence of the fact that this metric space is compact if X is finite.

Lemma 1 (One of the problems in the first chapter of [13], related to Theorem 1.2.4). *Let u_i be a finite or an infinite sequence over a finite set X for all $i \in \mathbb{N}$. If $\lim_{i\to\infty} |u_i| = \infty$, then the sequence $(u_i)_{i\in\mathbb{N}}$ has a subsequence that converges to a right-infinite sequence.*

Similarly, a sequence $(u_i)_{i\in\mathbb{N}}$ of shifted finite or bi-infinite sequences over X *converges* to a bi-infinite sequence w (or, in other words, has a *limit* w) if for all $k \in \mathbb{Z}$ there exists a bound L such that $w(k) = u_i(k)$ for all $i \geq L$. The following lemma is analogous to Lemma 1, and the proof is essentially the same.

Lemma 2. *Let $u_i : A_i \to X$ be shifted finite or a bi-infinite sequences over a finite set X and let $\lim_{i\to\infty} \min A_i = -\infty$ and $\lim_{i\to\infty} \max A_i = \infty$. The sequence $(u_i)_{i\in\mathbb{N}}$ has a subsequence that converges to a bi-infinite sequence.*

Problem Statement and Notation Used: In this paper, Σ and Γ are fixed alphabets which are finite and nonunary. We also use $\mathrm{alph}(w)$ to denote the minimal sized alphabet for which w is defined over.

Factor f of a (finite or infinite) word w is a word such that $w = ufv$ for some words u and v. We denote the set of length-n factors of a word w by $\mathrm{Fact}_n(w)$, and the set of all nonempty factors by $\mathrm{Fact}_+(w)$. Furthermore if f is a factor of w such that $w = fv$ for some v, then f is called a *prefix* of the word w. Similarly, if $w = uf$ for some word u then f is a *suffix* of the word w.

Any equality where a word $w \in \Sigma^*$ is written in terms of two or more shorter words, e.g. $w = w_1 w_2 \cdots w_k$ for some words $w_i \in \Sigma^*$ is called a *factorization* of w. Often when talking about factorization of some fixed word, we mean some particular factorization that is clear from the context.

Let $v \in \Sigma^+$ be a word and $p, q \in \mathbb{N}$ with $q = |v|$. By v^ω we denote the infinite word $vvv\cdots$ and the word $v^{\frac{p}{q}}$ then denotes the prefix of length p of v^ω. With this notation we have, for example, $(ab)^3 = (ab)^{\frac{6}{2}} = (abab)^{\frac{6}{4}} = (abab)^{\frac{3}{2}} = ababab$ and $(abc)^{\frac{4}{3}} = abca$.

For a nonempty word $u \in \Sigma^+$, its *integer exponent* $\mathrm{IE}(u)$ and *fractional exponent* $\mathrm{E}(u)$ are defined by:

$$\mathrm{IE}(u) = \sup\{n \in \mathbb{Z} \mid \exists v \in \Sigma^+ : u = v^n\},$$
$$\mathrm{E}(u) = \sup\{r \in \mathbb{Q} \mid \exists v \in \Sigma^+ : u = v^r\}.$$

We may also use the term *exponent* for $\mathrm{E}(u)$ (in the literature, exponent is sometimes used for integer exponent and sometimes for fractional exponent). If $\mathrm{IE}(x) = 1$, then we say that x is *primitive*. Note that if $w = x^{\mathrm{E}(w)}$, then x is primitive since $(y^n)^r = y^{rn}$ if $n \in \mathbb{N}$. If $w = x^{\mathrm{IE}(w)}$, then we say that x is a *primitive root* of w. We say that non-empty words u, v are *conjugate*, if $u = pq$ and $v = qp$ for some words p, q. Equivalently words u and v are conjugate if $|u| = |v|$ and v is a factor of uu. Following three well-known facts can be found in the first few pages of [12];

1. Two words u and v commute, i.e. satisfy $uv = vu$, only if words u and v are integer powers of a common word,
2. If words x and y are conjugate, then their primitive roots are conjugate and
3. Conjugates of a primitive word are primitive. (This follows from the fact 2.)

We use these facts without explicitly stating so, for example by declaring a contradiction if we show that a conjugate of a primitive word is nonprimitive.

For an infinite word $w \in \Sigma^{\mathbb{N}}$, its *critical exponent* $\mathrm{CE}(w)$ and *asymptotic critical exponent* $\mathrm{ACE}(w)$ are defined by:

$$\mathrm{CE}(w) = \sup\{\mathrm{E}(u) \mid u \in \mathrm{Fact}_+(w)\},$$
$$\mathrm{ACE}(w) = \limsup_{n \to \infty}\{\mathrm{E}(u) \mid u \in \mathrm{Fact}_n(w)\}.$$

A function $h : \Sigma^* \to \Gamma$ is a *morphism* if $h(uv) = h(u)h(v)$ for all words $u, v \in \Sigma^*$. We only give the images of letters when defining morphisms as those determine images of all words.

Let $\mathcal{I}$ be the set of all injective morphisms $\Sigma^* \to \Gamma^*$.

Consider the following variations of the different exponents that were defined above:

$$\mathrm{IE}_{\mathcal{I}}(u) = \sup\{\mathrm{IE}(h(u)) \mid h \in \mathcal{I}\},$$
$$\mathrm{E}_{\mathcal{I}}(u) = \sup\{\mathrm{E}(h(u)) \mid h \in \mathcal{I}\},$$
$$\mathrm{CE}_{\mathcal{I}}(w) = \sup\{\mathrm{CE}(h(w)) \mid h \in \mathcal{I}\},$$
$$\mathrm{ACE}_{\mathcal{I}}(w) = \sup\{\mathrm{ACE}(h(w)) \mid h \in \mathcal{I}\}.$$

Of these, $\mathrm{IE}_{\mathcal{I}}$ has previously been studied in [15] by Saarela, and $\mathrm{CE}_{\mathcal{I}}$ is not interesting because it is infinite for all w. In this article, we study $\mathrm{E}_{\mathcal{I}}$ and $\mathrm{ACE}_{\mathcal{I}}$, which we call *mapped (fractional) exponent* and *mapped asymptotic critical exponent*, respectively.

2 Finite Words

The main theorem of this section is the classification of words $w \in \Sigma^*$ such that $\mathrm{E}_{\mathcal{I}}(w) = \infty$. There are also noteworthy example words of low mapped exponent and high but finite mapped exponent in Theorems 2 and 4.

2.1 Lemmas for Finite Words

We present some needed results for the study of mapped exponent. The first result is the famous theorem of Fine and Wilf.

Theorem 1 (Fine and Wilf [7]). *If a power of u and a power of v have a common prefix of length $|u| + |v| - \gcd(|u|, |v|)$, then u and v are integer powers of a common word.*

The following is a folklore lemma that we prove for completeness.

Lemma 3. *For any rational r and primitive word x, x can be a factor of x^r only in a trivial way, that is, if $x^r = txu$, then $t \in x^*$.*

Proof. Let n be an integer bigger than r. If x can be a factor of x^n only in a trivial way, then the same is true for x^r. Assume for a contrary that x is a factor of x^n in a nontrivial way. Then x is a nontrivial factor of x^2, that is, $xx = txu$ for some nonempty words t and u. We see that t is a prefix of x and u is a suffix of x. We also see that $|tu| = x$, so together we know that $x = tu$. We get that $xx = tutu = txu = ttuu$ and then by canceling we have $ut = tu$. Thus $u = z^k$ and $t = z^l$ for some word z and positive integers k and l and then $x = z^{k+l}$, contradicting the fact that x is primitive. □

Lemma 4. *For any rational r and primitive word $x = pq$, if $x^r = tqpu$, then $t \in (pq)^*p$ and $u = (qp)^{r'}$ for some rational r'.*

Proof. We have $qtqpu = qx^r = q(pq)^r = (qp)^\gamma$ for some rational γ. By Lemma 3, $qt = (qp)^n$ for some positive integer n and thus $t = (pq)^{n-1}p$, and then $u = (qp)^{\gamma-n-1}$. □

Lemma 5. *Let $w \in \Sigma^+$, $h \in \mathcal{I}$ and $h(w) = x^r$ for $x \in \Gamma^+$ of minimal length and $r \in \mathbb{Q}$. If $|h(a)| \geq |x|$ for some $a \in \Sigma$ and $ava, aua \in \mathrm{Fact}_+(w)$ for some words $v \in (\Sigma \setminus \{a\})^*$ and $u \in (\Sigma \setminus \{a\})^*$, then $v = u$.*

Proof. We can write $w = u_1auau_2 = v_1avav_2$ for some words $u_1, u_2, v_1, v_2 \in \Sigma^*$. Because $h(a)$ is a factor of x^r, there must be p, q such that $x = pq$ and $h(a)$ begins with qp (that is, a prefix of $h(a)$ is a conjugate of X). By Lemma 4, $h(u_1), h(u_1au), h(v_1), h(v_1av) \in (pq)^*p$, and thus $h(au), h(av) \in (qp)^*$. It follows that $h(auav) = h(avau)$ and, by the injectivity of h, $auav = avau$. Since a does not occur in u or v, this leads to $u = v$. □

Part of the next lemma was already seen in [15].

Lemma 6. *Let $w \in \Sigma^*$ be a word. If there exists an injective morphism $h \in \mathcal{I}$ such that $h(w) = x^r$ for some $x \in \Gamma^*$ and $r \geq 1$ and where $|x|_a = 1$ for some $a \in \Gamma$, then $\mathrm{E}_{\mathcal{I}}(w) = \infty$.*

Proof. Since $|x|_a = 1$, we can write $x = uav$ for some words $u, v \in (\Gamma \setminus \{a\})^*$. Then $h(w) = (uav)^r = uavuav \cdots uavu'$, where u' is some prefix of uav. For all $n \in \mathbb{N}$, we can define an injective morphism $h' : \Gamma^* \to \Gamma^*$ by $h'(a) = a(vua)^{n-1}$ and $h'(b) = b$ for all letters $b \in \Gamma \setminus \{a\}$. We then obtain:

$$\begin{aligned} h'(h(w)) &= ua(vua)^{n-1}vua(vua)^{n-1}v \ldots ua(vua)^{n-1}vh'(u'), \\ &= (uav)^{n\lfloor r \rfloor}h'(u'). \end{aligned}$$

It then remains to show that $h'(u') = (uav)^{r'}$, where $r' \in \mathbb{Q}$.

- If $a \notin \mathrm{alph}(u')$, then $h'(u') = u'$, and as u' is a prefix of uav, it follows that $h'(u')$ is a prefix of uav.
- If $a \in \mathrm{alph}(u')$, then $u' = uam$, where $m \in (\Gamma \setminus \{a\})^*$ is a prefix of v. Then,

$$\begin{aligned} h(u') &= ua(vua)^{n-1}m, \\ &= ua(vua)^{n-2}vuam, \\ &= (uav)^{n-1}uam. \end{aligned}$$

As m is a prefix of v, it follows that uam is a prefix of uav.

□

2.2 Mapped Fractional Exponent

While computing $\mathrm{E}_{\mathcal{I}}$ is difficult for most words, there are some words for which it is relatively easy. We demonstrate this with a small example. Let $a, b \in \Sigma$ and consider a word $(aabb)^2 \in \Sigma^*$ and morphism $h \in \mathcal{I}$. We know $h(aabb)$ is primitive (see [15]). Let $h((aabb)^2) = x^r$ for some primitive word $x \in \Gamma^*$ and $r \geq 2$. Since a power of $h(aabb)$ and a power of x have a common prefix $h((aabb)^2)$, by Theorem of Fine and Wilf 1 we see that x and $h(aabb)$ are integer powers of a common word. We can apply Theorem of Fine and Wilf since $|h((aabb)^2)| \geq |h(aabb)x|$, which holds since $r \geq 2$. By primitivity this leads to $x = h(aabb)$ and thus $r = 2$. Since $h \in \mathcal{I}$ was arbitrary, we see that $\mathrm{E}_{\mathcal{I}}((aabb)^2) = \mathrm{E}((aabb)^2) = 2$. The next theorem shows that $\mathrm{E}_{\mathcal{I}}(w)$ can be arbitrarily close to 1 using similar ideas.

Theorem 2. *Let $a, b \in \Sigma$ be different letters. For all integers $n \geq 2$, $\mathrm{E}_{\mathcal{I}}((ab)^n ba) = 1 + \frac{2}{n-1}$.*

Proof. Let $w = (ab)^n ba$ with $n \geq 2$. Let h be an injective morphism and let $h(w) = v^r$ for some word v and rational number r. We first prove that $r \leq 1 + \frac{2}{n-1}$. We may assume v to be primitive.

A power of $h(ab)$ and a power of v have a common prefix of length $|h(ab)^n|$. If this is at least $|h(ab)v|$, it follows from Theorem of Fine and Wilf 1 that

v and $h(ab)$ are powers of a common word. Since v is primitive this leads to $h(ab) = v^m$ for some $m \in \mathbb{N}$. Now by length argument, $h(ba) = v^m = h(ab)$, which contradicts the injectivity of h. This shows that $|h(ab)^n| < |h(ab)v|$ and thus $|v| > |h(ab)^{n-1}|$. Consequently,

$$r = \frac{|h(w)|}{|v|} < \frac{n+1}{n-1} = 1 + \frac{2}{n-1}.$$

As $h \in \mathcal{I}$ was arbitrary, $\mathrm{E}_{\mathcal{I}}(w) \leq 1 + \frac{2}{n-1}$.

For the converse, let $h_k \in \mathcal{I}$ be defined by $h_k(a) = (cd)^k c$ and $h_k(b) = dc$ for each integer $k \in \mathbb{N}$ and for some different letters $c, d \in \Gamma$. Then

$$\begin{aligned} h_k(w) &= ((cd)^k cdc)^n dc(cd)^k c \\ &= ((cd)^k cdc)^{n-1}(cd)^k cdcdc(cd)^k c \\ &= [((cd)^k cdc)^{n-1}cd](cd)^k cdc(cd)^k c \\ &= [((cd)^k cdc)^{n-1}cd]^{1+\frac{4k+4}{(2k+3)(n-1)+2}}, \end{aligned}$$

where the square brackets show a period of $h_k(w)$. So

$$\mathrm{E}_{\mathcal{I}}(w) \geq \lim_{k\to\infty} \mathrm{E}(h_k(w)) \geq \lim_{k\to\infty} 1 + \frac{4k+4}{(n-1)(2k+3)+2} = 1 + \frac{2}{n-1}.$$

□

We say two words $u \in \Sigma^*$ and $v \in \Sigma^*$ are *prefix-comparable* if there exists a word $s \in \Sigma^*$ such that u is a prefix of the word vs. Similarly the words u and v are *suffix-comparable* if there exists a word $p \in \Sigma^*$ such that u is a suffix of pv.

Theorem 3. *For a word $w \in \Sigma^+$, the following are equivalent:*

1. $\mathrm{E}_{\mathcal{I}}(w) = \infty$.
2. $\mathrm{E}_{\mathcal{I}}(w) > |w|$.
3. *There exists an injective morphism $h \in \mathcal{I}$ and a letter $a \in \mathrm{alph}(w)$ such that $h(w) = x^r$ for some primitive $x \in \Gamma^*$, $r \in \mathbb{Q}$ and $|h(a)| \geq |x|$.*
4. *There exists an integer $k \geq 0$, a letter $a \in \Sigma$, words $w_1, w_2, w_3 \in (\Sigma \setminus \{a\})^*$, and an injective morphism h such that $w = w_1(aw_2)^k aw_3$, $h(w_1)$ and $h(w_2)$ are suffix-comparable, and $h(w_2)$ and $h(w_3)$ are prefix-comparable.*

Proof. We prove the claim as an implication chain 1. $\Longrightarrow$ 2. $\Longrightarrow$ 3. $\Longrightarrow$ 4. $\Longrightarrow$ 1.. Clearly 1. $\Longrightarrow$ 2., so we can start at 2. $\Longrightarrow$ 3..

Assume $w \in \Sigma^+$ has $\mathrm{E}_{\mathcal{I}}(w) > |w|$. By definition, there exists an injective morphism $h \in \mathcal{I}$ such that $h(w) = x^r$ with $r \geq |w|$. If for all $a \in \mathrm{alph}(w)$ we have $|h(a)| < |x|$, then $|h(w)| < |w|\,|x| \leq |x^r|$, a contradiction. Thus 2. $\Longrightarrow$ 3..

Next, 3. $\Longrightarrow$ 4.. By Lemma 5, between consecutive appearances of a in w we always have the same word. This happens exactly when w admits the factorization $w = w_1(aw_2)^k aw_3$ for words $w_1, w_2, w_3 \in (\Sigma \setminus \{a\})^*$ and integer $k \in \mathbb{N}$. If $k = 0$, we can freely choose $w_2 = \varepsilon$ so the comparability claim of

4. holds. If $k > 0$, then since $|h(a)| \geq |x|$ and $h(a)$ is a factor of a repetition of x, there are $p, q \in \Gamma^*$ such that $x = pq$ and $h(a)$ begins with qp (that is, a prefix of $h(a)$ is a conjugate of x). By Lemma 4, $h(w_1), h(w_1 a w_2) \in (pq)^* p$. Thus $h(w_1)$ and $h(w_2)$ are suffix-comparable. Similarly, by Lemma 4, $h(aw_2aw_3)$ and $h(aw_3)$ are both prefixes of some word in $(qp)^*$, and thus $h(w_2)$ and $h(w_3)$ are prefix-comparable.

Lastly we show that 4. $\implies$ 1.. We assume a factorization $w = w_1(aw_2)^k aw_3$ with $k \in \mathbb{N}$, words $w_1, w_2, w_3 \in (\Sigma \setminus \{a\})^*$ and an injective morphism h such that $h(w_1)$ and $h(w_2)$ are suffix-comparable and $h(w_2)$ and $h(w_3)$ are prefix-comparable. By definition, there exists words p, s such that $h(w_1)$ is a suffix of $ph(w_2)$ and $h(w_3)$ is a prefix of $h(w_2)s$. Define an injective morphism h' with $h'(b) = h(b)$ for all $b \neq a$ and $h'(a) = scp$ for some new letter $c \notin \Gamma$. Now, since $|w_1|_a = |w_2|_a = |w_3|_a = 0$, we have

$$h'(w) = h(w_1)s(cph(w_2)s)^k cph(w_3).$$

By suffix-comparability, $ph(w_2) = th(w_1)$ for some word t,

$$h'(w) = (h(w_1)sct)^k h(w_1)scph(w_3).$$

Similarly by prefix-comparability $ph(w_3)$ is a prefix of $ph(w_2)s = th(w_1)s$, finally we have

$$h'(w) = (h(w_1)sct)^r$$

for some $r < k+2$. Now $|h(w_1)sct|_c = 1$, so by Lemma 6 it follows that $\mathrm{E}_{\mathcal{I}}(w) = \infty$ (by using $\Gamma \cup \{c\}$ as Σ). This completes the implication chain. □

For binary words, the fourth statement of Theorem 3 holds already if the factorization part holds. However, in general, the factorization $w = w_1(aw_2)^k aw_3$ is not enough to ensure unbounded $\mathrm{E}_{\mathcal{I}}(w)$. This can be seen with the word $w = bbccabcbca$, where $w_1 = bbcc$ and $w_2 = cbcb$ cannot be injectively mapped to suffix-comparable words.

Lastly, the next theorem shows that $\mathrm{E}_{\mathcal{I}}(w)$ can be large without being infinite for a word with low $\mathrm{E}(w)$.

Theorem 4. *For all n, there exists a $2n$-ary word w_n of length $6n$ with $\mathrm{E}(w_n) = 1$ such that*

$$n - \frac{1}{6} < \mathrm{E}_{\mathcal{I}}(w_n) < \infty.$$

Proof. For $n = 1$, the lower bound is obtained by identity morphism and it is enough to note that $w_1 = aababb$ is a binary word which by Theorem 3 has $\mathrm{E}_{\mathcal{I}}(w_1) < \infty$. So let $n \geq 2$. Let $a_1, \ldots, a_n, b_1, \ldots b_n$ be distinct letters in Σ and let

$$w_n = \prod_{i=1}^{n} (a_i a_i b_i a_i b_i b_i).$$

Since w_n does not admit a factorization as in fourth statement in Theorem 3, $\mathrm{E}_{\mathcal{I}}(w_n) < \infty$. Let $h \in \mathcal{I}$ be the morphism defined by

$$h(a_i) = c^{i-1} a c^{n-i}, \qquad h(b_i) = c^{i-1} b c^{n-i}$$

for all $i \leq n$ and for some distinct letters $a, b, c \in \Gamma$.

We have

$$h(w_n) = (ac^{n-1}ac^{n-1}bc^{n-1}ac^{n-1}bc^{n-1}bc^n)^{n-\frac{n}{6n+1}},$$

so $\mathrm{E}_{\mathcal{I}}(w_n) \geq n - \frac{n}{6n+1} > n - \frac{1}{6}$. □

3 Infinite Words

We now change our focus to mapped asymptotic critical exponent and infinite words. The binary case has its own subsection after the investigation of the general alphabets, since in binary we are able to prove an exact upper bound of $\mathrm{ACE}_{\mathcal{I}}$ in terms of ACE. We start with some lemmas and needed definitions.

3.1 Lemmas for Infinite Words

In this section we present and prove some needed results about injective morphisms, ACE and periodic words. Infinite word w is said to be *eventually periodic* if $w = puuuu\cdots$ for some non-empty word u. If p is empty, then w is said to be *periodic*. Similarly, if w is a bi-infinite word, then it is said to be periodic if $w = \cdots uuu \cdots$ for some non-empty word u.

If w is infinite word that is not eventually periodic or bi-infinite word that is not periodic, it is said to be *aperiodic*.

Theorem 5 (Morse and Hedlund [14]). *For all infinite words $w \in \Sigma^{\mathbb{N}}$, let $\rho_n(w) = |\mathrm{Fact}_n(w)|$. The following are equivalent:*

1. *w is eventually periodic.*
2. *$\rho_n(w) \leq C$ for some constant $C \in \mathbb{R}$ for all $n \in \mathbb{N}$.*
3. *$\rho_n(w) \leq n$ for some $n \in \mathbb{N}$.*

Lemma 7. *Let $h \in \mathcal{I}$ and $u \in \Sigma^{\mathbb{N}}$ be an infinite word. If $h(u)$ is eventually periodic, then so is u.*

Proof. Assume $h(u)$ is eventually periodic as in the statement of the theorem. Let $F_n = \{h(f) \mid f \in \mathrm{Fact}_+(u) \text{ and } |f| \leq n\}$. The words of F_n are factors of $h(u)$ and have length at most $n \max\{|h(a)|\}$, where a is a letter. Then by Theorem 5 we have $|F_n| \leq Cn \max\{|h(a)|\}$ for some constant $C \in \mathbb{R}$. On the other hand, since h is injective, each $f \in \mathrm{Fact}_+(u)$ with $|f| \leq n$ gives a unique element to F_n, and thus, using the notation of Theorem 5, $|F_n| = \sum_{i=1}^{n} \rho_i(u)$. If u is aperiodic, then by Theorem 5 we would have $\rho_i(u) \geq i + 1$, which would lead to $|F_n|$ growing faster than linearly, a contradiction. Thus u is eventually periodic. □

Lemma 8. *Let $w \in \Sigma^{\mathbb{N}}$ be an infinite word and $h \in \mathcal{I}$. If there exists a word $x \neq \varepsilon$ such that x^l is a factor of $h(w)$ for all $l \in \mathbb{N}$, then $\mathrm{ACE}(w) = \infty$.*

Proof. Recall the definition of limits of words and subsequences of type $w \circ g$ defined in the preliminaries. Let $\sigma : \mathbb{Z} \to \mathbb{Z}$ be the shift map defined by $\sigma(n) = n + 1$. For σ and its iterates σ^n, we write (as is more common) $\sigma^n(f) :- f \circ \sigma^n$. Since x^l is a factor of $h(w)$ for all $l \in \mathbb{N}$, there exists a sequence of natural numbers $t_1, t_2, \ldots$ such that $\lim_{i\to\infty} \sigma^{t_i}(h(w)) = x^\omega$. Let $M = \max\{|h(a)| \mid a \in \text{alph}(w)\}$. Then for each i, there exists $p_i \leq M$ such that $\sigma^{t_i+p_i}(h(w)) = h(\sigma^{t'_i}(w))$ for some t'_i. Since the p_i are bounded, we have an infinite sequence of natural numbers j_i such that $p_{j_i} = p_{j_1}$. Finally, by possibly considering a subsequence of j_i, we may assume that $\lim_{i\to\infty} \sigma^{t'_{j_i}}(w)$ exists by 1. Now

$$
\begin{aligned}
&\sigma^{p_{j_1}}(\lim_{i\to\infty} \sigma^{t_{j_i}}(h(w))) \\
=& \lim_{i\to\infty} \sigma^{t_{j_i}+p_{j_1}}(h(w)) \\
=& \lim_{i\to\infty} \sigma^{t_{j_i}+p_{j_i}}(h(w)) \\
=& \lim_{i\to\infty} h(\sigma^{t'_{j_i}}(w)) \\
=& h(\lim_{i\to\infty} \sigma^{t'_{j_i}}(w))
\end{aligned}
$$

where the last equality is by the continuity of h in the metric space of sequences, or by following: Let $a \in \Sigma$ be the first letter in the infinite word $\lim_{i\to\infty} \sigma^{t'_{j_i}}(w)$. If $h(a)$ is then not a prefix of $\lim_{i\to\infty} h(\sigma^{t'_{j_i}}(w))$, it must be that infinite number of the words $\sigma^{t'_{j_i}}(w)$ begin with a letter $b \neq a$. This contradicts the assumption that a is a first letter of $\lim_{i\to\infty} \sigma^{t'_{j_i}}(w)$. Similar argument works on arbitrary large prefixes of $\lim_{i\to\infty} \sigma^{t'_{j_i}}(w)$.

Now we see that a sequence of shifts of w has a limit point that is a preimage of a periodic point. By Lemma 7, we have that this preimage must be also eventually periodic. But since this preimage limit point has only factors that are also factors of w (as the limit is over shifted versions of w), we see that $\text{ACE}(w) = \infty$. □

Let $X \subseteq \Sigma^*$ be a finite set of finite words. If $w : \mathbb{Z} \to \Sigma$ is bi-infinite sequence (written here as a concatenation of words)

$$w = \cdots x_{-2}x_{-1}p|qx_1x_2\cdots,$$

where $x_i \in X$ for all i, $pq \in X$, $q \neq \varepsilon$, then

$$(\ldots, x_{-2}, x_{-1}, p|q, x_1, x_2, \ldots)$$

is an *X-factorization* of w.

The combinatorial rank of a finite set of words X is defined as the size of the smallest set Y such that $X \subseteq Y^*$.

Theorem 6 (Karhumäki, Maňuch and Plandowski [11]). *Let $X \subset \Sigma^+$ be a finite set of words and let w be a bi-infinite word over Σ. If w has two different X-factorizations, then w is periodic or the combinatorial rank of X is at most $|X| - 1$.*

Corollary 1. *Let $h : \{a,b\}^* \to \{a,b\}^*$ be an injective morphism and let $w \in \{a,b\}^{\mathbb{Z}}$ be a bi-infinite word. If w can be $\{h(a),h(b)\}$-factored in two different ways, then w is periodic.*

Proof. By Theorem 6, since w can be $\{h(a),h(b)\}$-factored in two different ways, either w is periodic, or the combinatorial rank of $\{h(a),h(b)\}$ is 1. Since h is injective, the latter is not the case and thus w is periodic. □

An *X-interpretation* of a finite word $w \in \Sigma^*$ is a factorization $w = w_1w_2w_3\cdots w_n$ such that w_1 is a suffix of a word in X, w_n is a prefix of a word in X and $w_i \in X$ for all $1 < i < n$. Two X-interpretations $(w_1, w_2, w_3, \ldots, w_n)$ and $(v_1, v_2, v_3, \ldots, v_m)$ of the same word are *disjoint* if $w_1w_2\cdots w_i \neq v_1v_2\cdots v_j$ for all $i < n$ and $j < m$. If two X-interpretations are not disjoint, we say that they are *linked.*

The *X-degree* of a word w is the maximal number of pairwise disjoint X-interpretations w has.

Theorem 7 (Theorem 8.3.1 in [12]). *Let $w \in \Sigma^*$ be a word and $X \subseteq \Sigma^*$ a finite set of words. Let $r = \mathrm{E}(w)$ and $w = x^r$. If $|x| > \max\{|v| \mid v \in X\}$, then the X-degree of w is at most $|X|$.*

3.2 Mapped Asymptotic Critical Exponent over an Arbitrary Alphabet

The significance of Theorem 8 is that for all words w with finite $\mathrm{ACE}(w)$, the mapped asymptotic critical exponent is also finite.

Theorem 8. *For all words $w \in \Sigma^{\mathbb{N}}$,*

$$\mathrm{ACE}_{\mathcal{I}}(w) \leq (|\Sigma|+1) + |\Sigma|\,(|\Sigma|+1)(\lfloor \mathrm{ACE}(w) \rfloor + 1).$$

Proof. Let $N = (|\Sigma|+1) + |\Sigma|\,(|\Sigma|+1)(\lfloor \mathrm{ACE}(w) \rfloor + 1)$ and assume that $\mathrm{ACE}_{\mathcal{I}}(w) > N$ for some word $w \in \Sigma^{\mathbb{N}}$. By the definition of $\mathrm{ACE}_{\mathcal{I}}(w)$, there exists $h \in \mathcal{I}$ such that for all $n \in \mathbb{N}$, there exists $u_n^N \in \mathrm{Fact}_+(h(w))$ with $|u_n| > n$.

By Theorem 7, all long enough factors of $h(w)$ have $h(\Sigma)$ -degree of at most $|\Sigma|$, or there exists arbitrary long factors $f = x^{\mathrm{E}(f)} \in \mathrm{Fact}_+(h(w))$ such that $|x| \leq \max\{|h(a)| \mid a \in \Sigma\}$. From the latter, Lemma 8 gives us $\mathrm{ACE}(w) = \infty$ and the proof ends. So assume that the former is the case.

In the factor u_n^N, by the size of N, we can choose $|\Sigma|+1$ occurrences of u_n such that each occurrence pair chosen has a factor $u_n^{(|\Sigma|+1)(\lfloor \mathrm{ACE}(w) \rfloor+1)}$ in between them. In the chosen occurrences of u_n, at least two of the $h(\Sigma)$-interpretations we get from the natural $h(\Sigma)$-factorization of $h(w)$ are linked, because otherwise the $h(\Sigma)$-degree of v would be at least $|\Sigma|+1$, contradicting Theorem 7. This means that in the factor u_n^N, there is a subfactor $h(x_n) = v^k$ such that $x_n \in \mathrm{Fact}_+(w)$, v is a conjugate of u_n and $k \geq (|\Sigma|+1)(\lfloor \mathrm{ACE}(w) \rfloor + 1)$.

In the repetition v^k, consider the first $|\Sigma|+1$ occurrences of v. Again, we can conclude that two of the $h(\Sigma)$-interpretations of these must be linked as in the previous step. This means that we can write $x_n = ry^m t$ and $v = pq$ so that y is primitive, $h(r) \in (pq)^* p$, $h(y^m) \in (qp)^*$, $h(t) \in q(pq)^*$ and $|h(y^m)| \leq |v^{|\Sigma|+1}|$.

Now $h(rt), h(ry^m t) \in (pq)^*$, so $h(rt)h(ry^m t) = h(ry^m t)h(rt)$. By injectivity of h, $rtry^m t = ry^m trt$ and thus $try^m = y^m tr$. It follows that tr and y^m have the same primitive root, namely, y. Because $x_n = ry^m t$ is a conjugate of try^m, we can write $x_n = z^j$, where z is a conjugate of y.

We have

$$\mathrm{E}(x_n) = j = \frac{|h(z^j)|}{|h(z)|} = \frac{|v^k|}{|h(y)|} \geq \frac{|v^{(|\Sigma|+1)(\lfloor \mathrm{ACE}(w) \rfloor + 1)}|}{|v^{|\Sigma|+1}|} = \lfloor \mathrm{ACE}(w) \rfloor + 1$$

and

$$\mathrm{ACE}(w) \geq \limsup_{n \to \infty} \mathrm{E}(x_n) \geq \lfloor \mathrm{ACE}(w) \rfloor + 1,$$

a contradiction. This means that the original claim must hold: $\mathrm{ACE}_{\mathcal{I}}(w) \leq N$. □

We end this subsection by showing that, given large enough alphabet, $\mathrm{ACE}_{\mathcal{I}}$ can be large compared to ACE. For this, we need to construct words with low ACE. It follows from Beck's nonconstructive result [1] that there are infinite binary words with ACE equal to 1. Cassaigne in [2] gives an explicit construction of such a word.

Theorem 9 **([1] or [2]).** *There exists an infinite binary word u with* $\mathrm{ACE}(u) = 1$.

Theorem 10. *For all n, there exists a $2n$-ary infinite word w_n with* $\mathrm{ACE}(w_n) = 1$ *such that* $\mathrm{ACE}_{\mathcal{I}}(w_n) \geq n$.

Proof. Let u be a binary word with $\mathrm{ACE}(u) = 1$ as in Theorem 9. For $n = 1$, this word already proves the statement, so let $n \geq 2$. Let $a_1, \ldots, a_n, b_1, \ldots b_n$ be distinct letters. Let $u_i \in \{a_i, b_i\}^{\mathbb{N}}$ be a version of u defined over $\{a_i, b_i\}$ for all $i \leq n$ (that is, there is a renaming of letters such that u_i turns into u). Let $u_i = u_{i,1} u_{i,2} u_{i,3} \ldots$ where $|u_{i,j}| = j$. Now let

$$w_n = \prod_{j=1}^{\infty} u_{1,j} u_{2,j} u_{3,j} \ldots u_{n,j},$$

which is $2n$-ary word.

Let (x_l) be a sequence of factors of w_l such that $\lim_{l \to \infty} |x_l| = \infty$ and $\lim_{l \to \infty} \mathrm{E}(x_n) = \mathrm{ACE}(w_n)$. If for all large l the word x_l is a factor of some u_{i_l} for some i_l, then $\mathrm{ACE}(w_n) \leq \max \mathrm{ACE}(u_i) = 1$. Also, if for all large l the word x_l has $\mathrm{E}(x_l) = 1$, then $\mathrm{ACE}(w_n) = 1$. So assume that x_l has alphabets from two different sets $\{a_i, b_i\}$ and $\{a_j, b_j\}$ and $\mathrm{E}(x_l) > 1$ for all large l. Let $x_{l,i}$ be the largest subsequence of x_l such that $x_{l,i} \in \{a_i, b_i\}^*$, which is then a factor of u_i.

By $\mathrm{E}(x_l) > 1$, some binary alphabet $\{a_j, b_j\}$ appears in x_l twice with all the other alphabets in between these appearances. Then we have $\lim_{l \to \infty} |x_{l,i}| = \infty$ for all $i \neq j$. If $\lim_{l \to \infty} |x_{l,j}| \neq \infty$, it follows that $\lim_{l \to \infty} \mathrm{E}(x_l) = 1$, and thus

$\text{ACE}(w_n) = 1$. So we assume that $x_{l,i}$ gets arbitrarily large for all $i \leq n$ as l increases.

Let $x_l = y_l^{\text{E}(x_l)}$ for some word y_l and let $y_{l,i}$ be the largest subsequence of y_l with $y_{l,i} \in \{a_i, b_i\}^*$. Then from $x_l = y_l^{\text{E}(x_l)}$ it follows that $x_{l,i} = y_{l,i}^{r_{l,i}}$ for some rational $r_{l,i}$. Now

$$\begin{aligned}\text{ACE}(w_n) &= \lim_{l\to\infty} \text{E}(x_n) = \lim_{l\to\infty} \frac{|x_l|}{|y_l|} = \lim_{l\to\infty} \frac{\sum_{i=1}^{n} |x_{l,i}|}{|y_l|} \\ &= \lim_{l\to\infty} \frac{\sum_{i=1}^{n} \frac{|x_{l,i}|}{|y_{l,i}|} |y_{l,i}|}{|y_l|} \leq \lim_{l\to\infty} \max_{i\leq n} \left\{\frac{|x_{l,i}|}{|y_{l,i}|}\right\} \frac{\sum_{i=1}^{n} |y_{l,i}|}{|y_l|} = 1,\end{aligned}$$

so $\text{ACE}(w_n) = 1$.

Let $h \in \mathcal{I}$ be the morphism defined by

$$h(a_i) = c^{i-1}ac^{n-i}, \qquad h(b_i) = c^{i-1}bc^{n-i}$$

for all $i \leq n$. We have by construction of h (similarly to Theorem 4)

$$h(u_{1,j}u_{2,j}u_{3,j}\dots u_{n,j}) = (h(u_{1,j})c)^{n-\frac{1}{j}},$$

So $\text{ACE}_{\mathcal{I}}(w_n) \geq n$. □

Unfortunately the upper bound given in 8 is quadratic in the alphabet size, and the lower bound obtained from the examples is linear. In the next section, we get tighter results over the binary alphabet.

3.3 Binary Case

In this subsection, we will show that if in the word u all large enough factors are so called synchronizing words of $h \in \mathcal{I}$, then $\text{ACE}(h(u))$ is not too far from $\text{ACE}(u)$. For binary words the complement case leads to arbitrary long repetitions. This means that for binary words, one cannot increase ACE too much with injective morphism. Theorem 12 can be seen as complement result to upper bound given in Corollary 2.

In [5], Dvořáková, Ochem and Opočenská studied ACE of words with a few distinct palindromes. As a tool, they showed sufficient conditions for a morphism h and a word u to satisfy $\text{ACE}(h(u)) = \text{ACE}(u)$. In Theorem 11, we drop one of the conditions on the word u and weaken the conclusion, making the statement interesting in studying $\text{ACE}_{\mathcal{I}}$.

We use slightly adjusted notation from [5]. Let X be a finite set of finite words. We say that a word w is a *synchronizing word (of X)* if there exist words w_1, w_2 such that $w = w_1w_2$ and for all words $v \in X^*$, if $v = pws$ for some words p and s, then $v = v_1v_2$ and $v_1 = pw_1 \in X^*$ and $v_2 = w_2s \in X^*$ for some words v_1 and v_2. We also then say that w_1 is a synchronizing prefix and w_2 is a synchronizing suffix. While we do not need any old results on synchronizing words, we want to note that synchronizing words have appeared in different forms for a long time in coding theory, earliest explicit mention we found being from 1965 by Schützenberger [16].

Let $h : \Sigma^* \to \Gamma^*$ be a morphism. If a word $w \in \Gamma^*$ is a synchronizing word of $h(\Sigma)$, then we also say that it is a synchronizing word of h.

Theorem 11. *Let h be an injective morphism, $u \in \Sigma^{\mathbb{N}}$ an infinite word and L a natural number. If all factors f of $h(u)$ with $|f| \geq L$ are synchronizing words of h and $\mathrm{ACE}(u) \in \mathbb{R}$, then $\mathrm{ACE}(h(u)) \leq \lceil \mathrm{ACE}(u) \rceil$. Moreover, if u is over a binary alphabet and $\mathrm{ACE}(u) \notin \mathbb{N}$, then there exists $\lambda_u > 0$ such that for all h and L pairs, $\mathrm{ACE}(h(u)) \leq \lceil \mathrm{ACE}(u) \rceil - \lambda_u$.*

Proof. Theorem 3.16 in the full version of this paper [8] (preprint). □

Lemma 9. *Let X be a finite set of words and let w_i be words such that $\lim_{i\to\infty} |w_i| = \infty$. If every w_i has an X-interpretation but is not a synchronizing word of X, then we can write $w_i = u_i v_i$ so that a subsequence of $(u_i|v_i)_{i\in\mathbb{N}}$ converges to a bi-infinite word with two different X-factorizations.*

Proof. Let w_i have an X-interpretation $(x_{i,0}, \ldots, x_{i,m_i})$. Let $k_i = \lfloor m_i/2 \rfloor$ and

$$u_i = x_{i,0} \cdots x_{i,k_i-1}, \qquad v_i = x_{i,k_i} \cdots x_{i,m_i}.$$

Because u_i is not a synchronizing prefix of w_i, w_i must have an X-interpretation $(y_{i,0}, \ldots, y_{i,n_i})$ such that for some l_i we have $y_{i,l_i} = p_i q_i$, $p_i \neq \varepsilon \neq q_i$ and

$$u_i = y_{i,0} \cdots y_{i,l_i-1} p_i, \qquad v_i = q_i y_{i,l_i+1} \cdots y_{i,n_i}.$$

By Lemma 2, there exists a function g such that the g-subsequences of the sequences

$$((x_{i,0}, \ldots, x_{i,k_i-1}, \varepsilon | x_{i,k_i}, \ldots, x_{i,m_i}))_{i\in\mathbb{N}}, \qquad ((y_{i,0}, \ldots, y_{l_i-1}, p_i | q_i, y_{l_i+1}, \ldots, y_{i,n_i}))_{i\in\mathbb{N}}$$

both have a limit. Let these limits be S and T, respectively. It can be verified that both S and T are X-factorizations of the bi-infinite word that is the limit of the g-subsequence of $(u_i|v_i)_{i\in\mathbb{N}}$. Moreover, $S(-1) = \varepsilon \neq T(-1)$, so S and T are different. □

Lemma 10. *Let h be an injective morphism and w infinite binary word. If there exists infinite list $f_1, f_2, f_3, \ldots$ of factors of $h(w)$ that are not synchronizing words of h and $\lim_{i\to\infty} |f_i| = \infty$, then $\mathrm{ACE}(h(w)) = \mathrm{ACE}(w) = \infty$.*

Proof. By Lemma 9 there exists a bi-infinite word u that has at least two different $\{h(a), h(b)\}$-factorizations such that $\mathrm{Fact}_+(u) \subseteq \mathrm{Fact}_+(h(w))$. By Corollary 1 we see that u is then periodic, $u = \cdots xxx \cdots$ for some $x \in \Gamma^*$. Thus $h(w)$ has a factor x^l for each $l \in \mathbb{N}$. By Lemma 8, we see that $\mathrm{ACE}(w) = \mathrm{ACE}_{\mathcal{I}}(w) = \infty$. □

Corollary 2. *Let w be an infinite binary word with $\mathrm{ACE}(w) \in \mathbb{R}$. If $\mathrm{ACE}(w) \in \mathbb{N}$, then $\mathrm{ACE}_{\mathcal{I}}(w) = \mathrm{ACE}(w)$. Otherwise, $\mathrm{ACE}_{\mathcal{I}}(w) < \lceil \mathrm{ACE}(w) \rceil$.*

Proof. Let $h \in \mathcal{I}$. If $h(w)$ has arbitrary large factors that are not synchronizing, then by Lemma 10 we have $\mathrm{ACE}_{\mathcal{I}}(w) = \mathrm{ACE}(w) = \infty$.

Assume then that all large enough factors of $h(w)$ are synchronizing words. Then the claim follows directly from Theorem 11. □

Corollary 3. *There exists an infinite binary word w with $\mathrm{ACE}_{\mathcal{I}}(w) = 1$.*

Proof. Theorem 9 and Corollary 2. □

Theorem 12. *Let $n \in \mathbb{N}$, $n \geq 1$, $\lambda \in (0,1)$ and $\delta > 0$. There exists an infinite binary word $w_{(n,\lambda,\delta)}$ such that*

$$n < \mathrm{ACE}(w_{(n,\lambda,\delta)}) \leq n + \lambda$$

and

$$\mathrm{ACE}_{\mathcal{I}}(w_{(n,\lambda,\delta)}) \geq n + 1 - \delta.$$

Proof. Theorem 3.22 in the full version of this paper [8] (preprint). □

4 Conclusions and Open Problems

We investigated how much we can introduce repetitions to words via injective morphisms.

For finite binary words, the fourth statement in Theorem 3 reduces to a much simpler statement, because the suffix- and prefix-comparability conditions are automatically satisfied. Thus we think it is a good characterization of binary words w with $\mathrm{E}_{\mathcal{I}}(w) = \infty$. In the case of larger alphabets the situation is more complicated, which leads to the following question:

Open problem 1. Find a simpler form for the last condition of Theorem 3, or estimate the complexity of checking whether it holds for a given word.

Similarly, for infinite words we have powerful theorems concerning binary words that fail in the general case. In the binary case, by Corollary 2, an injective morphism cannot increase the asymptotic critical exponent by one or more. When the alphabet size is $2n$, then, by Theorem 10, the asymptotic critical exponent can sometimes increase at least by a factor of n. Theorem 8 shows that an increase by an arbitrarily large factor is not possible if the alphabet is fixed, but the upper bound here is most likely not very good. This makes the following question interesting:

Open problem 2. Let Σ be an alphabet with $|\Sigma| = n$. What is the value of

$$\sup_{\substack{w \in \Sigma^{\mathbb{N}} \\ \mathrm{ACE}(w) \in \mathbb{R}}} \frac{\mathrm{ACE}_{\mathcal{I}}(w)}{\mathrm{ACE}(w)}?$$

We conjecture that the answer is linear with respect to n.

References

1. Beck, J.: An application of Lovász local lemma: there exists an infinite 01-sequence containing no near identical intervals. In: Finite and infinite sets, Vol. I, II (Eger, 1981), vol. 37 of Colloq. Math. Soc. János Bolyai, pp. 103–107. North-Holland, Amsterdam (1984)
2. Cassaigne, J.: On extremal properties of the Fibonacci word. RAIRO-Theoret. Inf. Appl. **42**(4), 701–715 (2008)
3. Currie, J., Rampersad, N.: For each $\alpha>2$ there is an infinite binary word with critical exponent α. Electron. J. Comb. **15** (2008)
4. Dolce, F., Dvořáková, L., Pelantová, E.: On balanced sequences and their asymptotic critical exponent. In: International Conference on Language and Automata Theory and Applications, pp. 293–304. Springer (2021)
5. Dvořáková, L., Ochem, P., Opočenská, D.: Critical exponent of binary words with few distinct palindromes. Electron. J. Comb. **31**(2) (2024)
6. Entringer, R., Jackson, D., Schatz, J.: On nonrepetitive sequences. J. Comb. Theory Ser. A **16**(2), 159–164 (1974)
7. Fine, N., Wilf, H.: Uniqueness theorems for periodic functions. Proc. Am. Math. Soc. **16**(1), 109–114 (1965)
8. Foster, E., Saarela, A., Vanhatalo, A.: Mapped exponent and asymptotic critical exponent of words (2025). Preprint. arXiv:2506.04091
9. Harju, T., Nowotka, D.: Binary words with few squares. Bull. EATCS **89**, 164–166 (2006)
10. Karhumäki, J.: On cube-free ω-words generated by binary morphisms. Discret. Appl. Math. **5**(3), 279–297 (1983)
11. Karhumäki, J., Maňuch, J., Plandowski, W.: A defect theorem for bi-infinite words. Theoret. Comput. Sci. **292**(1), 237–243 (2003)
12. Lothaire, M.: Combinatorics on words. Encyclopedia of mathematics and its applications; 17. Cambridge University Press, Cambridge (1997)
13. Lothaire, M.: Algebraic combinatorics on words. Encyclopedia of mathematics and its applications; 90. Cambridge University Press, Cambridge (2002)
14. Morse, M., Hedlund, G.A.: Symbolic dynamics. Am. J. Math. **60**(4), 815–866 (1938)
15. Saarela, A.: Mapping words to powers by morphisms (2025). Preprint. arXiv:2503.00960
16. Schützenberger, M.-P.: Codes à longueur variable. Cours l'école d'été de l'OTAN sur les méthodes combinatoires en théorie de l'information et du codage, Royan, France (1965)
17. Thue, A.: Über unendliche Zeichenreihen. Norske Vid Selsk. Skr. I Mat-Nat Kl.(Christiana) **7**, 1–22 (1906)
18. Vandeth, D.: Sturmian words and words with a critical exponent. Theoret. Comput. Sci. **242**(1–2), 283–300 (2000)

Jumbled Scattered Factors

Pamela Fleischmann[1], Annika Huch[1], Melf Kammholz[1], and Tore Koß[2(✉)]

[1] Kiel University, Kiel, Germany
{fpa,ahu}@informatik.uni-kiel.de, stu209214@mail.uni-kiel.de
[2] University of Göttingen, Göttingen, Germany
tore.koss@cs.uni-goettingen.de

Abstract. In this work, we combine the research on (absent) scattered factors with the one of jumbled words. For instance, `wolf` is an absent scattered factor of `cauliflower` but since `lfow`, a jumbled (or abelian) version of `wolf`, is a scattered factor, `wolf` occurs as a jumbled scattered factor in `cauliflower`. A *jumbled scattered factor* u of a word w is constructed by letters of w with the only rule that the number of occurrences per letter in u is smaller than or equal to the one in w. We proceed to partition and characterise the set of jumbled scattered factors by the number of jumbled letters and use the latter as a measure. For this new class of words, we relate the folklore longest common subsequence (scattered factor) to the number of required jumbles. Further, we investigate the smallest possible number of jumbles alongside the jumbled scattered factor relation as well as Simon's congruence from the point of view of jumbled scattered factors and jumbled universality.

1 Introduction

In many areas of computer science, bioinformatics and data science, it is crucial to analyse data. One fundamental concept reoccurring in all these fields is the concept of *scattered factors* (also known as *scattered subwords* or *subsequences*), which are derived by selecting non-necessarily consecutive parts of a string while preserving their relative order. For example, the words `cute`, `outer` and `mute` are scattered factors of `computer`, while `route` and `tempo` are not since the letters occur in the wrong order. The study of scattered factors dates back to the 1970s, when Imre Simon introduced them in the context of piecewise-testable languages [46]. Words and their scattered factors have been studied in combinatorics on words [32,39], algorithms and stringology [4,5,8,19,37,48], logics [7,22,34,35], as well as language and automata theory [1,24,25,45,46,50] (see [36, Chapter 6, Subwords] for an overview). Since then, scattered factors played an important role also in a broader variety of fields such as bioinformatics [11,49], formal software verification [22,50] and database theory [3,20,30,31,43]. Further, scattered factors are also used to model corrupted data in the context of the theoretical problem of reconstruction [12,18,38]. The investigation of Simon's congruence [46], where two words are called *Simon congruent* if they

S.-K. Ko and F. Manea (Eds.): DLT 2025, LNCS 16036, pp. 261–277, 2026.
https://doi.org/10.1007/978-3-032-01475-7_18

have the same sets of scattered factors up to length $k \in \mathbb{N}$ is of ongoing research interest. This congruence has attracted recent interest in language theory [27,28] and pattern matching [17,29]. Closely related to Simon's congruence is the k-universality of words (also known as k-richness) that receives attention from combinatorial [6,10,14,44] and language theoretic perspectives [1,13]. A word is called k-universal if it contains all scattered factors of length k. This line of research is continued by characterising words that are not k-universal by their *absent scattered factors*, which are investigated in [2,15,16,32,47].

While lossy transmissions and absent scattered factors have been extensively studied, there is no line of research that allows the transmitted data to be scrambled [26,41]. A famous application in which the order of letters is not relevant is the *binary jumbled pattern matching* (BJPM) [9,21]: given a word over the binary alphabet $\{0,1\}$ and two natural numbers x and y, decide whether the word contains a factor with x zeros and y ones. Here, the order of the occurring letters is irrelevant (hence the name *jumbled* pattern matching). The description of a word by only the number of occurrences of each letter is given by its *Parikh vector* [40,42]. Therefore, we introduce and study the new notion of *jumbled* scattered factors by combining the research on (absent) subsequences and the one of jumbled (abelian) words. We define a word u to be a *jumbled scattered factor* of a word w if u can by constructed by the letters of w, i.e., u has a Parikh vector that is componentwise smaller than or equal to the Parikh vector of w. To distinguish those jumbled scattered factors in a more fine-grained way, we say that a jumbled scattered factor u of a word w has ℓ jumbles if at most ℓ letters of u occur at wrong places, i.e., $|u| - \ell$ letters occur in the correct order as a scattered factor in w. For instance, the words from before `route` and `tempo` are jumbled scattered factors of `computer` since all their letters occur in `computer`. Further, `route` is 1-jumbled since `oute` is a scattered factor of `computer`; `tempo` is 3-jumbled since `te` or `mp` are scattered factors of `computer` while the other letters occur in the wrong order. Even though we show that the minimal number of jumbles of a jumbled scattered factor in a word is closely related to the longest common scattered factor (lcsf) of those two words, this work is significantly different than the work on longest common scattered factors since we are taking a purely combinatorial perspective and partition the class of jumbled scattered factors by the number of jumbles where lcsf is just one part of this partition.

Own Contribution. In this work, we introduce jumbled scattered factors and start to explore the broad field of scattered factors in the jumbled case. First, we take a look at the *jumble index* of a word u in a word w which is the minimal amount of jumbles necessary to make u a scattered factor of w (Sect. 3). Then, we continue with a jumbled variant of Simon's congruence (which is only an equivalence relation) showing that there are finitely many equivalence classes of sizes either $1, |\Pi(w)|$, or ∞ and giving results on the structure of the classes where $\Pi(w)$ denotes the set of words with the same Parikh vector as w (Sect. 4). In Sect. 5, we investigate jumbled k-universality and search for the minimal number of jumbles such that the jumbled scattered factor set of a given word contains every k-length word.

2 Preliminaries

Let $\mathbb{N}$ denote the natural numbers starting with 1 and let $\mathbb{N}_0 = \mathbb{N} \cup \{0\}$. Further, let $[n] = \{1, \ldots, n\}$ for some $n \in \mathbb{N}$ and $[n]_0 = [n] \cup \{0\}$. An *alphabet* Σ is a finite, non-empty set whose elements are called *letters*. As Σ is finite, it can be assumed ordered and w.l.o.g. we take $\Sigma = \{\mathtt{a}_1, \ldots, \mathtt{a}_\sigma\}$ for some $\sigma \in \mathbb{N}$ with $\mathtt{a}_i < \mathtt{a}_{i+1}$ for all $i \in [\sigma - 1]$. A finite concatenation of letters from Σ is called a *word*. The set of words Σ^* with concatenation as binary operation and with the empty word ε, i.e., the word without any letters, as the neutral element forms a free monoid. For $M \subseteq \Sigma$, set $M^{\mathsf{C}} = \Sigma \backslash M$. The number of letters of a word $w \in \Sigma^*$ is called its *length* and denoted by $|w|$; this implies $|\varepsilon| = 0$. Let $\Sigma^k = \{w \in \Sigma^* \mid |w| = k\}$ and define $\Sigma^{\leq k}$ similarly for all $k \in \mathbb{N}_0$. The letter of $w \in \Sigma^*$ at position $i \in [|w|]$ is denoted by $w[i]$. For $w \in \Sigma^*$ and $\mathtt{a} \in \Sigma$ define $|w|_\mathtt{a} = |\{i \in [|w|] \mid w[i] = \mathtt{a}\}|$, i.e. the number of occurrences of $\mathtt{a}$ in w. Based on this define the *Parikh vector* of $w \in \Sigma^*$ by $p(w) = (|w|_{\mathtt{a}_1}, \ldots, |w|_{\mathtt{a}_\sigma}) \in \mathbb{N}_0^\sigma$. We denote the ith entry of $p(w)$ by $p(w)[i]$ for $i \in [\sigma]$. It is obvious that two words with the same Parikh vector have the same length and are permutations of each other like `flow` and `wolf`. Given two words $u, v \in \Sigma^*$, we say $p(u) \leq p(v)$ $(p(u) < p(v))$ if $|u|_{\mathtt{a}_i} \leq |v|_{\mathtt{a}_i}$ $(|u|_{\mathtt{a}_i} < |v|_{\mathtt{a}_i})$ for all $i \in [\sigma]$. If neither $p(u) \leq p(v)$ nor $p(v) \leq p(u)$ holds, $p(u)$ and $p(v)$ are said to be *incomparable*. We extend this notion by saying that u and v are incomparable (w.r.t. their Parikh vector) if $p(u)$ and $p(v)$ are incomparable. We use the maximum norm $\|\mathbf{x}\|_\infty = \max_{i \in [n]} |x_i|$ for all $\mathbf{x} \in \mathbb{Z}^n$ to refer to the absolute maximum component of a vector.

The main object of interest in our work are *scattered factors* also known as subsequences or subwords.

Definition 1. *Let $w \in \Sigma^*$. A word $u \in \Sigma^*$ is called a* scattered factor *of w if there exists $v_1, \ldots, v_{|u|+1} \in \Sigma^*$ such that $w = v_1 u[1] \ldots u[|u|] v_{|u|+1}$. We denote the set of all scattered factors of w by* ScatFact(w). *The largest k such that* ScatFact$(w) \supseteq \Sigma^k$ *is called the* universality index *of w, denoted by $\iota(w)$.*

Definition 2. *For a set $S \subseteq \Sigma^*$ and $k \in \mathbb{N}_0$ define $S_k = S \cap \Sigma^k$ and $S_{\leq k} = \bigcup_{i=0}^{k} S_i$.*

We mostly use Definition 2 for $S = \text{ScatFact}(w)$ for a given $w \in \Sigma^*$. Roughly spoken, a word u is a scattered factor of a word $w \in \Sigma^*$, if some of w's letters are deleted and the remaining ones form u. For instance, `allow` is a scattered factor of `cauliflower` but `owl` is not since the `l` occurs only before `ow`. The latter motivates the definition of *jumbled scattered factors*: not the word itself is a scattered factor but a permutation of it is.

Definition 3. *Let $w \in \Sigma^*$. A word $u \in \Sigma^*$ is called a* jumbled scattered factor *of w (denoted $u \mid_J w$) if there exists $v \in$* ScatFact(w) *such that $p(u) = p(v)$. If neither $u \mid_J v$ nor $v \mid_J u$ holds, we say that u and v are* incomparable *(w.r.t. jumbling). The set of jumbled scattered factors of w is denoted by* JScatFact(w).

Remark 4. Two words are incomparable w.r.t. their Parikh vector if and only if they are incomparable w.r.t. jumbling. Hence, in the following we will not distinguish between both incomparabilities.

Thus, in `cauliflower`, not only `owl` but also `wolf`, `recall`, `failure`, and `firewall` are jumbled scattered factors. Notice that Definition 3 implies that every scattered factor is also a jumbled scattered factor and $\mathrm{JScatFact}(w)$ is finite for all $w \in \Sigma^*$. Inspecting `recall` and `wolf` in more detail one notices that the longest scattered factor which `recall` and `cauliflower` have in common (`call`) is longer than the one `wolf` and `cauliflower` have in common (`lf`). Before we relate the well known concept of the *longest common scattered factor* and the jumbled scattered factors, we introduce some definitions for the number of required jumbles.

Definition 5. *Let* $w \in \Sigma^*$. *The word* $u \in \Sigma^*$ *is called a* scattered factor of w with ℓ jumbles *(or* u *is* ℓ*-jumbled in* w*) for* $\ell \in [|u|-1]_0$ *if* $p(u) \leq p(w)$ *and there exists* $v \in \mathrm{ScatFact}_{|u|-\ell}(u)$ *such that* $v \in \mathrm{ScatFact}(w)$ *holds. The set of scattered factors of* w *with* ℓ*-jumbles is denoted by* $\mathrm{JScatFact}(w, \ell)$.

Note that, for $w \in \Sigma^*$, $u \in \mathrm{JScatFact}(w, \ell)$ implies that every subsequence of u is at most ℓ-jumbled in w. By Definition 5 we have that `owl` is 1-jumbled and `wolf` is not only 2-jumbled but also 3-jumbled depending on which `l` in `cauliflower` is chosen. Having a measure in mind that a jumbled scattered factor is *better*, i.e., closer to being a scattered factor, the less jumbles are required (for example in DNA-sequencing or in data transmission), we fix the smallest possible number of needed jumbles.

Definition 6. *Let* $w \in \Sigma^*$ *and* $u \in \mathrm{JScatFact}(w)$. *We define the* jumble index of u w.r.t. w *by* $\delta_w(u) = \min\{\ell \in \mathbb{N}_0 \mid u \in \mathrm{JScatFact}(w, \ell)\}$. *Let* $\overline{\mathrm{JScatFact}}(w, \ell) = \{u \in \mathrm{JScatFact}(w, \ell) \mid \delta_w(u) = \ell\}$.

Definition 6 gives us $\delta_{\texttt{cauliflower}}(\texttt{wolf}) = 2$ and $\delta_{\texttt{cauliflower}}(\texttt{failure}) = 3$. Note that $w \in \mathrm{JScatFact}(w, \ell)$ for all $\ell \in [|w|-1]_0$ but $w \in \overline{\mathrm{JScatFact}}(w, \ell) \iff \ell = 0$ for $w \in \Sigma^*$.

Remark 7. For $w \in \Sigma^*, k \in \mathbb{N}$, every $u \in \mathrm{JScatFact}_k(w)$ with $\delta_w(u) > 0$ is of length at least $\iota(w) + 1$ since $|u| \leq \iota(w)$ implies $u \in \mathrm{JScatFact}(w, 0) = \mathrm{ScatFact}(w)$. The same argument gives us that unary jumbled scattered factors u yield $\delta_w(u) = 0$.

Definition 8. *Let* $u, v \in \Sigma^*$. *We define the set of* common scattered factors *by* $\mathrm{CSF}(u, v) = \mathrm{ScatFact}(u) \cap \mathrm{ScatFact}(v)$ *and* $\mathrm{LCSF}(u, v) = \{x \in \mathrm{CSF}(u, v) \mid \forall y \in \mathrm{CSF}(u, v) : |x| \geq |y|\}$ *as the set of* longest common scattered factors of u and v. *Their length is denoted by* $\mathrm{lcsf}(u, v)$.

Some relations between the LCSF and JScatFact are immediate.

Lemma 9. *(1) Let* $u, w \in \Sigma^*$, $\ell \in [|u| - 1]_0$ *and* $v \in \mathrm{CSF}(u, w)$. *If* $\ell + |v| = |u|$ *and* $p(u) \leq p(w)$, *then* $u \in \mathrm{JScatFact}(w, \ell)$.

(2) For $w \in \Sigma^*$ *and* $u \in \mathrm{JScatFact}(w, \ell)$ *there exists* $v \in \mathrm{CSF}(u, w)$ *such that* $\ell + |v| = |u|$.

(3) For $w \in \Sigma^*$, $\ell \in \mathbb{N}$, *we have* $u \in \overline{\mathrm{JScatFact}}(w, \ell)$ *iff* $\ell = |u| - \mathrm{lcsf}(u, w)$. *Thus,* $\delta_w(u) = |u| - \mathrm{lcsf}(u, w)$.

Proof. (1) follows immediately by $v \in \mathrm{ScatFact}(u)$. By definition of u, there exists $v \in \mathrm{ScatFact}(w)$ with $|v| = |u| - \ell$ and $v \in \mathrm{ScatFact}(u)$. Thus, $v \in \mathrm{CSF}(u, w)$ and $\ell + |v| = |u|$ holds which proves (2). If we choose a longest common scattered factor, we get immediately the duality between the jumble index and longest common scattered factors. This finishes the proof of (3). □

Based on Simon's congruence [46], we also define a variant for jumbled scattered factors.

Definition 10. *Let $v, w \in \Sigma^*$ and $k, \ell \in \mathbb{N}$. We say that v is* ℓ-jumble Simon k-equivalent *to w, for short $v \overset{J}{\sim}_{k,\ell} w$ iff* $\mathrm{JScatFact}_{\leq k}(v, \ell) = \mathrm{JScatFact}_{\leq k}(w, \ell)$. *We denote the equivalence class of w under $\overset{J}{\sim}_{k,\ell}$ by $[w]_{k,l} = \{v \in \Sigma^* \mid v \overset{J}{\sim}_{k,\ell} w\}$.*

Lemma 11. *For $k, \ell \in \mathbb{N}, w \in \Sigma^*$, $\mathrm{JScatFact}_k(u, \ell) = \mathrm{JScatFact}_k(v, \ell)$ implies $\mathrm{JScatFact}_{\leq k}(u, \ell) = \mathrm{JScatFact}_{\leq k}(v, \ell)$.*

Proof. Suppose $\mathrm{JScatFact}_k(u, \ell) = \mathrm{JScatFact}_k(v, \ell)$ and $\mathrm{JScatFact}_{\leq k}(u, \ell) \neq \mathrm{JScatFact}_{\leq k}(v, \ell)$. Thus, w.l.o.g., there exists a $k' < k$ (choose k' as the largest) and $x \in \Sigma^*$ with $x \in \mathrm{JScatFact}_{k'}(u, \ell)$ (in particular $|x| = k'$) and $x \notin \mathrm{JScatFact}_{k'}(v, \ell)$. Now there exist $y_1, \ldots, y_{k'+1} \in \Sigma^*$ of length $|y_1 \cdots y_{k'+1}| = k - k'$ such that $y_1 x[1] y_2 x[2] y_3 x[3] \cdots y_{k'} x[k'] y_{k'+1} \in \mathrm{JScatFact}_k(u, \ell)$. Thus, $y_1 x[1] y_2 x[2] y_3 x[3] \cdots y_{k'} x[k'] y_{k'+1} \in \mathrm{JScatFact}_k(v, \ell)$ which implies a contradiction: $x \in \mathrm{JScatFact}_{k'}(v, \ell)$. □

This leads to the problem of investigating the jumbled variant $\overset{J}{\sim}_{k,\ell}$ of Simon's congruence relation.

Problem 12. Given two words $v, w \in \Sigma^*$ and two integers k, ℓ is $v \overset{J}{\sim}_{k,\ell} w$?

Tightly related to the classical Simon's congruence with scattered factors is the notion of universality [6,14,44] that we extend to the jumbled scattered factors. Intuitively, one may think that the notion of k-universality can be extended to ℓ-jumble k-universality by requiring that a word contains all words of length k as scattered factors with exactly ℓ-jumbles, i.e., $\overline{\mathrm{JScatFact}_k}(w, \ell) = \Sigma^k$. This definition would invoke that no word is ℓ-jumble k-universal for $\ell > 0$: all scattered factors of length k are scattered factors with exactly 0 jumbles and thus are missing in $\overline{\mathrm{JScatFact}_k}(w, \ell)$ (cf. Remark 7). For this reason we relax the generalisation of universality to jumbled-universality by allowing *up to ℓ* jumbles. This idea is captured in the following definition.

Definition 13. *A word $w \in \Sigma^*$ is called* ℓ-jumble k-universal *if*

$$\bigcup_{\ell' \in [0, \ell]} \overline{\mathrm{JScatFact}_k}(w, \ell') = \Sigma^k.$$

This leads to the problem of determining the smallest value for ℓ to obtain universality by the union of all ℓ-jumbled scattered factors.

Problem 14. Given $w \in \Sigma^*$ and $k \in \mathbb{N}$ with $\iota(w) < k$. Determine the minimal $\ell \in \mathbb{N}$ such that $\bigcup_{\ell' \in [\ell]_0} \overline{\mathrm{JScatFact}_k(w, \ell')} = \Sigma^k$.

We conclude with a tool that has proven useful for the study of scattered factor universality: the arch factorisation that was defined by Hébrard [23].

Definition 15. *For a word $w \in \Sigma^*$ the* arch factorisation *is given by* $w = \mathrm{ar}_1(w) \ldots \mathrm{ar}_k(w) r(w)$ *for* $k \in \mathbb{N}_0$ *with*
- $\iota(\mathrm{ar}_i(w)) = 1$ *for all* $i \in [k]$,
- $\mathrm{ar}_i(w)[|\mathrm{ar}_i(w)|] \notin \mathrm{alph}(\mathrm{ar}_i(w)[1..|\mathrm{ar}_i(w)| - 1])$, *and*
- $\mathrm{alph}(r(w)) \subset \Sigma$.

The unique last letters of each arch $m(w) = \mathrm{ar}_1(w)[|\mathrm{ar}_1(w)|] \ldots \mathrm{ar}_k(w)[|\mathrm{ar}_k(w)|]$ *build the* model *of* w.

For a given word $w \in \Sigma^*$ the arch factorisation decomposes a word into uniquely defined arches: read w from left to right and every time the complete alphabet was read, close an arch. For instance, consider $w =$ `alfalfa` we have the decomposition `alf.alf.a` where the dots indicate the end of an arch.

3 The Jumble Index and Jumble Division

We start this section with some basic properties of the relation $|_J$ (in this section we write $u |_J w$ instead of $u \in \mathrm{JScatFact}(w)$ since we are studying the relation). It is immediate that $|_J$ is reflexive and transitive. In contrast to the scattered factor relation [36], $|_J$ is neither symmetric nor antisymmetric: from $u |_J v$ and $v |_J u$ it only follows $p(u) = p(v)$. Let $u, u', w, w' \in \Sigma^*$. If $u |_J w$ and $u' |_J w'$ then $uu' |_J ww'$. The relation $|_J$ is a preorder on Σ^* that is compatible with $\cdot$ on Σ^*.

Remark 4 motivates the following proposition, a similar results was shown for scattered factors in [36, Theorem 6.1.2]. The proof follows immediately by [33].

Proposition 16. *Let $S \subseteq \Sigma^*$ be a set of words that are not pairwise comparable with $|_J$. Then S is finite.*

A word u can be a scattered factor of w with a different number of jumbles. For example, consider $w =$ `abba` and $u =$ `ba`. Here, the number of jumbles can be either zero or one. The discrete interval $[\delta_w(u), |u| - 1]$ describes the set of all possible numbers of jumbles for which u is a jumbled scattered factor of w. By Lemma 9 we can choose any common scattered factor of u and w as long as we choose the number of jumbles appropriately. Or to express it differently, once we have determined $\delta_w(u)$, we can simply remove a character from a fixed longest common scattered of both words, thus increasing the number of jumbles that are used for u to be a jumbled scattered factor.

Corollary 17. *Let $w \in \Sigma^*$ and $u \in \mathrm{JScatFact}(w)$. Then, $u \in \mathrm{JScatFact}(w, \ell)$ for all $\ell \in [\delta_w(u), |u| - 1]$.*

Proof. Let $u \in \text{JScatFact}(w)$, thus we have $p(u) \leq p(w)$. Then there exists $v \in \text{ScatFact}(w)$ with $p(u) = p(v)$. All $v' \in \text{CSF}(u, v)$ provide by Lemma 9 an ℓ with $u \in \text{JScatFact}(w, \ell)$. The claim follows by $0 \leq |v'| \leq \text{lcsf}(u, v)$. □

This statement also holds for the length-restricted jumbled scattered factors.

Lemma 18. *Let $w \in \Sigma^*$, $k \leq |w|$, and $\ell \in [k-2]$. Then,* $\text{JScatFact}_k(w, \ell) \subseteq \text{JScatFact}_k(w, \ell + 1)$ *and* $\text{JScatFact}_k(w, i) = \emptyset$ *for all $i \geq k$.*

Proof. Let $\ell \in [0, k-2]$ and $u \in \text{JScatFact}_k(w, \ell)$. There exists $v \in \text{CSF}(u, w)$ such that $\ell + |v| = |u|$. Then, $v \neq \varepsilon$ since $|v| = |u| - \ell \geq 1$ holds. We can remove one character from v and get $v' = v[1..|v| - 1] \in \text{CSF}(u, w)$. Thus, u is $(\ell + 1)$-jumbled w.r.t. w, since $\ell + 1 + |v'| = |u|$ holds. Furthermore, since all quantities must be non-negative in $\ell + |v| = |u| = k$, we cannot choose $|v| < 0$ to accommodate for $\ell > k$. Suppose that $\text{JScatFact}(w, \ell) \neq \emptyset$ for some $\ell > |w|$, then there exists $u \in \text{JScatFact}(w, \ell)$ and thus a $v \in \text{CSF}(u, w)$ with $|v| + \ell = |u|$. However, $|w| < \ell = |u| - |v|$ contradicts $|u| \leq |w|$. This concludes the proof. □

Whereas in the field of scattered factors, we have that $u \in \text{ScatFact}(w)$ and $w \in \text{ScatFact}(u)$ implies $w = u$, in the field of jumbled scattered factors, it only implies that u and w are permutations of each other. This leads to the following result regarding the number of jumbles.

Lemma 19. *Let $u, w \in \Sigma^*$ such that $u \mid_J w$ and $w \mid_J u$. Then, $|u| = |w|$, $\delta_w(u) = \delta_u(w)$.*

Proof. The first claim follows immediately by the definition, in particular $p(u) \leq p(w)$ and $p(w) \leq p(u)$ imply $p(u) = p(w)$ which implies $|u| = |w|$. The second part follows from $\text{lcsf}(u, w) = \text{lcsf}(w, u)$: $\delta_u(w) = |w| - \text{lcsf}(w, u) = |u| - \text{lcsf}(u, w) = \delta_w(u)$. □

Another difference to *unjumbled* scattered factors is the concatenation of jumbled scattered factors. The words `life` and `low` are both scattered factors of $w =$ `cauliflower` and even the sum of their Parikh vectors is smaller than the Parikh vector of w but `lowlife` is not a scattered factor of w. In the scenario of jumbled scattered factors, we have that $\delta_{\texttt{cauliflower}}(\texttt{low} \cdot \texttt{life}) = 3$, thus the jumble index is higher than the sum of the jumble indices. The sum is the lower bound, e.g. $\delta_{\texttt{cauliflower}}(\texttt{ufo} \cdot \texttt{war}) = 1$.

Lemma 20. *Let $w \in \Sigma^*$ and $uv \in \text{JScatFact}(w)$. Then, $\delta_w(uv) \geq \delta_w(u) + \delta_w(v)$.*

Proof. Choose $\ell \in \mathbb{N}_0$ minimal with $uv \in \text{JScatFact}(w, \ell)$. Then, we have $p(uv) \leq p(w)$ and there exists

$$x \in \text{ScatFact}_{|uv| - \delta_w(uv)}(uv) \cap \text{ScatFact}(w).$$

Thus, there exist $x_u, x_v \in \Sigma^*$ with $x = x_u x_v$, $x_u \in \text{ScatFact}(u)$, and $x_v \in \text{ScatFact}(v)$. Moreover, we have

$$|x_u| + |x_v| = |x_u x_v| = |x| = |uv| - \delta_w(uv) = |u| + |v| - \delta_w(uv).$$

Thus, there exist ℓ_u and ℓ_v with $|x_u| = |u| - \ell_u$, $|x_v| = |v| - \ell_v$ and $\delta_w(uv) = \ell_u + \ell_v$. Accordingly, $x_u \in \mathrm{ScatFact}_{|u|-\ell_u}(u)$ and $x_v \in \mathrm{ScatFact}_{|v|-\ell_v}(v)$. Because x is a subsequence of w, x_u and x_v also are subsequences of w. It follows $\delta_w(u) \leq \ell_u$, $\delta_w(v) \leq \ell_v$ and, thereby, $\delta_w(uv) = \ell_u + \ell_v \geq \delta_w(u) + \delta_w(v)$. □

Moreover, for the concatenation of not only the jumbled scattered factors but also the words, we can bind the jumble index. Consider for instance $w = \mathtt{cba}$, $u = \mathtt{abc} \in \mathrm{JScatFact}(w)$, with $\delta_w(u) = 2$, and $w' = \mathtt{bcbaa}$, $v = \mathtt{ab} \in \mathrm{JScatFact}(w')$, with $\delta_{w'}(v) = 1$. In $ww' = \mathtt{cbabcbaa}$, we have $u\mathtt{a}$ as a scattered factor and thus $\delta_{ww'}(uv) = 1$. For the total number of jumbles, we had the advantage that u is a scattered factor of ww'. Generally, in this case we have the sum $\delta_w(u) + \delta_{w'}(v)$ as upper bound for $\delta_{ww'}(uv)$.

Lemma 21. *Let $u, v, w, w' \in \Sigma^*$. If $u \in \mathrm{JScatFact}(w)$ and $v \in \mathrm{JScatFact}(w')$ then $uv \in \mathrm{JScatFact}(ww')$ with $\min\{\delta_w(u), \delta_{w'}(v)\} \leq \delta_{ww'}(uv) \leq \delta_w(u) + \delta_{w'}(v)$.*

Proof. The fact that uv is a jumbled scattered factor of ww' follows directly from Definition 3. The inequality holds by the minimality of the jumbled index. □

We finish this section with an immediate result on the relation of the index of a word and its index when extending the word by one letter.

Remark 22. Let $w \in \Sigma^*$ and $u \in \mathrm{JScatFact}(w)$. Then, $\delta_w(ux) - \delta_w(u) \leq 1$ holds for all $x \in \Sigma$ with $p(ux) \leq p(w)$.

4 Jumbled Simon's Equivalence $\overset{J}{\sim}_{k,\ell}$

In this section, we investigate the extension of the famous Simon congruence [46] in the jumbled scenario. Note that there are subsets $M \subset \Sigma^{\leq k}$ such that for all $\ell \in \mathbb{N}_0$ and $S \subseteq \Sigma^*$ it holds $M \neq \bigcup_{w \in S} \mathrm{JScatFact}_{\leq k}(w, \ell)$ (this property is already true for $\ell = 0$, i.e., scattered factors). For instance, there exists no $w \in \Sigma^*, \ell \in \mathbb{N}_0$ such that $\mathrm{JScatFact}_{\leq k}(w, \ell) = \{\mathtt{a}, \mathtt{bb}\} \subset \Sigma^{\leq 2}$ - at least one of $\mathtt{ab}$ or $\mathtt{ba}$ needs to be a (0-jumbled) scattered factor. Moreover, if $|w| < k$ then $[w]_{k,\ell} = \{w\}$. Thus, we consider only values $\iota(w) \leq k \leq |w|$ and $0 \leq \ell < k$. We extend results from [36, 46] on $\sim_k$ to the jumbled case $\overset{J}{\sim}_{k,\ell}$ and continue with new results for the structure of jumbled Simon's equivalence classes (Problem 12).

Before we start with insights about $\overset{J}{\sim}_{k,\ell}$, we present some examples for the kinds of equivalence classes that may occur. First, consider some singleton classes $[\mathtt{ab}]_{\overset{J}{\sim}_{2,0}}$ or $[\mathtt{abcd}]_{\overset{J}{\sim}_{3,2}}$. Second, the class $[\mathtt{ab}]_{\overset{J}{\sim}_{2,1}} = \{\mathtt{ab}, \mathtt{ba}\}$ containing exactly the two words with the Parikh vector $(1, 1)$. And the infinite class $[\mathtt{aab}]_{\overset{J}{\sim}_{2,1}} = \{\mathtt{a}^n\mathtt{ba}^m \mid n, m \in \mathbb{N}_0, n + m \geq 2\}$ that all share the 1-jumbled scattered factor set of length 2, i.e. $\mathrm{JScatFact}_2(w, 1) = \{\mathtt{aa}, \mathtt{ab}, \mathtt{ba}\}$ for all $w \in [\mathtt{aab}]_{\overset{J}{\sim}_{2,1}}$.

Lemma 23. *$\overset{J}{\sim}_{k,\ell}$ is a equivalence relation on Σ^* of finite index.*

Proof. It is immediate that $\overset{J}{\sim}_{k,\ell}$ is an equivalence relation.

Further, the index of $\overset{J}{\sim}_{k,\ell}$ is equal to the number of distinct suitable choices for $\mathrm{JScatFact}_{\leq k}(w,\ell)$ for all $w \in \Sigma^*$ (each equivalence class is uniquely determined by the set of its jumbled scattered factors). We have

$$\left| \bigcup_{w\in\Sigma^*} \mathrm{JScatFact}_{\leq k}(w,\ell) \right| \leq |\mathcal{P}(\Sigma^{\leq k})| = 2^{|\Sigma^{\leq k}|}.$$

□

Note that $\overset{J}{\sim}_{k,\ell}$ is in contrast to $\sim_k$ not a congruence relation. As an example consider $w = \mathtt{cd}$ and $v = \mathtt{dc}$. We have $w \overset{J}{\sim}_{3,1} v$ but $\mathtt{ab}w \overset{J}{\not\sim}_{3,1} \mathtt{ab}v$ since $\mathtt{cda} \in \mathrm{JScatFact}_3(\mathtt{ab}w, 1)$ and $\mathtt{cda} \notin \mathrm{JScatFact}_3(\mathtt{ab}v, 1)$.

Lemma 24. *Let $\ell, k \in \mathbb{N}$, $\ell \leq k-2$ and $w, v \in \Sigma^*$ satisfy $w \overset{J}{\sim}_{k+1,\ell} v$. Then $w \overset{J}{\sim}_{k+1,\ell+1} v$, $w \overset{J}{\sim}_{k,\ell} v$, and $w \overset{J}{\sim}_{k,\ell+1} v$ also hold.*

Proof. First, let $w \overset{J}{\sim}_{k+1,\ell} v$ and assume $w \overset{J}{\not\sim}_{k+1,\ell+1} v$. Then, without loss of generality, there exists $u \in \Sigma^m$, with $m \leq k+1$, such that $\delta_v(u) = \ell + 1 < m$ and $\delta_w(u) \geq \ell + 2$. Let $u = u_1 \cdots u_m$, $\hat{u} = u_{i_1} \cdots u_{i_{m-\delta_v(u)}}$ be a (longest) common scattered factor of u and v and $j \notin \{i_1, \ldots, i_{m-\delta_v(u)}\}$. Then $\tilde{u} = u_1 \cdots u_{j-1} u_{j+1} \cdots u_m$ and v have the same common scattered factor $\hat{u}$. Accordingly, $\tilde{u} \in \mathrm{JScatFact}_{\leq k+1}(v,\ell) = \mathrm{JScatFact}_{\leq k+1}(w,\ell)$ and there exists a common scattered factor $\bar{u}$ of length $|\tilde{u}| - \ell$ of $\tilde{u}$ and w. Lastly, $\bar{u}$ is also a common scattered factor of u and w which implies $\delta_w(u) \leq |u| - |\bar{u}| = |\tilde{u}| + 1 - |\bar{u}| = \ell + 1$.

Second, $w \overset{J}{\sim}_{k,\ell} v$ holds because any jumbled scattered factor of length k of v and, respectively, w occurs as scattered factor in a jumbled scattered factor of length $k+1$ of v and, respectively, w.

The last statement follows as a combination of the previous two.

Let $v \in \mathrm{JScatFact}(w)$. First, assume that $v \overset{J}{\sim}_{k,\ell} w$ holds and suppose that there exists a $u \in \Sigma^*$ with $|u| \leq k$ and $\ell \geq \delta_w(u)$ such that $u \notin \mathrm{JScatFact}(v)$. By definition this implies $v \notin \mathrm{JScatFact}(w)$, a contradiction. Second, assume that for all $u \in \Sigma^*$, $|u| \leq k$ and $\ell \geq \delta_w(u)$ implies $u \in \mathrm{JScatFact}(w)$. We are going to show that $v \overset{J}{\sim}_{k,\ell} w$ by supposing the contrary. Suppose $v \overset{J}{\not\sim}_{k,\ell} w$. Since $v \in \mathrm{JScatFact}(w)$, there exists $u \in \mathrm{JScatFact}_{\leq k}(w)$ and $\ell \geq \delta_w(u)$ such that $u \notin \mathrm{JScatFact}(v)$, again a contradiction to the assumption. □

Lemma 25. *For all $u \in \Sigma^*$ and $\mathtt{a} \in \Sigma$ there exist $p, \ell \in \mathbb{N}_0$, $s \in \Sigma^*$ such that $u \overset{J}{\sim}_{p,\ell} u\mathtt{a}$, $s \in \mathrm{JScatFact}_p(u)$ with $\delta_u(s) \leq \ell$ and $s\mathtt{a} \notin \mathrm{JScatFact}(u)$.*

Proof. The existence of such p follows from Lemma 24 since $u \overset{J}{\sim}_{0,\ell} u\mathtt{a}$ and $u \overset{J}{\not\sim}_{|u|+1,\ell} u\mathtt{a}$ hold for every $\ell \in \mathbb{N}_0$. Thus, we may choose the greatest p with $u \overset{J}{\sim}_{p,\ell} u\mathtt{a}$ and $u \overset{J}{\not\sim}_{p+1,\ell} u\mathtt{a}$. Then the existence of $s \in \mathrm{JScatFact}_p(u)$ with $s\mathtt{a} \notin \mathrm{JScatFact}_p(u)$ follows. □

We now continue by a characterisation of the equivalence classes of $\overset{J}{\sim}_{k,\ell}$. Note that $\ell = 0$ implies by [36] that $\left|[w]_{\overset{J}{\sim}_{k,\ell}}\right| \in \{1, \infty\}$ since $\overset{J}{\sim}_{k,0}$ is the same as Simon's congruence relation. Nevertheless, the classes of infinite size differ if $\ell \geq 1$, e.g., $[\mathtt{aab}]_{\sim_3}$ is a singleton congruence class whereas $[\mathtt{aab}]_{\overset{J}{\sim}_{2,\ell}}$ is an infinite equivalence class for $\ell \geq 1$. One may think that singleton classes of $\overset{J}{\sim}_{k,\ell}$ only exist for $\ell = 0$, i.e., in the *classical* Simon's congruence. Consider $w = \mathtt{abcd} \in \{\mathtt{a}, \mathtt{b}, \mathtt{c}, \mathtt{d}\}^*$ providing a counterexample: we have $|\,\mathrm{JScatFact}_3(\mathtt{abcd}, 1)| = 20$ (missing the words $\mathtt{cba}, \mathtt{dba}, \mathtt{dca}, \mathtt{dcb}$); no word of length > 4 is in this class because $\mathtt{xxy} \notin \mathrm{JScatFact}_3(w, 1)$, for all $\mathtt{x}, \mathtt{y} \in \Sigma$, and similarly any word missing one of the four letters cannot be equivalent since we would lose jumbled scattered factors; lastly no permutation of $\mathtt{abcd}$ is in the class since there exists a jumbled scattered factor of any permutation of $\mathtt{abcd}$ that is contained in the set $\mathrm{JScatFact}_3(\mathtt{abcd}, 1)$ but needs more than 1 jumble.

Remark 26. Let $k, \ell \in \mathbb{N}$. Note, that two words $w, w' \in \Sigma^*$ with $\mathrm{alph}(w) = \mathrm{alph}(w')$ and $|w|_{\mathtt{a}} < k$ for all $\mathtt{a} \in \Sigma$ cannot be equivalent if there exists $\mathtt{b} \in \Sigma$ such that $|w|_{\mathtt{b}} \neq |w'|_{\mathtt{b}}$.

As a first step we consider infinite classes. The main idea on the characterisation of infinite equivalence classes is that a letter that already occurs more than or equal to k times in a sufficiently large block can also occur more often: all further occurrences are superfluous for k-length equivalence. Recall that we only need to investigate $0 \leq \ell < k$.

Proposition 27. *Let $w \in \Sigma^*$ and $k, \ell \in \mathbb{N}$. If there exist $\mathtt{a} \in \mathrm{alph}(w)$, $k' \geq k - \ell$, and $x, y \in \Sigma^*$ with $|w|_{\mathtt{a}} \geq k$ and $w = x\mathtt{a}^{k'}y$ then $|[w]_{\overset{J}{\sim}_{k,\ell}}| = \infty$.*

Proof. Let $\mathtt{a} \in \mathrm{alph}(w)$, $k' \geq k - \ell$ and $x, y \in \Sigma^*$ with $|w|_{\mathtt{a}} \geq k$ and $w = x\mathtt{a}^{k'}y$. We now want to show that $w \overset{J}{\sim}_{k,\ell} w'$ for $w' = x\mathtt{a}^{k'+1}y$, i.e., $\mathrm{JScatFact}_k(w, \ell) = \mathrm{JScatFact}_k(w', \ell)$.

Let $u \in \mathrm{JScatFact}_k(w, \ell)$. Thus, there exists $v \in \mathrm{ScatFact}_{k-\ell}(u)$ with $v \in \mathrm{ScatFact}(w)$. By construction, we have that $v \in \mathrm{ScatFact}(w')$. By $p(u) \leq p(w) \leq p(w')$, we have $u \in \mathrm{JScatFact}_k(w', \ell)$.

For the other inclusion, let $u' \in \mathrm{JScatFact}_k(w', \ell)$. So, there exists $v' \in \mathrm{ScatFact}_{k-\ell}(u')$ with $v' \in \mathrm{ScatFact}(w')$. Since $k' + 1 > k - \ell$ we obtain $p(u') \leq p(w)$, i.e., that at least one of the $\mathtt{a}$'s in the block of length k' must not be used in v'. Thus, we have $v' \in \mathrm{ScatFact}_{k-\ell}(w)$.

The claim follows by induction. □

Note that the restriction on the length of the $\mathtt{a}$-block is necessary. Consider for example $k = 6, \ell = 1$, $w = \mathtt{cba}^3\mathtt{cba}^3$ and $w' = \mathtt{cba}^4\mathtt{cba}^3$. One can verify that $\mathtt{a}^4\mathtt{bc} \in \mathrm{JScatFact}_6(w, 2)$ with $\delta_w(\mathtt{a}^4\mathtt{bc}) = 2$ and $\mathtt{a}^4\mathtt{bc} \in \mathrm{JScatFact}_6(w', 1)$. Now, we show that if there exist two equivalent words with a different Parikh vector then there always exists a word within the same equivalence class that contains the two equivalent words as its jumbled scattered factors. Following the idea given in Proposition 27 this word will be constructed by the addition of superfluous letters that occur more than k times anyway.

Corollary 28. *For all* $w, u \in \Sigma^*, k, \ell \in \mathbb{N}$ *with* $p(w) \neq p(u)$ *and all* $\mathtt{a}_i \in \Sigma$ *with* $p(w)[i]$, $p(u)[i] \geq k$, $\mathtt{a}_i^{k'} \in \mathrm{Fact}(w) \cap \mathrm{Fact}(u)$ *for* $k' \geq k - \ell$, *and* $w \overset{J}{\sim}_{k,\ell} u$ *there exists* $v \in \Sigma^* \setminus \{u, w\}$ *such that* $v \overset{J}{\sim}_{k,\ell} w, v \overset{J}{\sim}_{k,\ell} u$.

Proof. Let $w \overset{J}{\sim}_{k,\ell} u$ with $p(w) \neq p(u)$. First, we want to show that for all $i \in [\sigma]$ with $p(w)[i] \neq p(u)[i]$ it follows that $p(u)[i], p(w)[i] \geq k$. For the sake of contradiction suppose the contrary, i.e., that there exists $i \in [\sigma]$ with $p(w)[i] \neq p(u)[i]$ and with $p(w)[i] < k$ or $p(u)[i] < k$. W.l.o.g. let $p(w)[i] < p(u)[i]$. Thus, there exists $x \in \mathrm{JScatFact}_k(u, \ell)$ with $p(x)[i] = p(u)[i]$. Since $p(w)[i] < p(u)[i]$ and at least one of those values is smaller than k we obtain $x \notin \mathrm{JScatFact}_k(w, \ell)$ - a contradiction to $w \overset{J}{\sim}_{k,\ell} u$.

Thus, for all $i \in [\sigma]$ with $p(w)[i] \neq p(u)[i]$ we have that $p(w)[i] \geq k$ and $p(u)[i] \geq k$. Let $\mathtt{a}_i$ be a letter for which the Parikh vectors of u and w are different. W.l.o.g. let w be the word that contains a larger block of $\mathtt{a}_i$. By assumption we can write w by $w = x\mathtt{a}_i^{k'}y$ for some $x, y \in \Sigma^*$ and $k' \geq k - \ell$. Now we can construct a v following the proof of Proposition 27: Let $v = x\mathtt{a}_i^{k'+1}y$. Thus, $v \overset{J}{\sim}_{k,\ell} w \overset{J}{\sim}_{k,\ell} u$. It is left to show that $w, u \in \mathrm{JScatFact}(v)$. The fact that $w \in \mathrm{JScatFact}(v)$ is immediately visible. □

Note that we can also conclude $w, u \in \mathrm{JScatFact}(v)$ from the above corollary if $p(w) \leq p(v)$ or $p(u) \leq p(v)$ additionally holds for v.

Now we show when two equivalent words are abelian equivalent.

Proposition 29. *Let* $w, w' \in \Sigma^*, k \in \mathbb{N}_{>1}$ *with* $|w|_\mathtt{a}, |w'|_\mathtt{a} < k$ *for all* $\mathtt{a} \in \Sigma$. *We have* $w \overset{J}{\sim}_{k,k-1} w'$ *iff* $p(w) = p(w')$.

Proof. We have

$$\begin{aligned}
& w \overset{J}{\sim}_{k,k-1} w' \\
\iff & \{u \in \Sigma^k \mid (\exists v \in \mathrm{ScatFact}_1(u) : v \in \mathrm{ScatFact}(w)) \wedge p(u) \leq p(w)\} \\
& = \{u' \in \Sigma^k \mid (\exists v' \in \mathrm{ScatFact}_1(u') : v' \in \mathrm{ScatFact}(w')) \wedge p(u) \leq p(w')\} \\
\iff & \{u \in \Sigma^k \mid (\exists v \in \Sigma : v \in \mathrm{alph}(u) \cap \mathrm{alph}(w)) \wedge p(u) \leq p(w)\} \\
& = \{u' \in \Sigma^k \mid (\exists v' \in \Sigma : v' \in \mathrm{alph}(u') \cap \mathrm{alph}(w')) \wedge p(u') \leq p(w')\} \\
\iff & p(w) = p(w').
\end{aligned}$$

The last equivalence holds because we can choose $u = a_i^{p(w)[i]}$ and $u' = a_i^{p(w')[i]}$ for all $a_i \in \mathrm{alph}(w)$. □

Corollary 30. *Let* $w \in \Sigma^*, k \in \mathbb{N}_{>1}, \ell \in \mathbb{N}$ *with* $|w|_\mathtt{a} < k$ *for all* $\mathtt{a} \in \Sigma$. *We have* $[w]_{\overset{J}{\sim}_{k,\ell}} = \Pi(w)$ *iff* $\ell = k - 1$.

Whether other sizes of equivalence classes are possible stays an open problem for future research.

5 Jumbled Universality

In this section, we want to investigate Problem 14, i.e., reach k-universality w.r.t. some word with as few jumbles as possible. Recall that $\ell < |w|$ and $\iota(w) < k \leq |w|$. For instance, for $w = \mathtt{a}^2\mathtt{b}^2\mathtt{c}^2$, we have $\overline{\text{JScatFact}_2(w,0) \cup \text{JScatFact}_2(w,1)} = \Sigma^2$. Note that we can only get Σ^k if all letters occurs at least k times in w, i.e., $k \in [\iota(w), \min_{\mathtt{a}\in\Sigma} |w|_{\mathtt{a}}]$. Set $I = [\iota(w), \min_{\mathtt{a}\in\Sigma} |w|_{\mathtt{a}}]$ for abbreviation.

Lemma 31. *Let $w \in \Sigma^*$ and $k+1 \in I$. Then, we have for all $\ell \in [|w|]$*

$$\text{JScatFact}_k(w,\ell) = \Sigma^k \Rightarrow \text{JScatFact}_{k+1}(w,\ell+1) = \Sigma^{k+1}.$$

Proof. By definition, we have $\text{JScatFact}_{k+1}(w,\ell+1) \subseteq \Sigma^{k+1}$. Let $u \in \Sigma^{k+1}$, $v \in \Sigma^k$ and $x \in \Sigma$ with $u = vx$. Since $v \in \text{JScatFact}_k(w,\ell)$ and $p(u) \leq p(w)$, we can apply Remark 22 and get $\delta_w(u) \leq \delta_w(v) + 1 \leq \ell + 1$. By Lemma 18 we get $u \in \text{JScatFact}_{k+1}(w,\ell+1)$. □

With Lemma 31 we can show that we need at most $k - \iota(w)$ many jumbles.

Proposition 32. *For all $w \in \Sigma^*$ there exist $\ell \leq k - \iota(w)$ and $k \in I$, such that $\bigcup_{\ell' \in [0,\ell]} \text{JScatFact}_k(w,\ell') = \Sigma^k$.*

Proof. We only need to consider $\ell = k - \iota(w)$ since for any larger ℓ the union of jumbled scattered factors just becomes larger as well. We show this claim by induction over k. If we consider $k = \iota(w)$, then we can choose $\ell = 0$ and get $\text{JScatFact}_k(w,0) = \text{ScatFact}_k(w) = \Sigma^k$ since w is k-universal.

Now assume $\bigcup_{\ell' \in [0,\ell]} \text{JScatFact}_k(w,\ell') = \Sigma^k$ with $\ell = k - \iota(w)$ for some $k \in [\iota(w), \min_{\mathtt{a}\in\Sigma} |w|_{\mathtt{a}} - 1]$. By Lemma 31, we can conclude

$$\bigcup_{\ell' \in [0,\ell+1]} \text{JScatFact}_{k+1}(w,\ell') = \Sigma^{k+1} \quad \text{with } \ell + 1 = k + 1 - \iota(w). \quad \square$$

The next lemma shows that in each iteration, we get new scattered factors, if the jumbled universality is not reached yet.

Proposition 33. *Let $w \in \Sigma^*$, $k \in I$ and $\text{JScatFact}_k(w,\ell) \subset \Sigma^k$. Then, we have $\text{JScatFact}_k(w,\ell+1) \setminus \text{JScatFact}_k(w,\ell) \neq \emptyset$.*

Proof. We show this claim by *dejumbling* an arbitrary word that needs more than ℓ jumbles. Let $u \in \Sigma^k \setminus \text{JScatFact}_k(w,\ell)$. This implies $\delta_w(u) > \ell$. If $\delta_w(u) = \ell+1$, then we are done. Thus, we assume $\delta_w(u) > \ell + 1$. Let $x, y \in \Sigma^*, \mathtt{a} \in \Sigma$ such that $u = x\mathtt{a}y$ and $\delta_w(xy) = \delta_w(u) - 1$. Furthermore, let $\alpha, \beta \in \Sigma^*$ such that $w = \alpha\beta$ and $\text{lcsf}(x,\alpha) + \text{lcsf}(y,\beta) = \text{lcsf}(u,w)$. W.l.o.g. we assume $|\alpha|_{\mathtt{a}} > |x|_{\mathtt{a}}$ (otherwise consider β and y). Then, there exist $\mu, \nu \in \Sigma^*$ such that $x = \mu\nu$ and $\text{lcsf}(\alpha, \mu\mathtt{a}\nu) = \text{lcsf}(\alpha, x) + 1$. Now, we have $u' = \mu\mathtt{a}\nu y \notin \text{JScatFact}_k(w,\ell)$ with $|u'| = k$ and $\delta_w(u') = \delta(u) - 1$. We can repeat this process $\delta_w(u) - (\ell+1)$ times until we get a word that is $(\ell+1)$-jumbled. □

Note that $k - \iota(w)$ is not a tight bound for ℓ witnessed by $w' = \mathtt{a}^3\mathtt{b}^3$ resulting in $\overline{\text{JScatFact}_3(w,0)} \cup \overline{\text{JScatFact}_3(w,1)} = \Sigma^3$. The following part connects the classical k-universality with the jumbled k-universality. Similarly as in Sect. 4 we observe that the letters that occur more than k times play an important role in the following results.

Lemma 34. *For* $w \in \Sigma^*$ *with* $\iota(w) = k$ *and* $|w|_{\mathtt{a}} \geq k+1$ *for all* $\mathtt{a} \in \Sigma$, *we have* $\overline{\text{JScatFact}_{k+1}(w,1)} = \text{ScatFact}_k(w) \cdot \text{alph}(r(w))^{\complement} \setminus \text{ScatFact}_{k+1}(w)$.

Proof. Let $u \in \overline{\text{JScatFact}_{k+1}(w,1)}$. Then, $u[1..k] \in \text{ScatFact}_k(w)$ since w is k-universal. If $u[k+1]$ were in $\text{alph}(r(w))$, then $\delta_w(u) = 0 \neq 1$. □

Let $u \in \text{ScatFact}_k(w) \cdot \text{alph}(r(w))^{\complement}$ and $u \notin \text{ScatFact}_{k+1}(w)$. Then, u must have at least one jumble and it cannot have more than one, since $u[1..k] \in \text{ScatFact}_k(w)$. Furthermore,

$$\|p(u)\|_\infty \leq \|p(u[1..k])\|_\infty + \|p(u[k+1])\|_\infty = k+1.$$

By $|w|_{\mathtt{a}} \geq k+1$ for all $\mathtt{a} \in \Sigma$, each entry of $p(w)$ ist at least $k+1$. This concludes the proof. □

Proposition 35. *For* $w \in \Sigma^*$ *with* $\iota(w) = k$ *and* $|w|_{\mathtt{a}} \geq k+1$ *for all* $\mathtt{a} \in \Sigma$, *we have* $\overline{\text{JScatFact}_{k+1}(w,0)} \cup \overline{\text{JScatFact}_{k+1}(w,1)} = \Sigma^{k+1}$.

Proof. The following equation holds

$$\begin{aligned} &\overline{\text{JScatFact}_{k+1}(w,0)} \cup \overline{\text{JScatFact}_{k+1}(w,1)} \\ =&\text{ScatFact}_{k+1}(w) \cup (\text{ScatFact}_k(w) \cdot \text{alph}(r(w))^{\complement}) \setminus \text{ScatFact}_{k+1}(w) \\ =&\text{ScatFact}_{k+1}(w) \cup \text{ScatFact}_k(w) \cdot \text{alph}(r(w))^{\complement} \\ =&\text{ScatFact}_k(w) \cdot \text{alph}(r(w)) \cup \text{ScatFact}_k(w) \cdot \text{alph}(r(w))^{\complement} \\ =&\Sigma^{k+1}. \end{aligned}$$

□

Lemma 36. *Let* $w \in \Sigma^*$ *with* $\iota(w) = k$ *and* $|w|_{\mathtt{a}} \geq k+1$ *for all* $\mathtt{a} \in \Sigma$. *For all* $x \in \text{alph}(r(w))^{\complement}$ *there exists* $w_x \in \text{JScatFact}(w,1)$ *such that* $\text{ScatFact}_{k+1}(w) \cup \bigcup_{x \in \text{alph}(r(w))^{\complement}} \text{ScatFact}_{k+1}(w_x) = \Sigma^{k+1}$.

Proof. Let $w \in \Sigma^*$ be k-universal and $|w|_{\mathtt{a}} \geq k+1$ for all $\mathtt{a} \in \Sigma$. Let $x \in \text{alph}(r(w))^{\complement}$. Since $|w|_{\mathtt{a}} \geq k+1$ and w is just k-universal, there exists a smallest $i \in [k]$ with $|\text{ar}_i(w)|_x > 1$. Let $\alpha, \beta \in \Sigma^*$ with $x \notin \text{alph}(\alpha)$ such that $\text{ar}_i(w) = \alpha x \beta$. We define

$$w_x = \text{ar}_1(w) \cdots \text{ar}_{i-1}(w) \cdot \alpha \cdot \beta \cdot \text{ar}_{i+1}(w) \cdots \text{ar}_k(w) \cdot x \cdot r(w).$$

We can write $w = uxv \cdot r(w)$ and $w_x = uvx \cdot r(w)$ for some $u, v \in \Sigma^*$. Then, we have

$$\text{lcsf}(w, w_x) = \text{lcsf}(uxv \cdot r(w), uvx \cdot r(w)) = |u| + \text{lcsf}(xv, vx) + |r(w)|.$$

Note that $\text{lcsf}(xv, vx) \geq |v|$.
Supposition: $\text{lcsf}(xv, vx) = |xv|$
This is equivalent to $xv = vx$. By the Lyndon-Schützenberger Theorem, also known as xy-yx-lemma, we get $v = x^{|v|}$. This implies $i = k$ and $v = \beta$. Thus, $|\,\text{ar}_k(w)|_x = 2$ and $m(w)[|m(w)|] = x$ - a contradiction to the arch factorisation. Hence we have $\text{lcsf}(xv, vx) = |v|$ and consequently

$$\text{lcsf}(w, w_x) = |u| + |v| + |r(w)| = |w| - 1.$$

This implies $\delta_w(w_x) = 1$ as claimed.

For the second claim the $\subseteq$-inclusion is immediate. Let $u \in \Sigma^{k+1}$. If u is a scattered factor w, then the statement trivially holds. Thus, we consider the case that u is not a scattered factor of w. Let $v \in \Sigma^*$ and $x \in \Sigma$ such that $u = vx$. Then $u \in \text{ScatFact}_{k+1}(w_x)$ since either $\iota(w_x) = k + 1$, or $\iota(w_x) = k$ and $x \in \text{alph}(r(w_x))$ by construction. □

Note that w_x is not unique for a given $x \in \text{alph}(r(w))^{\mathsf{C}}$: given $\mathtt{aabc.bcbc}$ we may move the $\mathtt{a}$ in front of and at the end of $r(w)$. If x is the only absent letter in the rest, then putting it in front of the rest yields a shortest new arch. For Lemma 36 it suffices to pick one. Nevertheless, we believe that it is worth to apply this approach of moving letters on one such w_x for a given x. Before we define a refined factorisation, we give an example. Consider the word $\mathtt{a^3bcd.ab^3cd.cd^2cd}$ over $\Sigma = \{\mathtt{a}, \mathtt{b}, \mathtt{c}, \mathtt{d}\}$. The alphabet of the rest (in the first iteration) is $\Sigma_1 = \{\mathtt{c}, \mathtt{d}\}$ and we factorise the rest w.r.t. $\text{alph}(r(w))$ obtaining $r(w) = \mathtt{cd.dc.d}$ with $r^1(w) = \mathtt{d}$. Thus, $\mathtt{cd}$ is the first arch of $\text{r}(w)$ which we can fill up with *superfluous* letters from w's arches. Now, we iterate through Σ: considering $\mathtt{a}$, we notice that in $\text{ar}_i(w)$, we have three $\mathtt{a}$ and by Lemma 36 we can move them towards the end in order to increase the universality. Thus here, we get $\mathtt{abcd.ab^3cd.acd.adc.d}$. For $\mathtt{b} \in \Sigma$, we apply the same idea obtaining $\mathtt{abcd.abcd.bacd.badc.d}$ - a 4-universal word. Thus, w itself is jumbled 4-universal.

For easier readability we introduce first the definition of the *potential* using the observation that only $k \in [\iota(w), \min_{\mathtt{a} \in \Sigma} |w|_{\mathtt{a}}]$ needs to be considered.

Definition 37. *Define the potential of* $w \in \Sigma^*$ *by* $\nabla(w) = \min_{\mathtt{a} \in \Sigma} |w|_{\mathtt{a}} - \iota(w)$.

By the definition of k-universality, we have $\nabla(w) \geq 0$. If $\text{alph}(w) \subset \Sigma$ then we have $\nabla(u) = 0$.

Lemma 38. *Let* $u, v \in \Sigma^*$ *with* $\min_{\mathtt{a} \in \Sigma} |uv|_{\mathtt{a}} = \min_{\mathtt{a} \in \Sigma} |u|_{\mathtt{a}} + \min_{\mathtt{a} \in \Sigma} |v|_{\mathtt{a}}$. *Then,* $\nabla(u) + \nabla(v) \in \{\nabla(uv), \nabla(uv) - 1\}$.

Proof. There exists $\mathtt{a} \in \Sigma$ with $\min_{x \in \Sigma} |u|_x = |u|_{\mathtt{a}}$ and $\min_{x \in \Sigma} |v|_x = |v|_{\mathtt{a}}$. Thus, we have

$$\nabla(u) + \nabla(v) = |u|_{\mathtt{a}} - \iota(u) + |v|_{\mathtt{a}} - \iota(v) = |uv|_{\mathtt{a}} - (\iota(u) + \iota(v)).$$

Since $\mathtt{a}$ occurs minimally often in u and in v resp. it occurs minimally often in uv. The claim follows by [14]. □

In this section, we started to investigate the jumbled scattered factor universality. These results are only a beginning of the research in this area.

6 Conclusion

In this work, we introduced jumbled scattered factors as a combination of (absent) scattered factors and jumbled words. After fundamental properties on the jumbled index, we continued the investigation of the jumbled Simon's equivalence and the jumbled k-universality based on the respective versions of scattered factors. For the equivalence relation, it is left as future work to give a more fine-grained analysis of the classes that contain words with the same Parikh vector and to determine its number of equivalence classes which might lead to insights of the number of congruence classes for the original Simon's congruence. Regarding the universality, we are also interested in the following problem: Given $w \in \Sigma^*$ with $\ell \in [|w|]$. What is the maximal value k s.t. $\bigcup_{\ell' \in [0,\ell]} \mathrm{JScatFact}_k(w, \ell') = \Sigma^k$? In light of the recent interest of universality from a language perspective this and Problem 14 could be extended to the inspection of not only a single but a set of words.

Further, one can extend classical scattered factor problems from combinatorics on words to their jumbled equivalent, e.g., finding the longest common jumbled scattered factor for a given pair of words.

References

1. Adamson, D., Fleischmann, P., Huch, A., Koß, T., Manea, F., Nowotka, D.: k-universality of regular languages. In: ISAAC 2023. LIPIcs, vol. 283, pp. 4:1–4:21. Schloss Dagstuhl - Leibniz-Zentrum für Informatik (2023)
2. Adamson, D., Fleischmann, P., Huch, A., Wiedenhöft, M.: Tight bounds for the number of absent scattered factors. arXiv preprint arXiv:2407.18599 (2024)
3. Artikis, A., Margara, A., Ugarte, M., Vansummeren, S., Weidlich, M.: Complex event recognition languages: tutorial. In: DEBS, pp. 7–10 (2017)
4. Baeza-Yates, R.A.: Searching subsequences. Theor. Comput. Sci. **78**(2), 363–376 (1991)
5. Bannai, H., I, T., Kociumaka, T., Köppl, D., Puglisi, S.J.: Computing longest Lyndon subsequences and longest common Lyndon subsequences. Algorithmica **86**(3), 735–756 (2024)
6. Barker, L., Fleischmann, P., Harwardt, K., Manea, F., Nowotka, D.: Scattered factor-universality of words. In: Jonoska, N., Savchuk, D. (eds.) DLT 2020. LNCS, vol. 12086, pp. 14–28. Springer, Cham (2020). https://doi.org/10.1007/978-3-030-48516-0_2
7. Baumann, P., Ganardi, M., Thinniyam, R.S., Zetzsche, G.: Existential definability over the subword ordering. Log. Methods Comput. Sci. **19**(4) (2023)
8. Bringmann, K., Künnemann, M.: Multivariate fine-grained complexity of longest common subsequence. In: SODA 2018, pp. 1216–1235. SIAM (2018)
9. Burcsi, P., Cicalese, F., Fici, G., Lipták, Z.: Algorithms for jumbled pattern matching in strings. Int. J. Found. Comput. Sci. **23**(2), 357–374 (2012)
10. Day, J.D., Fleischmann, P., Kosche, M., Koß, T., Manea, F., Siemer, S.: The edit distance to k-subsequence universality. In: STACS 2021. LIPIcs, vol. 187, pp. 25:1–25:19. Schloss Dagstuhl - Leibniz-Zentrum für Informatik (2021)
11. Do, D.T., Le, T.T.Q., Le, N.-Q.-K.: Using deep neural networks and biological subwords to detect protein s-sulfenylation sites. Brief. Bioinform. **22**(3) (2021)

12. Dress, A.W.M., Erdős, P.L.: Reconstructing words from subwords in linear time. Ann. Comb. **8**, 457–462 (2005)
13. Fazekas, S.Z., Koß, T., Manea, F., Mercas, R., Specht, T.: Subsequence matching and analysis problems for formal languages. In: ISAAC 2024. LIPIcs, vol. 322, pp. 28:1–28:23 (2024)
14. Fleischmann, P., Germann, S.B., Nowotka, D.: Scattered factor universality–the power of the remainder. arXiv preprint arXiv:2104.09063 (published at RuFiDim) (2021)
15. Fleischmann, P., Haschke, L., Höfer, J., Huch, A., Mayrock, A., Nowotka, D.: Nearly k-universal words - investigating a part of Simon's congruence. Theor. Comput. Sci. **974**, 114113 (2023)
16. Fleischmann, P., Höfer, J., Huch, A., Nowotka, D.: α-β-factorization and the binary case of Simon's congruence. In: FCT 2023. LNCS, vol. 14292, pp. 190–204. Springer (2023)
17. Fleischmann, P., Kim, S., Koß, T., Manea, F., Nowotka, D., Siemer, S., Wiedenhöft, M.: Matching patterns with variables under Simon's congruence. In: RP 2023. LNCS, vol. 14235, pp. 155–170. Springer (2023)
18. Fleischmann, P., Lejeune, M., Manea, F., Nowotka, D., Rigo, M.: Reconstructing words from right-bounded-block words. Int. J. Found. Comput. Sci. **32**(6), 619–640 (2021)
19. Forrester, P.J., Mays, A.: Finite size corrections relating to distributions of the length of longest increasing subsequences. Adv. Appl. Math. **145**, 102482 (2023)
20. Frochaux, A., Kleest-Meißner, S.: Puzzling over subsequence-query extensions: disjunction and generalised gaps. In: AMW 2023 of CEUR Workshop Proceedings, vol. 3409. CEUR-WS.org (2023)
21. Gagie, T., Hermelin, D., Landau, G.M., Weimann, O.: Binary jumbled pattern matching on trees and tree-like structures. Algorithmica **73**(3), 571–588 (2015)
22. Halfon, S., Schnoebelen, P., Zetzsche, G.: Decidability, complexity, and expressiveness of first-order logic over the subword ordering. In:: LICS 2017, pp. 1–12. IEEE Computer Society (2017)
23. Hébrard, J.-J.: An algorithm for distinguishing efficiently bit-strings by their subsequences. Theor. Comput. Sci. **82**(1), 35–49 (1991)
24. Karandikar, P., Kufleitner, M., Schnoebelen, P.: On the index of Simon's congruence for piecewise testability. Inf. Process. Lett. **115**(4), 515–519 (2015)
25. Karandikar, P., Schnoebelen, P.: The height of piecewise-testable languages and the complexity of the logic of subwords. Log. Methods Comput. Sci. **15**(2) (2019)
26. Karhumäki, J., Puzynina, S., Rao, M., Whiteland, M.A.: On cardinalities of k-abelian equivalence classes. Theor. Comput. Sci. **658**, 190–204 (2017)
27. Kim, S., Han, Y.-S., Ko, S.-K., Salomaa, K.: On Simon's congruence closure of a string. Theor. Comput. Sci. **972**, 114078 (2023)
28. Kim, S., Han, Y.-S., Ko, S.-K., Salomaa, K.: On the Simon's congruence neighborhood of languages. In: DLT. LNCS, vol. 13911, pp. 168–181. Springer (2023)
29. Kim, S., Ko, S.-K., Han, Y.-S.: Simon's congruence pattern matching. Theor. Comput. Sci. **994**, 114478 (2024)
30. Kleest-Meißner, S., Sattler, R., Schmid, M.L., Schweikardt, N., Weidlich, M.: Discovering event queries from traces: laying foundations for subsequence-queries with wildcards and gap-size constraints. In: ICDT. LIPIcs, vol. 220, pp. 18:1–18:21 (2022)
31. Kleest-Meißner, S., Sattler, R., Schmid, M.L., Schweikardt, N., Weidlich, M.: Discovering multi-dimensional subsequence queries from traces - from theory to practice. In: BTW. LNI, vol. P-331, pp. 511–533 (2023)

32. Kosche, M., Koß, T., Manea, F., Siemer, S.: Absent subsequences in words. Fundam. Inf. **189**(3–4), 199–240 (2022)
33. Kruskal, J.B.: The theory of well-quasi-ordering: a frequently discovered concept. J. Comb. Theory, Ser. A **13**(3), 297–305 (1972)
34. Kuske, D.: The subtrace order and counting first-order logic. In: CSR, pp. 289–302. Springer (2020)
35. Kuske, D., Zetzsche, G.: Languages ordered by the subword order. In: FoSSaCS, pp. 348–364. Springer (2019)
36. Lothaire, M.: Combinatorics on Words. Cambridge University Press, Cambridge Mathematical Library (1997)
37. Maier, D.: The complexity of some problems on subsequences and supersequences. J. ACM **25**(2), 322–336 (1978)
38. Manuch, J.: Characterization of a word by its subwords. In: DLT 1999, pp. 210–219. World Scientific (1999)
39. Mateescu, A., Salomaa, A., Yu, S.: Subword histories and Parikh matrices. J. Comput. Syst. Sci. **68**(1), 1–21 (2004)
40. Parikh, R.: On context-free languages. J. ACM **13**(4), 570–581 (1966)
41. Puzynina, S.: Abelian properties of words. In: Mercaş, R., Reidenbach, D. (eds.) WORDS 2019. LNCS, vol. 11682, pp. 28–45. Springer, Cham (2019). https://doi.org/10.1007/978-3-030-28796-2_2
42. Salomaa, A.: Counting (scattered) subwords. Bull. EA Theor. Comput. Sci. **81**, 165–179 (2003)
43. Sattler, R., Kleest-Meißner, S., Lange, S., Schmid, M.L., Schweikardt, N., Weidlich, M.: Disces: systematic discovery of event stream queries. Proc. ACM Manag. Data **3**(1) (2025)
44. Schnoebelen, P., Veron, J.: On arch factorization and subword universality for words and compressed words. In: WORDS 2023. LNCS, vol. 13899, pp. 274–287. Springer (2023)
45. Simon, I.: Hierarchies of events with dot-depth one, PhD thesis, University of Waterloo (1972)
46. Simon, I.: Piecewise testable events. In: AFL, 2nd GI Conference, Kaiserslautern, May 20-23, 1975. LNCS, vol. 33, pp. 214–222. Springer (1975)
47. Tronícek, Z.: On problems related to absent subsequences. In: COCOA 2023. LNCS, vol. 14462, pp. 351–363. Springer (2023)
48. Wagner, R.A., Fischer, M.J.: The string-to-string correction problem. J. ACM **21**(1), 168–173 (1974)
49. Wang, C., Cho, K., Gu, J.: Neural machine translation with byte-level subwords. In: EAAI 2020, pp. 9154–9160. AAAI Press (2020)
50. Zetzsche, G.: The complexity of downward closure comparisons. In: ICALP 2016. LIPIcs, vol. 55, pp. 123:1–123:14. Schloss Dagstuhl – Leibniz-Zentrum für Informatik (2016)

Heuristic Universality Detection over Regular Expressions Specified by Systems

Florent Koechlin[1], Carine Pivoteau[2], and Pablo Rotondo[2](✉)

[1] CNRS, Université Sorbonne Paris Nord, LIPN, 93430 Villetaneuse, France
[2] Univ. Gustave Eiffel, CNRS, LIGM, 77454 Marne–la–Vallée, France
pablo.rotondo@univ-eiffel.fr

Abstract. In this article, we provide a computational approach for evaluating the expressivity of systems of equations for generating random regular expressions. More concretely, we obtain bounds on the proportion of universal expressions these systems produce. These bounds rely on the heuristic proposed by Koechlin et al., 2021, extended here to more general systems of several equations, which are harder to analyze. We have developed a tool that enables one to automatically test many of them. We discuss the results obtained for several systems from the literature.

Keywords: Universality · Regexp · Systems · Analytic combinatorics

1 Introduction

Regular expressions have a long history in computer science. They provide a simple way to describe regular languages in the form of a simple string. They are popular among programmers as a mean to specify pattern searches in text. For instance, they are used in the `grep` utility on Unix systems. Most of these programs translate regular expressions into finite automata, which can then be simulated efficiently. The construction of a minimal automaton from a regular expression is PSPACE-complete [18]. In practice many suboptimal constructions are used, such as the Thompson construction or the Glushkov automaton. The complexity of these algorithms depends on the size of the regular expressions. Hence, it is natural to first simplify the input regular expression into a smaller equivalent one using quick and simple rules (see e.g., [12,13]).

In the literature, these automata constructions have been compared on average, both experimentally and theoretically [6,19]. To conduct this type of study, the starting point is to define a proper random distribution on regular expressions. This is usually done by introducing a combinatorial system of trees and considering the uniform distribution. Generating uniform random expressions from such systems, even with several equations, can be done systematically using the recursive[1] method [20] or Boltzmann samplers [9]. On the theoretical side, these systems are amenable to the techniques of Analytic Combinatorics [10].

[1] For instance, `FAdo` implements this method for regular expressions [2].

S.-K. Ko and F. Manea (Eds.): DLT 2025, LNCS 16036, pp. 278–293, 2026.
https://doi.org/10.1007/978-3-032-01475-7_19

For example, the following one-equation system:

$$\mathcal{L} = a + b + \varepsilon + \overset{\star}{\underset{\mathcal{L}}{|}} + \overset{\bullet}{\underset{\mathcal{L}\ \mathcal{L}}{/\backslash}} + \overset{+}{\underset{\mathcal{L}\ \mathcal{L}}{/\backslash}}, \tag{1.1}$$

is the simplest description of the set of regular expressions on the letters $\{a, b\}$. This description introduces many redundancies semantically. Two consecutive stars are allowed, the associativity of $+, \bullet$ is not taken into account, etc. Thus this description may not be relevant for randomly testing algorithms that deal with regular expressions. A theoretical argument in support of this claim was given in [15], where the authors proved that there is an overabundance of universal expressions[2]. This phenomenon was also noticed by Kahrs and Runciman [13] when evaluating their *lifting method* experimentally.

More refined systems have been used in the literature [7,17] to avoid common redundancies. To give a small example, we can avoid two consecutive stars by considering the following two equations:

$$S = R + \overset{\star}{\underset{R}{|}}, \qquad R = a + b + \varepsilon + \overset{+}{\underset{S\ S}{/\backslash}} + \overset{\bullet}{\underset{S\ S}{/\backslash}}. \tag{1.2}$$

No matter how large the system is, there remains a risk of over-representing some languages; it was shown in [14] that, under certain technical conditions, the system is not suitable for random testing and average case analysis if it cannot prevent the occurrence of *all* universal expressions under a union-node. This very general result presents a serious challenge for anyone designing such systems, as verification of universality is PSPACE–complete. On the other hand, using this argument to demonstrate that a system is not well-designed is also difficult for the very same reason. To address this issue, in [15] we resorted to simple heuristics for universality detection, which we showed to be effective in the context of the single equation (1.1).

In this article, we present a computational method to bound the proportion of universal expressions produced by a system. This provides a concrete indicator for the expressivity of such systems for random generation or average case analysis. Our methodology relies on a systematic analysis of the heuristic of [15] on systems, that are sensibly harder to analyze than the one-dimensional case. We developed a tool implemented in Sagemath that allows to easily test many systems from the literature, including [17].

2 Generating Regular Expressions via Systems of Trees

The family of trees representing regular expressions we consider throughout this article are rooted and planar, with planar meaning that the trees $\overset{op}{\underset{T_1\ T_2}{/\backslash}}$ and $\overset{op}{\underset{T_2\ T_1}{/\backslash}}$ are not considered equal. They are formally defined as follows.

[2] An expression is universal if it recognizes every word of the alphabet.

Definition 1. *Given a finite alphabet $\Sigma = \{a_1, \ldots, a_k\}$, we define the class of regular expression trees $\mathcal{L}_{\mathcal{R}}(\Sigma)$ on Σ, written $\mathcal{L}_{\mathcal{R}}$ for short when the alphabet is clear, inductively from the equation*

$$\mathcal{L}_{\mathcal{R}} = \varepsilon + a_1 + \ldots + a_k + \underset{\mathcal{L}_{\mathcal{R}}}{\overset{\star}{|}} + \underset{\mathcal{L}_{\mathcal{R}}\ \mathcal{L}_{\mathcal{R}}}{\overset{\bullet}{/\backslash}} + \underset{\mathcal{L}_{\mathcal{R}}\ \mathcal{L}_{\mathcal{R}}}{\overset{+}{/\backslash}}\,. \tag{2.1}$$

The size $|T|$ of an expression tree $T \in \mathcal{L}_{\mathcal{R}}$ is defined as its number of nodes. In particular the leaves $a_1, \ldots, a_k$ and ε have size 1. We write $\overline{\Sigma} = \{\varepsilon, a_1, \ldots, a_k\}$.

2.1 Systems of Regular Expression Trees

As explained in the introduction, to avoid redundancies in the description of regular expressions, it is possible to define subsets (not necessarily strict) of $\mathcal{L}_R$ by using systems of several equations.

Definition 2 (System of expression trees). *A system of regular expression trees of dimension m is a vector of m equations*

$$\mathcal{Y}_i = \Phi_i(\varepsilon, a_1, \ldots, a_k; \mathcal{Y}_1, \ldots, \mathcal{Y}_m)\,,$$

where $a_1, \ldots, a_k$ are the letters of Σ, and each Φ_i is a formal sum of regular expression trees on the variables $\overline{\Sigma} \cup \{\mathcal{Y}_1, \ldots, \mathcal{Y}_m\}$, where we recall the notation $\overline{\Sigma} = \Sigma \cup \{\varepsilon\}$. By analogy with grammars, we call the symbols $\mathcal{Y}_i$ non-terminals, and the elements of $\overline{\Sigma}$ terminals. When the alphabet is clear from the context, we write $\boldsymbol{\mathcal{Y}} = \boldsymbol{\Phi}(\boldsymbol{\mathcal{Y}})$ where bold symbols indicate vectors, $\boldsymbol{\mathcal{Y}} = (\mathcal{Y}_1, \ldots, \mathcal{Y}_m)$ and $\boldsymbol{\Phi} = (\Phi_1, \ldots, \Phi_m)$.

Note that Definition 1 is an example of such a system, with variables $\overline{\Sigma} \cup \{\mathcal{L}_{\mathcal{R}}\}$, where $\overline{\Sigma} = \{\varepsilon, a_1, \ldots, a_k\}$. Equation 1.2 is also an example of such a system, with variables $\overline{\Sigma} \cup \{S, R\}$, where $\overline{\Sigma} = \{\varepsilon, a, b\}$.

Systems arising from combinatorics have been extensively studied under the name combinatorial systems. They verify additional conditions to be well-defined. In order to avoid a technical discussion on the weakest set of conditions (see e.g., [1,11,21]), we prefer to identify a sufficient set of conditions that ensure that there exists a unique combinatorial solution to the system and that are general enough to be applicable:

1. *Avoid cycles of unit rules.* Consider, for example, $\mathcal{Y}_1 = \mathcal{Y}_1 + a$. In this case the element a can be derived from the grammar-like equation in infinitely many ways. This is undesirable from a combinatorial point of view.
2. *Avoid emptysets.* Consider $\mathcal{Y}_1 = \underset{\mathcal{Y}_1\ \mathcal{Y}_1}{\overset{+}{/\backslash}}$. This equation defines $\mathcal{Y}_1 = \emptyset$. This is not desirable because the rule is then useless and can be removed.

2.2 Generating Series

The ordinary generating series associated to a combinatorial class $\mathcal{C}$ is the formal sum $\sum_{n \in \mathbb{N}} c_n x^n$ where c_n counts the number of objects in $\mathcal{C}$ of size n. In our setting, the size associated to a regular expression is the total number of nodes (operators and leaves) of its tree representation.

Formal system. Given a system $\boldsymbol{\mathcal{Y}} = \boldsymbol{\Phi}(\boldsymbol{\mathcal{Y}})$ of regular expressions, we define for each class $\mathcal{Y}_i$ its associated generating function $y_i(x) = \sum_{n\geq 0} c_{i,n}x^n$, where $c_{i,n}$ denotes the number of trees in $\mathcal{Y}_i$ having n nodes. When the system is unambiguous, following the symbolic method from [10], the system $\boldsymbol{\mathcal{Y}} = \boldsymbol{\Phi}(\boldsymbol{\mathcal{Y}})$ translates directly into a system of equations on $\mathbf{y}(x) = (y_1(x), \ldots, y_m(x))$. For instance, Eqs. 1.1 and 1.2 translate directly to the following two systems:

$$\{\ L(x) = 3x + xL(x) + 2xL(x)^2\ \}, \tag{2.2}$$

$$\{\ S(x) = R(x) + xR(x), \qquad R(x) = 3x + 2xS(x)^2\ \}. \tag{2.3}$$

These systems can be written more generally under the form $\mathbf{y}(x) = \boldsymbol{\phi}(x; \mathbf{y}(x))$. For instance, for Eq. 2.3, we have $(S(x), R(x)) = \boldsymbol{\phi}(x; S(x), R(x))$ with $\boldsymbol{\phi}(x; y_1, y_2) = (y_2 + xy_2, 3x + 2xy_1^2)$.

Asymptotics for the coefficients. When a generating series $y_i(x)$ has a positive radius of convergence $\rho_i > 0$, it is possible to interpret it as an analytic function (called generating function or GF for short) on $|z| < \rho_i$, and apply the analytic combinatorics methodology: the asymptotics of the coefficients is associated with the behavior of the GF near its dominant singularities (i.e., the ones with smallest modulus). We first recall that the radius of convergence $\rho > 0$ is always a dominant singularity (there might be others), when the coefficients are positive, due to Pringsheim's Theorem [10, Thm.IV.6]. Then the *Transfer Theorem* [10, Thm.VI.3] states that, under technical conditions, if a generating series $f(z)$ has a unique singularity at $\rho > 0$ on its radius of convergence, and if around this singularity $f(z) \sim_{z\to\rho} a - b \cdot (1 - z/\rho)^\alpha$ with $\alpha \notin \mathbb{Z}_{\geq 0}$, then $[z^n]f(z) \sim_{n\to\infty} \frac{\alpha \cdot b}{\Gamma(1-\alpha)} n^{-\alpha-1}\rho^{-n}$ where Γ is the well-known Gamma function, a generalization of the factorial to complex numbers. In the case of several singularities on the circle of convergence $|z| = \rho$, one must add the contribution of each singularity by applying similarly the same Transfer Theorem to each of them separately.

For example, solving Equation (2.2) we obtain $L(x) = \frac{1}{4x}\left(1 - x - \sqrt{\Delta(x)}\,\right)$, with $\Delta(x) = -23(1 - x/\rho)(1 - x/\hat{\rho})$ where $\rho = \frac{-1+2\sqrt{6}}{23}$ and $\hat{\rho} = \frac{-1-2\sqrt{6}}{23}$. The singularities are ρ and $\hat{\rho}$, while $x = 0$ is a false singularity. The only dominant singularity is $\rho \doteq 0.16952\ldots$ The Transfer Theorem implies, provided that we can find $b \neq 0$ such that $L(x) \sim a - b \cdot \sqrt{1 - x/\rho}$, that

$$L_n = [z^n]L(z) \sim \frac{b}{2\Gamma(\frac{1}{2})} n^{-3/2}\rho^{-n} = \frac{b}{2\sqrt{\pi}} n^{-3/2}\rho^{-n}.$$

The constant b can be determined by differentiation, since

$$L'(z) \sim \tfrac{b}{2\rho}(1 - z/\rho)^{-1/2}, \qquad \frac{b}{2\rho} = \lim_{z\to\rho^-} L'(z)(1 - z/\rho)^{1/2} = \frac{6^{\frac{1}{4}}\sqrt{2\sqrt{6}-1}\,23^{\frac{3}{2}}}{4(25-4\sqrt{6})}.$$

This analysis, apparently confined to simple systems, is true under more general conditions. We discuss the conditions and the result in the next section. We will later show, in Sect. 4.2, that we can obtain similar results for a refined version of the system, taking into account the subsets of $\overline{\Sigma}$ that are recognized.

2.3 Conditions on the Systems

In Sect. 2.1 we gave two conditions to ensure that the equation $\mathbf{y}(x) = \phi(x; \mathbf{y}(x))$ defines proper generating series. In this section we introduce stronger conditions, which imply the previous ones, to restrict ourselves to systems (which we call $\mathcal{D}$-analyzable[3]) for which the solution is of simple form (see Lemma 1). In practice, these requirements allow many systems from the literature.

In order to state the conditions, we introduce the concept of the dependency digraph of the system. This digraph expresses the dependencies between the non-terminals and thus plays an important role in our study.

Definition 3 (Dependency digraph). *Let $\boldsymbol{\mathcal{Y}} = \boldsymbol{\Phi}(\boldsymbol{\mathcal{Y}})$ be a system of regular expression trees. The dependency digraph associated to $\boldsymbol{\mathcal{Y}}$, denoted $\mathcal{G}(\boldsymbol{\Phi})$ is the directed graph whose vertices are the non-terminal symbols $\mathcal{Y}_1, \ldots, \mathcal{Y}_m$, and such that there is an edge $\mathcal{Y}_i \to \mathcal{Y}_j$ if and only if $\mathcal{Y}_j$ appears as a leaf inside some regular expression tree of $\Phi_i(\boldsymbol{\mathcal{Y}})$, and we say that $\mathcal{Y}_i$ depends on $\mathcal{Y}_j$.*

The unit dependency graph $\mathcal{G}_{unit}(\boldsymbol{\Phi})$ is defined similarly with an edge $\mathcal{Y}_i \to \mathcal{Y}_j$ if and only if the expression $\mathcal{Y}_j$ is a regular expression tree of $\Phi_i(\boldsymbol{\mathcal{Y}})$.

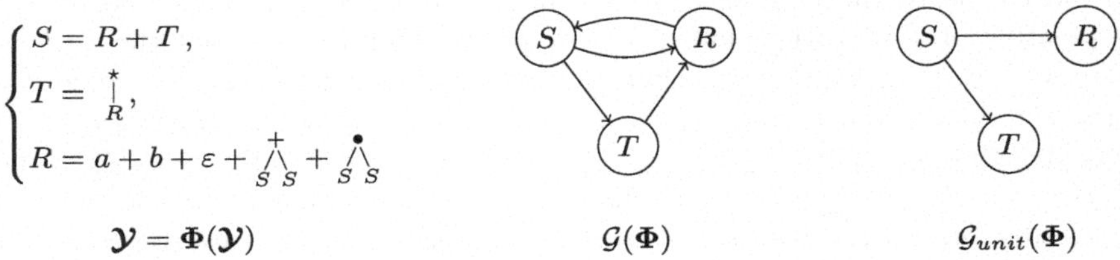

Fig. 1. Example of a system of regular expression trees, its dependency digraph and its unit dependency graph.

Definition 3 is illustrated in Fig. 1. Now we can define the precise conditions we consider for our systems.

Definition 4 ($\mathcal{D}$-analyzable system). *A system $\boldsymbol{\mathcal{Y}} = \boldsymbol{\Phi}(\boldsymbol{\mathcal{Y}})$ is said to be $\mathcal{D}$-*analyzable *if it satisfies the following properties:*

1. *The graph $\mathcal{G}(\boldsymbol{\Phi})$ is strongly connected, and the graph $\mathcal{G}_{unit}(\boldsymbol{\Phi})$ is acyclic.*
2. *The system is aperiodic: for n_0 large enough, for all $n \geq n_0$, every class $\mathcal{Y}_i$ contains an expression of size n.*
3. *The system is not linear, in the sense that $\boldsymbol{\Phi}$ contains at least one expression with two or more non-terminal symbols in its leaves*
4. *The system is unambiguous: every regular expression is described by the system in at most one way.*

[3] $\mathcal{D}$ stands for Drmota.

Remark 1 Conditions (1) and (2) imply the conditions of well-foundness mentioned in Sect. 2.1.

Let us discuss the new set of conditions. The *absence of cycle of unit rules* is not very constraining, as there are well-known procedures [1] to remove them. Note that the elimination procedure to remove unit-rule can however change the connectivity of the system. In practice, no system of regular expressions we have come through in the literature had a cycle of unit rule.

The condition of *strong connectivity* is actually quite strong, even if most systems encountered in the literature satisfy (or almost satisfy) this condition. In practice, we managed to weaken it a bit to treat more general systems.

The *aperiodicity* condition could be relaxed ; however system with periodicities can present empty classes for arbitrary large sizes n. In this case, defining the limit proportion of expressions detected as universal is more challenging, and we would have to restrict ourselves to certain subsequences.

The *unambiguity* condition could also be relaxed; in this case the system describes regular expressions with multiplicity. The uniform distribution on these trees of size n then takes into account these multiplicities.

2.4 Asymptotics for the Number of Expression Trees

The conditions defining $\mathcal{D}$-analyzable systems ensure that all of the hypotheses of Drmota's theorem [8, Thm. 1] are satisfied. Hence, a direct application (see Lemma 1) of Drmota's theorem yields the asymptotic behaviour of the number of regular expression trees described by the system. These estimates will be used for the calculation of the proportions of universals in Sect. 4.2.

Lemma 1 (Corollary of [8, Thm. 1]). *Let $\boldsymbol{\mathcal{Y}} = \boldsymbol{\Phi}(\boldsymbol{\mathcal{Y}})$ be a $\mathcal{D}$-analyzable system, and let $\mathbf{y}(z) = (y_1(z), \ldots, y_m(z))$ be the associated GFs. Then, the GFs $y_i(z)$ share a unique common dominant singularity $\rho > 0$, and moreover*

- *for each $i \in [m]$, there are functions $a_i(z)$ and $b_i(z)$, analytic in a neighborhood $\mathcal{N} \subseteq \mathbb{C}$ of ρ with $b_i(\rho) \neq 0$, such that for $z \in \mathcal{N} \setminus (\rho, +\infty)$,*

$$y_i(z) = a_i(z) - b_i(z) \cdot \sqrt{1 - z/\rho}\,.$$

- *for each $i \in [m]$, there is a constant $C_i > 0$ such that $[z^n]y_i(z) \sim C_i n^{-3/2} \rho^{-n}$.*

The following lemma characterizes the singularity ρ from Lemma 1, and in particular it yields a procedure to calculate it. Here $\mathtt{Jac}_{\mathbf{y}}[\phi](z; \mathbf{y})$ denotes the Jacobian matrix of the system, namely, $(\mathtt{Jac}_{\mathbf{y}}[\phi])_{i,j}(z; \mathbf{y}) = \partial_j \phi_i(z; \mathbf{y})$.

Lemma 2 (Corollary of [8, Thm. 1]). *Let $\boldsymbol{\mathcal{Y}} = \boldsymbol{\Phi}(\boldsymbol{\mathcal{Y}})$ be a $\mathcal{D}$-analyzable system, and let $\mathbf{y}(z) = (y_1(z), \ldots, y_m(z))$ be the associated GFs, which satisfy $\mathbf{y}(z) = \boldsymbol{\phi}(z; \mathbf{y}(z))$. Then the unique dominant singularity ρ and $\boldsymbol{\tau} = \mathbf{y}(\rho)$ verify the characteristic system*

$$\boldsymbol{\tau} = \Phi(\rho; \boldsymbol{\tau}), \qquad \det\left(\mathtt{Id}_{m\times m} - \mathtt{Jac}_{\mathbf{y}}[\Phi](\rho; \boldsymbol{\tau})\right) = 0\,. \tag{2.4}$$

There can be several solutions to the characteristic system, but in the case of polynomial systems, there are properties that distinguish $(\rho, \boldsymbol{\tau})$ from the other solutions of the system, see [4, Theorem 14].

3 Refinements of Systems Towards Universality Detection

In this section, we refine a system of expression trees in order to distinguish those in each class $\mathcal{Y}_i$ that are detected as universals by the heuristic in [15].

3.1 Letters, Empty Word and Partial Universality Detection

Following [15], we define a subclass $\mathcal{U}$ of universal expressions inductively:

- If $T = \overset{\star}{\underset{T_1}{|}}$, and T_1 recognizes every letter of the alphabet, then T belongs to $\mathcal{U}$. In particular, if $T_1 \in \mathcal{U}$, then $T \in \mathcal{U}$
- If $T = \overset{+}{\underset{T_1\ T_2}{\wedge}}$ where T_1 *or* T_2 belongs to $\mathcal{U}$, then $T \in \mathcal{U}$
- If $T = \overset{\bullet}{\underset{T_1\ T_2}{\wedge}}$, where $T_1 \in \mathcal{U}$ and T_2 recognizes ε (or the converse), then $T \in \mathcal{U}$.

It is clear that all elements of $\mathcal{U}$ are universal. Furthermore, deciding whether an expression belongs to $\mathcal{U}$ can be done in linear time. Indeed, given a regular expression tree T, it is easy to decide if T recognizes the empty word, by induction on the structure of the tree. The base case (a leaf a, b, ε) is immediate. If T is the star of a tree, then it always recognizes ε; and if T is a union (resp. concatenation) of two trees, then T recognizes ε if and only if one (resp. all) of its children recognizes it. Similarly we can also decide if an expression T recognizes a given letter (by using the ε detection for treating the concatenation case).

Note that this detection of universals is truly heuristic, in the sense that it cannot possibly recognize all universal expressions. For example, it does not recognize the tree associated with $L = \Sigma \cdot \Sigma^\star + \varepsilon$ as universal.

3.2 Refinement of a Combinatorial System of Trees

In this section we refine the system in order to follow the heuristic detection for universals described in the previous section.

Example 1. Consider the following system:

$$\mathcal{L}_0 = \mathcal{L}_1 + \mathcal{L}_2\,, \quad \mathcal{L}_1 = \overset{\star}{\underset{\mathcal{L}_2}{|}} + \varepsilon\,, \quad \mathcal{L}_2 = a + b + \overset{+}{\underset{\mathcal{L}_0\ \mathcal{L}_0}{\wedge}} + \overset{\bullet}{\underset{\mathcal{L}_0\ \mathcal{L}_1}{\wedge}}\,.$$

Suppose we decompose each class $\mathcal{L}_i$ into two classes $\mathcal{L}_{i,\varepsilon}$ and $\mathcal{L}_{i,\overline{\varepsilon}}$ corresponding to whether ε is recognized or not by the expression trees. The original system can be easily transformed into a system on $\mathcal{L}_{i,\varepsilon}$ and $\mathcal{L}_{i,\overline{\varepsilon}}$, following the procedure described on the previous section. Then we obtain the following system of combinatorial equations:

$$\begin{cases} \mathcal{L}_{0,\varepsilon} = \mathcal{L}_{1,\varepsilon} + \mathcal{L}_{2,\varepsilon}\,, \quad \mathcal{L}_{0,\overline{\varepsilon}} = \mathcal{L}_{1,\overline{\varepsilon}} + \mathcal{L}_{2,\overline{\varepsilon}}\,, \quad \mathcal{L}_{1,\varepsilon} = \underset{\mathcal{L}_{2,\overline{\varepsilon}}}{\overset{\star}{|}} + \underset{\mathcal{L}_{2,\varepsilon}}{\overset{\star}{|}} + \varepsilon\,, \quad \mathcal{L}_{1,\overline{\varepsilon}} = \emptyset\,, \\ \mathcal{L}_{2,\varepsilon} = \underset{\mathcal{L}_{0,\varepsilon}\ \mathcal{L}_{0,\varepsilon}}{\overset{+}{/\backslash}} + \underset{\mathcal{L}_{0,\overline{\varepsilon}}\ \mathcal{L}_{0,\varepsilon}}{\overset{+}{/\backslash}} + \underset{\mathcal{L}_{0,\varepsilon}\ \mathcal{L}_{0,\overline{\varepsilon}}}{\overset{+}{/\backslash}} + \underset{\mathcal{L}_{0,\varepsilon}\ \mathcal{L}_{1,\varepsilon}}{\overset{\bullet}{/\backslash}}\,, \\ \mathcal{L}_{2,\overline{\varepsilon}} = a + b + \underset{\mathcal{L}_{0,\overline{\varepsilon}}\ \mathcal{L}_{0,\overline{\varepsilon}}}{\overset{+}{/\backslash}} + \underset{\mathcal{L}_{0,\overline{\varepsilon}}\ \mathcal{L}_{1,\varepsilon}}{\overset{\bullet}{/\backslash}} + \underset{\mathcal{L}_{0,\varepsilon}\ \mathcal{L}_{1,\overline{\varepsilon}}}{\overset{\bullet}{/\backslash}} + \underset{\mathcal{L}_{0,\overline{\varepsilon}}\ \mathcal{L}_{1,\overline{\varepsilon}}}{\overset{\bullet}{/\backslash}}\,. \end{cases}$$

This example can also be generalized to detect at the same time which letters are recognized and, finally, whether the expression trees are detected as universal. In order to do this, we introduce the refinement into classes as follows: for each class $\mathcal{Y}_i$ of the system, and each subset $S \subseteq \overline{\Sigma}$, we introduce the class $\mathcal{Y}_i^{\langle S\rangle}$ which corresponds to the trees in $\mathcal{Y}_i$ recognizing every element of S and none from $\overline{\Sigma} \setminus S$. Observe that $\mathcal{Y}_i$ is the disjoint union of $\mathcal{Y}_i^{\langle S\rangle}$ for all $S \subseteq \overline{\Sigma}$.

We can similarly refine $\mathcal{Y}_i^{\langle\overline{\Sigma}\rangle}$ into two disjoint classes, according to whether they belong to $\mathcal{U}$ defined in Sect. 3.1 or to its complement $\mathcal{G} := \mathcal{L}_R \setminus \mathcal{U}$. We write, in an abuse of notation, $\mathcal{Y}_i^{\langle\mathcal{U}\rangle} := \mathcal{Y}_i^{\langle\overline{\Sigma}\rangle} \cap \mathcal{U}$ and $\mathcal{Y}_i^{\langle\mathcal{G}\rangle} := \mathcal{Y}_i^{\langle\overline{\Sigma}\rangle} \cap \mathcal{G}$. Hence each class $\mathcal{Y}_i$ decomposes into $2^{|\Sigma|}+1$ classes, namely $\mathcal{Y}_i^{\langle S\rangle}$ for all $S \subsetneq \overline{\Sigma}$, and $\mathcal{Y}_i^{\langle\mathcal{G}\rangle}, \mathcal{Y}_i^{\langle\mathcal{U}\rangle}$. These classes also constitute a system of regular expression trees equations.

Proposition 1. *If $\boldsymbol{\mathcal{Y}} = \boldsymbol{\Phi}(\boldsymbol{\mathcal{Y}})$ is a system of regular expression trees, then the associated classes $(\mathcal{Y}_i^{\langle S\rangle})_{S\subsetneq\overline{\Sigma}} \cup (\mathcal{Y}_i^{\langle\mathcal{G}\rangle}, \mathcal{Y}_i^{\langle\mathcal{U}\rangle})$ are also described by a system of regular expression trees, that we call the extended system. Furthermore, the extended system does not introduce ambiguities.*

Proof (sketch). We follow the inductive definition of Sect. 3.1 as in Example 1 to refine the system in successive steps. First, we start with the empty word detection, then we refine it with letters, such that we obtain the system for $(\mathcal{Y}_i^{\langle S\rangle})_{S\subseteq\overline{\Sigma}}$. Finally, we consider the refinement of $\mathcal{Y}_i^{\langle\overline{\Sigma}\rangle}$ into $\mathcal{Y}_i^{\langle\mathcal{U}\rangle}$ and $\mathcal{Y}_i^{\langle\mathcal{G}\rangle}$. □

It is important to note that the extended system does not introduce new ambiguities, as we will use the GFs for counting the different objects. In particular, their generating series satisfy the relation:

$$y_i(z) = y_i^{\langle\mathcal{U}\rangle}(z) + y_i^{\langle\mathcal{G}\rangle}(z) + \sum_{S\subsetneq\overline{\Sigma}} y_i^{\langle S\rangle}(z)\,. \tag{3.1}$$

3.3 The Underlying Digraph of the Extended System

The extended system of a $\mathcal{D}$-analyzable system may not (and need not) be again $\mathcal{D}$-analyzable. Indeed the extended system may fail to satisfy the strong connectivity and aperiodicity needed to be $\mathcal{D}$-analyzable. To see it, it is enough to consider the first step of the construction, namely, the refinement according to whether ε is recognized or not. Consider the system:

$$\begin{cases} S = a + b + \varepsilon + \overset{+}{\underset{T\ \ S}{\wedge}} + \overset{\bullet}{\underset{S\ \ S}{\wedge}} \\ T = a + b + \overset{\star}{\underset{S}{|}} \end{cases}$$

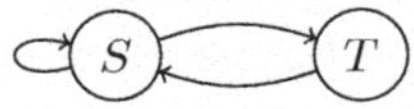

Then the dependency graph of its refined system $(S_\varepsilon, S_{\overline{\varepsilon}}, T_\varepsilon, T_{\overline{\varepsilon}})$ is not strongly connected (nor aperiodic!) anymore:

$$\begin{cases} S_\varepsilon = \varepsilon + \overset{+}{\underset{T_\varepsilon\ S_\varepsilon}{\wedge}} + \overset{+}{\underset{T_{\overline{\varepsilon}}\ S_\varepsilon}{\wedge}} + \overset{+}{\underset{T_\varepsilon\ S_{\overline{\varepsilon}}}{\wedge}} + \overset{\bullet}{\underset{S_\varepsilon\ S_\varepsilon}{\wedge}} & T_\varepsilon = \overset{\star}{\underset{S_\varepsilon}{|}} + \overset{\star}{\underset{S_{\overline{\varepsilon}}}{|}} \\ S_{\overline{\varepsilon}} = a + b + \overset{+}{\underset{S_{\overline{\varepsilon}}\ S_{\overline{\varepsilon}}}{\wedge}} + \overset{\bullet}{\underset{T_{\overline{\varepsilon}}\ S_{\overline{\varepsilon}}}{\wedge}} + \overset{\bullet}{\underset{T_{\overline{\varepsilon}}\ S_\varepsilon}{\wedge}} + \overset{\bullet}{\underset{T_\varepsilon\ S_{\overline{\varepsilon}}}{\wedge}} & T_{\overline{\varepsilon}} = a + b \end{cases}$$

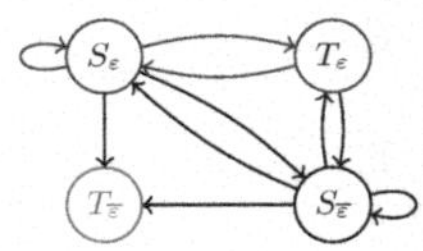

Another important issue that further distinguishes the extended system comes from the fact that some of its classes might be empty. See Example 1 above, where $\mathcal{L}_{1,\overline{\varepsilon}} = \emptyset$. In this system, the term $\overset{\bullet}{\underset{\mathcal{L}_{0,\varepsilon}\ \mathcal{L}_{1,\overline{\varepsilon}}}{\wedge}}$ appears in the system, but does not generate any actual expression trees. This does not express a true dependency, then, and we should remove it from the dependency digraph. This can be done simply by removing the empty classes and every rule involving them.

Definition 5 (Reduced digraph of the extended system). *The* reduced digraph *of the extended system is the dependency digraph of the system obtained after removing every empty class. Henceforth, we still call this the* dependency digraph *of the extended system for short.*

4 Limiting Proportion of Universals

Let $\mathcal{Y} = \Phi(\mathcal{Y})$ be a system of regular expression trees. For compactness, we rename each class $\mathcal{Y}_i^{\langle S \rangle}$ as $\mathcal{W}_j$, in some arbitrary order, to write the extended system under the form $\mathcal{W} = \Psi(\mathcal{W})$. Similarly, we write $(y_i^{\langle S \rangle}(z))$ as a vector $\mathbf{w}(z)$, so that the system of equations for the GFs is written as $\mathbf{w}(z) = \psi(z; \mathbf{w}(z))$. This section is dedicated to study the generating functions corresponding to the extended system, with the goal of estimating the proportion of universal expressions in the original system for $\mathcal{Y}$.

4.1 Analysis of the Extended Dependency DAG in Terms of Series

In Sect. 2.3, we have seen that although the extended system is a refinement of the original system, their underlying structure can be significantly different. In this case, Drmota's Theorem does not apply anymore, and we need to take into account the Directed Acyclic Graph (DAG) of the strongly connected components (scc for short) of the dependency graph. To do this, we follow the work of Banderier and Drmota [3].

The following simple lemma shows that the DAG structure of its dependency digraph is naturally related to an ordering of the convergence radius of the GFs. This is a classical extension of the proof of [10, Thm.VII.6] for a single scc.

Lemma 3. *If two classes lie in the same strongly connected component C, then their GFs have the same radius of convergence ρ_C. Moreover these radii do not decrease along the edges of the DAG.*

This means that if we decompose the extended dependency graph into the DAG of its strongly connected components, then we can associate to each component C its radius of convergence ρ_C. Moreover, if a strongly connected component C_1 has an arrow into another C_2, then $\rho_{C_1} \leq \rho_{C_2}$. By Lemma 1, the classes $\mathcal{Y}_i$ of the original system all have the same radius of convergence ρ. Hence from Eq. (3.1) we deduce that the radii ρ_C in the extended system can only be greater or equal to ρ.

The following proposition shows that the classes of the extended system that are defined in a scc C with $\rho_C > \rho$ represent a negligible proportion.

Proposition 2. *If $\mathcal{Y}_i^{\langle S\rangle}$ is in a scc C with $\rho_C > \rho$, then $\lim_{n\to\infty} \frac{[z^n]y_i^{\langle S\rangle}(z)}{[z^n]y_i(z)} = 0$. In particular, if both $\mathcal{Y}_i^{\langle \mathcal{U}\rangle}$ and $\mathcal{Y}_i^{\langle \mathcal{G}\rangle}$ are not in a scc with $\rho_C = \rho$, then the limiting proportion of universals in $\mathcal{Y}_i$ is zero.*

Proof. The denominator behaves like $\sim c\cdot n^{-3/2}\rho^{-n}$ by Lemma 1. By hypothesis $\rho_C > \rho$, and we can find an $\epsilon > 0$ such that $\frac{1}{\rho_C} + \epsilon < \rho^{-1}$. Then, by the Cauchy-Hadamard Theorem (see [10, Thm IV.7]), we have $[z^n]y_i^{\langle S\rangle}(z) \leq (\frac{1}{\rho_C} + \epsilon)^n$ for large enough n. Since $\frac{1}{\rho_C} + \epsilon < \rho^{-1}$, we deduce the first part of the proposition. The last part follows from the fact that the number of universal expressions in $\mathcal{Y}_i$ is between the number of expressions in $\mathcal{Y}_i^{\langle \mathcal{U}\rangle}$ and $\mathcal{Y}_i^{\langle \Sigma\rangle} = \mathcal{Y}_i^{\langle \mathcal{U}\rangle} \cup \mathcal{Y}_i^{\langle \mathcal{G}\rangle}$. □

4.2 Statement of the Main Result

Our main result extends Proposition 2 to all classes in the extended system. The proof of this theorem is given in Sect. 4.4.

Theorem 1 (Limiting probabilities). *There exists an integer $q \geq 1$ such that, for every remainder $k \in \{0, \ldots, q-1\}$, the limiting probabilities:*

$$\mathbb{P}_i^{\langle S\rangle}(k) := \lim_{n\to\infty} \frac{[z^{k+qn}]y_i^{\langle S\rangle}(z)}{[z^{k+qn}]y_i(z)},$$

exist for every $1 \leq i \leq m$, and every $S \subseteq \overline{\Sigma}$, $S = \mathcal{U}$ and $S = \mathcal{G}$. Furthermore, if $\mathcal{Y}_i^{\langle S\rangle}$ is in a scc with $\rho_C = \rho$, at least one of $(\mathbb{P}_i^{\langle S\rangle}(k))_{k=0}^{q-1}$ is non-null.

An important observation is the following corollary that allows to detect overabundance of universal expressions.

Corollary 1. *If $\mathcal{Y}_i^{\langle \mathcal{U}\rangle}$ is in a scc with $\rho_C = \rho$, then the system presents an overabundance of universal expressions for infinitely many sizes.*

Corollary 1 and Proposition 2 yield a simple procedure to determine the overabundance of the universal expression detected by our heuristic, without computing the limiting probabilities. The procedure relies on determining, in the DAG, which scc's have $\rho_C = \rho$ as singularity (see Fig. 2 for an example). We show in Sect. 4.3 how to detect this algorithmically.

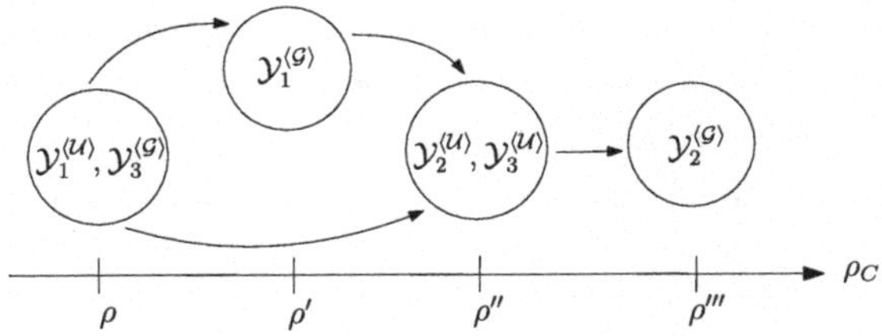

Fig. 2. Illustration of a generic reasoning on the DAG. In this example, $\mathcal{Y}_1^{\langle\mathcal{U}\rangle}$ is in a scc with radius ρ, so we can deduce that it presents an overabundance of universals. On the contrary, both $\mathcal{Y}_2^{\langle\mathcal{U}\rangle}$ and $\mathcal{Y}_2^{\langle\mathcal{G}\rangle}$ are in scc's with strictly larger radius, so we can deduce that they have a negligible proportion of universals. For the last class $\mathcal{Y}_3$, we cannot conclude using our heuristic.

4.3 Determining the Strongly Connected Components in ρ

In this section we show how to detect algorithmically if $\mathcal{Y}_i^{\langle\mathcal{U}\rangle}$ is in a scc C with $\rho_C = \rho$. This, along with Corollary 1, gives a simple procedure to determine the overabundance of universal expressions (see Fig. 2).

We show that there are two categories of scc's with a singularity at $z = \rho$, with two different roles for our calculations:

(*a*) the strongly connected components that depend only on scc's with larger radius of convergence. We call them *source* scc's.
(*b*) the scc's that *inherit* the singularity at $z = \rho$.

In order to define an *inherited singularity*, we need some notation. Let C be a scc, we denote by ψ_C and $\mathbf{w}_C$ the restriction of ψ and $\mathbf{w}$ to C.

Definition 6. (Inherited singularity). *A scc C with $\rho_C = \rho$ is said to inherit the singularity ρ if and only if* $\det(\mathtt{Id} - \mathtt{Jac}_{\mathbf{w}_C}[\psi_C](\rho; \mathbf{w}(\rho)) \neq 0$.

A key observation is that the scc's in ρ that are not sources necessarily inherit the singularity at $z = \rho$, and therefore there are no other categories. This fact simplifies the calculations: it suffices to locate the source scc's, which we prove are characterized by $\det(\mathtt{Id} - \mathtt{Jac}_{\mathbf{w}_C}[\psi_C](\rho; \mathbf{w}(\rho))\,) = 0$. The non-source scc's in ρ, i.e., those which inherit the singularity, can then be deduced by following the DAG structure.

We summarize this in the following proposition.

Proposition 3. *Let C be a scc with $\rho_C = \rho$. Then C is either a source scc in ρ or it inherits the singularity at $z = \rho$. Moreover, an scc C' is a source in ρ if and only if* $\det(\mathtt{Id} - \mathtt{Jac}_{\mathbf{w}_{C'}}[\psi_{C'}](\rho; \mathbf{w}(\rho)) = 0$.

4.4 Proof of Theorem 1

Let us now sketch the proof of Theorem 1. By Proposition 2 it is enough to study the classes that lie inside components C with $\rho_C = \rho$. We show that all scc's with $\rho_C = \rho$ have the same behavior as $z \to \rho$, namely:

Lemma 4. *Let C be a scc with $\rho_C = \rho$, and let $\mathcal{Y}_i^{\langle S \rangle}$ be a class in C. Then locally around $z = \rho$, we have $y_i^{\langle S \rangle}(z) = a_i^{\langle S \rangle}(z) - b_i^{\langle S \rangle}(z)\sqrt{1 - z/\rho}$ with $a_i^{\langle S \rangle}(z)$ and $b_i^{\langle S \rangle}(z)$ analytic at $z = \rho$, and $b_i^{\langle S \rangle}(\rho) \neq 0$.*

Proof (sketch). We rely on the admissible exponents for solutions of positive systems from [3] combined with the fact that the solutions for the original system has a square root behaviour by Lemma 1. □

An important observation is that scc's that inherit the singularity ρ (see Def. 6) do not introduce new singularities on the circle $z = |\rho|$. This means that they do not introduce periodicities.

Periodicities. So far we have discussed the behavior around $z = \rho$, the radius of convergence. However, unlike Lemma 1, there may be other singularities on the circle of convergence that must be taken into account for the asymptotics.

We apply the general techniques of the proof of Thm 5 in [3] to our particular system. We show that for the source scc's in ρ, the singularities on the circle are well-spaced due to the Daffodil Lemma [10], which classically leads to a periodic behavior in the asymptotics. For the remaining components, we prove that the scc's that inherit the singularity ρ do not introduce periodicities.

Lemma 5. *There is $q \geq 1$ such that every singularity of the extended system on the circle $|z| = \rho$ is of the form $\varrho = \rho\omega^r$, where $\omega = \exp(2\pi i/q)$, for some integer r. Moreover, around $z = \varrho$, for each i and $S \subseteq \overline{\Sigma}$, $S = \mathcal{U}$ or $S = \mathcal{G}$, such that $\mathcal{Y}_i^{\langle S \rangle} \in C$ for a scc of the extended system with $\rho_C = \rho$, then*

$$y_i^{\langle S \rangle}(z) = a_{i,r}^{\langle S \rangle}(z) - b_{i,r}^{\langle S \rangle}(z)\sqrt{1 - z/\varrho},$$

with $a_{i,r}^{\langle S \rangle}(z), b_{i,r}^{\langle S \rangle}(z)$ analytic at $z = \varrho$.

We can now finish the proof-sketch of Theorem 1. Lemma 5 and the Transfer Theorem imply that $[z^n]y_i^{\langle S \rangle}(z) = c \cdot \rho^{-n} n^{-3/2}(\sum_{r=0}^{q-1} b_{i,r}^{\langle S \rangle}(\rho\omega^r)\omega^{nr} + o(1))$ for some $c > 0$. Due to Lemma 1, which applies to $[z^n]y_i(z)$, we are done if we can prove that the polynomial $g(z) = \sum_{r=0}^{q-1} b_{i,r}^{\langle S \rangle}(\rho\omega^r) \cdot z^r$ is non-zero for some $z = \omega^0, \ldots, \omega^{q-1}$. Being of degree $q - 1$, if this were not the case, it would be identically zero, but we know that the coefficient $b_{i,0}^{\langle S \rangle}(\rho) = b_i^{\langle S \rangle}(\rho)$ is non-zero.

4.5 Computation of the Limiting Probabilities

We summarize the strategy stemming from the analysis of $\mathcal{D}$-analyzable systems to derive numerical estimates on the proportion of universals.

1. First, compute the radius of convergence ρ from the original system by using the *characteristic system* in Eq. (2.4).

2. Determine the source scc's C with $\rho_C = \rho$ in the extended system. It can be shown that these are characterized solely by $\det(\mathtt{Id} - \mathtt{Jac}_{\mathbf{w}_C}[\psi_C](\rho; \mathbf{w}(\rho))) = 0$. A scc C admits ρ as singularity if and only if it has some descendant source scc in ρ (maybe C itself). Hence, the classes that have an overabundance of universals $\mathcal{U}$ can be determined without further calculations by Corollary 1.
3. Compute a period $q \geq 1$ of the extended system, by taking the least common multiple of the periods of the sources. For the sources, computing the period reduces to computing its span. For the span, apply the procedure given in [5, Prop. 2.9].
4. Compute $b_{i,r}^{\langle\mathcal{U}\rangle}(\rho\omega^r)$ and $b_{i,r}^{\langle\mathcal{G}\rangle}(\rho\omega^r)$, in Lemma 5, for each i and $r = 0, \ldots, q-1$. Finally, for the denominators, compute $b_i(\rho)$, in Lemma 1.

Note that this method is numerical; testing the equality in step 2 depends on a numerical tolerance.

Implementation. We give a partial implementation[4] of this method in SageMath. The detection of overabundance (steps 1 and 2) has been completely implemented. For the numeric computation of the proportions, since the systems we encountered did not present periodicities, the code assumes for the moment that the extended system remains aperiodic. In particular, we managed to handle many of them (see Sect. 5).

5 Application to Systems from the Literature

Several systems have been introduced in the literature in order to limit the most obvious syntactic redundancies of regular expressions. It is natural to expect more refined systems to present fewer universal expressions; however, our calculations show surprisingly that this is not always the case. Indeed, reducing redundancies decreases both the number of universal and non-universal expressions; it is a matter of the proportion for each, as in the following example.

Example 2. Consider the following systems (1) and (2). System (1) avoids consecutive stars and takes into account the associativity of $+$, while system (2) also considers the associativity of $\bullet$.

$$(1)\begin{cases} S = a + b + \varepsilon + \overset{\star}{\underset{S_s}{|}} + \overset{+}{\underset{S\ S_p}{\wedge}} + \overset{\bullet}{\underset{S\ S}{\wedge}} \\ S_s = a + b + \varepsilon + \overset{+}{\underset{S\ S_p}{\wedge}} + \overset{\bullet}{\underset{S\ S}{\wedge}} \\ S_p = a + b + \varepsilon + \overset{\bullet}{\underset{S\ S}{\wedge}} + \overset{\star}{\underset{S_s}{|}} \end{cases} \qquad (2)\begin{cases} S = a + b + \varepsilon + \overset{\star}{\underset{S_s}{|}} + \overset{+}{\underset{S\ S_p}{\wedge}} + \overset{\bullet}{\underset{S\ S_c}{\wedge}} \\ S_s = a + b + \varepsilon + \overset{+}{\underset{S\ S_p}{\wedge}} + \overset{\bullet}{\underset{S\ S_c}{\wedge}} \\ S_p = a + b + \varepsilon + \overset{\star}{\underset{S_s}{|}} + \overset{\bullet}{\underset{S\ S_c}{\wedge}} \\ S_c = a + b + \varepsilon + \overset{\star}{\underset{S_s}{|}} + \overset{+}{\underset{S\ S_p}{\wedge}} \end{cases}$$

In both cases, the extended system remains strongly connected and aperiodic. Our calculations give, after renaming $S = \mathcal{Y}_1$:

$$(1)\ \ \mathbb{P}_1^{\langle\mathcal{U}\rangle} = 0.16\ldots, \quad \mathbb{P}_1^{\langle\overline{\Sigma}\rangle} = 0.27\ldots\,; \quad (2)\ \ \mathbb{P}_1^{\langle\mathcal{U}\rangle} = 0.38\ldots, \quad \mathbb{P}_1^{\langle\overline{\Sigma}\rangle} = 0.51\ldots.$$

[4] https://gitlab.com/carine.pivoteau/universality-detection.

Note that $\mathbb{P}_1^{\langle \mathcal{U} \rangle}$ is a lower-bound, while $\mathbb{P}_1^{\langle \overline{\Sigma} \rangle}$ is an upper-bound, for the proportion of universals. This shows that (2) has (proportionally) more universals than (1).

Example 3. The following elaborate system comes from Lee and Shallit [17, p.17]. It avoids several more redundancies of regular expressions. For instance, it takes into account the redundancy coming from $(L_1 + \ldots + L_k)^* \equiv L_1^* + \ldots + L_k^*$.

$$A = \overset{+}{\varepsilon \; A_B} + \overset{+}{C \; A_C} + \overset{+}{B \; A_B} + \overset{+}{D \; A_D} + \varepsilon \qquad A' = \overset{+}{B \; A_B} + \overset{+}{D \; A_D} \qquad A_C = C + \overset{+}{C \; A_C} + A_B$$

$$A_B = B + \overset{+}{B \; A_D} + A_D \qquad A_D = D + \overset{+}{D \; A_D} \qquad B = \overset{\bullet}{B_0 \; B_0} + \overset{\bullet}{B_0 \; B}$$

$$B_0 = A + C + D \qquad C = \overset{\star}{A'} + \overset{\star}{B} + \overset{\star}{D} \qquad D = a + b + \varepsilon\,.$$

We calculated that the proportion of universals in this system, starting from an extra axiom symbol $S = A + B + C$, is comprised between $\mathbb{P}_1^{\langle \mathcal{U} \rangle} = 0.30\ldots$ and $\mathbb{P}_1^{\langle \overline{\Sigma} \rangle} = 0.44\ldots$ Despite the complexity of the system, these bounds remain close to those obtained in [15] for the system of one equation, Eq. (1.1). We recall that in this case the limiting proportion of universals belongs to $[0.31, 0.46]$.

6 Conclusion

We have shown how to systematically estimate the proportion of universal expressions in systems of expression trees that are $\mathcal{D}$-analyzable. Our code (in Sagemath) is available online for anyone interested to test their own systems.

The analysis can be generalized to some systems which are not strongly-connected. For example, if in the original system there is only one strongly-connected component C in ρ, which we call the dominant component, Theorem 1 still applies to all $\mathcal{Y}_i$ in C. As a matter of fact, in Example 3, the system is not strongly-connected, but we could carry out our procedure as there is only one dominant component.

Our experiments and calculations reveal that the relationship between redundancy and universality can be surprising and counter-intuitive, which stresses the difficulty to design and predict the behavior of systems of regular expressions.

Here, the heuristic for universal detection relies on an under-approximation of universal expressions to detect redundancies. It would be interesting to study how randomized approximation algorithms (e.g., [16]), developed to estimate universality by *over-approximation*, could be interfaced with our approach.

References

1. Aho, A.V., Ullman, J.D.: The Theory of Parsing, Translation, and Compiling. Prentice-Hall Inc, USA (1972)
2. Almeida, A., Almeida, M., Alves, J., Moreira, N., Reis, R.: FAdo and GUItar: tools for automata manipulation and visualization. In: Maneth, S. (ed.) CIAA 2009. LNCS, vol. 5642, pp. 65–74. Springer, Heidelberg (2009). https://doi.org/10.1007/978-3-642-02979-0_10
3. Banderier, C., Drmota, M.: Formulae and asymptotics for coefficients of algebraic functions. Comb. Probab. Comput. **24**(1), 1–53 (2015)
4. Bell, J.P., Burris, S., Yeats, K.A.: Characteristic points of recursive systems. Electr. J. Comb. **17**(1) (2010)
5. Bell, J.P., Burris, S., Yeats, K.A.: Spectra and systems of equations. Contemp. Math. AMS **558**, 43–96 (2011)
6. Broda, S., Machiavelo, A., Moreira, N., Reis, R.: On the average state complexity of partial derivative automata: an analytic combinatorics approach. Int. J. Found. Comput. Sci. **22**, 1593–1606 (2011)
7. Broda, S., Machiavelo, A., Moreira, N., Reis, R.: On the uniform distribution of regular expressions. In: Han, YS., Ko, SK. (eds.) Descriptional Complexity of Formal Systems. DCFS 2021. LNCS, vol. 13037, pp. 13–25. Springer, Cham (2021)
8. Drmota, M.: Systems of functional equations. Random Struct. Algorithms **10**(1–2), 103–124 (1997)
9. Duchon, P., Flajolet, P., Louchard, G., Schaeffer, G.: Boltzmann samplers for the random generation of combinatorial structures. Comb. Probab. Comput. **13**(4–5), 577–625 (2004)
10. Flajolet, P., Sedgewick, R.: Analytic Combinatorics. Cambridge University Press (2009)
11. Flajolet, P., Steyaert, J.-M.: A complexity calculus for recursive tree algorithms. Math. Syst. Theory **19**(4), 301–331 (1987)
12. Gruber, H., Gulan, S.: Simplifying regular expressions. In: Dediu, AH., Fernau, H., Martín-Vide, C. (eds.) Language and Automata Theory and Applications. LATA 2010. LNCS, vol. 6031, pp. 285–296. Springer, Berlin, Heidelberg (2010)
13. Kahrs, S., Runciman, C.: Simplifying regular expressions further. J. Symb. Comput. **109**, 124–143 (2022)
14. Koechlin, F., Nicaud, C., Rotondo, P.: Simplifications of uniform expressions specified by systems. Int. J. Found. Comput. Sci. **32**(06), 733–760 (2021)
15. Koechlin, F., Rotondo, P.: Analysis of an efficient reduction algorithm for random regular expressions based on universality detection. In: Santhanam, R., Musatov, D. (eds.) Computer Science – Theory and Applications, pp. 206–222. Springer, Cham (2021)
16. Konstantinidis, S., Mastnak, M., Moreira, N., Reis, R.: Approximate NFA universality and related problems motivated by information theory. Theoret. Comput. Sci. **972**, 114076 (2023)
17. Lee, J., Shallit, J.: Enumerating regular expressions and their languages. In: Domaratzki, M., Okhotin, A., Salomaa, K., Yu, S., (eds.) Implementation and Application of Automata, pp. 2–22. Springer, Berlin, Heidelberg (2005)
18. Meyer, A.R., Stockmeyer, L.J.: The equivalence problem for regular expressions with squaring requires exponential space. In: Proceedings of the 13th Annual Symposium on Switching and Automata Theory (Swat 1972), SWAT '72, pp. 125–129, USA. IEEE Computer Society (1972)

19. Nicaud, C.: On the average size of Glushkov's automata. In: Dediu, A.-H., Ionescu, A.-M., Martín-Vide, C. (eds.) Language and Automata Theory and Applications, Third International Conference, LATA 2009, Tarragona, Spain, April 2–8, 2009. Proceedings. LNCS, vol. 5457, pp. 626–637. Springer (2009)
20. Nijenhuis, A., Wilf, H.S.: Combinatorial algorithms for computers and calculators. In: Computer Science and Applied Mathematics, 2nd edn.. Academic Press, New York 1978 (1978)
21. Pivoteau, C., Salvy, B., Soria, M.: Algorithms for combinatorial structures: well-founded systems and Newton iterations. J. Comb. Theory Ser. A **119**(8), 1711–1773 (2012)

Sorting Circular Suffixes in Linear Time

Nicola Cotumaccio(✉)

Department of Computer Science, University of Helsinki, Helsinki, Finland
nicola.cotumaccio@helsinki.fi

Abstract. In a well-studied variant of the dictionary matching problem, the strings in the dictionary are circular. Hon et al. [ISAAC 2011] showed how to solve circular dictionary queries efficiently within compressed space under some technical assumptions on the dictionary. Several papers studied the problem of building efficiently the required data structures, until Bannai et al. finally achieved linear time [CPM 2021] by presenting an algorithm for building the bijective Burrows-Wheeler transform. More precisely, to support circular dictionary matching queries efficiently, Hon et al. rely on Mantaci et al.'s eBWT, so Boucher et al. described a direct linear-time algorithm for building the eBWT [SPIRE 2021]. Very recently, it was shown how to solve circular dictionary queries in full generality by introducing a suffix array for arbitrary dictionaries [CPM 2025]. To build the suffix array and support circular dictionary queries, one needs to consider a problem more general than constructing the eBWT, because it is necessary to compute the *ordered partition* of all (infinite) suffixes with respect to the lexicographic order.

In this paper, we provide a linear-time construction algorithm that extends Ko and Aluru's algorithm for building the suffix array. As a by-product, we show that induced sorting yields a new linear-time algorithm to compute the primitive root of a string.

Keywords: Circular dictionary matching · eBWT · induced sorting · dictionary matching · Burrows-Wheeler transform

1 Introduction

In the *dictionary matching problem*, we simultaneously match the elements of a dictionary against a pattern. The classical solution to the problem relies on Aho-Corasick automata [1]. More recently, it was shown that the dictionary matching problem can be solved within compressed space [9,35]. Several variations have been studied, including dictionary matching in the streaming model [17], dynamic dictionary matching [31], internal dictionary matching [15] and dictionary matching with gaps [5,6,36] or mismatches [30].

In this paper, we consider the *circular dictionary matching problem*, in which after reading the last character of a string in the dictionary one can jump back to the beginning of the string. The problem is motivated by some applications in computational geometry and bioinformatics [40]. In computational geometry, a

S.-K. Ko and F. Manea (Eds.): DLT 2025, LNCS 16036, pp. 294–310, 2026.
https://doi.org/10.1007/978-3-032-01475-7_20

polygon can be stored by listing its vertices in clockwise order. In bioinformatics, the genome of some viruses, such as the herpes simplex virus (HSV-1), exists as circular strings [51], and some microbial samples have circular chromosomes [27, 50]. Approximate variants [16] and average-case analysis of the circular matching problem [39] have also been considered in the literature.

Given an alphabet of size σ and a dictionary with d strings of total length n, Hon et al. showed how to solve solve circular dictionary matching queries in $O((m+occ)\log^{1+\epsilon} n)$ time for every $\epsilon > 0$ (where m is the length of the pattern) by storing a compressed data structure of $n \log \sigma(1 + o(1)) + O(n) + O(d \log n)$ bits [37]. This result is valid under the assumption that the elements of the dictionary are distinct primitive strings of similar length (Hon et al. introduce the technical notion of *bounded aspect ratio*) such that no string is a conjugate (i.e., a circular rotation) of another string, even though subsequent papers [33,34] tried to remove these constraints. The two key ingredients of the data structures are Mantaci et al.'s eBWT [44] and an extension of Sadakane's compressed suffix tree [49] to the circular setting.

A chain of papers considered the problem of building Hon et al.'s data structures. In [34], it was shown that, given the eBWT, it is possible to build the (circular) compressed suffix tree in $O(n \log n)$ time within $O(n \log \sigma + d \log n)$ bits of working space, even though it is not too difficult to show that, if we do not insist on achieving the space bound, we can reduce the time bound to $O(n)$. As a consequence, the problem of constructing the data structure of Hon et al. reduces to building the eBWT. A construction algorithm for the eBWT running in $O(n \log n)$ time was presented in [33]. Bonomo et al. [11] [Sect. 6] showed that the problem of building the eBWT can be reduced in linear time to the problem of building the bijective BWT, and Bannai et al. [8] finally achieved linear-time construction of the eBWT by describing a linear-time construction algorithm for the bijective BWT. In 2021, Boucher et al. gave a direct linear-time construction algorithm for the eBWT [13], which can also be used to r-index the eBWT [14]. Both Bannai et al.'s algorithm and Boucher et al.'s algorithm use *induced sorting*, the recursive paradigm of Farach's linear-time algorithm for building the suffix tree [28]. More precisely, both algorithms generalize SAIS algorithm [46] for building the suffix array. An additional eBWT construction algorithm [47] is based on Baier's algorithm for building the suffix array [7].

2 Our Contribution

Very recently, it was shown that there exists a data structure achieving the same space bound $n \log \sigma(1 + o(1)) + O(n) + O(d \log n)$ and supporting circular dictionary matching queries on *arbitrary* dictionaries in $O((m + occ) \log n)$ time [22]. This result, which does not assume any constraint on the dictionary, is obtained by defining a more general suffix array for arbitrary dictionaries and drawing inspiration from a recent research line that has extended the suffix array, the Burrows-Wheeler transform and the FM-index to labeled graphs and automata [2–4,18,19,21,23–26,29,52,53]. Consequently, to build the

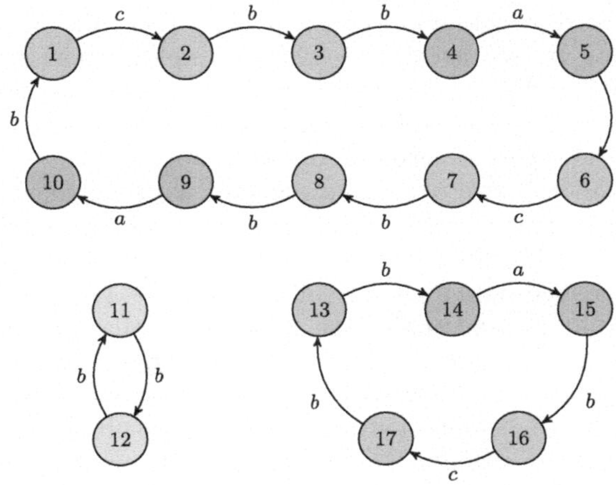

Fig. 1. A disjoint union of cycles $G = (V, E)$. The graph G encodes the circular nature of the (circular) dictionary $\mathcal{T} = (cbbabcbbab, bb, babcb)$. Different colors correspond to different values of $\tau(u)$ (see Definition 1).

new data structure we will need to solve a problem more general than building the original eBWT. To illustrate the problem, we will again use labeled graphs. Broadly speaking, the original paper on the eBWT [44] extended the Burrows-Wheeler transform to a collection of strings through the powerful and non-obvious trick of using an order on infinite strings to sort the suffixes of finite strings. The original paper on the circular dictionary matching problem [37] established a systematic connection between the eBWT, Sadakane's compressed suffix tree and pattern matching. The recent paper on the circular dictionary matching problem [22] employed labeled graphs and their properties to visualize and study the cyclic nature of the problem. Consider the dictionary $\mathcal{T} = (cbbabcbbab, bb, babcb)$. Every string in the dictionary is circular (after reading the last character, we can move back to the first character), so we can capture the circular nature of the dictionary by drawing a cycle for each string (see Fig. 1). Every node identifies the corresponding position in the concatenation *cbbabcbbabbbbabcb* of the three strings of the dictionary (for example, node 13 corresponds to the starting position of *babcb* in the concatenation). For every node u, let $\mathcal{S}_u$ be the infinite string that can be read starting from node u in the graph (for example, $\mathcal{S}_2 = (bbabc)^\omega$). Our goal is to sort the $\mathcal{S}_u$'s lexicographically keeping track of which nodes share the same string $\mathcal{S}_u$, thus building the $\mathcal{S}$*-partition* $(\{4,9,14\}, \{3,8,13\}, \{2,7,17\}, \{11,12\}, \{5,10,15\}, \{1,6,16\})$ (see Fig. 2 for details and see Sect. 3 for the formal definition of the $\mathcal{S}$-partition). Intuitively, each element of the $\mathcal{S}$-partition consists of all nodes u sharing the same string $\mathcal{S}_u$ (for example, in Fig. 2 we have $\mathcal{S}_u = (abcbb)^\omega$ if and only if $u \in \{4,9,14\}$, so $\{4,9,14\}$ is an element of $\mathcal{S}$-partition), and the elements of $\mathcal{S}$-partition are sorted according to the lexicographic order of the corresponding $\mathcal{S}_u$'s (for example, in Fig. 2 the element $\{4,9,14\}$ must occur before $\{3,8,13\}$

in the $\mathcal{S}$-partition because $(abcbb)^\omega$ is lexicographically smaller than $(babcb)^\omega$). Differently from Bannai et al.'s algorithm and Boucher et al.'s algorithm, we do not need to output a permutation of the text positions (as in the standard suffix array), but an *ordered partition* (note that a permutation can be seen as an ordered partition consisting of singletons). Intuitively, in the general setting of [22] we need an ordered partition because, if the strings in the dictionary can be distinct or non-primitive, or if some strings in the dictionary can be circular rotations of some other strings, then multiple positions in the dictionary may correspond to the same circular suffix (or equivalently, from a graph-theoretical perspective, multiple nodes u may correspond to the same $\mathcal{S}_u$).

$\mathcal{S}_u$											u	$\tau(u)$
a	b	c	b	b	a	b	c	b	b	...	4	3
a	b	c	b	b	a	b	c	b	b	...	9	3
a	b	c	b	b	a	b	c	b	b	...	14	3
b	a	b	c	b	b	a	b	c	b	...	3	1
b	a	b	c	b	b	a	b	c	b	...	8	1
b	a	b	c	b	b	a	b	c	b	...	13	1
b	b	a	b	c	b	b	a	b	c	...	2	1
b	b	a	b	c	b	b	a	b	c	...	7	1
b	b	a	b	c	b	b	a	b	c	...	17	1
b	b	b	b	b	b	b	b	b	b	...	11	2
b	b	b	b	b	b	b	b	b	b	...	12	2
b	c	b	b	a	b	c	b	b	a	...	5	3
b	c	b	b	a	b	c	b	b	a	...	10	3
b	c	b	b	a	b	c	b	b	a	...	15	3
c	b	b	a	b	c	b	b	a	b	...	1	1
c	b	b	a	b	c	b	b	a	b	...	6	1
c	b	b	a	b	c	b	b	a	b	...	16	1

Fig. 2. The strings $\mathcal{S}_u$'s (see Sect. 2) and the values $\tau(u)$'s (see Definition 2) for the graph of Fig. 1. The strings $\mathcal{S}_u$'s are sorted lexicographically, and the nodes u in the same block have the same $\mathcal{S}_u$. The $\mathcal{S}$-partition is $(\{4,9,14\},\{3,8,13\},\{2,7,17\},\{11,12\},\{5,10,15\},\{1,6,16\})$ because $\mathcal{S}_4 = \mathcal{S}_9 = \mathcal{S}_{14} \prec \mathcal{S}_3 = \mathcal{S}_8 = \mathcal{S}_{13} \prec \mathcal{S}_2 = \mathcal{S}_7 = \mathcal{S}_{17} \prec \mathcal{S}_{11} = \mathcal{S}_{12} \prec \mathcal{S}_5 = \mathcal{S}_{10} = \mathcal{S}_{15} \prec \mathcal{S}_1 = \mathcal{S}_6 = \mathcal{S}_{16}$.

Consider a polynomial alphabet (that is, of size $O(|V|^{O(1)})$, where V is the set of all nodes). The main result of the paper is the following.

Theorem 1. *Let $G = (V, E)$ be a disjoint union of cycles. We can build the $\mathcal{S}$-partition of V in $O(|V|)$ time.*

Our algorithm is also based on induced sorting. We do not generalize SAIS, but Ko and Aluru's algorithm [41], another linear-time construction algorithm for the suffix array. Ko and Aluru's algorithm has already been extended to a superlinear algorithm to compute the Burrows-Wheeler transform of a deterministic automaton [20]. However, in the setting of this paper, we can achieve linear time.

We prove our main result in the graph setting because providing the intuition behind the main steps will be much easier. Still, we do not need to explicitly build the graph to compute the $\mathcal{S}$-partition of the positions of a dictionary $\mathcal{T}$ because we can use $\mathcal{T}$ as an implicit representation of the graph. The algorithms based on induced sorting are generally rather complex, and our goal is to propose not only a clear explanation but also a rigorous proof (while, e.g., Boucher et al. provide an outline of the main differences between SAIS and their approach and present some techniques to enhance the practical performance of their algorithm). Note that from Theorem 1 one can immediately obtain that the suffix array of a string T can be built in linear time, because it is sufficient to consider the dictionary $\mathcal{T} = (T\$)$ (which contains only the string T suffixed by a new character \$ smaller than all the other characters in the alphabet), build the corresponding graph (which consists of a single cycle) as in Fig. 1 and compute the $\mathcal{S}$-partition (which consists of singletons due to the presence of \$).

Theorem 1 implies as a special case that the eBWT (as defined by Bannai et al. and Boucher et al.) can be built in linear time, so our algorithm generalizes the results on the eBWT construction mentioned in the introduction. Our algorithm has an additional advantage. While Boucher et al.'s algorithm is the first algorithm to provide a direct, linear-time construction of the eBWT, it still requires a linear-time preprocessing to compute the *root* of each string in the dictionary $\mathcal{T}$. Recall that a finite string $T \in \Sigma^*$ is primitive if for every string $S \in \Sigma^*$ and for every integer $k \geq 1$, if $T = S^k$, then $k = 1$ and $T = S$. It is well-known that for every $T \in \Sigma^*$ there exists exactly one primitive string $R \in \Sigma^*$ and exactly one integer $z \geq 1$ such that $T = R^z$ (see [42, Prop. 1.3.1]); we say that R is the root of T. Boucher et al. first compute the primitive root of each string in $\mathcal{T}$, and then show that it is sufficient to build of eBWT of $\mathcal{T}'$, where $\mathcal{T}'$ is the set of all primitive roots of some string in $\mathcal{T}$ (Boucher et al. remark that, as a matter of fact, the original definition of the eBWT [44] only addresses multisets of primitive strings).

In our algorithm, we do not need this preprocessing step, because via induced sorting we can implicitly compute the root of each string. In fact, our algorithm yields a new algorithm for this problem:

Corollary 1. *Let $T \in \Sigma^*$ be a string. We can compute the root of T in $O(|T|)$ time.*

Corollary 1 follows immediately from Theorem 1: given $T \in \Sigma^*$, we consider the dictionary $\mathcal{T} = (T)$ (which consists of a single string), we (implicitly) build the corresponding graph (which consists of a single cycle) as in Fig. 1, and we compute the $\mathcal{S}$-partition. Then, the root of T is the prefix of T of length ℓ, where ℓ is the number of elements in the $\mathcal{S}$-partition. For example, if $T = ababab$, the $\mathcal{S}$-partition is $(\{1, 3, 5\}, \{2, 4, 6\})$ and so the root of T is $T[1, 2] = ab$. We remark that the classical algorithms for computing the root of a string are simpler and already achieve linear time in the comparison model (while Corollary 1 is only true for polynomial alphabets). However, the classical algorithms are very different in flavor: they are based on the border array used in the Knuth-Morris-Pratt (KMP) algorithm (as explained by Boucher et al. [13]) or the related

Z-array of the Z-algorithm (see [32]). In other words, Corollary 1 confirms the power of (suffix) sorting in solving classical pattern matching and stringology problems. As an additional example, in 1975 Manacher showed that the problem of computing the longest palindromic substrings of a string can be solved in linear time [43], but later it was shown that the same problem can also be solved in linear time via a very different approach based on building the suffix tree of the string [32]. Manacher's algorithm achieves linear time in the comparison model, while achieving linear-time construction of the suffix tree requires considering polynomial alphabets [28].

Theorem 1 also implies that the lexicographically smallest circular rotation of a string can be computed in linear time (for polynomial alphabets). The same problem was solved, e.g., by Booth in 1980 (in the comparison model) [12]. Conversely, Booth's algorithm can be used to infer Theorem 1 from Boucher et al.'s algorithm: after a first linear-time preprocessing step to compute the primitive root of each string in the dictionary, it is possible to run a second linear-time preprocessing step to determine which strings in the dictionary have the same root or have the same root up to a circular rotation. To this end, it is sufficient to compute the lexicographically smallest circular rotations of the primitive roots, and then sort them in linear time [38,48].

3 Preliminaries

Ordered Partitions. In the paper, we consider q-ples $\mathcal{I} = (I_1, I_2, \dots, I_q)$, where $q \geq 0$ and each I_i is a nonempty set (note that e.g., $(\{2,3\}, \{4,1,5\})$ and $(\{3,2\}, \{5,4,1\})$ are the same 2-ple). The case $q = 0$ corresponds to the empty q-ple, denoted by $\perp$. We say that I_1, I_2, ..., I_q are the *elements* of $\mathcal{I}$. We can scan the element of the q-ple $\mathcal{I}$ from left to right (that is, following the order I_1, I_2, ..., I_q) or from right to left (that is, following the order I_q, I_{q-1}, ..., I_1). For every $0 \leq r \leq q$, we say that $(I_1, I_2, \dots, I_r)$ is a *prefix* of $\mathcal{I}$ (the case $r = 0$ corresponds to the empty prefix $\perp$) and for every $1 \leq s \leq q+1$ we say that $(I_s, I_{s+1}, \dots, I_q)$ is a *suffix* of $\mathcal{I}$ (the case $s = q+1$ corresponds to the empty suffix $\perp$).

Given a q_1-ple $\mathcal{I} = (I_1, I_2, \dots, I_{q_1})$ and a q_2-ple $\mathcal{J} = (J_1, J_2, \dots, J_{q_2})$, let $\mathcal{I} \circ \mathcal{J}$ be the $(q_1 + q_2)$-ple $(I_1, I_2, \dots, I_{q_1}, J_1, J_2, \dots, J_{q_2})$ obtained by concatenating $\mathcal{I}$ and $\mathcal{J}$. Notice for every q-ple $\mathcal{I}$ we have $\mathcal{I} \circ \perp = \perp \circ \mathcal{I} = \mathcal{I}$. Moreover, if $\mathcal{I}_i$ is a q_i-ple for $1 \leq i \leq 3$, we have $(\mathcal{I}_1 \circ \mathcal{I}_2) \circ \mathcal{I}_3 = \mathcal{I}_1 \circ (\mathcal{I}_2 \circ \mathcal{I}_3)$ (that is, concatenation is associative), so the $(q_1 + q_2 + q_3)$-ple $\mathcal{I}_1 \circ \mathcal{I}_2 \circ \mathcal{I}_3$ is unambiguously defined.

Let $\mathcal{I} = (I_1, I_2, \dots, I_q)$ be a q-ple, and define $|\mathcal{I}| = \sum_{i=1}^{q} |I_i|$. If each I_i is a set of integers, we store $\mathcal{I}$ by using two arrays $C_\mathcal{I}[1, |\mathcal{I}|]$ and $D_\mathcal{I}[1, |\mathcal{I}|]$. The array $C_\mathcal{I}[1, |\mathcal{I}|]$ is obtained by concatenating the elements in I_1, the elements in I_2, ..., the elements in I_q, in this order (the elements in each I_i can be enumerated in an arbitrary order). The array $D_\mathcal{I}[1, |\mathcal{I}|]$ is a bit array that contains exactly q 1 s, and each 1 marks the end of a set I_i (see Fig. 3). Alternatively, we can use the variant of $D_\mathcal{I}[1, |\mathcal{I}|]$ where we mark the beginning of each set I_i, and we can build one variant from the other in linear time. For $q = 0$ we have the empty q-ple $\perp$, and $C_\mathcal{I}[1, |\mathcal{I}|]$ and $D_\mathcal{I}[1, |\mathcal{I}|]$ are the empty arrays, which can be remembered by using null pointers. We can scan all elements of $\mathcal{I}$ from left to right or from right to left in linear time. More generally, if for every $1 \le j \le t$ we store the q_j-ple $\mathcal{I}_j$, then we can go through the elements of $I_1 \circ I_2 \circ \ldots I_t$ from left to right or from right to left in $O(t + \sum_{j=1}^{t} |\mathcal{I}_j|)$ time (we write $O(t + \sum_{j=1}^{t} |\mathcal{I}_j|)$ and not $O(\sum_{j=1}^{t} |\mathcal{I}_j|)$ because we may have $I_j = \perp$ for some j's).

A q-ple $\mathcal{I} = (I_1, I_2, \ldots, I_q)$ is an *ordered partition* if $I_i \cap I_j = \emptyset$ for every $1 \le i < j \le q$. Moreover, if $\mathcal{I} = (I_1, I_2, \ldots, I_q)$ is an ordered partition and X is a set, then $\mathcal{I}$ is an *ordered partition of* X if $X = \bigcup_{i=1}^{q} I_i$. Note that the unique ordered partition of the empty set is the empty ordered partition $\perp$. With a slight abuse of notation, we can see an ordered partition $\mathcal{I} = (I_1, I_2, \ldots, I_q)$ as a set endowed with a total order $\le_{\mathcal{I}}$. The elements of $\mathcal{I}$ are $I_1, I_2, \ldots, I_q$ (so for $1 \le i \le q$ we can write $I_i \in \mathcal{I}$) and $\le_{\mathcal{I}}$ is the total order on the set $\mathcal{I}$ such that $I_r \le_{\mathcal{I}} I_s$ if and only if $r \le s$.

k	1	2	3	4	5
$C_{\mathcal{I}}[k]$	2	3	4	1	5
$D_{\mathcal{I}}[k]$	0	1	0	0	1

Fig. 3. How to represent the 2-ple $\mathcal{I} = (\{2,3\}, \{4,1,5\})$.

Strings and Graphs. Let Σ be a finite alphabet. Let Σ^* be the set of all finite strings on Σ (including the empty string ϵ), and let Σ^ω be the set of all infinite strings on Σ. For example, if $\Sigma = \{a, b\}$, then *baaba* is a string in Σ^* and *ababab*... is a string in Σ^ω. Let $\alpha \in \Sigma^* \cup \Sigma^\omega$. If $i \ge 1$, we denote the i-th character of α by $\alpha[i]$, and if $j \ge i$, we define $\alpha[i,j] = \alpha[i]\alpha[i+1]\ldots\alpha[j-1]\alpha[j]$ (if $\alpha \in \Sigma^*$, we assume $i, j \le |\alpha|$, where $|\alpha|$ is the length of α). If $j < i$, we define $\alpha[i,j] = \epsilon$.

We consider a fixed total order $\preceq$ on Σ and we extend it to $\Sigma^* \cup \Sigma^\omega$ *lexicographically.* If $\alpha, \beta \in \Sigma^* \cup \Sigma^\omega$, we write $\alpha \prec \beta$ if $(\alpha \preceq \beta) \wedge (\alpha \ne \beta)$. In our examples, $\preceq$ is the standard order on the English alphabet.

Let $G = (V, E)$ be a directed, edge-labeled graph on Σ, where V is the set of all nodes and $E \subseteq V \times V \times \Sigma$ is the set of all edges. We say that G is a *disjoint union of cycles* if every node has exactly one outgoing edge and exactly one incoming edge (which implies $|E| = |V|$), as in Fig. 1.

Let $G = (V, E)$ be a disjoint union of cycles. An *occurrence* of $\alpha \in \Sigma^\omega$ starting at $u \in V$ is a sequence of nodes $(u_i)_{i \ge 1}$ of V such that (i) $u_1 = u$ and (ii) $(u_i, u_{i+1}, \alpha[i]) \in E$ for every $i \ge 1$. For every $u \in V$, there exists exactly one occurrence starting at u and there exists exactly one infinite string having an occurrence starting at u; we will denote such a string by $\mathcal{S}_u$ (for example, $\mathcal{S}_{14} = (abcbb)^\omega$ in Fig. 1). Note that the unique edge leaving u is labeled with $\mathcal{S}_u[1]$. Fix $u \in V$, and let $(u_i)_{i \ge 1}$ be the occurrence of $\mathcal{S}_u$ starting at u. Note that for every $i \ge 1$ we have that $(u_j)_{j \ge i}$ is the occurrence of $\mathcal{S}_{u_i}$ starting at u_i and $\mathcal{S}_u = \mathcal{S}_u[1, i-1]\mathcal{S}_{u_i}$ (in particular $\mathcal{S}_u[i] = \mathcal{S}_{u_i}[1]$). For every $u \in V$, let c_u be the length of the cycle (i.e., the number of nodes in the cycles, or equivalently, the number of edges in the cycles) to which u belongs (for example, $c_7 = 10$ in Fig. 1). Notice that, if $(u_i)_{i \geq 1}$ is the occurrence starting at u, then for every $i' \geq i \geq 1$ we have $u_{i'} = u_i$ if and only if $i' - i$ is a multiple of c_u.

Let $G = (V, E)$ be a disjoint union of cycles, and let $V' \subseteq V$. The *$\mathcal{S}$-partition* of V' (with respect to G) is the ordered partition $(I_1, I_2, \ldots, I_q)$ of V' such that, for every $u, v \in V'$, (i) if u and v are in the same I_j, then $\mathcal{S}_u = \mathcal{S}_v$ and (ii) if $u \in I_j$, $v \in I_{j'}$ and $j < j'$, then $\mathcal{S}_u \prec \mathcal{S}_v$ (see Fig. 2 for an example). The main result in this paper is that the $\mathcal{S}$-partition of V can be built in $O(|V|)$ time (Theorem 1).

Let $G = (V, E)$ be a disjoint union of cycles. When we implement G, we assume $V = \{1, 2, \ldots, |V|\}$ and $\Sigma = \{1, 2, \ldots, |\Sigma|\}$, with $1 \prec 2 \prec \cdots \prec |\Sigma|$, and E is given through its adjacency list. Since each node has exactly one outgoing edge, we can implement the adjacency list through two arrays $W[1, |V|]$ and $Z[1, |V|]$ where for every $1 \leq i \leq |V|$, if (i, j, k) is the edge leaving node i, we have $W[i] = j$ and $Z[i] = k$ (that is, $W[i]$ is the node reachable from i and $Z[i]$ is the label of the corresponding edge). For notational clarity, we typically denote nodes using u, v, u', v' and characters using a, b, c. This means that, if $u \in V$, then u is an integer between 1 and $|V|$, and if $a \in \Sigma$, then a is an integer between 1 and $|\Sigma|$. Note that for every $1 \leq i \leq |V|$ we have $\mathcal{S}_i[1] = Z[i]$. In our algorithm, we will also need to navigate G following edges in a backward fashion. To this end, in $O(|V|)$ time we can build from W and Z the adjacency list of the *reverse* graph G^R obtained by swapping the direction of each edge of G. More precisely, let $W^*[1, |V|]$ and $Z^*[1, |V|]$ be the arrays such that for every $1 \leq j \leq |V|$, if (i, j, k) is the edge entering node j, we have $W^*[j] = i$ and $Z^*[j] = k$. We can build $W^*[1, |V|]$ and $Z^*[1, |V|]$ in $O(|V|)$ time by setting $W^*[W[i]] \leftarrow i$ and $Z^*[W[i]] \leftarrow Z[i]$ for every $1 \leq i \leq |V|$.

Let $G = (V, E)$ be a disjoint union of cycles. We mentioned that Theorem 1 holds for polynomial alphabets, that is, we assume $|\Sigma| = O(|V|^{O(1)})$. In the rest of the paper, we can assume without loss of generality that the alphabet Σ is *effective*, that is, every character in Σ labels some edge. Indeed, given a polynomial alphabet, we can sort all the $|V|$ edge labels in $O(|V|)$ time [38,48] and we can replace each label with its rank in the sorted list of all labels, which yields an effective alphabet and does not alter the $\mathcal{S}$-partition. In particular, we can assume without loss of generality that $|\Sigma| \leq |V|$.

Outline of the Paper. Let us sketch our (recursive) strategy to prove Theorem 1. In Sect. 4, we classify each $u \in V$ into one of three types: $\tau(u) = 1$, $\tau(u) = 2$ or $\tau(u) = 3$ (the definition of $\tau(u)$ will be given later and it is not necessary to follow the general strategy of the proof). Let $V_i = \{u \in V \mid \tau(u) = i\}$ for $1 \leq i \leq 3$. Since V is the disjoint union of V_1, V_2 and V_3, then at least one of the following is true: (i) $|V_1| \leq |V|/2$; (ii) $|V_3| \leq |V|/2$. Assume that $|V_3| \leq |V|/2$ (the case $|V_1| \leq |V|/2$ requires minor modifications). In Sect. 5, we show that, if we have the $\mathcal{S}$-partition of V_3, we can compute the $\mathcal{S}$-partition of V in $O(|V|)$ time. In Sect. 6, we show that we can reduce in $O(|V|)$ time the problem of computing the $\mathcal{S}$-partition of V_3 to the problem of computing the $\mathcal{S}$-partition of $\overline{V}$ with respect to a new graph $\overline{G} = (\overline{V}, \overline{E})$ for which $|\overline{V}| = |V_3|$, and we show that $\overline{G}$ can be built in $O(|V|)$ time from G. Since $|V_3| \leq |V|/2$, we obtain the recurrence $T(|V|) = T(|V|/2) + O(|V|)$ and so the algorithm runs in $O(|V|)$ time.

Due to space constraints and the complexity of the argument, we focus on the intuition behind the proof of Theorem 1 and we give an example, while the detailed proofs (and some auxiliary results) can be found in the extended version on arXiv.

4 The Array *T*

Following [20], we classify the strings in Σ^ω into three types.

Definition 1. *Let* $\alpha \in \Sigma^\omega$, *and let* $a \in \Sigma$ *and* $\alpha' \in \Sigma^\omega$ *such that* $\alpha = a\alpha'$. *Then, we define* $\tau(\alpha)$ *as follows: (i)* $\tau(\alpha) = 1$ *if* $\alpha' \prec \alpha$, *(ii)* $\tau(\alpha) = 2$ *if* $\alpha' = \alpha$, *and (iii)* $\tau(\alpha) = 3$ *if* $\alpha \prec \alpha'$.

For example, $\tau((bac)^\omega) = 1$, $\tau(a^\omega) = 2$ and $\tau((abc)^\omega) = 3$.

To describe our recursive approach, we extend Definition 1 from strings to nodes.

Definition 2. *Let* $G = (V, E)$ *be a disjoint union of cycles.*

- *For every* $u \in V$, *define* $\tau(u) = \tau(\mathcal{S}_u)$.
- *Let* $T[1, \ldots, |V|]$ *be the array such that* $T[i] = \tau(i)$ *for every* $1 \leq i \leq |V|$.

For example, in Fig. 1 we have $T = [1, 1, 1, 3, 3, 1, 1, 1, 3, 3, 2, 2, 1, 3, 3, 1, 1]$.

To understand how we can use $T[1, \ldots, |V|]$ to compute the $\mathcal{S}$-partition, let us give a definition.

Definition 3. *Let* $G = (V, E)$ *be a disjoint union of cycles.*

- *Let* $\mathcal{A}$ *be the* $\mathcal{S}$*-partition of* V.
- *For every* $c \in \Sigma$ *and for* $t \in \{1, 2, 3\}$, *let* $V_{c,t} = \{u \in V \mid \mathcal{S}_u[1] = c, \tau(u) = t\}$, *and let* $\mathcal{A}_{c,t}$ *be the* $\mathcal{S}$*-partition of* $V_{c,t}$.

Note that some $V_{c,t}$ may be empty (and in this case $\mathcal{A}_{c,t}$ is the empty $\mathcal{S}$-partition $\perp$). In Fig. 2, we have $\mathcal{A}_{a,1} = \perp$, $\mathcal{A}_{a,2} = \perp$, $\mathcal{A}_{a,3} = (\{4, 9, 14\})$, $\mathcal{A}_{b,1} = (\{3, 8, 13\}, \{2, 7, 17\})$, $\mathcal{A}_{b,2} = (\{11, 12\})$, $\mathcal{A}_{b,3} = (\{5, 10, 15\})$, $\mathcal{A}_{c,1} = (\{1, 6, 16\})$, $\mathcal{A}_{c,2} = \perp$, $\mathcal{A}_{c,3} = \perp$. Notice that $\mathcal{A}$ can be obtained by concatenating these ordered partitions, which suggests the following result.

Lemma 1. *Let* $G = (V, E)$ *be a disjoint union of cycles. Then,* $\mathcal{A} = \mathcal{A}_{1,1} \circ \mathcal{A}_{1,2} \circ \mathcal{A}_{1,3} \circ \mathcal{A}_{2,1} \circ \mathcal{A}_{2,2} \circ \mathcal{A}_{2,3} \circ \cdots \circ \mathcal{A}_{|\Sigma|,1} \circ \mathcal{A}_{|\Sigma|,2} \circ \mathcal{A}_{|\Sigma|,3}$.

Lemma 1 shows that we can reduce the problem of computing $\mathcal{A}$ to the problem of computing the $\mathcal{A}_{c,t}$'s. In particular, we need to know $\tau(u)$ for every $u \in V$, so the rest of this section is devoted to computing the array $T[1, \dots, |V|]$ in $O(|V|)$ time. We start with a simple lemma: if two consecutive edges of G have distinct labels, we can immediately determine one value $\tau(u)$.

Lemma 2. *Let $G = (V, E)$ be a disjoint union of cycles. Let $u, v \in V$ be such that there is an edge from u to v.*

1. *If $\mathcal{S}_v[1] \prec \mathcal{S}_u[1]$, then $\tau(u) = 1$. Equivalently, if $\tau(u) \geq 2$, then $\mathcal{S}_u[1] \preceq \mathcal{S}_v[1]$.*
2. *If $\mathcal{S}_u[1] \prec \mathcal{S}_v[1]$, then $\tau(u) = 3$. Equivalently, if $\tau(u) \leq 2$, then $\mathcal{S}_v[1] \preceq \mathcal{S}_u[1]$.*

In Fig. 1, one cycle only contains nodes of type 2, and the other two cycles contain no nodes of type 2 but contain both nodes of type 1 and nodes of type 3. Lemma 3 below shows that these are the only two possible cases.

Lemma 3. *Let $G = (V, E)$ be a disjoint union of cycles. Let $u \in V$, and let $(u_i)_{i \geq 1}$ be the occurrence of $\mathcal{S}_u$ starting at u. Then, exactly one of the following holds true:*

- *$\tau(u_i) = 2$ and $\mathcal{S}_{u_i} = \mathcal{S}_u$ for every $i \geq 1$.*
- *$\tau(u_i) \neq 2$ for every $i \geq 1$, $\tau(u_i) = 1$ is true for infinitely many i's, and $\tau(u_i) = 3$ is true for infinitely many i's.*

The next lemma shows that, when we have consecutive equally-labeled edges, the corresponding nodes must be of the same type and, if we know the type, we can also immediately determine their mutual order in the $\mathcal{S}$-partition (for example, in Fig. 1 the labels of the two edges leaving nodes 7 and 8 are the same, and we have $\tau(7) = \tau(8) = 1$ and $\mathcal{S}_8 \prec \mathcal{S}_7$).

Lemma 4. *Let $G = (V, E)$ be a disjoint union of cycles. Let $u \in V$, and let $(u_i)_{i \geq 1}$ be the occurrence of $\mathcal{S}_u$ starting at u. Let $k \geq 2$ such that $\mathcal{S}_{u_1}[1] = \mathcal{S}_{u_2}[1] = \dots = \mathcal{S}_{u_k}[1]$. Then:*

1. *$\tau(u_i) = \tau(u_k)$ for every $1 \leq i \leq k$. Moreover: (i) if $\tau(u_k) = 1$, then $\mathcal{S}_{u_k} \prec \mathcal{S}_{u_{k-1}} \prec \dots \prec \mathcal{S}_{u_1}$; (ii) if $\tau(u_k) = 2$, then $\mathcal{S}_{u_1} = \mathcal{S}_{u_2} = \dots = \mathcal{S}_{u_k}$; (iii) if $\tau(u_k) = 3$, then $\mathcal{S}_{u_1} \prec \mathcal{S}_{u_2} \prec \dots \prec \mathcal{S}_{u_k}$.*
2. *If $u_k = u$, then $\tau(u_i) = 2$ for every $i \geq 1$.*

We now have all the elements to compute the array $T[1, |V|]$ in $O(|V|)$ time. Intuitively, we proceed as follows. We scan all nodes i. If we have not computed $\tau(i)$ yet, we navigate the graph starting from node i as long as all the edges have the same label. By Lemma 4, all the visited nodes have the same value $\tau(u)$. If we find an edge with a distinct label, we use Lemma 2 to determine whether $\tau(u) = 1$ or $\tau(u) = 3$. If we visit the whole cycle, by Lemma 4 we have $\tau(u) = 2$. If we encounter a node v for which we have already computed $\tau(v)$, we set $\tau(u) = \tau(v)$.

Lemma 5. *Let $G = (V, E)$ be a disjoint union of cycles. We can compute the array $T[1, |V|]$ in $O(|V|)$ time.*

5 Computing the $\mathcal{S}$-Partition

In this section, we show that if we have (recursively) computed the $\mathcal{S}$-partition of $\{u \in V \mid \tau(u) = 3\}$, then in $O(|V|)$ time we can compute $\mathcal{S}$-partition $\mathcal{A}$ of V.

First, let us show that, by Lemma 1, we only need to compute the $\mathcal{A}_{c,t}$'s.

Lemma 6. *Let $G = (V, E)$ be a disjoint union of cycles. Given $\mathcal{A}_{c,t}$ for every $c \in \Sigma$ and for every $t \in \{1, 2, 3\}$, we can compute $\mathcal{A}$ in $O(|V|)$ time.*

Let us show that we can compute all the values $|V_{c,t}|$'s in $O(|V|)$ time by using $T[1, |V|]$. This is helpful to allocate the arrays $C_{\mathcal{A}_{c,t}}[1, |V_{c,t}|]$'s and $D_{\mathcal{A}_{c,t}}[1, |V_{c,t}|]$'s.

Lemma 7. *Let $G = (V, E)$ be a disjoint union of cycles. Given $T[1, |V|]$, we can compute all the $3|\Sigma|$ values $|V_{c,t}|$'s in $O(|V|)$ time.*

We can easily compute all the $\mathcal{A}_{c,2}$'s in linear time. Indeed, to compare the nodes u's for which $\tau(u) = 2$, we only need to compare the $\mathcal{S}_u[1]$'s.

Lemma 8. *Let $G = (V, E)$ be a disjoint union of cycles. Given $T[1, |V|]$ and all the $|\Sigma|$ values $|V_{c,2}|$'s, we can compute all the $|\Sigma|$ ordered partitions $\mathcal{A}_{c,2}$'s in $O(|V|)$ time.*

Now assume that we have (recursively) computed the $\mathcal{S}$-partition of $\{u \in V \mid \tau(u) = 3\}$. We must show how to compute the $\mathcal{A}_{c,3}$'s and the $\mathcal{A}_{c,1}$'s in $O(|V|)$ time. The next lemma shows that, given the $\mathcal{S}$-partition $\mathcal{A}'$ of $\{u \in V \mid \tau(u) = 3\}$, we can easily build all the $\mathcal{A}_{c,3}$'s in $O(|V|)$ time because Lemma 1 implies that $\mathcal{A}' = \mathcal{A}_{1,3} \circ \mathcal{A}_{2,3} \circ \cdots \circ \mathcal{A}_{|\Sigma|,3}$.

Lemma 9. *Let $G = (V, E)$ be a disjoint union of cycles. Given the $\mathcal{S}$-partition of $\{u \in V \mid \tau(u) = 3\}$ and all the $|\Sigma|$ values $|V_{c,3}|$'s, we can compute all the $|\Sigma|$ ordered partitions $\mathcal{A}_{c,3}$'s in $O(|V|)$ time.*

We are now ready to show how to build all the $\mathcal{A}_{c,1}$'s in $O(|V|)$ time, given the $\mathcal{A}_{c,3}$'s. Let us give some intuition on our algorithm based on Fig. 1 and Fig. 2. We will consider three ordered partitions $\mathcal{A}^*_{a,1}$, $\mathcal{A}^*_{b,1}$ and $\mathcal{A}^*_{c,1}$ that will be modified throughout the algorithm. At the beginning of the algorithm we have $\mathcal{A}^*_{a,1} = \mathcal{A}^*_{b,1} = \mathcal{A}^*_{c,1} = \perp$ and at the end of the algorithm we will have $\mathcal{A}^*_{a,1} = \mathcal{A}_{a,1}$, $\mathcal{A}^*_{b,1} = \mathcal{A}_{b,1}$ and $\mathcal{A}^*_{c,1} = \mathcal{A}_{c,1}$. We process $\mathcal{A}^*_{a,1}$, $\mathcal{A}_{a,3}$, $\mathcal{A}^*_{b,1}$, $\mathcal{A}_{b,3}$, $\mathcal{A}^*_{c,1}$, $\mathcal{A}_{c,3}$, in this order. Then, we scan each of these ordered partitions from left to right. For each scanned I, we compute the set $\psi(I)$ of all nodes preceding I in G and, if the nodes $u \in \psi(I)$ satisfy $\tau(u) = 1$, we add $\psi(I)$ to $\mathcal{A}^*_{x,1}$, where $x = \mathcal{S}_u[1]$ for every $u \in \psi(I)$. In our example, at the beginning of the algorithm we have $\mathcal{A}^*_{a,1} = \perp$, so we move to $\mathcal{A}_{a,3} = (\{4, 9, 14\})$. We scan $I = \{4, 9, 14\}$. Let $\psi(I) = \{3, 8, 13\}$. Since for every $u \in \psi(I)$ we have $\tau(u) = 1$ and $\mathcal{S}_u[1] = b$, we update $\mathcal{A}^*_{b,1}$ from $\mathcal{A}^*_{b,1} = \perp$ to $\mathcal{A}^*_{b,1} = (\{3, 8, 13\})$. We have finished processing $\mathcal{A}_{a,3}$, so we move to $\mathcal{A}^*_{b,1} = (\{3, 8, 13\})$. We scan $I = \{3, 8, 13\}$. Let $\psi(I) = \{2, 7, 17\}$. Since for every $u \in \psi(I)$ we have $\tau(u) = 1$ and $\mathcal{S}_u[1] = b$, we update

$\mathcal{A}^*_{b,1}$ from $\mathcal{A}^*_{b,1} = (\{3,8,13\})$ to $\mathcal{A}^*_{b,1} = (\{3,8,13\},\{2,7,17\})$. The next I in $\mathcal{A}^*_{b,1}$ to scan is the set $\{2,7,17\}$ that we have just added. Let $\psi(I) = \{1,6,16\}$. Since for every $u \in \psi(I)$ we have $\tau(u) = 1$ and $\mathcal{S}_u[1] = c$, we update $\mathcal{A}^*_{c,1}$ from $\mathcal{A}^*_{c,1} = \bot$ to $\mathcal{A}^*_{c,1} = (\{1,6,16\})$. We have finished processing $\mathcal{A}^*_{b,1}$, so we move to $\mathcal{A}_{b,3} = (\{5,10,15\})$. We scan $I = \{5,10,15\}$. Let $\psi(I) = \{4,9,14\}$. Since for every $u \in \psi(I)$ we have $\tau(u) = 3 \neq 1$, we do not do anything. We have finished processing $\mathcal{A}_{b,3}$, so we move to $\mathcal{A}^*_{c,1} = (\{1,6,16\})$. We scan $I = \{1,6,16\}$. Let $\psi(I) = \{5,10,15\}$. Since for every $u \in \psi(I)$ we have $\tau(u) = 3 \neq 1$, we do not do anything. We have finished processing $\mathcal{A}^*_{c,1}$, so we move to $\mathcal{A}_{c,3} = \bot$, and we have no element to scan. At the end of the algorithm, we have correctly $\mathcal{A}^*_{a,1} = \mathcal{A}_{a,1} = \bot$, $\mathcal{A}^*_{b,1} = \mathcal{A}_{b,1} = (\{3,8,13\},\{2,7,17\})$ and $\mathcal{A}^*_{c,1} = \mathcal{A}_{c,1} = (\{1,6,16\})$. In general, we have the following result.

Lemma 10. *Let $G = (V,E)$ be a disjoint union of cycles. Given $T[1,|V|]$, all the $|\Sigma|$ ordered partitions $\mathcal{A}_{c,3}$'s and all the $|\Sigma|$ values $|V_{c,1}|$'s, we can compute all the $|\Sigma|$ ordered partitions $\mathcal{A}_{c,1}$'s in $O(|V|)$ time.*

6 The Recursive Step

In this section we show that we can reduce in $O(|V|)$ time the problem of computing the $\mathcal{S}$-partition of $\{u \in V \mid \tau(u) = 3\}$ to the problem of computing the $\mathcal{S}$-partition of $\overline{V}$ with respect to a new graph $\overline{G} = (\overline{V},\overline{E})$ for which $|\overline{V}| = |\{u \in V \mid \tau(u) = 3\}|$, and we show that $\overline{G}$ can be built in $O(|V|)$ time from G. We start with a definition.

Definition 4. *Let $G = (V,E)$ be a disjoint union of cycles. Let $u \in V$ be such that $\tau(u) = 3$ and let $(u_i)_{i \geq 1}$ be the occurrence of $\mathcal{S}_u$ starting at u. Let (i) ℓ_u be the smallest integer $k \geq 2$ such that $\tau(u_k) = 3$, define (ii) $\phi_u = u_{\ell_u}$, and define (iii) $\gamma_u = \mathcal{S}_u[1,\ell_u]$.*

Notice that ℓ_u is well defined because $u_{1+c_u} = u_1 = u$ and so $\tau(u_{1+c_u}) = \tau(u) = 3$. As an example, in Fig. 1 we have $\tau(5) = 3$, $\ell_5 = 5$, $\phi_5 = 9$ and $\gamma_5 = \mathcal{S}_5[1,\ell_5] = \mathcal{S}_5[1,5] = bcbba$.

Intuitively, the next lemma is needed to understand how to define the alphabet Σ' of $\overline{G} = (\overline{V},\overline{E})$ and a convenient total order $\preceq'$ on Σ'.

Lemma 11. *Let $G = (V,E)$ be a disjoint union of cycles. Let $u, v \in V$ be such that $\tau(u) = \tau(v) = 3$. Assume that one of the following is true: (i) γ_u is not a prefix of γ_v and $\gamma_u \prec \gamma_v$; (ii) γ_v is a strict prefix of γ_u. Then, $\mathcal{S}_u \prec \mathcal{S}_v$.*

Equivalently, if $\mathcal{S}_u \preceq \mathcal{S}_v$, then one of the following is true: (a) γ_u is not a prefix of γ_v and $\gamma_u \prec \gamma_v$; (b) γ_v is a strict prefix of γ_u; (c) $\gamma_u = \gamma_v$.

We are now ready to build a new graph $\overline{G} = (\overline{V},\overline{E})$ such that $|\overline{V}| = |\{u \in V \mid \tau(u) = 3\}|$ (see Fig. 4). In particular, every $u \in V$ such that $\tau(u) = 3$ will have a corresponding node $\overline{u}$ in $\overline{G}$. In the rest of this section, when we write $\overline{u}$, we implicitly imply that $\overline{u} \in \overline{V}$ corresponds to a node $u \in V$ for which $\tau(u) = 3$.

Definition 5. *Let $G = (V, E)$ be a disjoint union of cycles.*

- *Let $\Sigma' = \{\gamma_u \mid u \in V, \tau(u) = 3\}$.*
- *Let $\preceq'$ be the total order on Σ' such that, for every distinct $\alpha, \beta \in \Sigma'$, we have $\alpha \prec' \beta$ if and only if one of the following is true: (i) α is not a prefix of β and $\alpha \prec \beta$; (ii) β is a strict prefix of α.*
- *Let $\overline{G} = (\overline{V}, \overline{E})$ be the disjoint union of cycles on $(\Sigma', \preceq')$ defined as follows: (i) $\overline{V} = \{\overline{u} \mid u \in V, \tau(u) = 3\}$; (ii) $\overline{E} = \{(\overline{u}, \overline{\phi_u}, \gamma_u) \mid u \in V, \tau(u) = 3\}$.*

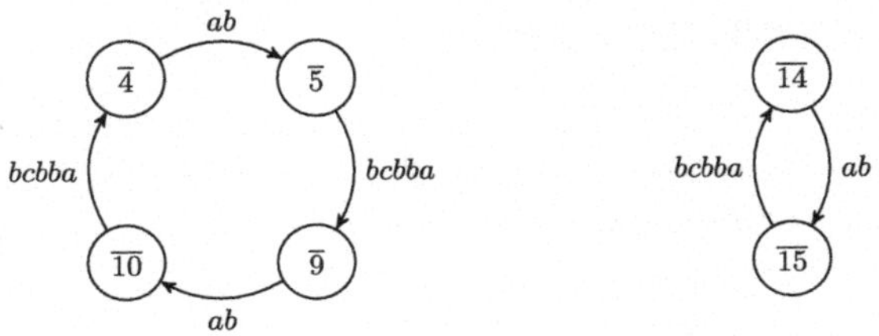

Fig. 4. The graph $\overline{G}$ of Definition 5 obtained from the graph G of Fig. 1.

Notice that $\preceq'$ and $\overline{G}$ are well defined. Indeed (i) $\preceq'$ is well defined because it is the variant of the (total) lexicographic order on Σ' in which a string is smaller than every strict prefix of it; (ii) $\overline{E}$ is well defined because if $u \in V$ and $\tau(u) = 3$, then $\tau(\phi_u) = 3$ by the definition of ℓ_u, so $\overline{\phi}_u \in \overline{V}$; (iii) $\overline{G}$ is a disjoint union of cycles: every $\overline{u} \in \overline{V}$ has exactly one outgoing edge by definition and every $\overline{u} \in \overline{V}$ has exactly one incoming edge because $\overline{\phi}_u = \overline{\phi}_v$ implies $u = v$ (G being a disjoint union of cycles). Notice also that the alphabet Σ' is effective, and if we pick two consecutive edges in $\overline{G}$, then the last character of the first edge label is equal to the first character of the second edge label (because in Definition 4 we have $\gamma_u = \mathcal{S}_u[1, \ell_u]$ and not $\gamma_u = \mathcal{S}_u[1, \ell_u - 1]$.)

In the example of Fig. 4, we have $|\Sigma'| = 2$, and we write $ab = 1$ and $bcbba = 2$ to highlight that ab is smaller than $bcbba$ w.r.t the total order $\preceq'$. Notice that the $\mathcal{S}$-partition of $\overline{V}$ (w.r.t $\overline{G}$) is $(\{\overline{4}, \overline{9}, \overline{14}\}, \{\overline{5}, \overline{10}, \overline{15}\})$, and the $\mathcal{S}$-partition of $\{u \in V \mid \tau(u) = 3\}$ (w.r.t G) is $(\{4, 9, 14\}, \{5, 10, 15\})$. This correspondence is ensured by the following lemma.

Lemma 12. *Let $G = (V, E)$ be a disjoint union of cycles. Let $u, v \in V$ be such that $\tau(u) = \tau(v) = 3$.*

1. *If $\mathcal{S}_u = \mathcal{S}_v$, then $\mathcal{S}_{\overline{u}} = \mathcal{S}_{\overline{v}}$.*
2. *If $\mathcal{S}_u \prec \mathcal{S}_v$, then $\mathcal{S}_{\overline{u}} \prec' \mathcal{S}_{\overline{v}}$.*

As explained in Sect. 3, when we implement $\overline{G} = (\overline{V}, \overline{E})$, we assume $\overline{V} = \{1, 2, \ldots, |\overline{V}|\}$ and $\Sigma' = \{1, 2, \ldots, |\Sigma'|\}$, with $1 \prec' 2 \prec' \cdots \prec' |\Sigma'|$. Consequently, the correspondence between the γ_u's and the set of integers $\{1, 2, \ldots, |\Sigma'|\}$ is uniquely determined and can be computed by sorting the γ_u's w.r.t. $\preceq'$ (for example, in Fig. 4 we have the correspondence $ab \mapsto 1$ and $bcbba \mapsto 2$). To implement $\overline{G}$, we need to store its adjacency list (through the two arrays $\overline{W}[1,|\overline{V}|]$ and $\overline{Z}[1,|\overline{V}|]$ described in Sect. 3). To this end, we must establish a correspondence between $\overline{V}$ and the set of integers $\{1,2,\dots,|\overline{V}|\}$. For example, if we want to store the edge $(\overline{5},\overline{9},bcbba)$ of Fig. 4, we know that $bcbba$ will be stored as 2, but we still have to decide how to store $\overline{5}$ and $\overline{9}$ (they must correspond to integers from the set $\{1,2,\dots,6\}$ because the graph of Fig. 4 has six nodes). Computationally, this correspondence should be as simple as possible because, given the $\mathcal{S}$-partition of $\overline{V}$ (w.r.t $\overline{G}$), we want to retrieve the $\mathcal{S}$-partition of $\{u \in V \mid \tau(u) = 3\}$ (w.r.t G). For this reason, we will use the correspondence obtained by sorting the $\overline{u}$'s according to the value u. In Fig. 4, we have $\overline{4} \mapsto 1$, $\overline{5} \mapsto 2$, $\overline{9} \mapsto 3$, $\overline{10} \mapsto 4$, $\overline{14} \mapsto 5$, $\overline{15} \mapsto 6$. In particular, the edge $(\overline{5},\overline{9},bcbba)$ will be stored as $(2,3,2)$. When we recursively compute the $\mathcal{S}$-partition of $\overline{V}$, the output will be $(\{1,3,5\},\{2,4,6\})$, which corresponds to $(\{\overline{4},\overline{9},\overline{14}\},\{\overline{5},\overline{10},\overline{15}\})$.

In the next lemma, we formally prove that in $O(|V|)$ time we can store $\overline{G} = (\overline{V},\overline{E})$ and, if we have recursively computed the $\mathcal{S}$-partition of $\overline{V}$ (w.r.t $\overline{G}$), in $O(|V|)$ time we can retrieve the $\mathcal{S}$-partition of $\{u \in V \mid \tau(u) = 3\}$ (w.r.t G). The main technical ingredient to achieve $O(|V|)$ time is to sort the elements of Σ' w.r.t $\preceq'$, which can be done in $O(|V|)$ time because $\sum_{u\in V,\tau(u)=3} |\gamma_u| = O(|V|)$ and $\preceq'$ is only a slight variant of the lexicographic order.

Lemma 13. *Let $G = (V,E)$ be a disjoint union of cycles. Given $T[1,|V|]$, in $O(|V|)$ time we can store $\overline{G} = (\overline{V},\overline{E})$ in such a way that, given the $\mathcal{S}$-partition of $\overline{V}$ (w.r.t $\overline{G}$), in $O(|V|)$ time we can compute the $\mathcal{S}$-partition of $\{u \in V \mid \tau(u) = 3\}$ (w.r.t G).*

We have now completed all steps to prove Theorem 1.

7 Conclusions and Future Work

In this paper, we have shown that the suffixes of a disjoint union of cycles can be sorted in linear time (Theorem 1). This algorithm extends previous construction algorithms for the eBWT and shows that the key component of the data structure in [22] to solve circular dictionary matching queries on arbitrary dictionaries can be built in linear time. As a by-product, we have obtained an algorithm to compute the primitive root of a string in linear time (Corollary 1), which is very different in flavor from the algorithms that are already known for this problem.

The data structure for solving circular dictionary matching queries requires $n\log\sigma(1+o(1))+O(n)+O(d\log n)$ bits, so the natural open problem is whether we can design a linear-time construction algorithm that only uses $O(n\log\sigma + d\log n)$ bits of working space by extending the (advanced) results in [10,45] for building the compressed suffix tree of a string in $O(n)$ time and $O(n\log\sigma)$ working bits. We conjecture that this is possible.

Acknowledgments. Funded by the Helsinki Institute for Information Technology (HIIT).

References

1. Aho, A.V., Corasick, M.J.: Efficient string matching: an aid to bibliographic search. Commun. ACM **18**(6), 333–340 (1975). https://doi.org/10.1145/360825.360855
2. Alanko, J., Cotumaccio, N., Prezza, N.: Linear-time minimization of Wheeler DFAs. In: 2022 Data Compression Conference (DCC), pp. 53–62. IEEE (2022)
3. Alanko, J., D'Agostino, G., Policriti, A., Prezza, N.: Regular languages meet prefix sorting. In: Proceedings of the Fourteenth Annual ACM-SIAM Symposium on Discrete Algorithms, pp. 911–930. SIAM (2020)
4. Alanko, J.N., Cenzato, D., Cotumaccio, N., Kim, S.H., Manzini, G., Prezza, N.: Computing the LCP array of a labeled graph. In: 35th Annual Symposium on Combinatorial Pattern Matching (CPM 2024). Schloss Dagstuhl–Leibniz-Zentrum für Informatik (2024)
5. Amir, A., Kopelowitz, T., Levy, A., Pettie, S., Porat, E., Shalom, B.R.: Mind the gap! Online dictionary matching with one gap. Algorithmica **81**, 2123–2157 (2019)
6. Amir, A., Levy, A., Porat, E., Shalom, B.R.: Dictionary matching with a few gaps. Theoret. Comput. Sci. **589**, 34–46 (2015)
7. Baier, U.: Linear-time suffix sorting-a new approach for suffix array construction. In: 27th Annual Symposium on Combinatorial Pattern Matching (CPM 2016), pp. 23:1–23:12. Schloss Dagstuhl–Leibniz-Zentrum für Informatik (2016)
8. Bannai, H., Kärkkäinen, J., Köppl, D., Piątkowski, M.: Constructing the bijective and the extended Burrows-Wheeler transform in linear time. In: 32nd Annual Symposium on Combinatorial Pattern Matching (CPM 2021). Schloss Dagstuhl-Leibniz-Zentrum für Informatik (2021)
9. Belazzougui, D.: Succinct dictionary matching with no slowdown. In: Amir, A., Parida, L. (eds.) Combinatorial Pattern Matching, pp. 88–100. Springer, Berlin, Heidelberg (2010)
10. Belazzougui, D., Cunial, F., Kärkkäinen, J., Mäkinen, V.: Linear-time string indexing and analysis in small space. ACM TALG **16**(2), 1–54 (2020)
11. Bonomo, S., Mantaci, S., Restivo, A., Rosone, G., Sciortino, M.: Sorting conjugates and suffixes of words in a multiset. Int. J. Found. Comput. Sci. **25**(08), 1161–1175 (2014)
12. Booth, K.S.: Lexicographically least circular substrings. Inf. Process. Lett. **10**(4–5), 240–242 (1980)
13. Boucher, C., Cenzato, D., Lipták, Z., Rossi, M., Sciortino, M.: Computing the original eBWT faster, simpler, and with less memory. In: Lecroq, T., Touzet, H. (eds.) String Processing and Information Retrieval, pp. 129–142. Springer, Cham (2021)
14. Boucher, C., Cenzato, D., Lipták, Z., Rossi, M., Sciortino, M.: r-indexing the eBWT. Inf. Comput. **298**, 105155 (2024)
15. Charalampopoulos, P., Kociumaka, T., Mohamed, M., Radoszewski, J., Rytter, W., Waleń, T.: Internal dictionary matching. Algorithmica **83**(7), 2142–2169 (2021)
16. Charalampopoulos, P., et al.: Approximate circular pattern matching. In: ESA 2022-30th Annual European Symposium on Algorithms (2022)
17. Clifford, R., Fontaine, A., Porat, E., Sach, B., Starikovskaya, T.: Dictionary matching in a stream. In: Algorithms-ESA 2015: 23rd Annual European Symposium, Patras, Greece, September 14–16, 2015, Proceedings. pp. 361–372. Springer (2015)
18. Conte, A., Cotumaccio, N., Gagie, T., Manzini, G., Prezza, N., Sciortino, M.: Computing matching statistics on Wheeler DFAs. In: 2023 Data Compression Conference (DCC), pp. 150–159 (2023). https://doi.org/10.1109/DCC55655.2023.00023

19. Cotumaccio, N.: Graphs can be succinctly indexed for pattern matching in $O(|E|^2 + |V|^{5/2})$ time. In: 2022 Data Compression Conference (DCC), pp. 272–281 (2022). https://doi.org/10.1109/DCC52660.2022.00035
20. Cotumaccio, N.: Prefix sorting DFAs: a recursive algorithm. In: 34th International Symposium on Algorithms and Computation (ISAAC 2023). Schloss-Dagstuhl-Leibniz Zentrum für Informatik (2023)
21. Cotumaccio, N.: A Myhill-Nerode theorem for generalized automata, with applications to pattern matching and compression. In: 41st International Symposium on Theoretical Aspects of Computer Science (STACS 2024). Schloss Dagstuhl–Leibniz-Zentrum für Informatik (2024)
22. Cotumaccio, N.: Improved circular dictionary matching. In: 36th Annual Symposium on Combinatorial Pattern Matching (CPM 2025), pp. 18–1. Schloss Dagstuhl–Leibniz-Zentrum für Informatik (2025)
23. Cotumaccio, N., D'Agostino, G., Policriti, A., Prezza, N.: Co-lexicographically ordering automata and regular languages - part I. J. ACM **70**(4) (2023). https://doi.org/10.1145/3607471
24. Cotumaccio, N., Gagie, T., Köppl, D., Prezza, N.: Space-time trade-offs for the LCP array of Wheeler DFAs. In: Nardini, F.M., Pisanti, N., Venturini, R. (eds.) String Processing and Information Retrieval, pp. 143–156. Springer, Cham (2023)
25. Cotumaccio, N., Prezza, N.: On indexing and compressing finite automata. In: Proceedings of the 2021 ACM-SIAM Symposium on Discrete Algorithms (SODA), pp. 2585–2599. SIAM (2021)
26. Cotumaccio, N., Trubiani, C.: Convex Petri nets. In: Köhler-Bussmeier, M., Moldt, D., Rölke, H. (eds.) Proceedings of the International Workshop on Petri Nets and Software Engineering 2024 co-located with the 45th International Conference on Application and Theory of Petri Nets and Concurrency (PETRI NETS 2024), June 24–25, 2024, Geneva, Switzerland. CEUR Workshop Proceedings, vol. 3730. CEUR-WS.org (2024)
27. Eisen, J.A.: Environmental shotgun sequencing: its potential and challenges for studying the hidden world of microbes. PLoS Biol. **5**(3), e82 (2007)
28. Farach, M.: Optimal suffix tree construction with large alphabets. In: Proceedings 38th Annual Symposium on Foundations of Computer Science, pp. 137–143. IEEE (1997)
29. Gagie, T., Manzini, G., Sirén, J.: Wheeler graphs: a framework for BWT-based data structures. Theoret. Comput. Sci. **698**, 67–78 (2017)
30. Gawrychowski, P., Starikovskaya, T.: Streaming dictionary matching with mismatches. Algorithmica, 1–21 (2019)
31. Golan, S., Kociumaka, T., Kopelowitz, T., Porat, E.: Dynamic dictionary matching in the online model. In: Friggstad, Z., Sack, J.-R., Salavatipour, M.R. (eds.) WADS 2019. LNCS, vol. 11646, pp. 409–422. Springer, Cham (2019). https://doi.org/10.1007/978-3-030-24766-9_30
32. Gusfield, D.: Algorithms on Strings, Trees, and Sequences - Computer Science and Computational Biology. Cambridge University Press (1997). https://doi.org/10.1017/CBO9780511574931
33. Hon, W.K., Ku, T.H., Lu, C.H., Shah, R., Thankachan, S.V.: Efficient algorithm for circular Burrows-Wheeler transform. In: Kärkkäinen, J., Stoye, J. (eds.) Combinatorial Pattern Matching, pp. 257–268. Springer, Berlin, Heidelberg (2012)
34. Hon, W.K., Ku, T.H., Shah, R., Thankachan, S.V.: Space-efficient construction algorithm for the circular suffix tree. In: Fischer, J., Sanders, P. (eds.) Combinatorial Pattern Matching, pp. 142–152. Springer, Berlin, Heidelberg (2013)

35. Hon, W.K., Ku, T.H., Shah, R., Thankachan, S.V., Vitter, J.S.: Faster compressed dictionary matching. Theoret. Comput. Sci. **475**, 113–119 (2013). https://doi.org/10.1016/j.tcs.2012.10.050
36. Hon, W.K., Lam, T.W., Shah, R., Thankachan, S.V., Ting, H.F., Yang, Y.: Dictionary matching with a bounded gap in pattern or in text. Algorithmica **80**, 698–713 (2018)
37. Hon, W.-K., Lu, C.-H., Shah, R., Thankachan, S.V.: Succinct indexes for circular patterns. In: Asano, T., Nakano, S., Okamoto, Y., Watanabe, O. (eds.) ISAAC 2011. LNCS, vol. 7074, pp. 673–682. Springer, Heidelberg (2011). https://doi.org/10.1007/978-3-642-25591-5_69
38. Hopcroft, J.E., Ullman, J.D., Aho, A.V.: Data Structures and Algorithms, vol. 175. Addison-wesley, Boston, MA, USA (1983)
39. Iliopoulos, C.S., Pissis, S.P., Rahman, M.S.: Searching and indexing circular patterns. In: Algorithms for Next-Generation Sequencing Data: Techniques, Approaches, and Applications, pp. 77–90 (2017)
40. Iliopoulos, C.S., Rahman, M.S.: Indexing circular patterns. In: Nakano, S.I., Rahman, M.S. (eds.) WALCOM: Algorithms and Computation, pp. 46–57. Springer, Berlin, Heidelberg (2008)
41. Ko, P., Aluru, S.: Space efficient linear time construction of suffix arrays. J. Disc. Algorithms **3**(2–4), 143–156 (2005)
42. Lothaire, M.: Combinatorics on Words, vol. 17. Cambridge University Press (1997)
43. Manacher, G.: A new linear-time on-line algorithm for finding the smallest initial palindrome of a string. J. ACM (JACM) **22**(3), 346–351 (1975)
44. Mantaci, S., Restivo, A., Rosone, G., Sciortino, M.: An extension of the Burrows-Wheeler transform. Theoret. Comput. Sci. **387**(3), 298–312 (2007)
45. Munro, J.I., Navarro, G., Nekrich, Y.: Space-efficient construction of compressed indexes in deterministic linear time. In: Proceedings of the Twenty-Eighth Annual ACM-SIAM Symposium on Discrete Algorithms, pp. 408–424. SIAM (2017)
46. Nong, G., Zhang, S., Chan, W.H.: Two efficient algorithms for linear time suffix array construction. IEEE Trans. Comput. **60**(10), 1471–1484 (2011). https://doi.org/10.1109/TC.2010.188
47. Olbrich, J., Ohlebusch, E., Büchler, T.: Generic non-recursive suffix array construction. ACM Trans. Algorithms **20**(2), 1–42 (2024)
48. Paige, R., Tarjan, R.E.: Three partition refinement algorithms. SIAM J. Comput. **16**(6), 973–989 (1987)
49. Sadakane, K.: Compressed suffix trees with full functionality. Theor. Comp. Sys. **41**(4), 589–607 (2007). https://doi.org/10.1007/s00224-006-1198-x
50. Simon, C., Daniel, R.: Metagenomic analyses: past and future trends. Appl. Environ. Microbiol. **77**(4), 1153–1161 (2011)
51. Strang, B.L., Stow, N.D.: Circularization of the herpes simplex virus type 1 genome upon lytic infection. J. Virol. **79**(19), 12487–12494 (2005)
52. Cotumaccio, N.: Fast pattern matching with epsilon transitions. In: Fernau, H., Zhu, B. (eds.) International Workshop on Combinatorial Algorithms 2025. LNCS, vol. 15885, pp. 242–255. Springer, Cham (2025). https://doi.org/10.1007/978-3-031-98740-3_18
53. Becker, R., Cotumaccio, N., Kim, S.-H., Prezza, N., Tosoni, C.: Encoding Co-Lex orders of finite-state automata in linear space. In: 36th Annual Symposium on Combinatorial Pattern Matching (CPM 2025), pp. 15–1. Schloss Dagstuhl–Leibniz-Zentrum für Informatik (2025)

Author Index

S.-K. Ko and F. Manea (Eds.): DLT 2025, LNCS 16036, pp. 311–312, 2026.
https://doi.org/10.1007/978-3-032-01475-7

The manufacturer's authorised representative in the EU is Springer
Nature Customer Service Centre GmbH, Europaplatz 3, 69115 Heidelberg,
Germany. If you have any concerns regarding our products, please
contact ProductSafety@springernature.com

Printed and bound by CPI Group (UK) Ltd, Croydon, CR0 4YY
30/09/2026
02225560-0008